GREEN FLUORESCENT PROTEIN

Properties, Applications, and Protocols

Edited by

MARTIN CHALFIE
Columbia University
New York, New York

STEVEN KAIN
CLONTECH Laboratories
Palo Alto, California

A JOHN WILEY & SONS, INC., PUBLICATION
New York • Chichester • Weinheim • Brisbane • Singapore • Toronto

This book is printed on acid-free paper. ♾

Published simultaneously in Canada.

Cover Image GFP structure solved by Fan Yang and George N. Phillips Jr. of Rice University and Larry Moss of Tufts University School of Medicine. Figure designed and rendered by Tod D. Romo of Rice University.

Library of Congress Cataloging-in-Publication Data:
Green fluorescent protein: Properties, Applications, and Protocols / edited by Martin Chalfie, Steven Kain.
p. cm.
"A Wiley–Liss publication."
Includes index.
ISBN 0-471-17839-X (pbk. : alk. paper)
1. Green fluorescent protein. I. Chalfie, Martin. II. Kain, Steven
QP552.G73G47 1998 97-38102
572.6–dc21

Printed in the United States of America.

10 9 8 7 8 6 5 4 3 2

GREEN FLUORESCENT PROTEIN

Contents

Preface

Now it is such a bizarrely improbable coincidence that anything so mind-bogglingly useful could have evolved purely by chance that some thinkers have chosen to see it as a final and clinching proof of the nonexistence of God.

Douglas Adams, Hitchhikers Guide to the Galaxy

In 1955, Davenport and Nicol reported that the light-producing cells of the jellyfish *Aequorea victoria* fluoresced green when animals were irradiated with long-wave ultraviolet. Five years later, Shimomura et al. (1962) described a protein extract from this jellyfish that could produce this fluorescence. Independently, Morin and Hastings (1971) found the same protein a few years later. This protein, now called the Green Fluorescent Protein (GFP), was studied for many years in virtual obscurity. However, with the cloning and expression of *A. victoria* GFP (Prasher et al., 1992; Chalfie et al., 1994), interest in this protein has grown enormously. To steal a phrase from a recent movie, GFP has gone from "zero to hero." As of January, 1998 at least 500 scientific publications have been published with the term "GFP" in their titles or abstracts. In the last 3 years hundreds of people have used GFP to mark proteins, cells, and organisms in a wide range of prokaryotic and eukaryotic species. They have used GFP to investigate fundamental questions in cell biology, developmental biology, neurobiology, and ecology. The interest in GFP goes beyond its utility as a biological marker. The protein is intrinsically intriguing, and investigators have sought to understand its structure, fluorescent properties, and biochemistry. This increased interest in GFP, serves as an important reminder of the usefulness of studying the biology of organisms that are not among the chosen "model" systems.

The usefulness of GFP as a biological marker derives from the finding that the protein's fluorescence requires no other cofactor: The fluorophore forms from the cyclization of the peptide backbone. This feature makes the molecule a virtually unobtrusive indicator of protein position in cells. Indeed, use of GFP as a tag suggests that the protein does not alter the normal function or localization of the fusion partner. Because permeabilization for substrate entry and fixation are not needed to localize GFP, proteins, organelles, and cells marked with this protein can be examined in living tissue. This ability to examine processes in living cells has permitted biologists to study the dynamics of cellular and developmental processes in intact tissues.

In addition to the broad impact of GFP technology on basic research, several companies have also incorporated this important reporter into more applied efforts such as high throughput drug screening, evaluation of viral vectors for human gene therapy, biological pest control, and monitoring genetically altered microbes in the environment. Most notable on this list are applications for GFP in drug discovery, here the potential for real time kinetics, ease of use, and cost savings provided by this reporter are leading to the replacement of other markers such as firefly luciferase and β-galactosidase. As the development of GFP technology continues to expand, the instrument companies are introducing new and better instruments for detecting GFP fluorescent. Finally, two U.S. patents have issued (as of July, 1997) on GFP and its variants, with many more certain to appear in the next few years.

As editors we find ourselves in the exciting, yet frustrating, position of producing a book that, in some aspects, will be out of date as it is published. The excitement comes from seeing the wealth of information being discovered about GFP and the many uses that people are finding for this molecule. The frustration results from the same source: New applications and information about GFP are published weekly, and no book on this subject can remain current. For example, as we write this preface, two papers have appeared on single molecule fluorescence of GFP (Dickson et al., 1997; Pierce et al., 1997), three on modifying GFP to measure calcium levels (Miyawaki et al., 1997; Persechini et al., 1997; Romoser et al., 1997), two on conditions that make GFP fluoresce red (Elowitz et al., 1997; Sawin and Nurse, 1997), and one on converting GFP to a voltage indicator (Siegel and Isacoff, 1997). We feel, however, that the contents of this volume serve as an important foundation for strategies that utilize GFP, and should guide the reader in using the marker in his or her system. We are in a period of rapid development of GFP as a tool for the biological sciences as people adapt the molecule for use in different organisms, generate variants with altered properties, and discover new ways that the protein can be used.

Despite the intrinsic incompleteness of this enterprise, we have asked our colleagues to summarize the state of GFP research and they have done an admirable job. We are grateful that so many of the initial investigators that pioneered the study and use of GFP consented to write chapters for this volume. We have organized this book into four sections. We start with two introductory chapters by Osamu Shimomura on the discovery of GFP and by Woody Hastings and James Morin on bioluminescence and biofluorescence in nature.

The second section describes the biochemistry and molecular biology of GFP. Bill Ward has written a very useful description of the biochemistry of GFP, pointing out both the gaps in our knowledge and the importance of physical chemical criteria for evaluating new variants of GFP. George Phillips then discusses the structure of GFP and implications of this structure for its function as a fluorescent molecule. In the last chapter in this section, Roger Tsien and Douglas Prasher describe many of these variants, their uses, and how they were derived.

The third section documents various biological applications of GFP. The people we asked to contribute these chapters are the major developers of GFP in the various organisms described. As described above these chapter are incomplete in that new information and application are developing at a very rapid rate. Nonetheless, each of these chapters provides insights into how GFP is being applied to particular species. We urge readers not to look only at the organism they love best, since approaches used for one organism may prove important when applied to others. For example, the use of species-specific codon usage, presumably by allowing greater production of protein, has been very important for GFP expression in mammalian cells. Also Andy Fire has found that GFP (and β-galactosidase) expression is elevated in the nematode *Caenorhabditis elegans* when artificial introns are interspersed in the cDNA sequence. Both of these observations may be important for those considering optimizing GFP expression in their organisms. Finally, we asked the contributors in the third section to provide protocols on the purification of GFP and its application in various organisms and Bill Ward to contribute information on purifying GFP. Sharyn Endow and David Piston have admirably taken on the formidable task of collecting, editing, and adding to this material for the fourth section of this book. In particular, they have provided outstanding protocols for visualizing and recording GFP fluorescence.

We are just beginning to learn about and use GFP, and, as always, many questions remain. Much still needs to be learned about the chemistry of fluorophore formation and the role of the protein structure in this formation. Additional variants are needed. In particular, variants with spectra that do not significantly overlap with those of existing variants would be very useful. Such variants could be used in multiple labeling experiments, but they may have an even greater potential. Specifically, the use of fluorescence resonance energy transfer between two fluorescent proteins would enable the generation of a system analogous to the yeast two-hybrid system (Fields and Song, 1989) to look at protein : protein interactions. The advantage of such a system is that it would not require transcription as a readout of the interaction, and could therefore be used anywhere in the cell (e.g., cytosol, plasma membrane, mitochondria). Morever, suitably marked molecules would allow the testing of protein interactions in situ in a variety of organisms. Finally, as we learn more about the properties of these protein, we need to take advantage of this information to optimize GFP fluorescence intensity, excitation and emission spectra, and protein and message stability for different uses. In the next few years, we will undoubtedly see many more uses for this protein. The future does look bright for GFP.

The editors of a book have, perhaps, the easiest jobs; everyone contributes to an effort that they get the credit for. As this was the first volume that either of us had edited, we are particularly grateful for all the help that we have been given. Foremost we wish to thank the contributors who graciously consented to write chapters and then put up with our requests for rewrites and for "just a little more information" with great good humor. We are indebted to David Ades and Kaaren Janssen for starting us on this endeavor. We will get even. Finally, we are most obligated to Colette Bean, our editor at John Wiley, for showing us the ropes, keeping us on schedule, and getting us over the anxieties of producing this volume.

Martin Chalfie
Steven Kain

REFERENCES

Chalfie, M., Tu, Y., Euskirchen, G., Ward, W. W. and Prasher, D. C. (1994). Green fluorescent protein as a marker for gene expression. *Science* 263:802–805.

Davenport, D. and Nicol, J. A. C. Luminescence in Hydromedusae. *Proc. R. Soc. London Ser. B* 144:399–411.

Dickson, R. M., Cubitt, A. B., Tsien, R. Y., and Moener, W. E. (1997). On/off blinking and switching behavior of single molecules of green fluorescent protein. *Nature (London)* 388:355–358.

Elowitz, M. B., Surette, M. G., Wolf, P. E., Stock, J., and Leibler, S. (1997). Photoactivation turns green fluorescent protein red. *Curr. Biol.* 7:809–812.

Fields, S. and Song, O. K. (1989). A novel genetic system to detect protein-protein interactions. *Nature (London)* 340:245–246.

Miyawaki, A., Llopis, J., Heim, R., McCaffery, J. M., Adams, J. A., Ikura, M., and Tsien, R. Y. (1997). Fluorescent indicators for Ca^{2+} based on green fluorescent proteins and calmodulin. *Nature* 388:882–887.

Morin, J. G. and Hastings, J. W. (1997). Biochemistry of the bioluminescence of colonial hydroids and other coelenterates. *J. Cell. Physiol.* 77:305–312.

Persechini, A., Lynch, J. A., and Romoser, V. A. (1997). Novel fluorescent indicator proteins for monitoring free intracellular Ca^{2+}. *Cell Calcium* 22:209–216.

Pierce, D. W., Hom-Booher, N., and Vale, R. D. (1997). Imaging individual green fluorescent proteins. *Nature (London)* 388:338.

Prasher, D. C., Eckenrode, V. K., Ward, W. W., Prendergast, F. G., and Cormier, M. J. (1992). Primary structure of the *Aequorea victoria* green-fluorescent protein. *Gene* 111:229–233.

Romoser, V. A., Hinkle, P. M., Persechini, A. (1997). Detection in living cells of Ca^{2+}-dependent changes in the fluorescence emission of an indicator composed of two green fluorescent protein variants linked by a calmodulin-binding sequence. A new class of fluorescent indicators *J. Biol. Chem.* 272:13270–13274.

Sawin, K. E., and Nurse, P. (1997). Photoactivation of green fluorescent protein. *Curr. Biol.* 7:R606–R607.

Shimomura, O., Johnson, F. H., and Saiga, Y. (1962). Extraction, purification, and properties of aequorin, a bioluminescent protein from the luminous hydromedusan, Aequorea. *J. Cell. Comp. Physiol.* 59:223–239.

Contributors

JOOHONG AHNN
Carnegie Institute of Washington
Department of Embryology
115 West University Parkway
Baltimore, MD 21210

ADAM AMSTERDAM
Department of Biology
Massachusetts Institute of Technology
77 Massachusetts Avenue, E17-341
Cambridge, MA 02139

MARTIN CHALFIE
Department of Biological Sciences
1012 Fairchild Center
Columbia University
1212 Amsterdam Avenue
New York, NY 10027

BRENDAN P. CORMACK
Department of Microbiology and Immunology
Stanford University School of Medicine
Stanford, CA 94305

SHARYN A. ENDOW
Department of Microbiology and Immunology
Duke University Medical Center
Research Drive, Jones Building
Durham, NC 27710

STANLEY FALKOW
Rocky Mountain Laboratories
National Institute of Allergy and Infectious Diseases
Hamilton, MT 59840

ANDREW FIRE
Carnegie Institute of Washington
Department of Embryology
115 West University Parkway
Baltimore, MD 21210

BRIAN D. HARFE
Carnegie Institute of Washington
Department of Embryology
115 West University Parkway
Baltimore, MD 21210

JIM HASELOFF
Division of Cell Biology
MRC Laboratory of Molecular Biology
Hills Road, Cambridge CB2 2QH
United Kingdom

J. WOODLAND HASTINGS
Department of Molecular and Cellular Biology
16 Divinity Ave,
Harvard University
Cambridge, MA 02138

TULLE HAZELRIGG
Department of Biological Sciences
602 Fairchild, Columbia University,
New York, NY 10027

NANCY HOPKINS
Department of Biology
Massachusetts Institute of Technology
77 Massachusetts Avenue, E17-341
Cambridge, MA 02139

JENNY HSIEH
Carnegie Institute of Washington
Department of Embryology
115 West University Parkway
Baltimore, MD 21210

MEI HSU
Carnegie Institute of Washington
Department of Embryology
115 West University Parkway
Baltimore, MD 21210

JASON A. KAHANA
Department of Biological Chemistry and Molecular Pharmacology
Harvard Medical School
Dana Farber Cancer Institute
Boston, MA 02115

STEVEN R. KAIN
CLONTECH Laboratories, Inc.
1020 E. Meadow Circle
Palo Alto, CA 94303

WILLIAM G. KELLY
Carnegie Institute of Washington
Department of Embryology
115 West University Parkway
Baltimore, MD 21210

STEPHEN A. KOSTAS
Carnegie Institute of Washington
Department of Embryology
115 West University Parkway
Baltimore, MD 21210

JENNIFER LIPPINCOTT-SCHWARTZ
Department of Cell Biology and Metabolism
NICHD, NIH
Building 18T
Bethesda, MD 20892-0001

JAMES G. MORIN
Section of Ecology and Systematics
Cornell University
Ithaca, NY

GEORGE N. PHILLIPS, JR.
Department of Biochemistry and Cell Biology
Rice University
Houston, TX 77005-1892

DAVID W. PISTON
Department of Molecular Physics and Biophysics
Vanderbilt University
702 Light Hall
Nashville, TN 37232-0615

DOUGLAS PRASHER
USDA, APHIS
Building 1398
Otis ANGB, MA 02542

OSAMU SHIMOMURA
Marine Biological Laboratory
Woods Hole, MA 02543

KIRBY R. SIEMERING
Division of Cell Biology
MRC Laboratory of Molecular Biology
Hills Road
Cambridge CB2 2QH
United Kingdom

PAMELA A. SILVER
Department of Biological Chemistry and Molecular Pharmacology
Harvard Medical School
Dana Farber Cancer Institute
Boston, MA 02115

ROGER TSIEN
Department of Pharmacology
University of California, San Diego
310 Cell and Molecular Medicine, M-047
La Jolla, CA 92093-0647

RAPHAEL H. VALDIVIA
Department of Microbiology and Immunology
Stanford University School of Medicine
Stanford, CA 94305

WILLIAM W. WARD
Department of Biochemistry and Microbiology
Rutgers University
Cook College
New Brunswick, NJ 08903

SI-QUN XU
Carnegie Institute of Washington
Department of Embryology
115 West University Parkway
Baltimore, MD 21210

PART ONE

Background

1

The Discovery of Green Fluorescent Protein

OSAMU SHIMOMURA
Marine Biological Laboratory, Woods Hole, MA

1.1 PROLOGUE

It was early July in 1961. Dr. Frank Johnson and I were studying the bioluminescence of the jellyfish *Aequorea* aequorea (see Section 1.2 for the discussion on the species name) at the Friday Harbor Laboratories of the University of Washington, located on a small island near Victoria, British Columbia, Canada. Since early morning of that day, we were trying to develop a practical method to extract the active light emitting matter of the jellyfish, a substance later named "aequorin" (cf. Shimomura et al., 1962; Shimomura, 1995a); the day before, we had found the basic principle of solubilizing and extracting this substance. In the course of our experiments, however, I was deeply puzzled when I realized that the light emitted from the extract was clearly blue, contrary to our expectation of green light identical to the luminescence of live specimens.

A mature specimen of A. *aequorea* looks like a transparent, hemispherical umbrella, with its mouth at the underside of the body (Fig. 1.1). Average specimens measure 7–10 cm in diameter. Due to the high transparency of the body, the jellyfish can function as a magnifier lens when the mouth is fully open. The light organs, consisting of about 10-doz tiny granules, are distributed evenly along the edge of the umbrella, making a full circle. Soaking a specimen of the jellyfish in a dilute potassium chloride (KCl) solution in a dark-room, causes the light

Green Fluorescent Protein: Properties, Applications, and Protocols, Edited by Martin Chalfie and Steven Kain
ISBN 0-471-17839-X

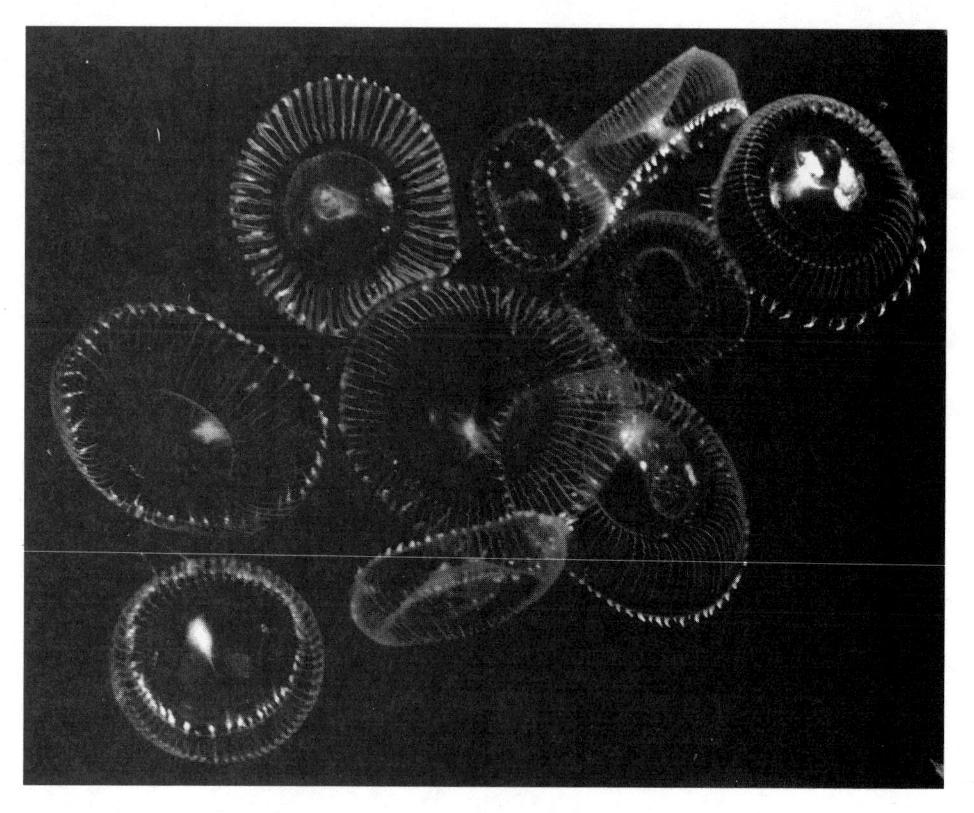

FIG. 1.1 Mid-summer specimens of *Aequorea* aequorea photographed in their natural habitat, at the University of Washington's Friday Harbor Laboratories.

organs to luminesce, exhibiting a ring of green light in the darkness. If a specimen is soaked in distilled water, a dimmer green ring is first observed, which gradually changes into blue, probably due to the cytolysis of cells. Under an ultraviolet (UV) light, a specimen of fresh jellyfish shows a bright ring of green fluorescence, similar to the luminescence caused by KCl.

The margin of the umbrella containing the light organs can be cut off with a pair of scissors, yielding a 2–3 mm wide strip called the "ring." When the rings obtained from 20–30 jellyfish were squeezed through a rayon gauze, a dimly luminescent, turbid liquid called the "squeezate" is obtained. The granules of light organs in the squeezate can be collected by filtration or centrifugation. When the granules are mixed with neutral buffer solutions, they are cytolyzed and emit light. When mixed with a pH 4.0 buffer, however, the granules are cytolyzed without light emission, preserving the light-emitting activity in the solution. After the removal of cell debris by centrifugation, the pH 4.0 cell-free solution can be luminesced by the addition of a neutral buffer solution containing Ca^{2+}. These are the experiments we were doing on that day in July 1961, and I saw that the luminescence of the neutralized solution was blue, contrary to our expectation. I doubled, then tripled, the number of the jellyfish used in each experiment in order to make the final luminescence stronger and clearer, but these efforts only helped to confirm my observation. My question concerning the seeming discrepancy remained in my consciousness, until we found an explanation more than 10 years later.

After returning to Princeton University with the jellyfish extracts, we purified the light-emitting substance. The substance obtained was a protein capable of emitting light in the presence of Ca^{2+}; the protein was named aequorin. During the purification of aequorin, we noticed the existence of a green fluorescent protein in the jellyfish extract. Although the presence of a green fluorescent substance in the light organs was previously known (Davenport and Nicol, 1955), it was the first time that the substance was recognized as a protein. Our observation was mentioned in our first full paper on the purification and characterization of aequorin (Shimomura et al., 1962), in a footnote, as follows:

> A protein giving solutions that look slightly greenish in sunlight though only yellowish under tungsten lights, and exhibiting a very bright, greenish fluorescence in the ultraviolet of a Mineralite, has also been isolated from the squeezates. No indications of a luminescent reaction of this substance could be detected.

The first measurements of the luminescence spectrum of aequorin and the fluorescence spectrum of the green protein were reported shortly (Johnson et al., 1962). The luminescence spectrum of aequorin was broad, with a peak at 460 nm. The fluorescence spectrum of the green protein was sharp, with a peak at 508 nm. Apparently, the light organs of the jellyfish contain these two kinds of protein, that is, aequorin and the "green protein", of which the former emits blue light in the presence of Ca^{2+} and the latter emits green fluorescence when excited. The green protein was later called green fluorescent protein (GFP) (Hastings and Morin, 1969).

How can a protein, aequorin, luminesce just by the addition of Ca^{2+}, even in the absence of oxygen? Why is the luminescence of a live jellyfish green, while aequorin emits blue light? To answer the first question, we would need to understand the mechanism of the intramolecular chemical reaction that takes place when Ca^{2+} is added, a formidable task at the time. For the second question, it would be necessary to consider two possibilities: (1) a filtering effect by the green protein or something else that shifts the emission maximum of aequorin luminescence to longer wavelength, and (2) an energy transfer from aequorin molecules to the green protein by a certain mechanism. Considering that the brightly fluorescent protein was created by nature presumably on purpose, the possibility of an energy transfer would be more likely. In those days, however, we were not concerned with the details of energy-transfer mechanism; we merely thought that the green protein absorbed the blue light of aequorin, then reemitted that absorbed energy as green light (i.e., an energy transfer by the trivial mechanism). We deferred the studies of these subjects for the next 5 years, because of various difficulties.

1.2 SPECIES NAME OF THE JELLYFISH FROM WHICH AEQUORIN AND GFP WERE EXTRACTED

A brief discussion concerning the names of *Aequorea* species is included here in consideration of the problems and confusions induced by the recent common use of the species name *Aequorea* victoria in place of A. *aequorea* (and *Aequorea* forskalea). The species names A. *aequorea* (Forskal, 1775) and A. *forskalea* (Peron and Lesueur, 1809) are synonymous and both names are commonly used; the decision of priority between them appears to be a matter of opinion. The species of A. *aequorea* is highly variable in both form and color (Mayer, 1910), and distributed very widely—Mediterranean; Atlantic coasts, from Norway to South Africa and Cape Cod to Florida; northeastern Pacific; east coast of Australia; and Iranian Gulf (Kramp, 1968). According to Mayer (1910), A. *victoria* (Murbach and Shearer, 1902) from the northeastern Pacific is probably a variety of A. *aequorea.* He stated "I can not distinguish this medusa from *Aequorea* forskalea of the Atlantic and Mediterranean. Were it described from the Atlantic I would not hesitate to designate it A. *forskalea.*" Mayer's opinion has been overwhelmingly accepted until recently (Russell, 1953; Kramp, 1965, 1968), thus the jellyfish we collected in the Friday Harbor area have been called A. *aequorea* for a long time. The situation changed, however, after Arai and Brinckmann-Voss (1980) reported their conclusion to separate A. *victoria* from the species A. *aequorea,* based on their study of about 40 specimens collected from more than 10 different areas around Vancouver Island (Friday Harbor included). Their reasons were that A. *victoria* has much more regularly serrated mouth lobes and a much larger, almost hemispherical lens in the stomach region, when compared with A. *aequorea* from the Mediterranean.

It is not clear in the Arai and Brinckmann-Voss paper why the conclusion to separate A. *victoria* from A. *aequorea* was made on the basis of the comparison between the former (from British Columbia and Puget Sound) and the latter from only the Mediterranean; their use of only a few specimens per study area, collected probably on a single occasion, brings about another problem. It has been well documented that a wide intraspecific variation of A. *aequorea* by geography exists (Mayer, 1910; Russell, 1953; Kramp, 1959, 1965). The Mediterranean form of A. *aequorea* is only one of many variations of this species. Therefore, the difference between A. *victoria* and A. *aequorea* cannot be fully determined by only the comparison between the former and the Mediterranean form of the latter; to determine the difference, A. *victoria* should be compared with various other forms of A. *aequorea.* To establish that A. *victoria* is a separate species, one must prove that A. *victoria* is not identical to any variety of A. *aequorea;* if an identical variety is found, the A. *aequorea* name should have priority over A. *victoria.* It seems particularly intriguing to compare A. *victoria* with the varieties of A. *aequorea* obtained from the northern and western Atlantic. In our record, we collected a large number of A. *aequorea* at Woods Hole, MA, in the summer of 1987, when there was a strong easterly wind; those medusae appeared to be indistinguishable from the average specimens of *Aequorea* obtained at Friday Harbor in both form and aequorin composition.

Most specimens of *Aequorea* used for biochemical research in the past have been collected around Friday Harbor. If all those medusae were *A. victoria* as implied by Arai and Brinckmann-Voss (1980), it seems that *A. victoria* must have a very wide variation, like *A. aequorea*, to be consistent with our observation in the past. We have collected over 1 million specimens of *Aequorea* in the vicinity of Friday Harbor in 17 summers between 1961 and 1989. All the specimens were mature (>7 cm in diameter), and they were collected, handled, and excised individually. More than several times during our operation, we observed pronounced changes in the form of the jellyfish that came to our collecting sites. The jellyfish can drift far and widely by current, tide and wind in groups, and the changes that we observed usually lasted for only a few days but occasionally continued for several weeks. The bell-height of the medusae were sometimes markedly higher than usual (relative to the diameter), and sometimes much flatter. In one of these occasions, we thought that the jellyfish we had collected were a wrong species because they were too flat; we suspended our operation until we had an assurance by a jellyfish expert in the lab that they were indeed a variety of *A. aequorea* (as known then). If all those medusae at Friday Harbor are the variations of *A. victoria*, the situation would be very confusing because *A. aequorea*, from which *A. victoria* was to be separated, also has wide variations and it would be very difficult to make a clear distinction between the two species.

Despite the high intraspecific variability that tends to cause confusions and difficulties, the species names *A. aequorea* and *A. forskalea* have been accepted and used by the majority of researchers for at least 60 years until 1980. To avoid further confusion, the name *A. victoria* should be used only as a synonym, not as the name of a separate species, until an overall difference between *A. victoria* and *A. aequorea* (or *A. forskalea*) is firmly established. Until that time, the species name *A. aequorea* (or *A. forskalea*) has priority.

1.3 ISOLATION OF THE GREEN FLUORESCENT PROTEIN

In 1967, Ridgway and Ashley reported the first successful application of aequorin bioluminescence. They used microinjected aequorin to monitor changes in Ca^{2+} concentration during the contraction of barnacle muscle single fibers; the study clearly demonstrated the importance and usefulness of this photoprotein in the studies of intracellular calcium. For the efficient and productive use of aequorin, it would be highly desirable to have a full knowledge concerning the properties of aequorin and the chemistry involved. Thus, we decided to study the chemistry of aequorin luminescence, an undertaking that seemed extremely difficult and almost unachievable at the time. We made a strenuous effort for several years, and we had the luck to uncover a large part of the intramolecular chemistry involved in the Ca^{2+} triggered luminescence of aequorin, including the chemical structure of the chromophore and also the means to regenerate spent aequorin into the original, active aequorin (Shimomura and Johnson, 1969, 1972, 1973, 1975).

During the same period, it was found that a number of other bioluminescent coelenterates also contain a green fluorescent protein that is similar to *Aequorea* GFP and functions as the light emitter of *in vivo* bioluminescence (Hastings and Morin, 1969; Morin and Hastings, 1971a,b; Wampler et al., 1971, 1973; Cormier et al., 1973, 1974; Morin, 1974). The genera of the coelenterates containing a GFP are as follows:

Class Hydrozoa

- The jellyfish *Aequorea*
- The jellyfish *Mitrocoma* (synonym *Halistaura*)
- The hydroid *Obelia*
- The jellyfish *Phialidium* (hydroid *Clytia*)

Class Anthozoa

- *Acanthoptilum*
- The sea cactus *Cavernularia*
- The sea pansy *Renilla*
- The sea pen *Ptilosarcus*
- *Stylatula*

Green fluorescent protein was not found in the jellyfish of scyphozoa (*Pelagia*) and ctenophora (*Mnemiopsis*). With regard to the mechanism of energy transfer from the excited state of a photoprotein to a GFP, Morin and Hastings (1971b) suggested the possible involvement of the Förster-type energy-transfer mechanism for the first time in the bioluminescence of coelenterates.

To clarify the mechanism of energy transfer involved in the emission of green light from the jellyfish *Aequorea*, we isolated and purified the green fluorescent protein of the jellyfish, and then studied its properties (Morise et al., 1974). The purified *Aequorea* GFP was easily crystallized by decreasing the ionic strength of the solvent (see Fig. 1.2). We studied the energy transfer from aequorin to GFP during the Ca^{2+} triggered luminescence reaction under two sets of conditions: one set contained high concentrations of GFP and the other set contained relatively low concentrations of GFP.

In the presence of high concentrations of GFP, an energy transfer by the trivial (radiative) mechanism apparently takes place, at least to some extent. Thus, the luminescence of aequorin (emission max 460 nm) is absorbed by GFP (absorption max 480 nm), then the absorbed energy is reemitted as the fluorescence of GFP (emission max 509 nm). The extent of energy transfer and the spectral shape of emitted light are dependent on the GFP concentration. It is clear, however, that GFP cannot absorb all the light energy emitted from aequorin, because the luminescence emission of aequorin extends to about 600 nm on the long wavelength side whereas GFP can absorb light up to only about 510 nm. Therefore, a complete energy transfer by the trivial mechanism is clearly impossible. In any event, a very high concentration of GFP (i.e., a very high level of absorbance) is required to obtain a significant extent of energy transfer by the trivial mechanism. Under such a condition, the self-absorption

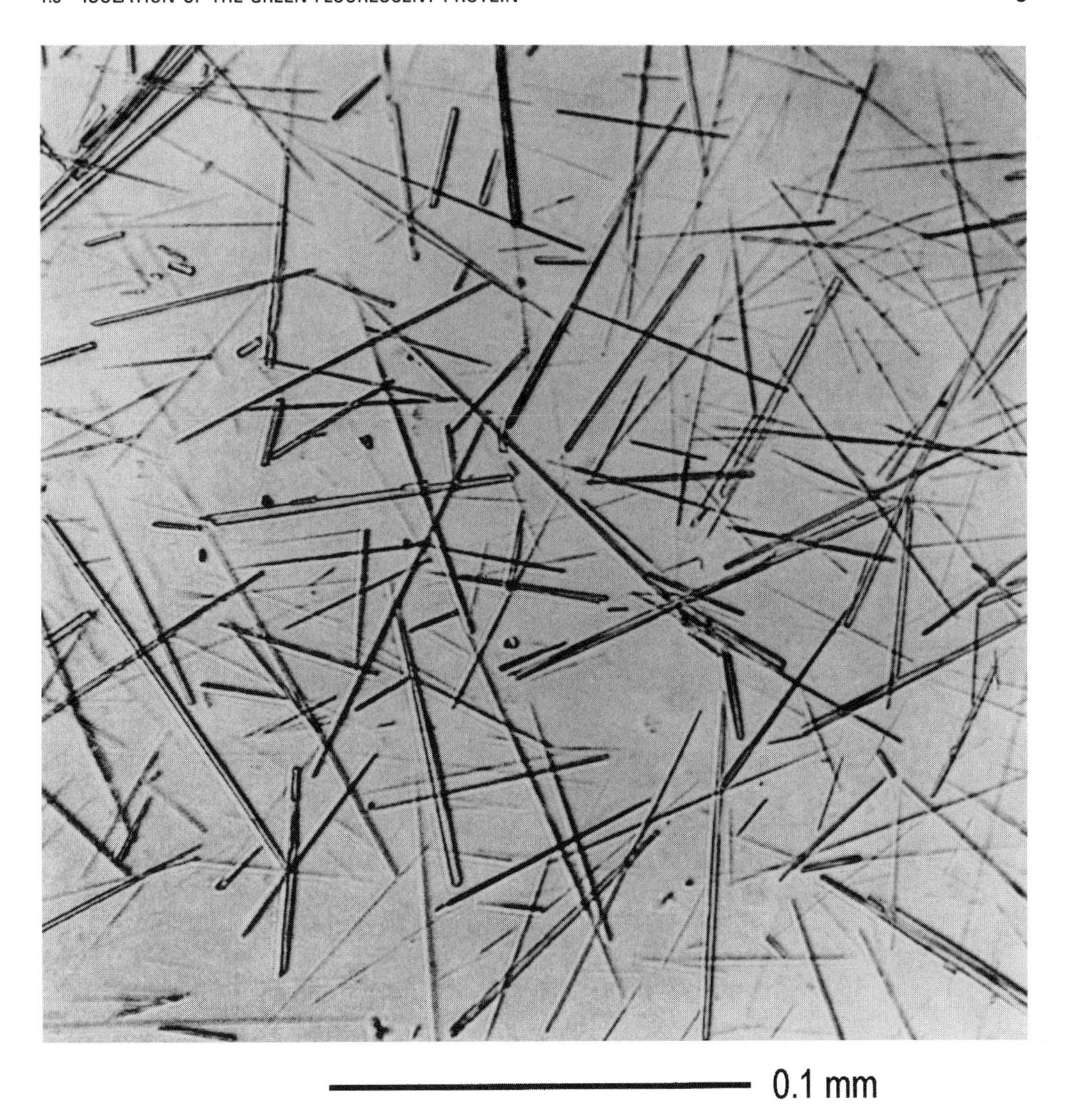

FIG. 1.2 The photomicrograph of the first crystals of *Aequorea* GFP obtained in 1973.

of GFP would strongly affect the spectral shape of the fluorescence emitted from GFP, in two manners: (1) a very steep decrease in the light intensities below 510 nm, and (2) a small red shift of the fluorescence peak position. The actual luminescence spectrum should be the sum of the aequorin luminescence unabsorbed by GFP and the GFP fluorescence distorted by self-absorption; it would be highly unlikely that such a spectrum coincides with the true, undistorted spectrum of GFP fluorescence or with the luminescence from the live *Aequorea*.

When aequorin was luminesced with a trace of Ca^{2+} in the presence of relatively low concentrations of GFP in a low ionic strength buffer, the emission spectrum of aequorin was little affected by GFP. However, when a small amount of fine particles of diethylaminoethylcellulose DEAE-cellulose or DEAE Sephadex (anion exchangers) was mixed in advance to the same solution, the Ca^{2+}-triggered luminescence of the clouded mixture became spectrally identical with the *in vivo* bioluminescence of *Aequorea*, indicating the occurrence of an efficient energy transfer from aequorin light emitter to GFP. It should be pointed out that the amounts of aequorin and GFP, as well as the volume used, were kept practically equal (i.e., the overall concentrations and the absorbance values were equal) in the above experiments; the only difference was the DEAE material added in the latter experiment.

Under the conditions used, the DEAE cellulose particles had coadsorbed GFP and aequorin by anion exchange mechanism, greatly increasing the local concentrations of the two proteins around the particles. The coadsorption perhaps made the distance between the GFP molecules and the aequorin molecules sufficiently short (roughly 30 Å) to make the Förster-type (radiationless) energy transfer workable. Thus, the result observed was the green light that spectrally matches with the *in vivo* luminescence and the fluorescence emission of GFP. Because the radiationless process is not significantly influenced by the concentration of GFP and does not require a very high concentration of GFP, the energy transfer can take place without being significantly affected by the absorbance of GFP. The above discussion strongly suggests that the energy transfer involved in the emission of green light from live *Aequorea* is mostly, if not entirely, a radiationless process.

The light organs of *Aequorea* contain tiny particles (Davenport and Nicol, 1955) that contain high concentrations of aequorin and GFP. In the particles, aequorin molecules and GFP molecules must be very closely arranged, if they are not directly bound to each other, to allow an efficient energy transfer by a radiationless process.

Some investigators seem to favor radiative process of energy transfer in the luminescence of *Aequorea* on the basis of quantum yields (Ward, 1979; Ward et al., 1980), presumably for the reason that the quantum yield of the coadsorbed aequorin–GFP (Morise et al., 1974) was not significantly higher than that of aequorin alone. It seems that they are not paying sufficient attention to the fact that the fluorescence quantum yield of the aequorin light emitter at the moment of light emission is not available, just like the corresponding data in many other bioluminescence systems. In the luminescence reaction, aequorin is converted into apoaequorin, coelenteramide plus carbon dioxide, a mixture called the blue fluorescent protein "BFP" (Shimomura and Johnson, 1970). The light emitter is the amide anion of coelenteramide (Hori et al., 1973; Shimomura, 1995b) that exists in BFP in a apoaequorin-bound form. The fluorescence quantum yield of BFP measured a few seconds after the light emission cannot be the quantum yield of the light emitter, because (1) BFP is an easily dissociable, equilibrium complex of coelenteramide, apoaequorin and calcium ions, and the fluorescence of coelenteramide is extremely sensitive to the environment (it is practically nonfluorescent in water) as discussed by Morise et al. (1974), and (2) the conformational change of apoaequorin bound. If the fluorescence quantum yield of the aequorin light emitter at the moment of light emission were as high as that of GFP (0.72), the overall quantum yield after radiationless energy transfer would not necessarily increase. Without data on the fluorescence quantum yield of the aequorin light emitter, a meaningful discussion of energy-transfer mechanism on the basis of quantum yields is difficult.

Another kind of green fluorescent protein, the GFP of the sea pansy *Renilla*, was purified to homogeneity, then physicochemically characterized (Ward and Cormier, 1979). The fluorescence emission peak of *Renilla* GFP (509 nm) was identical with that of *Aequorea* GFP, but the absorption spectrum (peak max 498 nm) was markedly different. The components that are directly involved in the *Renilla* bioluminescence are coelenterazine (the luciferin), a luciferase, a GFP and oxygen, whereas the bioluminescence system of *Aequorea* involves only aequorin, Ca^{2+} and a GFP. In the case of *Renilla*, the addition of coelenterazine into a solution containing the luciferase in the presence of oxygen results in the emission of blue light. However, if the GFP has been added to the luciferase solution before the addition of coelenterazine, green luminescence is emitted with a significant increase in the quantum yield, clearly indicating the occurrence of radiationless energy transfer (Ward and Cormier, 1979).

Thus, in the case of *Renilla*, there must be a sufficiently strong binding affinity between the molecules of the luciferase and the GFP, to make the distance between the chromophores of these two proteins sufficiently short and energy transfer by radiationless process workable. It is apparent that the affinity between *Renilla* luciferase and *Renilla* GFP is much greater than that between aequorin and *Aequorea* GFP.

Regarding the nature of the chromophore, it is believed that the GFPs of *Aequorea*, *Renilla* and many other coelenterates contain an identical chromophore (Ward and Cormier, 1978; Ward et al., 1980); the only exception presently known is the GFP of the jellyfish *Phialidium* that showed a blue-shifted fluorescence emission peak at 497 nm (Levine and Ward, 1982).

1.4 DISCOVERY OF THE STRUCTURE OF GFP CHROMOPHORE

In 1979, I was interested in the chemical structure of the chromophore of *Aequorea* GFP, which had never been studied before. From a papain digest of heat-denatured GFP, I isolated a small peptide containing the chromophore. I synthesized a model compound of the chromophore, and deduced the structure of the GFP chromophore (**D** in Fig. 1.3) based on the clear resemblance between this model compound and the chromophore of the peptide (Shimomura, 1979). It might look as though I were very lucky in my guesswork, because the data obtained from the peptide were clearly insufficient to elucidate the structure of the chromophore. In fact, several people questioned me as to how I could guess the imidazolone structure. The truth is that I was certainly lucky, but not by guesswork.

I was studying the structure of *Cypridina* luciferin in the late 1950s. The techniques for structure determination available at the time were not as sophisticated as at present. The techniques that would give clear-cut information, such as nuclear magnetic resonance (NMR), high-resolution mass spectroscopy, and high-performance liquid chromatography (HPLC), were not available, thus we often had to make a presumption in the course of study. In an early stage of study on the structure of *Cypridina* luciferin, we arrived at a tentative structure that

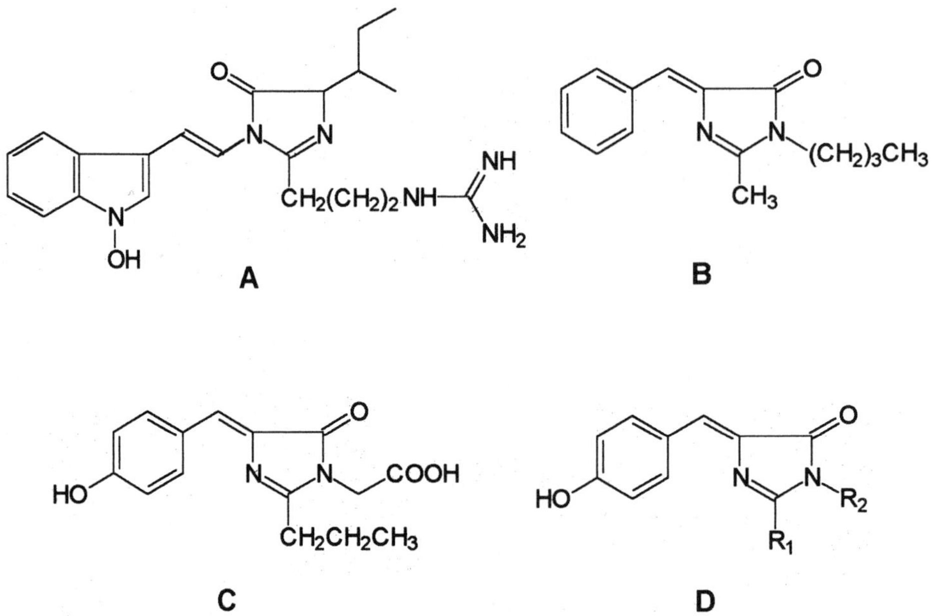

FIG. 1.3 A tentative structure of *Cypridina* luciferin proposed in 1959 (**A**), one of model compounds synthesized to test the feasibility of the structure **A** (**B**), a model compound synthesized for the GFP chromophore (**C**), and the chromophore of GFP proposed in 1979 (**D**). Both R_1 and R_2 are peptide residues.

contained an imidazolone ring (**A**) (Hirata et al., 1959). To test the absorption spectrum of this tentative structure, we synthesized various imidazolone compounds that contained one double bond conjugated with the imidazolone ring (Shimomura and Eguchi, 1960), though the results eventually showed that the structure **A** was incorrect. One of the imidazolones synthesized at that time was **B**. When I obtained the chromophore-bearing peptide from *Aequorea* GFP in 1979, I immediately noticed a close resemblance in spectroscopic and other properties between the chromophore derived from GFP and the imidazolone chromophore **B** that was synthesized some 20 years before. A small difference found in the wavelength of absorption peak was thought to be the effect of a phenolic OH, based on the evidence that the acid hydrolysis of the peptide yielded *p*-hydroxy benzaldehyde. I synthesized a new model compound **C**, of which the spectroscopic properties were in satisfactory agreement with those of the peptide. Thus, structure **D** was proposed as the chromophore of GFP (Shimomura, 1979). The chromophore structure was recently confirmed to be correct, although the side chains were different (Cody et al., 1993).

I learned in 1979 that W. W. Ward of Rutgers University, the pioneer of the isolation of the photosensitive ctenophore photoproteins (Ward and Seliger, 1974a,b), had been working on *Aequorea* GFP in addition to *Renilla* GFP. I thought my role was over and decided to discontinue my work on GFP. Since then, the work on *Aequorea* GFP by Ward and others has steadily progressed, finally developing into the successful cloning of GFP (Prasher et al., 1992), a memorable event that made the foundation of the present volume.

ACKNOWLEDGMENTS

Our work on the jellyfish *Aequorea* was initiated by the late Professor Frank H. Johnson, whose contribution to the project was immeasurable. I thank all the people who contributed directly or indirectly to our work described in this chapter. The work was made possible by the excellent facilities of the Friday Harbor Laboratories, University of Washington, and research grants from the National Science Foundation and National Institutes of Health.

REFERENCES

Arai, M. N. and Brinckman-Voss, A. (1980). Hydromedusae of British Columbia and Puget Sound. *Can. Bull. Fish. Aqua. Sci.* Bulletin 204:1–181.

Cody, C. W., Prasher, D. C., Westler, W. M., Prendergast, F. G., and Ward, W. W. (1993). Chemical structure of the hexapeptide chromophore of the *Aequorea* green-fluorescent protein. *Biochemistry* 32:1212–1218.

Cormier, M. J., Hori, K., and Anderson, J. M. (1974). Bioluminescence in coelenterates. *Biochim. Biophys. Acta* 346:137–164.

Cormier, M. J., Hori, K., Karkhanis, Y. D., Anderson, J. M., Wampler, J. E., Morin, J. G., and Hastings, J. W. (1973). Evidence for similar biochemical requirements for bioluminescence among the coelenterates. *J. Cell. Physiol.* 81:291–298.

Davenport, D. and Nicol, J. A. C. (1955). Luminescence of hydromedusae. *Proc. R. Soc. London, Ser. B* 144:399–411.

Forskal, P. (1775). Descriptiones animalium avium, amphibiorum, piscium, insectorum, vermium: quae in itinere orientali observavit Petrus Forskal. Post mortem auctoris edidit Carsten Niebuhr. 164 pages. *Ex Officina Moller* Hauniae (Copenhagen).

Hastings, J. W. and Morin, J. G. (1969). Comparative biochemistry of calcium-activated photoproteins from the ctenophore, Mnemiopsis and the coelenterates *Aequorea*, Obelia, Pelagia and *Biol. Bull.* 137:402.

Hirata, Y., Shimomura, O., and Eguchi, S. (1959). The structure of Cypridina luciferin. *Tetrahedron Lett.* 4–9.

Hori, K., Wampler, J. E., and Cormier, M. J. (1973). Chemiluminescence of *Renilla* (sea pansy) luciferin and its analogues. *Chem. Commun.* 492–493.

Johnson, F. H., Shimomura, O., Saiga, Y., Gershman, L. C., Reynolds, G. T., and Waters, J. R. (1962). Quantum Efficiency of *Cypridina* luminescence, with a note on that of *Aequorea. J. Cell. Comp. Physiol.* 60:85–104.

Kramp, P. L. (1959). The hydromedusae of the Atlantic ocean and adjacent waters. Dana-Report No. 46. Carlsberg Foundation, Copenhagen, Denmark.

Kramp, P. L. (1965). The hydromedusae of the Pacific and Indian Oceans. Dana Report No. 63. Carlsberg Foundation, Copenhagen, Denmark.

Kramp, P. L. (1968). The hydromedusae of the Pacific and Indian oceans, sections II and III. Dana-Report No. 72. Carlsberg Foundation, Copenhagen, Denmark.

Levine, L. D. and Ward, W. W. (1982). Isolation and characterization of a photoprotein, "phialidin", and a spectrally unique green-fluorescent protein from the bioluminescent jellyfish *Phialidium gregarium*. *Comp. Biochem. Physiol.* 72B:77–85.

Mayer, A. G. (1910). Medusae of the world. Vol. II (Hydromedusae). Carnegie Institute Washington Publication, Washington, D. C., pp. 231–498.

Morin, J. G. (1974). Coelenterate bioluminescence. *In* Coelenterate Biology. Reviews and Perspectives. Muscatine, L. and Lenhoff, H. M. Eds., Academic, New York, pp. 397–438.

Morin, J. G. and Hastings, J. W. (1971a). Biochemistry of the bioluminescence of colonial hydroids and other coelenterates. *J. Cell. Physiol.* 77:305–311.

Morin, J. G. and Hastings, J. W. (1971b). Energy transfer in a bioluminescent system. *J. Cell. Physiol.* 77:313–318.

Morise, H., Shimomura, O., Johnson, F. H., and Winant, J. (1974). Intermolecular Energy Transfer in the bioluminescent system of *Aequorea. Biochemistry* 13:2656–2662.

Murbach, L. and Shearer, C. (1902). Preliminary report on a collection of medusae from the coast of British Columbia and Alaska. *Ann. Mag. Nat. Hist. Ser. 7.* 9:71–73.

Peron, F. and Lesueur, C. A. (1809). Tableau des caracteres generiques et specifiques de toutes les especes de Meduses connues jusqu'a ce jour. *Ann. Mus. Hist. Nat. Paris* 14:325–366.

Prasher, D. C., Eckenrode, V. K., Ward, W. W., Prendergast, F. G., and Cormier, M. J. (1992). Primary structure of the *Aequorea victoria* green fluorescent protein. *Gene* 111:229–233.

Ridgway, E. B. and Ashley, C. C. (1967). Calcium transients in single muscle fibers. *Biochem. Biophys. Res. Commun.* 29:229–234.

Russell, F. S. (1953). The Medusae of the British Isles. Vol. I: Anthomedusae, Leptomedusae, Limnomedusae, Trachymedusae and Narcomedusae. Cambridge University Press, London, 530 p.

Shimomura, O. (1979). Structure of the chromophore of *Aequorea* green fluorescent protein. *FEBS Lett.* 104:220–222.

Shimomura, O. (1995a). A short story of aequorin. *Biol. Bull.* 189:1–5.

Shimomura, O. (1995b). Cause of spectral variation in the luminescence of semisynthetic aequorins. *Biochem. J.* 306:537–543.

Shimomura, O. and Eguchi, S. (1960). Studies on 5-imidazolone. I–II. *Nippon Kagaku Zasshi* 81:1434–1439.

Shimomura, O. and Johnson, F. H. (1969). Properties of the bioluminescent protein aequorin. *Biochemistry* 8:3991–3997.

Shimomura, O. and Johnson, F. H. (1970). Calcium binding, quantum yield, and emitting molecule in aequori bioluminescence. *Nature (London)* 227:1356–1357.

Shimomura, O. and Johnson, F. H. (1972). Structure of the light-emitting moiety of aequorin. *Biochemistry* 11:1602–1608.

Shimomura, O. and Johnson, F. H. (1973). Chemical nature of light emitter in bioluminescence of aequorin. *Tetrahedron Lett.* 2963–2966.

Shimomura, O. and Johnson, F. H. (1975). Regeneration of the photoprotein aequorin. *Nature (London)* 256:236–238.

Shimomura, O. Johnson, F. H., and Saiga, Y. (1962). Extraction, purification and properties of aequorin, a bioluminescent protein from the luminous hydromedusan, Aequorea. *J. Cell. Comp. Physiol.* 59:223–239.

Wampler, J. E., Hori, K., Lee, J. W., and Cormier, M. J. (1971). Structured bioluminescence. Two emitters during both the in vitro and the *in vivo* Bioluminescence of the sea pansy, *Renilla. Biochemistry* 10:2903–2909.

Wampler, J. E., Karkhanis, Y. D., Morin, J. G., and Cormier, M. J. (1973). Similarities in the bioluminescence from the Pennatulacea. *Biochim. Biophys. Acta* 314:104–109.

Ward, W. W. (1979). Energy transfer processes in bioluminescence. *In* Photochemical and Photobiological Reviews. Vol. 4., Smith, K. C., Ed. Plenum, New York, pp. 1–57.

Ward, W. W. and Cormier, M. J. (1978). Energy transfer via protein–protein interaction in *Renilla* bioluminescence. *Photochem. Photobiol.* 27:389–396.

Ward, W. W. and Cormier, M. J. (1979). An energy transfer protein in coelenterate bioluminescence. Characterization of the *Renilla* green fluorescent protein. *J. Biol. Chem.* 254:781–788.

Ward, W. W. and Seliger, H. H. (1974a). Extraction and purification of calcium-activated photoproteins from ctenophores. *Biochemistry* 13:1491–1499.

Ward, W. W. and Seliger, H. H. (1974b). Properties of mnemiopsin and berovin, calcium-activated photoproteins. *Biochemistry* 13:1500–1510.

Ward, W. W. Cody, C. W., Hart, R. C., and Cormier, M. J. (1980). Spectrophotometric identity of the energy transfer chromophores in *Renilla* and *Aequorea* green fluorescent proteins. *Photochem. Photobiol.* 31:611–615.

2

Photons for Reporting Molecular Events: Green Fluorescent Protein and Four Luciferase Systems

J. WOODLAND HASTINGS
Department of Molecular and Cellular Biology, Harvard University, Cambridge, MA

JAMES G. MORIN
Section of Ecology and Systematics, Cornell University, Ithaca, NY

2.1 INTRODUCTION

During the course of evolution, bioluminescence has repeatedly appeared where it serves biological functions important to the organism. Functions may differ among organisms and a given organism may utilize luminescence in more than one way (Morin, 1983; Hastings, 1983; Hastings and Morin, 1991). The different specific recognized functions may be classed under three major rubrics: defensive (to help deter predators), offensive (to aid in predation), and communication (e.g., for courtship or mating). Within each category a number of different specific strategies are recognized; for example, luminescence may be used defensively as a decoy to divert, as a flash to frighten, or as ventral luminescence to camouflage the silhouette.

In terms of the total number of different species, the emission of bioluminescence is rather rare, but it occurs in many phylogenetically different groups (Table 2.1; Harvey, 1952; Herring, 1978). In those groups that do emit light,

Green Fluorescent Protein: Properties, Applications, and Protocols, Edited by Martin Chalfie and Steven Kain
ISBN 0-471-17839-X

TABLE 2.1 Representatives of the major bioluminescent organisms

Type of Organism	Representative Genera	Luciferins and Other Factors (Emission Max. (nm))	Displays and Functions
Bacteria	Photobacterium	Reduced flavin and long luciferase	Steady bright glow
	Vibrio	chain aldehyde (475–540)	Autoinduction of luciferase
	Xenorrhabdus	YFP and LUMP as accessory emitters	Symbiosis
Mushrooms	Panus, Armillaria	Unknown	Steady dim glow; to attract insects
	Pleurotus	(535)	
Dinoflagellates	Gonyaulax	Linear tetrapyrrole	pH change causes short
	Pyrocystis	Cell organelles (scintillons)	(0.1 s) bright flashes;
	Noctiluca	(470)	To frighten or deter
Cnidaria			
Jellyfish	Aequorea	Ca^{2+}, coelenterazine/aequorin	Bright flash or train of flashes;
Hydroid	Obelia	Imidazo pyrazine nucleus	To frighten or deter
Sea Pansy	Renilla	(460–510), GFP as accessory emitter	
Ctenophores	Mnemiopsis; Beroe	Ca^{2+}, coelenterazine (460)	Bright flashes; frighten or deter
Annelids			
Earthworms	Diplocardia	*N*-isovalyeryl-3 amino propanal	Cellular exudates or intracellular
Chaetopterid worm	Chaetopterus	Unknown (465)	flashes sometimes very bright.
Syllid fireworm	Odontosyllis	Unknown (510)	To divert or deter; courtship
Scale worm	Acholoe	Unknown (530)	To divert or deter
Molluscs			
Limpet	Latia	Aldehyde	Exuded luminescence in all three.
Clam	Pholas	Clam luciferin, structure?, Cu^{2+}	Photophores and symbiotic bacteria
Squid	Heteroteuthis	Bacterial symbionts (490)	in some squid. Functions: diversion, decoy, camouflage, probably others
Crustacea			
Ostracod	Vargula; Cypridina	Imadazopyrazine nucleus (465)	Squirts enzyme and substrates
			Diversion, decoy, courtship
Shrimp	Meganyctiphanes	Linear tetrapyrrole (470)	Photophores; camouflage
Copepods and others	Gaussia	Unknown	Deter predators

TABLE 2.1 (contd)

Type of Organism	Representative Genera	Luciferins and Other Factors (Emission max. (nm))	Displays and Functions
Insects			
Coleopterids (beetles)			
firefly	Photinus, Photuris	Benzothiazole, ATP, Mg^{2+}	Flashes, specific kinetic patterns
click beetles	Pyrophorus	Similar chemistry in all coleoptera	Deter predators; courtship, mating
railroad worm	Phengodes, Phrixothrix	(most 550–580)	
Diptera (flies)	Arachnocampa	Biochemistry unknown (~480)	Lure to attract prey
Echinoderms			
Brittle stars	Ophiopsila	Biochemistry unknown, Ca^{2+}	Trains of rapid flashes; frighten, divert
Chordates			
Tunicates	Pyrosoma	Cell organelles evolved from bacteria (480-500)	Brilliant trains of flashes; function unknown Stimulated by light and other factors
Fishes			
Cartilaginous Fishes	Isistius	Unknown	Ventral glow; camouflage
Bony Fishes			
Ponyfish	Leiognathus	Symbiotic luminous bacteria (~490)	Ventral luminescence and flashes;
Flashlight fish	Photoblepharon	Symbiotic luminous bacteria (~490)	camouflage, attract and capture prey, courtship, deter predators, communication
Angler fish	Cryptopsaras	Symbiotic luminous bacteria (~490)	
Midshipman	Porichthys	Self-Luminous, Vargula type luciferin, Nutritionally obtained (485)	Ventral luminescence; camouflage, courtship
Midwater fishes			Camouflage, courtship, deterrence, capture prey
	Cyclothone	Self-luminous, biochemistry unknown	Many photophores, ventral and lateral
	Neoscopelus	Self-luminous, biochemistry unknown	Photophores: lateral, on tongue
	Tarletonbeania	Self-luminous, biochemistry unknown	Sexual dimorphism; males have dorsal (police car) photophores

the biochemical and physiological mechanisms responsible for it are often very different, as are its several functional roles. Indeed, luciferase (the enzyme) and luciferin (the substrate) are generic terms, and quite different in the different groups. Thus, the organism from which they are obtained must be specified.

Bioluminescence is thus *not* an evolutionarily conserved function; in the different groups of organisms the genes and proteins involved are mostly unrelated, and evidently originated and evolved independently. How many times this may have occurred is difficult to say, but it has been estimated that present day luminous organisms come from as many as 30 different evolutionarily distinct origins (Hastings, 1983; Hastings and Morin, 1991). Cnidarian luminescence, with green fluorescent protein (GFP) present as an accesssory emitter in some but not all species, is thus only one of a wide array of luminescent systems and biochemistries. The genes and proteins of several of the other systems (notably bacterial and firefly) have been used for many different analytical and reporter purposes (DeLuca, 1978; Hastings et al., 1997). Their diversity allows for many different possible applications in which photons are the reporter, but none of the other proteins or systems possesses the unique features of GFP.

The biochemistries of only four of the different luminous systems are known in detail (Table 2.1), namely, bacteria, dinoflagellates, cnidarians, and fireflies (Hastings and Morin, 1991). Although some information is known for another half-dozen or so, we confine our review to the four best described. While these systems differ in the structures of the luciferins (Fig. 2.1) and luciferases, all systems have some features in common at the chemical level. All known luciferases are oxygenases that utilize molecular oxygen to oxidize the associated luciferin, giving an intermediate enzyme-bound peroxide, whose breakdown then results in the production of an intermediate or product directly in its excited singlet state.

In most systems, emission occurs from the luciferase-bound substrate-derived excited molecule (see Fig. 2.1), but an accessory secondary emitter occurs in certain cnidarians and some bacteria. With GFP in cnidarians, the mechanism responsible was postulated and later confirmed to involve Förster-type energy transfer (Morin and Hastings, 1971b; Morise et al., 1974; Ward et al., 1980). GFP is unusual in that its chromophore is a part of the (modified) primary structure of the protein, thus not subject to dissociation (Cody et al., 1993; Heim et al., 1994). Its use as a transgene reporter, pioneered by Chalfie et al. (1994), relies on this feature and its fluorescence alone.

2.2 CNIDARIANS, CTENOPHORES AND GFP

Luminescence is common and widely distributed in these groups (Morin, 1974; Cormier, 1981; Herring, 1978). In the ctenophores (comb jellies), they comprise over one-half of all genera, whereas in the cnidarians it is about 6%. These organisms are mostly sessile, sedentary, or planktonic, and upon stimulation emit light as flashes. Hydroids such as *Obelia* occur as plant-like growths, typically adhering to rocks and kelp below low-tide level in many of the world's oceans. Upon stimulation, a conducted scintillating emission emanates as a wave along the colony from individual photocytes (cells specialized for light emission);

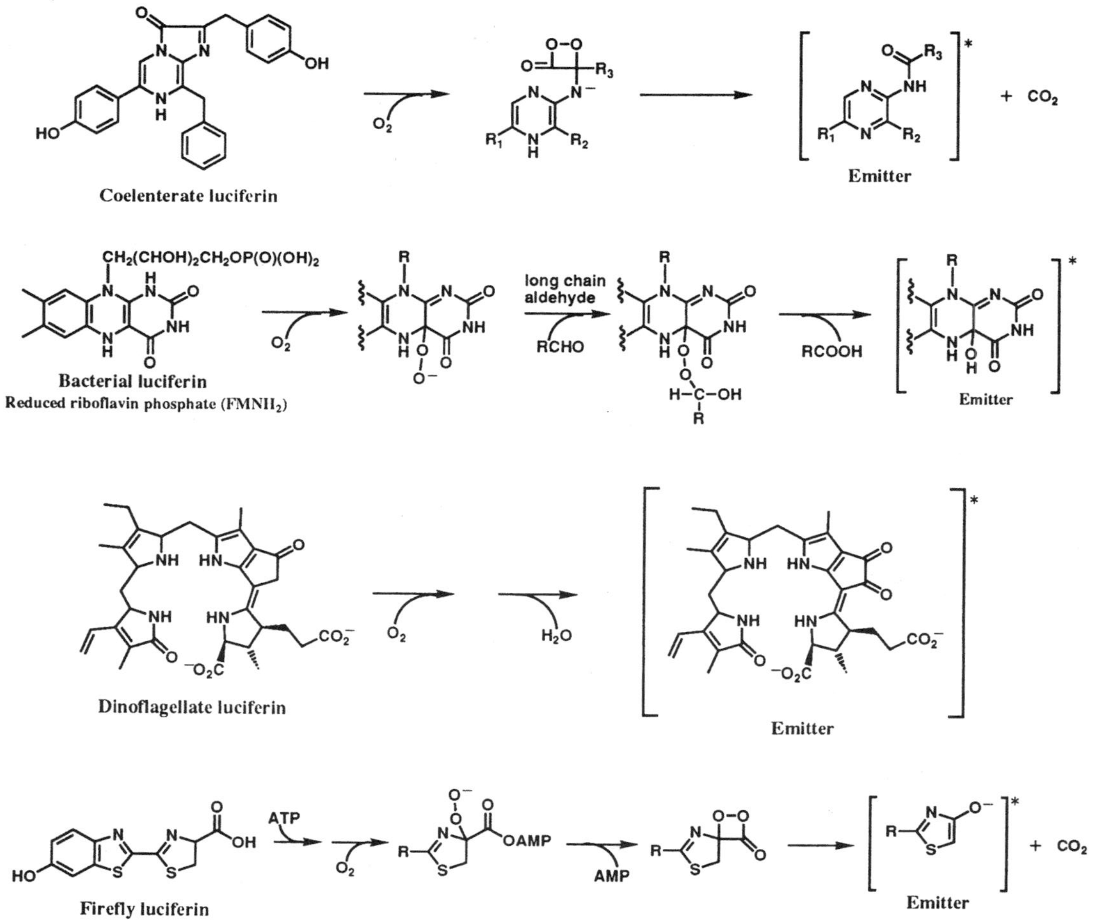

Coelenterate luciferin

O_2

Emitter

+ CO_2

$CH_2(CHOH)_2CH_2OP(O)(OH)_2$

Bacterial luciferin

Reduced riboflavin phosphate ($FMNH_2$)

O_2

long chain aldehyde

RCHO

RCOOH

Emitter

Dinoflagellate luciferin

O_2

H_2O

Emitter

Firefly luciferin

ATP

O_2

AMP

Emitter

+ CO_2

FIG. 2.1 Structures of four different luciferins, oxygen-containing intermediates, and postulated emitters. (see Table 2.1).

repetitive waves may occur from a single stimulus. *Aequorea*, a hydromedusan that is very abundant in the San Juan Islands region of the northwest United States has been extensively used for biochemical studies (Shimomura and Johnson, 1975; Cormier et al., 1989). The biochemistry of the sea pansy, *Renilla*, which occurs near shore on sandy bottoms, has also been elucidated (Cormier, 1981).

Early observations, attributable to what we now know as GFP, were reported by several investigators, including an emission spectrum of the bioluminescence of the sea pen *Pennatula phosphorea* showing a narrow bandwidth emission in the green (Nicol, 1958), which is now known to be characteristic of GFP. Shimomura et al. (1962) later noted that the bioluminescence of *Aequorea* extracts was blue while that of the intact organism was green, attributing this to a protein that fluoresced green in extracts (Johnson et al., 1962). More suggestive of a relationship between luminescence and fluorescence was the observation of green fluorescence in cells located in the vicinity of bioluminescence activity by Titschack (1964) in the pennatulacean *Veretillum cynomorium*. In none of these studies, however, was the relationship of the green fluorescence to the bioluminescence clearly established.

We discovered GFP quite independently while we were examining both the mechanisms controlling luminescent flashes and the biochemical underpinnings of the luminescence in the colonial hydroid *Obelia geniculata*. Our biochemical studies quickly expanded to studying calcium activated photoproteins in a variety of coelenterates, including several species of hydrozoans (*Obelia, Aequorea, Clytia*), pennatulaceans (the sea pens *Renilla* and *Ptilosarcus*), the scyphozoan jellyfish *Pelagia*, and the ctenophore *Mnemiopsis* (Morin et al., 1968; Hastings and Morin 1969a,b). Measurements of emission spectra of reactions in extracts gave wide bandwidth curves peaking in the blue, but somewhat different in different species; the blue emissions of the luciferase systems of *Aequorea, Obelia,* and *Renilla* exhibit maxima at 460, 472, and 486 nm, respectively (Fig. 2.2). *In vivo* luminescent spectra from all three, however, were narrow, peaking in the green at about 508 nm, matching fluorescence emission spectrum of GFP (Morin and Hastings, 1971a). So while the underlying biochemistry may differ somewhat, emissions are all in the green peaking at about 508 nm so long as the luciferases interact appropriately with GFP in order to transfer excitation energy. GFP is thus an accessory emitter protein of the cnidarian luminescent system, deriving its excitation by nonradiative energy transfer in association with the luciferase reaction, which in the absence of GFP emits blue light (Morin and Hastings, 1971b).

The biochemical studies provided critical information for our physiological and morphological studies on the colonial hydroid *O. geniculata,* where action potentials propagate through electrically excitable epithelial cells. These action potentials spread incrementally through a colony and repetitively elicit flashes from individual photocytes (Fig. 2.3), which are located along the length of the stems and pedicels but not in the polyps themselves (see color Fig. 2.4; Morin et al., 1968; Morin and Reynolds, 1969; Morin and Cooke, 1971a,b). Indeed, it was the green fluorescence of the protein that allowed us, in conjunction with image intensification, to identify the photocytes by the co-localization of the fluorescence and bioluminescence of the photocytes and to establish irrefutably the

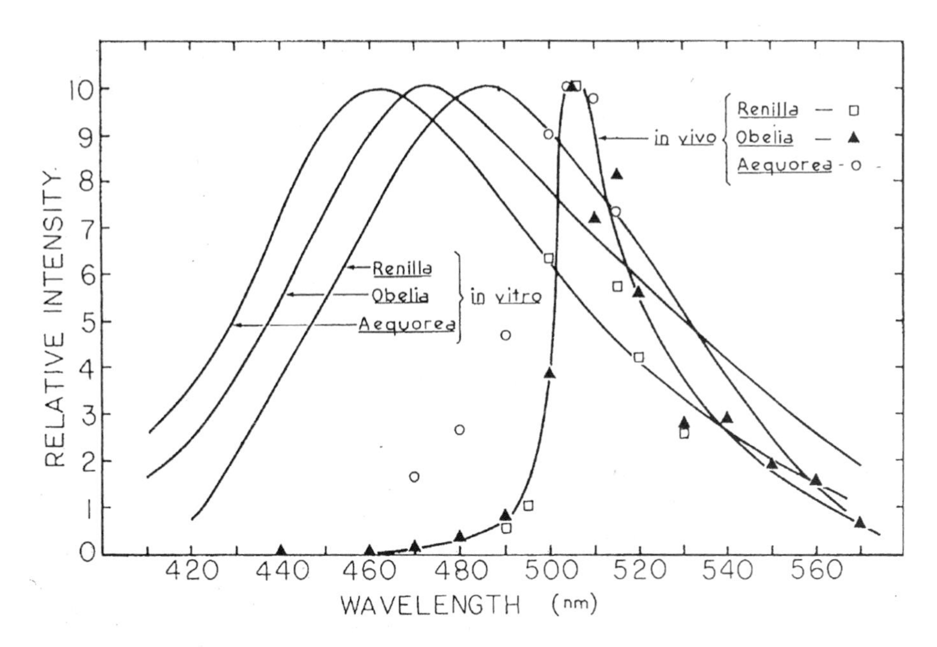

FIG. 2.2 Emission spectra for bioluminescence *in vivo* for *Obelia, Aequorea,* and *Renilla* compared with spectra from emission of *in vitro* reactions isolated from these same organisms. [From Morin and Hastings, 1971b].

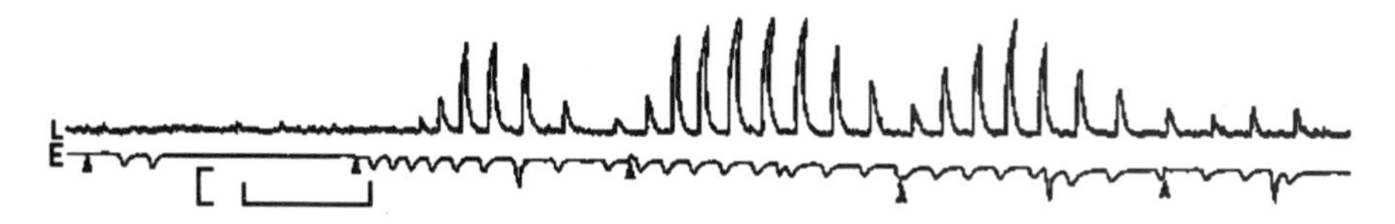

FIG. 2.3 Luminescent flash (*L*) and luminescent potential (*E*) bursts recorded concurrently from an *O. geniculata* photocyte and hydranth respectively. Responses are to a train of stimuli applied about once every 2 s, indicated by solid triangles beneath the lower trace (horizontal bar = 1 s); lower (*E*) ordinate, vertical bar = 1 mV; upper (*L*) ordinate, light intensity in relative units. [From Morin and Cooke 1971a].

association of GFP with the luminescent system (Fig. 2.5; Morin and Reynolds, 1969, 1970, 1974).

In different species, GFP is confined to discrete photocytes (~10–20 μm in diameter), which are either dispersed [Fig. 2.4a] or clumped [Fig. 2.4B) in specific locations within the gastrodermis of the colonies (Morin and Reynolds, 1969, 1970, 1974; Morin, 1974). In measuring *in vivo* bioluminescence from single *Obelia* photocytes, GFP allowed us to identify their location prior to stimulation, so as to record photometrically via a fine-tipped (0.5 mm) light guide (Fig. 2.3). Trains of action potentials, termed luminescent potentials, initiated by single stimuli, propagate via electrically excitable epithelial cells (rather than neurons). These action potentials spread incrementally through a colony and repetitively excite the photocytes, and also couple to other neuroid conducting

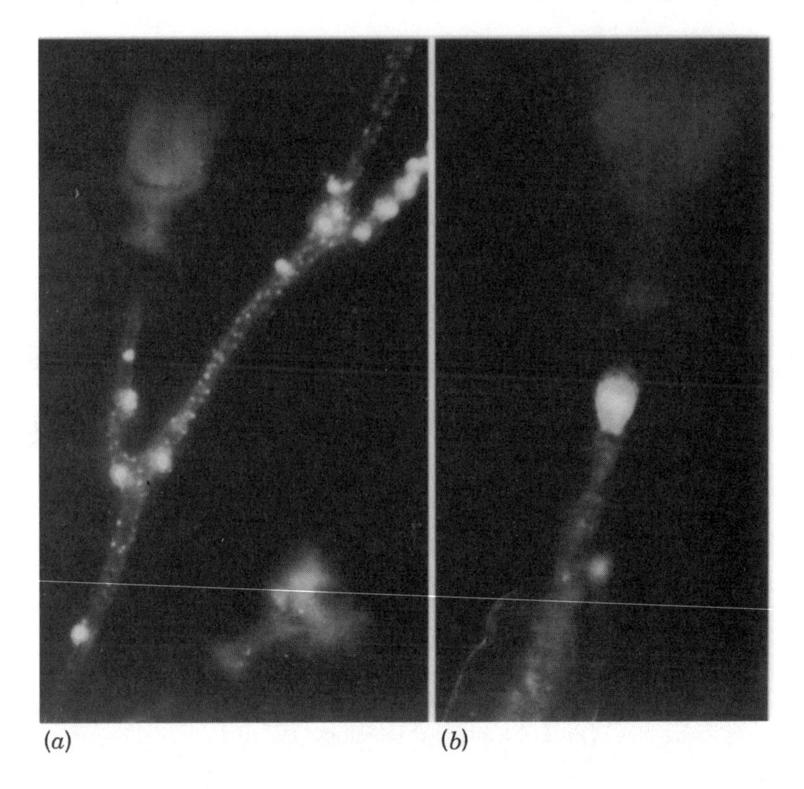

FIG. 2.4 Fluorescence micrograph of a living colony of *Obelia* species showing photocytes visualized by GFP. Height of field shown is about 3 mm in (*a*) and 1.5 mm in (*b*). (*a*) Dispersed photocytes (bright green spots) in an upright of *O. geniculata* (two polyps also shown at lower right and upper left). (*b*) Concentrated photocytes at the tip of a pedicel below the base of a hydranth of *O. bidentata* (=*bicuspidata*). Figure also appears in color section.

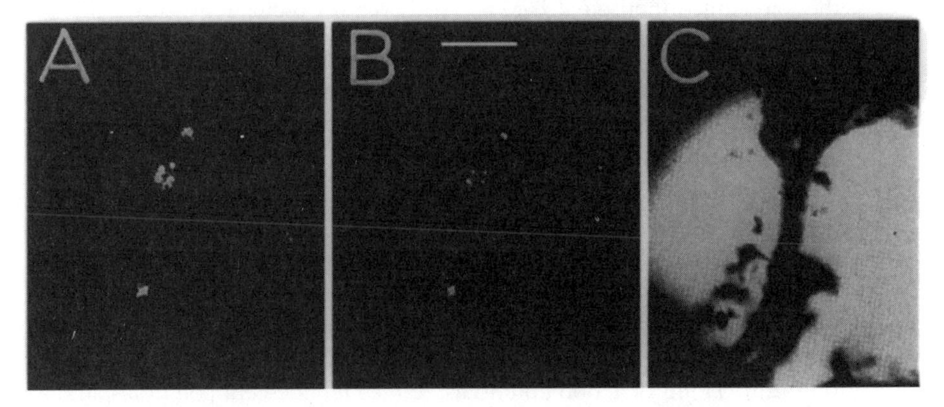

FIG. 2.5 Bioluminescence (A), fluorescence (B), and rear illumination (C) pictures of an *Obelia geniculata* upright. Note that the luminescent and fluorescent spots (six in each) directly superimpose. The scale bar indicates 200 μm. [From Morin and Reynolds 1974.]

systems such as those governing polyp contraction. Based on the number of photocytes, as determined with the aid of GFP, we were able to calculate from photometric measurements that each cell could emit about 1–2 × 10^8 quanta/cell (Shorey and Morin, 1974).

Studies of the GFP in the hydroid *Obelia* were also instrumental in providing the first demonstration that gap junctions can pass chemical signals in excitable tissues (Dunlap et al., 1987; Brehm et al., 1989). They showed that calcium actually enters neighboring nonluminescent, but electrically excitable, epithelial cells via voltage-dependent calcium channels and then into the photocytes, which are nonexcitable, via secondary calcium diffusion through gap junctions. This calcium then triggers the luminescence from the calcium activated photoprotein (obelin) with subsequent energy transfer to and emission from the GFP. Finally, by using GFP fluorescence as a reporter for the spatial distribution of luminescent cells in pennatulaceans (sea pens; see color Fig. 2.6) and photometry to measure the temporal aspects of the light emission, we have been able to infer that luminescence in sea pens and probably all cnidarians functions as an aposematic signal to deter damage to the colonies by potential aggressors or predators such as fishes and crustaceans (Morin, 1976, 1983). This inference has been experimentally verified for both pennatulaceans and brittle stars by Grober (1988a,b).

At the biochemical level, the luciferin (coelenterazine) is the same in different cnidarian luminescent systems. Coelenterazine possesses an imidazopyrazine skeleton (Fig. 2.1) and is notable for its widespread phylogenetic distribution (Thomson et al., 1997), but whether the reason is nutritional or genetic (hence, possible evolutionary relatedness) has not yet been elucidated. But there are differences between the cnidarian anthozoan and hydrozoan systems with regard to the site of calcium action. In the anthozoan, coelenterazine is sequestered by a Ca^{2+}-sensitive binding protein, and Ca^{2+} causes its release, thus triggering the *in vivo* flash. The *Renilla* luciferase reaction (EC 1.13.12.5) does not itself require calcium (Lorenz et al., 1991). In the hydrozoan *Aequorea* calcium reacts instead at the luciferase stage, namely with aequorin, a luciferase- bound hydroperoxy coelenterazine intermediate, poised for the completion of the reaction.

Aequorin was isolated by Shimomura et al. (1962) from the jellyfish *Aequorea* [in the presence of ethylenediaminetetraacetate (EDTA) to chelate calcium] and shown to emit light simply upon the addition of Ca^{2+}, which is presumably the trigger *in vivo* (Hastings and Morin 1971; Blinks et al., 1982; Cormier et al., 1989). It was postulated (Hastings and Gibson, 1963) that *in vivo* luciferin coelenterazine reacts with oxygen, catalyzed by its luciferase (EC1.13.12.5), to form the hydroperoxide in a calcium-free compartment (the photocyte), where it is stored. Excitation allows Ca^{2+} to enter and bind to the protein (which possesses homology with calmodulin; Lorenz et al., 1991), changing its conformation so that the reaction continues, but without the need for free oxygen at this stage. It had been reported in the early literature (Harvey, 1952) that coelenterates could inexplicably emit bioluminescence without oxygen. The explanation is now evident.

Aequorin luminescence has been widely used for the detection and measurement of calcium, most especially in living cells, into which aequorin can be microinjected (Blinks et al., 1982). The first such experiment was reported by

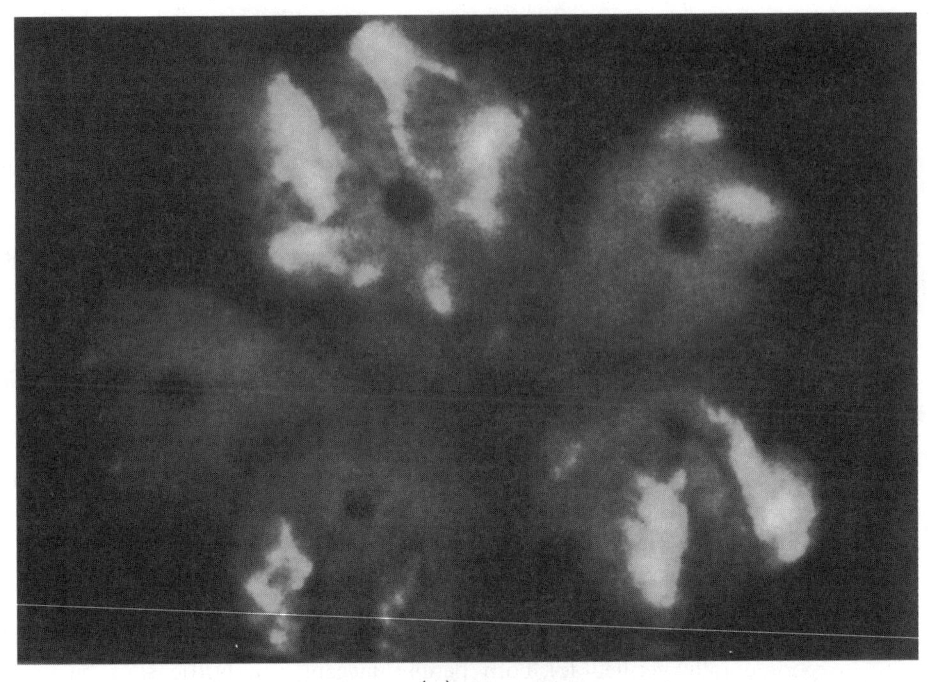

(*a*)

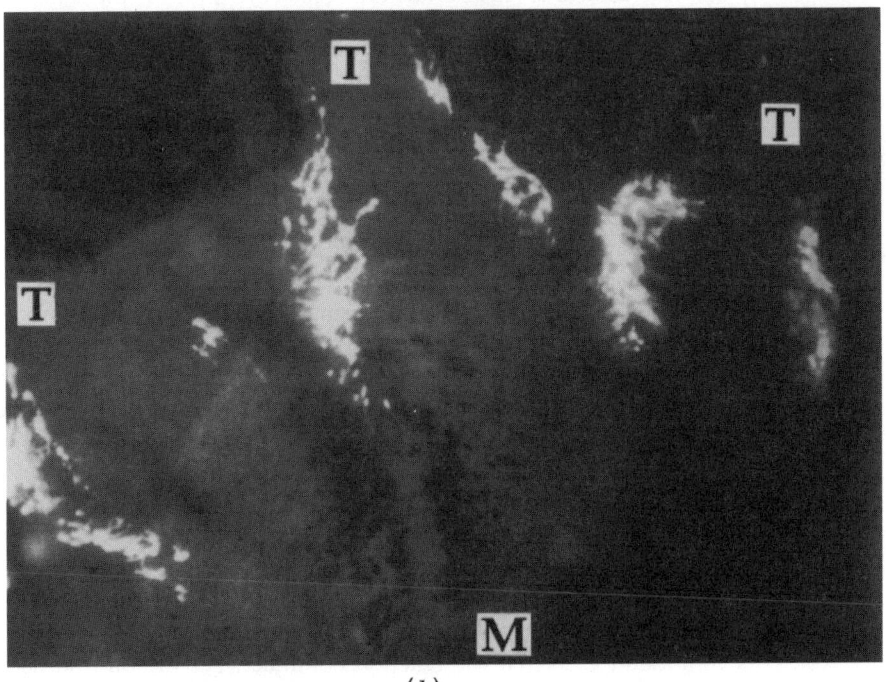

(*b*)

FIG. 2.6 Fluorescence micrographs of photocytes visualized by GFP in living pennatulacean (sea pen) colonies. Width of field shown is about 0.8 mm in (*a*) and 1.6 mm in (*b–d*). (*a*). Photocytes in a cluster of five siphonozooids (water pumping polyps) of *Renilla kollikeri*. (*b*) Photocytes clustered in the lateral-axial region of the tentacles and oral disk of an autozooid (feeding polyp) of *Renilla kollikeri* (mouth [M] and base of three of eight tentacles [T] shown).

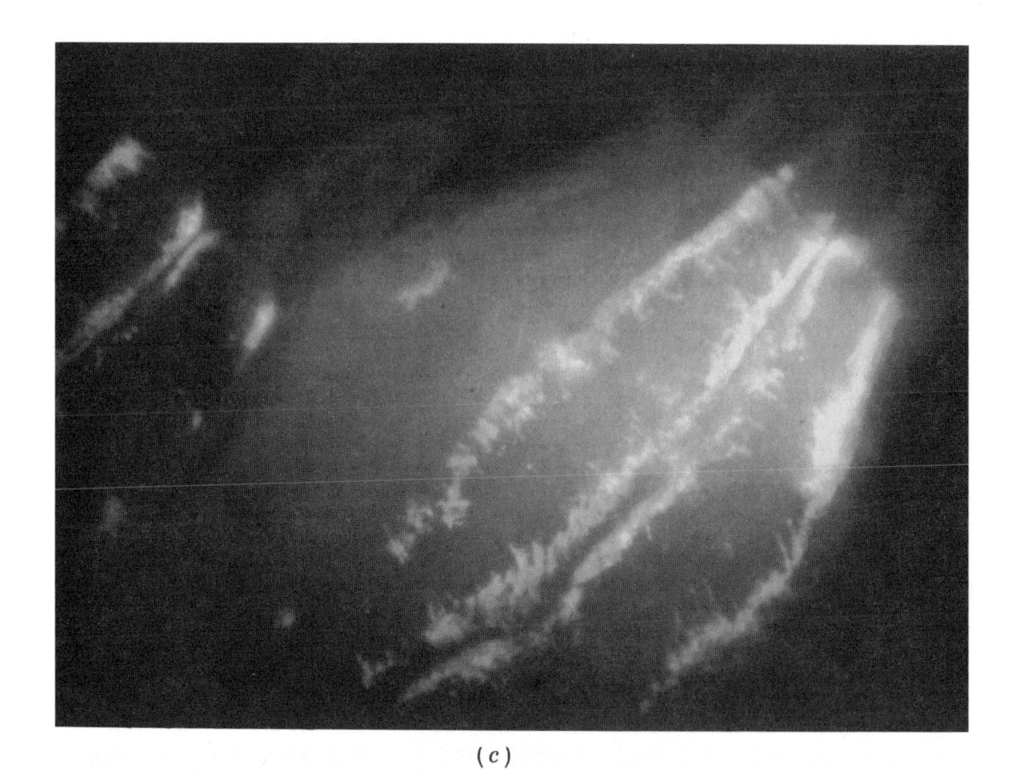

(c)

(d)

FIG. 2.6 (continued) (c) photocytes clustered in only the two outer (of the eight) chambers within the calyx of the column (and not the tentacles) of an autozooid of *Acanthoptilum gracile*. (d) photocytes clustered laterally along the length of each of the eight tentacles (T) of an autozooid of *Ptilosarcus guerneyi* (M = mouth). Figure also appears in color section.

Ridgeway and Ashly (1968), in which they detected a calcium transient accompanying the contraction of single muscle fibers. Since then there have been many analogous applications (Blinks et al., 1982), making aequorin an important tool in analytical biochemistry, physiology, and developmental biology.

Apoaequorin, which functions as the luciferase in this system, has been cloned and expressed in other cell types (Inouye et al., 1986, 1989; Tanahashi et al., 1990) where, in the presence of exogenously added coelenterazine, it serves to monitor intracellular calcium levels. For example, expressed as a transgene in *Dictyostelium*, it was used to monitor intracellular calcium changes in response to cyclic adenosinemonophosphate (cAMP) stimulation (Saran et al., 1994). In tobacco and *Arabidopsis* plants, the expressed transgene revealed circadian oscillations in free cytosolic calcium (Johnson et al., 1995); when targeted to the chloroplast, circadian chloroplast rhythms were likewise observed.

2.3 BACTERIA

Luminous bacteria (see color Fig. 2.7) occur ubiquitously in sea water samples. A primary habitat where most species abound is in association with another (higher) organism, dead (saprophytes) or alive (parasites or symbionts), where growth and propagation occur. Specific associations involve specialized light organs (e.g., in fish and squid; Ruby, 1996) in which a pure culture of luminous bacteria is maintained at a high density and at high light intensity (Nealson and Hastings, 1991). Parasitic and commensal relationships are also known. Terrestrial luminous bacteria are rare, the best described being those harbored by nematodes that are parasitic on insects such as caterpillars.

Luminous bacteria emit light continuously, peaking at about 490 nm if no accessory protein is present. When strongly expressed, a single bacterium may emit 10^4–10^5 photons s^{-1}. The luciferase is a flavin mixed-function monooxygenase (EC 1.14.14.3), and its presence is diagnostic for a bacterial symbiotic involvement in the luminescence of a higher organism. The pathway constitutes a shunt of cellular electron transport at the level of flavin; reduced flavin mononucleotide (FMN) (Fig. 2.1) reacts with oxygen in the presence of bacterial luciferase to produce an intermediate peroxy flavin, which then reacts with a long-chain aldehyde (tetradecanal) to form the acid and the luciferase-bound hydroxy flavin in its excited state (Hastings et al., 1985; Baldwin and Zeigler, 1992). Although there are two substrates in this case, the flavin can claim the name luciferin on etymological grounds, since it forms (bears) the emitter. The bioluminescence quantum yield has been estimated to be about 30%, the same as the fluorescence quantum yield of FMN. Curiously, no other flavin monoxygenases have been found to emit light, even at very low quantum yields, and no genes with significant sequence similarities have been recorded in any of the data bases.

Bacterial luciferases are heterodimeric (alpha–beta) proteins (~80 kDa) in all species; they are relatively simple, having no metals, disulfide bonds, prosthetic groups, or non-amino acid residues. The X-ray structure has recently been reported (Fisher et al., 1995). An interesting feature of the reaction is its inherent slowness: At 20°C the time required for a single catalytic cycle is about 20 s. The luciferase peroxy flavin itself has a long lifetime; at low temperatures

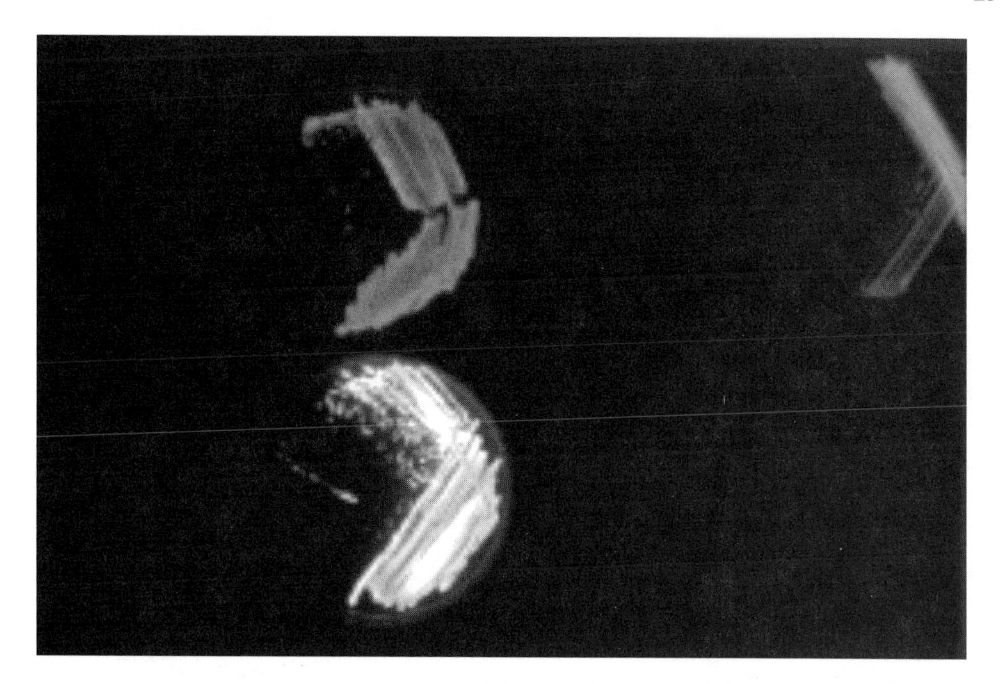

FIG. 2.7 Streaks of luminescent bacteria photographed by their own light, showing two strains of *Photobacterium fischeri*, one of which emits yellow light by virtue of having YFP (yellow fluorescent protein). The other lacks YFP, emitting only blue light. Figure also appears in color section.

(0 to −20°C) it has been isolated, purified, and characterized (Hastings et al., 1973). It can be further stabilized by long-chain alcohols and amines, which bind at the aldehyde site.

Two major operons contain genes for the luciferase and other proteins associated with the luminescent system, including enzymes that serve to maintain the supply of myristic aldehyde (Fig. 2.8a; Meighen, 1991). There are also genes that specifically control the development and expression of luminescence. This fascinating mechanism is called "autoinduction" (Nealson et al., 1970), in which the transcription of the luciferase and aldehyde synthesis genes of the *lux* operon is regulated by genes of the operon itself. A substance produced by the cells called autoinducer (a homoserine lactone; Eberhard et al., 1981; Fig. 2.8b) is a product of the *lux* I gene. The ecological implications are evident: In planktonic bacteria, a habitat where luminescence has no apparent value, autoinducer cannot accumulate, and no luciferase synthesis occurs (Nealson and Hastings, 1991). However, in the confines of a light organ, high autoinducer levels are reached and the luciferase genes are transcribed. Interestingly, it has recently been discovered that an autoinduction-type mechanism, now dubbed quorum sensing (Fuqua et al., 1994), similarly controls the expression of other specific genes in several different groups of bacteria.

Bacterial *lux* genes have been used as reporters in numerous instances (Chatterjee and Meighen, 1995), such as for visualizing gene expression in *Streptomyces* (Schauer et al., 1988) and following the circadian regulation of

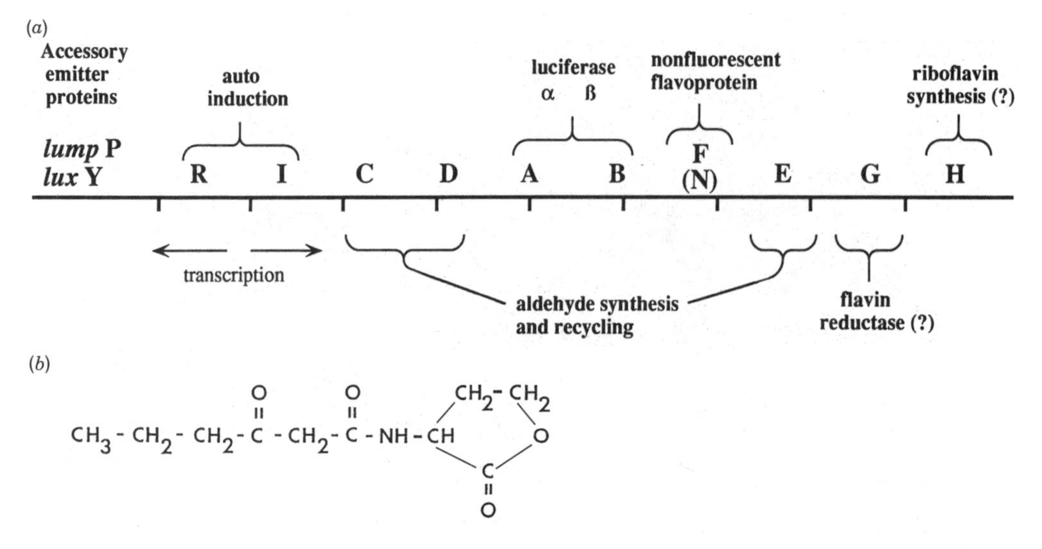

FIG. 2.8 Organization of the *lux* genes (*a*) and the homoserine structure of autoinducer (*b*) in *Vibrio fischeri*. The operon on the right, transcribed from the 5′ to the 3′ end, carries genes for synthesis of autoinducer (*lux* I), for luciferase α and β peptides (*lux* A and B), and for aldehyde production (*lux* C, D, and E). *Lux* R, transcribed from the operon on the left, codes for a receptor molecule that binds autoinducer, controlling the transcription of the right operon. Other genes, *lux* F (N), G, and H (right), are associated with the operon but with still uncertain functions. The genes coding for accessory fluorescent proteins Lump and YFP are located to the left.

transcription in cyanobacteria (Kondo et al., 1994), to name only two. In these cases it is necessary to supply exogenous aldehyde (as a vapor), but the reduced flavin substrate need not be added, since it is generally present in all cells. Bacterial luciferase is also useful in many analytical applications, where flavin or aldehyde, or any enzyme linked to nicotinamide adenine dinucleotide (NAD) or NAD phosphate (NADP) can be assayed (Hastings et al., 1997).

Several species and strains of luminous bacteria also contain accessory proteins that, like GFP, serve as secondary emitters. These include both blue- and red-shifted emissions. As with GFP, light is emitted from the luciferase reaction alone (in the absence of a second emitter protein), with the emission peaking at about 490 nm. The blue-shifted emission, due to a lumazine protein (LUMP), peaks at about 475–480 nm; the dissociable chromophore is identified as 6,7-dimethyl-8-ribitylumazine (Small et al., 1980; Petushkov et al., 1995a). A yellow emission peaking at 540 nm in a strain of *V. fischeri* is due to an analogous yellow fluorescent protein (YFP; recently shown to have homologies with LUMP) in which the chromophore is flavin mononucleitide (FMN) or riboflavin (Hastings et al., 1985; Macheroux et al., 1987; Karatani and Hastings, 1993; Petushkov et al., 1995b). In the YFP system, evidence has been obtained that energy transfer alone cannot account for the yellow emission (Eckstein et al., 1990). In that case a direct population of the excited state of the accessory emitter may occur, without the intermediacy of the luciferase-bound excited state.

2.4 DINOFLAGELLATES

Dinoflagellates occur ubiquitously in the oceans as planktonic forms, and contribute substantially to the bioluminescence commonly seen at night (especially in summer) when the water is disturbed. They occur primarily in surface waters and many species are photosynthetic. In the phosphorescent bays (e.g., in Puerto Rico and Jamaica), high densities of a single species (*Pyrodinium bahamense*) usually occur. The so called red tides are blooms of dinoflagellates, and some of these are bioluminescent.

About 6% of all dinoflagellate genera contain luminous species, but since there are no luminous dinoflagellates among the fresh water species, the proportion of luminous forms in the ocean is higher. As a group, dinoflagellates are important as symbionts, notably for contributing photosynthesis and carbon fixation in animals, especially corals. But unlike bacteria, no luminous dinoflagellates are known from symbiotic niches. Bioluminescent flashing is postulated to help reduce predation either by directly diverting predators or by revealing the location of the predators to their predators (Buskey et al., 1983; Hastings and Morin, 1991; Mensinger and Case, 1992; Fleisher and Case, 1995).

Luminescence in dinoflagellates is emitted from many small (~0.5 μm) cortical locations. The structures have been identified as novel organelles, termed the scintillons (flashing units). They occur as outpocketings of the cytoplasm into the cell vacuole, like balloons, with their necks remaining connected (Fig. 2.9). Scintillons contain only two major proteins, dinoflagellate luciferase (LCF) and luciferin binding protein (LBP) (Desjardins and Morse, 1993); the latter sequesters luciferin and prevents it from reacting with luciferase. Ultrastructurally, these proteins can be identified by immunolabeling (Nicolas et al., 1987, 1991; Fritz et al., 1990), and visualized with image intensification by their bioluminescent flashing following stimulation (Johnson et al., 1985), as well as by the fluorescence of luciferin, the emission spectrum of which is the same as the bioluminescence. Dinoflagellate luciferin is a novel tetrapyrrole related to chlorophyll (Fig. 2.1).

Activity (quantum yield, 0.2) can be obtained in extracts made at pH 8 simply by shifting the pH from 8 to 6; it occurs in both soluble and particulate (scintillon) fractions (Fogel and Hastings, 1971, 1972). The existence of activity in both fractions is explained by the rupture of some scintillons during extraction, while others seal off at the neck and form closed vesicles. With the scintillon fraction, the *in vitro* activity occurs as a flash (~100 ms), very similar to that of the living cell, and the kinetics are independent of the dilution of the suspension. For the soluble fraction, the kinetics depend on dilution, as in enzyme reactions.

A distinctive feature of the reaction is that the binding of luciferin to LBP is pH dependent, being bound at pH 8 and free at pH 6. Thus, the flashing of dinoflagellates *in vivo* is postulated to result from a transient pH change in the scintillons, triggered by an action potential in the vacuolar membrane which, while sweeping over the scintillon, opens ion channels that allow protons from the acidic vacuole to enter (Fig. 2.9).

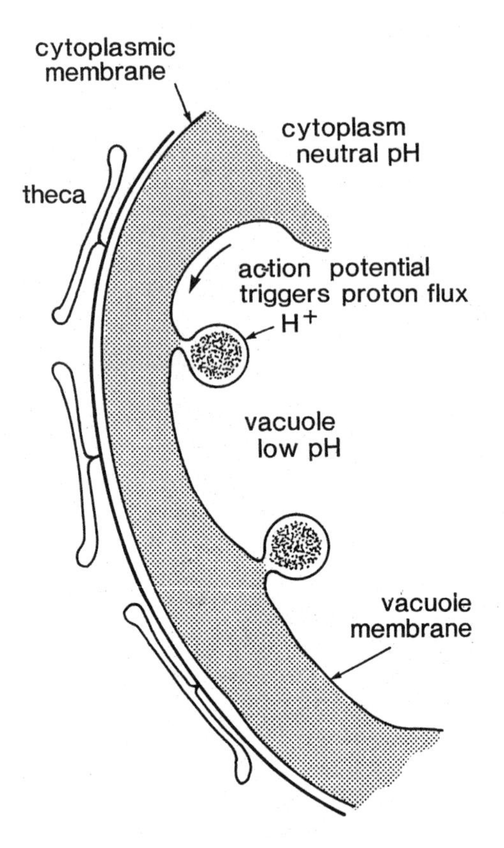

FIG. 2.9 A cartoon depicting scintillons of dinoflagellate, the organelles responsible for flashing light emission. They are formed as outpocketings of the cytoplasm projecting into the acidic vacuole.

The genes for the two dinoflagellate luminescence proteins have been cloned and sequenced (Lee et al., 1993; Bae and Hastings, 1994; Li et al., 1997); there are no introns in either gene. Both proteins are synthesized and destroyed each day, mediated translationally for LBP by proteins that bind to its mRNA 3′ untranslated region (Johnson et al., 1984; Morse et al., 1989; Mittag et al, 1994). Both of the cloned genes produce active proteins; when expressed *in vitro*, LBP exhibits a pH dependent binding of luciferin while LCF catalyses the oxidation of luciferin to give light.

LCF has an interesting and unusual feature (Li et al., 1997). The approximately 140 kDa protein has three tandem repeat domains (~377, 377, and 375 aa long, with no spacer sequences between). Recombinant proteins expressed from the three individual domains of the messenger ribonucleic acid (mRNA) are all separately active as luciferases. This means that in the scintillon three different sites in the molecule could be concurrently contributing to the activity of the luciferase: a three-ring circus with the same act in all three rings.

Genes for LBP and LCF have no homologies or similarities with other luciferases or other sequences in any of the data bases. This distinctiveness is consistent with the hypothesis that luciferases have arisen independently in evolution. However, the 5′ ends of both genes are about 50% homologous over a 90-nt region; for the luciferase, this constitutes the entire remainder of its sequence outside the three repeat regions. Both proteins bind luciferin, but since this region is not needed for luciferase activity, it must have some other function. It might be a sequence for targeting the proteins to the vacuolar membrane in the formation of scintillons.

2.5 FIREFLIES AND OTHER INSECTS

Out of a total of approximately 75,000 insect genera, there are only about 100 classed as luminous. But where seen, their luminescence is impressive, most notably in the many species of beetles: the fireflies and their relatives. Fireflies themselves possess ventral light organs on posterior segments, the South American railroad worm, *Phrixothrix*, has paired green lights on the abdominal segments and red head lights, while the click and fire beetles, Pyrophorini, have both "running lights" (dorsal) and "landing lights" (ventral). From ceiling perches, the dipteran cave glow worms (true flies, not beetles; they occur in New Zealand and Australia) use their light to attract flying prey, which are then entrapped.

In fireflies, communication in courtship is the major function of luminescence in fireflies; one sex emits a flash as a signal, to which the other responds, usually in a species-specific pattern (Lloyd, 1977, 1980; Case, 1984). The time delay between the two may be a signaling feature; for example, it is precisely 2 s in some North American species. But the flashing pattern (e.g., trains distinctive in duration and/or intensity) is also important in some cases, as is the kinetic character of the individual flash (duration; onset and decay kinetics). In some species, flickering occurs within the flashes, sometimes at very high frequencies (~40 Hz). Fireflies in Southeast Asia are particularly noteworthy for their synchronous flashing; congregations of many thousands form in single trees, where the males produce an all-night-long courtship display of synchronous flashing (Buck and Buck, 1976).

The adult firefly light organ comprises a series of photocytes arranged in rosettes, positioned radially around a central well, through which run nerves and trachea, the later carrying oxygen to the cells (Ghiradella, 1977). Within the photocytes, organelles containing luciferase have been identified with peroxisomes on the basis of immunochemical labeling (Hanna et al., 1976). This identification is supported by the presence of a C-terminal peroxisomal signal sequence in luciferase (Conti et al., 1996).

Although flashing is initiated by a nerve impulse that travels to the light organ, the nerve terminals in the light organ are not on photocytes but on tracheolar cells, which regulate the supply of oxygen (Case and Strause, 1978), suggesting that these cells control the flash. In support of this theory, there is a strong positive relationship between the flashing ability and the extent of the tracheal supply system in different species. On the other hand, rapid kinetics, complex waveforms, multiple flashes, and high-frequency flickering all seem unlikely to be regulated by a gas in solution. However, although oxygen might diffuse slowly, it reacts very rapidly in this system; the half rise-time of luminescence with the anaerobic enzyme intermediate (luciferase–luciferyl adenylate) is less than 10 ms (Hastings et al., 1953). Also, possibilities alternate to oxygen seem unlikely. The flash is not directly triggered by an action potential, and none of the ions typically gated by membrane potential changes (Na^+, K^+ and Ca^{2+}) appear to be candidates for controlling firefly luminescence chemistry.

The firefly system was the first in which the biochemistry was extensively studied. It had been known since before 1900 that cell-free extracts could continue to emit light for several minutes or hours, and that after the complete decay of the light, emission could be restored by adding a second extract, prepared by using boiling water to extract the cells (cooled before adding). The enzyme luciferase was assumed to be in the first (cold water) extract (with all the luciferin substrate being used up during the emission), whereas the enzyme would be denatured by the hot-water extraction, leaving some substrate intact. This test was referred to as the luciferin–luciferase reaction, and it was already known in the first part of this century that luciferins and luciferases from the different major groups would not cross react, indicative of their independent evolutionary origins (Harvey, 1952).

McElroy (1947) discovered that the addition of adenosine triphosphate (ATP) to an "exhausted" cold-water extract resulted in bioluminescence. This showed that luciferin had not actually been used up in the cold-water extract. But ATP could not be the emitter, since it does not have the appropriate fluorescence. It was thus discovered that firefly luciferin, which is a unique benzothiazole (Fig. 2.1), was still present in large amounts in the "exhausted" cold-water extract, and that it was ATP that was used up, but available in the hot-water extract.

With the elucidation of the luciferin structure, ATP was shown to be required to form the luciferyl adenylate intermediate, which with the adenylate as the leaving group then reacts with oxygen to form a cyclic luciferyl peroxy species (Fig. 2.1). This breaks down to yield CO_2 and an excited state of the carbonyl product (McElroy and DeLuca, 1978; Wood, 1995). A remarkably high quantum yield of 0.88 was reported (Seliger and McElroy, 1960).

In reactions in which luminescence has decreased to a low level (this may continue for days), it was found that emission is greatly increased by coenzyme A (CoA), but the reason for this was obscure. The recent discovery that long-chain acyl-CoA synthetase (EC 6.2.1.3) has homologies with firefly luciferase (EC 1.13.12.7) both explains this observation and indicates the evolutionary origin of the gene (Wood, 1995).

Firefly luciferase has been cloned and expressed in other organisms, including *Escherichia coli* and tobacco and its crystal structure has recently been determined (Conti et al., 1996). To visualize expression, luciferin must be added exogenously; tobacco "lights up" when the roots are dipped in luciferin (Ow et al., 1986; see color Fig. 2.10). Luciferase catalyzes both the luciferin activation with ATP and the subsequent steps leading to the excited product. There are some beetles in which the light from different organs is a different color, and there is additional color variation between individuals of the same species. In *Pyrophorus plagiophthalamus*, the same ATP dependent luciferase reaction with the same luciferin occurs in the different organs, but no accessory emitter pro-

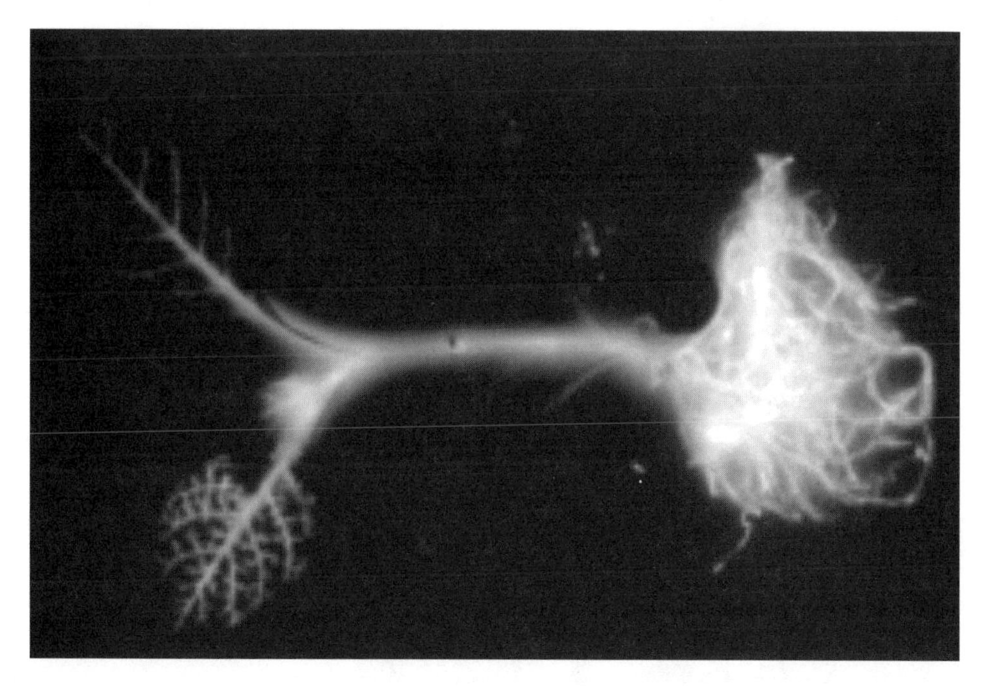

FIG. 2.10 Transgenic tobacco plant carrying the firefly luciferase gene photographed by its own light. The continuous luminescence occurs following the uptake of luciferin by the roots. [From Ow et al., 1986]. Figure also appears in color section.

teins have been implicated in any of these cases. Instead, differences in the luciferases appear to responsible. Different (but closely homologous) genes from a single organism have been cloned in *E. coli* and shown to fall into four color classes (see color Fig. 2.11; Wood et al., 1989; Wood, 1995). The chemical basis for the color differences remains to be elucidated (McCapra, 1997).

Firefly luciferase has been extensively used in analytical applications for the measurement of ATP (Brolin and Wettermark, 1992; Hastings et al., 1997). The cloned gene has also been used as a reporter gene in a number of studies, most recently to monitor circadian regulation of transcription of genes in higher plants (Millar et al., 1995). The use of the genes that result in different colors of luminescence has also been explored. More recently, the simultaneous use of two different luminous systems, for example, firefly and *Renilla*, for assays of two different substances, has been reported (Sherf et al., 1997).

2.6 CONCLUSION

Bioluminescence occurs in many different species in phylogenetically diverse groups. Among the different groups, the type and method of display of the light, its color, and its function may be very different. In two groups (and only two), bacteria and cnidaria, some of the luminous species possess accessory proteins carrying chromophores, which may serve as secondary emitters and

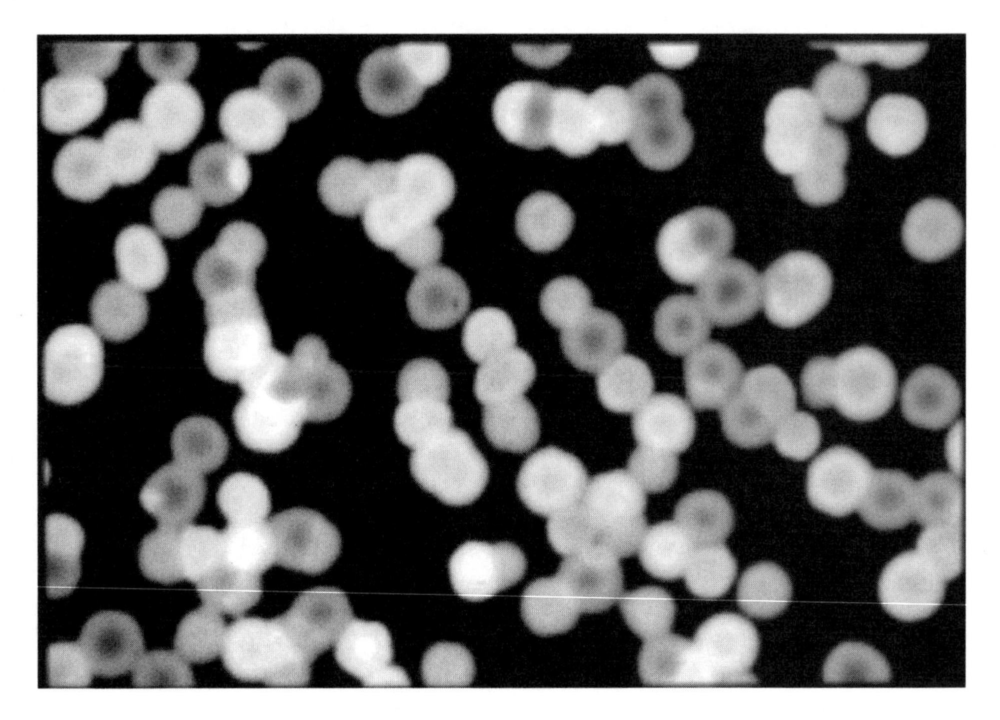

FIG. 2.11 Bacterial colonies carrying four different beetle luciferase genes cloned from the ventral organ, distinguished by their different luminescence colors: green, yellow-green, yellow and orange. [Wood et al., 1989.] Figure also appears in color section.

shift the spectrum of the light. The diversity of luminous organisms is indicative of what has been firmly established over the past several decades: Bioluminescent systems in different major groups are not evolutionarily conserved, so that the genes coding for the proteins (e.g., luciferases and accessory proteins) are not homologous. The consequent biochemical diversity offers a marvelous menu for many different specific analytical and reporter applications (Hastings et al., 1997), featuring noninvasive reporting by light emission, as exemplified by GFP.

REFERENCES

Bae, Y.-M. and Hastings, J. W. (1994). Cloning, sequencing and expression of dinoflagellate luciferase DNA from a marine alga, *Gonyaulax polyedra*. *Biochem. Biophys. Acta*, 1219: 449–456.

Baldwin, T. and Ziegler, M. (1992). The biochemistry and molecular biology of bacterial bioluminescence. *In* Chemistry and Biochemistry of Flavoenzymes, Vol. III, Müller, F. ed., CRC Press, Boca Ratan, FL, pp. 467–530.

Blinks, J. R., Wier, W. G., Hess, P., and Prendergast, F. G. (1982). Measurement of Ca^{++} concentrations in living cells. *Prog. Biophys. Mol. Biol.* 40:1–114.

Brehm, P., Lechleiter, J., Smith, S., and Dunlap, K. (1989). Intercellular signaling as visualized by endogenous calcium-dependent bioluminescence. *Neuron* 3: 191–198.

Brolin, S. and Wettermark, G. (1992). Bioluminescence Analysis. VCH Publishers, Inc., New York, 151 pp 71:176–182.

Buck, J. B. and Buck, E. (1976). Synchronous fireflies. *Sci. Am.* 234:74–85.

Buskey, E. J., Mills, L., and Swift, E. (1983). The effects of dinoflagellate bioluminescence on the swimming behavior of a marine copepod. *Limnol. Oceanogr.* 28:575–579.

Case, J. (1984). Firefly behavior and vision. *In* Insect Communication, Lewis, T. Ed. Harcourt, Brace, Jovanovich, London, pp. 195–222.

Case, J. F. and Strause, L. G. (1978). Neurally controlled luminescent systems. *In* P. J. Herring, Ed. *Bioluminescence in Action.* Academic Press, New York, pp. 331–366.

Chalfie, M., Tu, Y., Euskirchen, G., Ward, W. W., and Prasher, D. C. (1994). Green fluorescent protein as a marker for gene expression. *Science* 263:802–805.

Chatterjee, J. and Meighen, E. A. (1995). Biotechnological applications of bacterial bioluminescence (*lux*) genes. *Photochem. Photobiol.* 62:651–650.

Cody, C. W., Prasher, D. C., Westler, W. M., Prendergast, F. G., and Ward, W. W. (1993). Chemical structure of the hexapeptide chromophore of the *Aequorea* green-fluorescent protein. *Biochemistry* 32:1212–1218.

Conti, E. Franks, N. T. and Brick, P. (1996) Crystal structure of firefly luciferase throws light on a super family of adenylate-forming enzymes. *Structure* 4:287–298.

Cormier, M. J. (1981). *Renilla* and *Aequorea* bioluminescence. *In* Bioluminescence and Chemiluminescence, DeLuca, M. and McElroy, W. D. Eds., Academic, New York, pp. 225–233.

Cormier, M. J., Prasher, D. C., Longiaru, M. and McCann, R. O. (1989). The enzymology and molecular biology of the Ca^{++}-activated photoprotein, aequorin. *Photochem. Photobiol.* 49:509–512.

DeLuca, M. Ed., (1978). Bioluminescence and Chemiluminescence. *Methods Enzymol.* 57: 1–653.

Desjardins, M. and Morse, D. (1993). The polypeptide components of scintillons, the bioluminescence organelles of the dinoflagellate *Gonyaulax polyedra. Biochem. Cell Biol.* 71:176–182.

Dunlap, K., Takeda, K., and Brehm, P. (1987). Activation of a calcium-dependent photoprotein by chemical signalling through gap junctions. *Nature* (*London*) 325:60–62.

Eberhard, A., Burlingame, A. L., Eberhard, C., Kenyon, G. L., Nealson, K. H., and Oppenheimer, N. H. (1981). Structural Identification of Autoinducer of *Photobacterium fischeri* Luciferase. *Biochemistry* 20:2444–2449.

Eckstein, J., Cho, K. W., Colepicolo, P., Ghisla, S., Hastings, J. W., and Wilson, T. (1990). A time-dependent bacterial bioluminescence emission spectrum in an *in vitro* single turnover system: energy transfer alone cannot account for the yellow emission of *Vibrio fischeri* Y-1. *Proc. Natl. Acad. Sci.* 87: 1466–1470.

Fisher, A. J., Rauschel, F. M., Baldwin, T. O., and Rayment, I. (1995). The three-dimensional structure of bacterial luciferase from *Vibrio harveyi* at 2.4 Å resolution. *Biochemistry* 34:6581–6586.

Fleisher, K. J. and Case, J. F. (1995). Cephalopod predation facilitated by dinoflagellate luminescence. *Biol. Bull.* 189:263–271.

Fogel, M. and Hastings, J. W. (1971). A substrate binding protein in the *Gonyaulax* bioluminescence reaction. *Arch. Biochem. Biophys.* 142:310–321.

Fogel, M. and Hastings, J. W. (1972). Bioluminescence: Mechanism and mode of control of scintillon activity. *Proc. Nal. Acad. Sci.* 69:690–693.

Fritz, L., Morse, D., and Hastings, J. W. (1990). The circadian bioluminescence rhythm of *Gonyaulax* is related to daily variations in the number of light emitting organelles. *J. Cell. Sci.* 95:321–328.

Fuqua, W. C., Winans, S. C., and Greenberg, E. P. (1994). Quorum sensing in bacteria: the LuxR-Lufamily of cell density-responsive transcriptional regulators. *J. Bacteriol.* 176:269–275.

Ghiradella, H. (1977). Fine structure of the tracheoles of the lantern of a photurid firefly. *J. Morphol.* 153:187–204.

Grober, M. S. (1988a). Brittle-star bioluminescence functions as an aposematic signal to deter crustacean predators. *Anim. Behav.* 36:493–501.

Grober, M. S. (1988b). Responses of tropical reef fauna to brittle-star luminescence. *J. Exp. Mar. Biol. Ecol.* 115:157–168.

Hanna, C. H., Hopkins, T. A., and Buck, J. (1976). Peroxisomes of the firefly lantern. *J. Ultrastr. Res.* 57:150–162.

Harvey, E. N. (1952). Bioluminescence. Academic, New York.

Hastings, J. W. (1983). Biological diversity, chemical mechanisms and evolutionary origins of bioluminescent systems. *J. Molec. Evol.* 19:309–321.

Hastings, J. W., Balny, C., Le Peuch, C., and Douzou, P. (1973). Spectral properties of an oxygenated luciferase-flavin intermediate isolated by low-temperature chromatography. *Proc. Natl. Acad. Sci. USA* 70:3468–3472.

Hastings, J. W. and Gibson, Q. H. (1963). Intermediates in the bioluminescent oxidation of reduced flavin mononucleotide. *J. Biol. Chem.* 238:2537–2554.

Hastings, J. W., McElroy, W. D. and Coulombre, J. (1953). The effect of oxygen upon the immobilization reaction in firefly luminescence. *J. Cell and Comp. Physiol.* 42:137–150.

Hastings, J. W., Kricka, L., and Stanley, P. Eds. (1997). Bioluminescence and Chemiluminescence: Molecular reporting by photons. J. Wiley, Chichester, UK.

Hastings, J. W. and Morin, J. G. (1969a). Calcium-triggered light emission in *Renilla*. A unitary biochemical scheme for coelenterate bioluminescence. *Biochem. Biophys. Res. Commun.* 37:493–498.

Hastings, J. W. and Morin, J. G. (1969b). Comparative biochemistry of calcium activated photoproteins from the ctenophore, *Mnemiopsis* and the coelenterates *Aequorea, Obelia, Pelagia and Renilla. Biol. Bull.* 137:402.

Hastings, J. W. and Morin, J. G. (1971). Kinetics of calcium-triggered luminescence of aequorin and other photoproteins. *In* Contractility of Muscle Cells and Related Processes, Podolsky, R. J. Ed., Prentice-Hall, Engelwood, NJ, pp. 99–104.

Hastings, J. W. and Morin, J. G. (1991). Bioluminescence. *In* Neural and Integrative Animal Physiology, Prosser, C. L. ed., Wiley-Interscience, New York, pp. 131–170.

Hastings, J. W., Potrikus, C. J., Gupta, S., Kurfürst, M., and Makemson, J. C. (1985). Biochemistry and physiology of bioluminescent bacteria. *Adv. Microbial. Physiol.* 26:235–291.

Heim, R., Prasher, D. C., and Tsien, R. Y. (1994). Wavelength mutations and posttranslational autoxidation of green fluorescent protein. *Proc. Natl. Acad. Sci. USA* 9:12501–12504.

Herring, P. J. (Ed.) (1978). Bioluminescence in Action. Academic, New York, 570 pp.

Inouye, S., Aoyama, S., Miyata, T., Tsuji, F., and Sakaki, Y. (1989). Overexpression and Purification of the Recombinant Ca^{2+}-Binding Protein, Apoaequorin, *J. Biochem.* 105:473–477.

Inouye, S. Sakaki, Y., Goto, T., and Tsuji, F. I. (1986). Expression of apoaequorin complementary DNA in *Escherichia coli*. *Biochemistry* 25:8425–8429.

Johnson, C. H., Inoué, S., Flint, A., and Hastings, J. W. (1985). Compartmentation of algal bioluminescence: autofluorescence of bioluminescent particles in the dinoflagellate *Gonyaulax* as studied with the image-intensified video microscopy and flow cytometry. *J. Cell Biol.* 100:1435–1446.

Johnson, C. H., Knight, M. R., Kondo, T., Masson, P., Sedbrook, J., Haley, A., and Trewavas, A. (1995). Circadian oscillations of cytosolic and chloroplastic free calcium in plants. *Science* 269:1863–1865.

Johnson, C. H., Roeber, J., and Hastings, J. W. (1984). Circadian changes in enzyme concentration account for rhythm of enzyme activity in *Gonyaulax*. *Science* 223:1428–1430.

Johnson, F. H., Shimomura, O., Saiga, Y., Gershman, G., Reynolds, G. T., and Waters, J. R. (1962). Quantum efficiency of *Cypridina* luminescence with a note on that of *Aequorea*. *J. Cell. Comp. Physiol.* 60:85–103.

Karatani, H. and Hastings, J. W. (1993). Two active forms of the accessary yellow fluorescence protein of the luminous bacterium *Vibrio fischeri* strain Y1. *J. Photochem. Photobiol.* B. 18:227–232.

Kondo, T., Tsinoremas, N., Golden, S., Johnson, C., Kutsuna, S., and Ishiura, M. (1994). Circadian clock mutants of cyanobacteria. *Science* 266: 1233–1236.

Lee, D.-H., Mittag, M., Sczekan, S., Morse, D., and Hastings, J. W. (1993). Molecular cloning and genomic organization of a gene for luciferin-binding protein from the dinoflagellate *Gonyaulax polyedra*. *J. Biol. Chem.* 268:8842–8850.

Li, L., Hong, R. and Hastings, J. W. (1997). Three functional luciferase domains in a single polypeptide chain. *Proc. Natl. Acad. Sci. USA* 94:8954–8958.

Lloyd, J. E. (1977). Bioluminescence and communication. *In* How Animals Communicate, Sebeok, T. A. Ed., Indiana University Press, Bloomington, IN., pp. 164–183.

Lloyd, J. E. (1980). Male *Photuris* fireflies mimic sexual signals of their females' prey. *Science* 210:669–671.

Lorenz, W. W., McCann, R. O., Longiaru, M., and Cormier, M. J. (1991). Isolation and expression of cDNA encoding *Renilla reniformis* luciferase. *Proc. Natll. Acad. Sci. USA* 88:4438–4442.

Macheroux, P., Steinerstauch, P., Ghisla, S., Buntic, R., Colepiciolo, P., and Hastings, J. W. (1987). Purification of the yellow fluorescent protein from *V. fischeri* and identity of the flavin chromophore. *Biochem. Biophys. Res. Commun.* 146:101–106.

McCapra, F. (1997). Mechanisms in chemiluminescence and bioluminescence—unfinished business. *In* Hastings, J. W., Kricka, L . J. and Stanley, P. E., Ed. *Bioluminescence and Chemiluminiscence*. John Wiley and Sons, Chichester, pp. 7–15.

McElroy, W. D. (1947) The energy source for bioluminescence in an isolated system. *Proc. Natl. Acad. Sci. USA* 33:342–346.

McElroy, W. D. and DeLuca, M. (1978). Chemistry of Firefly Luminescence. *In* Herring, P. J., Ed., *Bioluminescence in Action*. Academic Press, New York, pp. 109–128.

Meighen, E. A. (1991). Molecular biology bacterial bioluminescence. *Microbiol. Rev* 55:123–142.

Mensinger, A. F. and Case, J. F. (1992). Dinoflagellate luminescence increases susceptibility of zooplankton to teleost predation. *Mar. Biol.* 112:207–210.

Millar, A. J., Carré, I. A., Strayer, C. A., Chua, N.-H., and Kay, S. A. (1995). Circadian clock mutants in *Arabidopsis* identified by luciferase imaging. *Science* 267: 1161–1166.

Mittag, M., Lee, D.-H., Hastings, J. W. (1994). Circadian expression of the luciferin-binding protein correlates with the binding of a protein to its 3′ untranslated region. *Proc. Natl. Acad. Sci. USA* 91:5257–5261.

Morin, J. G. (1974). Coelenterate bioluminescence. *In* Coelenterate Biology: Reviews and New Perspectives, Muscatine, L. and Lenhoff, H. Eds., Academic, New York, pp. 397–438.

Morin, J. G. (1976) Probable functions of bioluminescence in the Pennatulacea (Cnidaria, Anthozoa). *In* Coelenterate Ecology and Behavior, Mackie, G. O. Ed., Plenum, New York, pp. 629–638.

Morin, J. G. (1983). Coastal bioluminescence: patterns and functions. *Bull. Mar. Sci.* 33:787–817.

Morin, J. G. and Cooke, I. M. (1971a). Behavioural physiology of the colonial hydroid *Obelia.* II. Stimulus-initiated electrical activity and bioluminescence. *J. Exp. Biol.* 54:707–721.

Morin, J. G. and Cooke, I. M. (1971b). Behavioural physiology of the colonial hydroid *Obelia.* III. Characteristics of the bioluminescent system. *J. Exp. Biol.* 54:723–735.

Morin, J. G. and Hastings, J. W. (1971a). Biochemistry of the bioluminescence of colonial hydroids and other coelenterates. *J. Cell. Physiol.* 77:305–311.

Morin, J. G. and Hastings, J. W. (1971b). Energy transfer in a bioluminescent system. *J. Cell. Physiol.* 77:313–318.

Morin, J. G. and Reynolds, G. T. (1969). Fluorescence and time distribution of photon emission of bioluminescent photocytes in *Obelia geniculata. Biol. Bull.* 137:410.

Morin, J. G. and Reynolds, G. T. (1970). Luminescence and related fluorescence in coelenterates. *Biol. Bull.* 139:430–431.

Morin, J. G. Reynolds, G. T. (1974). The cellular origin of bioluminescence in the colonial hydroid *Obelia. Biol. Bull.* 147:397–410.

Morin, J. G., Reynolds, G. T., and Hastings, J. W. (1968). Excitatory physiology and localization of bioluminescence in *Obelia. Biol. Bull.* 135:429–430.

Morise, H., Shimomura, O., Johnson, F. H., and Winant, J. (1974). Intermolecular energy transfer in the bioluminescent system. *Biochemistry* 13:2656–2662.

Morse, D., Milos, P. M., Roux, E., and Hastings, J. W. (1989). Circadian regulation of the synthesis of substrate binding protein in the *Gonyaulax* bioluminescent system involves translational control. *Proc. Natl. Acad. Sci. USA* 86: 172–176.

Nealson, K. and Hastings, J. W. (1991). The luminous bacteria. *In* The Prokaryotes 2nd ed., Vol. I, Part 2, Chapter 25, Balows, A. Trüper, H. G., Dworkin, M., Harder, W., and Schleifer, K. H., Eds., Springer-Verlag, New York, pp. 625–639.

Nealson, K., Platt, T., and Hastings, J. W. (1970). The cellular control of the synthesis and activity of the bacterial luminescent system. *J. Bacteriol.* 104:313–322.

Nicol, J. A. C. (1958). Observations on the luminescence of *Pennatula phosphorea,* with a note on the luminescence of *Virgularia mirabilis. J. Mar. Biol. Ass.,* UK 37:551–563.

Nicolas, M-T., Morse, D., Bassot, J.-M., and Hastings, J. W. (1991). Colocalization of luciferin binding protein and luciferase to the scintillons of *Gonyaulax polyedra* revealed by immunolabeling after fast-freeze fixation. *Protoplasma* 160:159–166.

Nicolas, M-T., Nicolas, G., Johnson, C. H., Bassot, J-M., and Hastings, J. W. (1987). Characterization of the bioluminescent organelles in *Gonyaulax polyedra* (dinoflagellates) after fast–freeze fixation and antiluciferase immunogold staining. *J. Cell Biol.* 105:723–735.

Ow, D. W., Wood, K. V., DeLuca, M., de Wet, J. R., Helinski, D. R., and Howell, S. H. (1986). Transient and stable expression of the firefly luciferase gene in plant cells and transgenic plants. *Science* 234: 856–859.

Petushkov, V. N., Gibson, B. G., and Lee J. (1995a). Properties of recombinant fluorescent proteins from *Photobacterium leiognathi* and their interaction with luciferase intermediates. *Biochemistry* 34:3300–3309.

Petushkov, V. N., Gibson, B. G., and Lee J. (1995b). The yellow bioluminescence bacterium, *Vibrio fischeri* Y1, contains a bioluminescence active riboflavin protein in addition to the yellow fluorescence FMN protein. *Biochem. Biophys. Res. Commun.* 211;774–779.

Ridgeway, E. B. and Ashley, C. C. (1968). Simultaneous recording of membrane potential, calcium transient and tension in single muscle fibres. *Nature (London)* 219:1168–1169.

Ruby, E. G. (1996). Lessons from a cooperative, bacterial-animal association: the *Vibrio fischeri-Euprymna scolopes* light organ symbiosis. *Annu. Rev. Microbiol.* 50:591–624.

Saran, S., Nakao, H., Tasaki, M., Iida, H., Tsuji, F. I., Nanjundiah, V., and Takeuchi, I. (1994). Intracellular free calcium level and its response to cAMP stimulation in developing *Dictyostelium* cells transformed with jellyfish apoaequorin cDNA. *FEBS Lett.* 337:43–47.

Schauer, A., Ranes, M., Santamaria, R., Guijarro, J., Lawlor, E., Mendez, C., Chater, K., and Losick R. (1988). Gene expression in the filamentous bacterium *Streptomyces coelicolor. Science* 240:768–772.

Seliger, H. H. and McElroy, W. D. (1960). Spectral emission and quantum yield of firefly bioluminescence. *Arch. Biochem. Biophys.* 88:136.

Sherf, B., Navarro, S., Hannah, R. and Wood, K. V. (1997). Co-reporter technology integrating firefly and *Renilla* luciferase assays. *In* J. W. Hastings, L. J. Kricka and P. E. Stanley [Ed.] *Co-reporter technology integrating firefly and Renilla luciferase assays*. John Wiley and Sons, Chichester, pp. 228–231.

Shimomura, O. and Johnson, F. H. (1975). Regeneration of the photoprotein aequorin. *Nature (London)* 256:236–238.

Shimomura, O., Johnson, F. H., and Saiga, Y. (1962). Extraction, purification and properties of aequorin, a bioluminescent protein from the luminous hydromedusan, *Aequorea. J. Cell. Comp. Physiol.* 59:223–240.

Shorey, J. and Morin, J. G. (1974). Quantification of light produced from hydrozoan photocytes. *Biol. Bull.* 147:499.

Small, E. D., Koka, P., and Lee, J. (1980). Lumazine protein from the bioluminescent bacterium *Photobacterium phosphoreum. J. Biol. Chem.* 255:8804–8810.

Tanahashi, H., Ito, T., Inouye, S., Tsuji, F. I:, and Sakaki, Y. (1990). Photoprotein aequorin: use as a reporter enzyme in studying gene expression in mammalian cells. *Gene* 96:249–255.

Thomson, C. M., Herring, P. J., and Campbell, A. K. (1997). The widespread occurrence and tissue distribution of the imidazolopyrazine luciferins. *J. Biolum. Chemilum.* 12:87–91.

Titschack, H. (1964). Untersuchungen Über das Leuchten der Seefeder *Veretillum cynomorium* (Pallas). *Vie Milieu* 15:547–563.

Ward, W. W., Cody, C. W., Hart, R. C., and Cormier, M. J. (1980). Spectrophotometric identity of the energy transfer chromophores in *Renilla* and *Aequorea* green-fluorescent proteins. *Photochem. Photobiol.* 31:611-615.

Wood, K. V. (1995). The chemical mechanism and evolutionary development of beetle bioluminescence. *Photochem. Photobiol.* 62:662–673.

Wood, K. V., Lam, Y. A., Seliger, H. H., and McElroy, W. D. (1989). Complementary DNAs encoding click beetle luciferases can elicit bioluminescence of different colors. *Science* 244:700–702.

PART TWO

Biochemistry and Molecular Biology of Green Fluorescent Protein

3

Biochemical and Physical Properties of Green Fluorescent Protein

WILLIAM W. WARD
Department of Biochemistry and Microbiology, Rutgers University, Cook College, New Brunswick, NJ

3.1 INTRODUCTION

The recent popularity of green fluorescent protein (GFP) as a research tool in cellular and developmental biology (Chalfie, 1995; Hassler, 1995; Kain et al., 1995; Prasher, 1995; Stearns, 1995) requires that we look very carefully at the chemical and physical properties of the GFP molecule and its chromophore. Unfortunately, the chemical and physical characterizations of native and recombinant forms of GFP and numerous mutants of the original *Aequorea victoria* derived clone (Chalfie et al., 1994) have not, and cannot, keep pace with the proliferation of GFP mutants and the accelerating pace in GFP applications. After 30 years of research on the prototype native GFP molecules from the jellyfish, *Aequorea victoria* (Morin and Hastings, 1971; Morise et al., 1974; Prendergast and Mann, 1978) and the sea pansy, *Renilla reniformis* (Wampler et al, 1971, 1973; Morin, 1974; Cormier et al., 1974; Ward, 1979; Ward and Cormier, 1979), these proteins are still incompletely characterized; much less is known about the chemical and physical properties of the available mutants of GFP. In this chapter, the known chemical and physical properties of GFP are summarized to provide a sound basis for the qualitative and quantitative interpretations of data generated in its applications. Nonetheless, because so much

Green Fluorescent Protein: Properties, Applications, and Protocols, Edited by Martin Chalfie and Steven Kain
ISBN 0-471-17839-X

remains unknown about the biochemical properties of these molecules, users of GFP should cautiously interpret their GFP derived data.

3.2 BIOLOGICAL FUNCTION OF GREEN FLUORESCENT PROTEIN

Biologically, GFP acts to shift the color of bioluminescence from blue to green in luminous coelenterates (jellyfish, hydroids, sea pansies, and sea pens) and to increase the quantum yield of light emission (Ward, 1979). All coelenterates utilize the same luciferin (coelenterate-type luciferin or coelenterazine) in their bioluminescence reactions (Cormier et al., 1973; Hori and Cormier, 1973; Hori et al., 1973; Wampler et al., 1973; Ward and Cormier, 1975; Inoue et al., 1977a,b; Shimomura and Johnson, 1979; Shimomura et al., 1980), producing a protein-bound oxyluciferin (Hori et al., 1973, 1975, 1977) that emits blue light in the absence of GFP. In the presence of GFP, however, the emitted light is green and identical in spectral properties to the fluorescence emission spectrum of GFP (Ward, 1979) when excited directly by exogenous radiation. Such spectral shifts are known to occur in spectroscopy by one of two general mechanisms: (1) radiative (trivial) energy transfer in which the donor molecule emits light that is subsequently absorbed and reemitted by the acceptor and (2) radiationless (often called Förster type) energy transfer in which excitation energy is transferred, without photon emission, from the donor molecule to the acceptor (Ward, 1979). Efficient trivial transfer from blue-emitting oxyluciferin to GFP requires a relatively high concentration of GFP and a sufficiently long pathlength. In a 1-cm pathlength fluorometric cuvette, for example, 90% of the incident blue light could be absorbed by wild-type GFP at a concentration of 5–10 mg mL^{-1} (absorbance = 1.0 at 480 nm). But, with a very short pathlength as would be seen in animal cells (10 μm diameter), the GFP concentration would need to be 1000X as great for trivial transfer to operate efficiently. Clearly, intracellular protein concentrations on the order of 10,000 mg mL^{-1} cannot be achieved. Even if all oxyluciferin emission were absorbed by GFP, the quantum yield in a trivial transfer system can be no greater than the product of the quantum yields of donor and acceptor. In the coelenterate systems, the bioluminescence quantum yield for oxyluciferin is about 0.10 (Hori et al., 1973; Hart et al., 1979) and the fluorescence quantum yield for GFP is about 0.80 (Ward and Cormier, 1979). Thus the maximum overall quantum yield, by a trivial mechanism, would be $0.10 \times 0.80 = 0.08$, a slight decrease from oxyluciferin emission alone. But, a radiationless system traps excitation energy directly by resonance transfer, so long as donor and acceptor molecules are relatively close to each other (<100 Å). No light is actually emitted by the donor. Close proximity of donor and acceptor can be achieved by high concentrations of donor and acceptor or by chemical interactions that favor heterodimer formation at much lower concentrations.

In the *Renilla reniformis* system, luciferase-bound oxyluciferin and GFP are brought into intimate contact via protein–protein interaction facilitating radiationless energy transfer at submicromolar protein concentrations (<0.1 mg mL^{-1} GFP). Such radiationless energy transfer, unlike trivial transfer, can result in a potential enhancement in overall quantum yield. Thus, in the coelenterates, if the yield of excited-state oxyluciferin is 100%, transfer of excitation energy to GFP could result in overall quantum yield of 0.80—10-fold greater than the maximum achievable by trivial transfer. In practice, an enhancement in radiative quantum yield of 3X has been measured *in vitro* in the *R. reniformis* system using native coelenterate-type luciferin (Ward and Cormier, 1976, 1978a, 1979). With certain low quantum yield luciferin analogues (Hart et al., 1979) an enhancement of 200X has been measured. Efficient energy transfer of this sort occurs in the *Renilla* system at low concentrations of GFP (10^{-6} M) and low concentrations of salt (<0.1 M NaCl) implicating electrostatic protein–protein association between dimeric *Renilla* GFP and *Renilla* luciferase in the mechanism (Ward and Cormier, 1978a,b). Radiationless energy transfer between protein-bound oxyluciferin and GFP has been demonstrated in this system by five independent physical means including intraphyletic cross-reactions (Ward and Cormier, 1978a).

However, at similarly low-protein concentration levels, the formation of a complex of *Aequorea* GFP and aequorin (the blue light emitting photoprotein of *Aequorea*) in aqueous solution has not been demonstrated. This lack of association led to our premature conclusion that energy transfer in *Aequorea* occurs by a trivial mechanism (Ward and Cormier, 1978a), while earlier Morise et al. (1974), who coimmobilized aequorin and *Aequorea* GFP to an ion exchange gel, had reached the opposite conclusion. Years later, we demonstrated that radiationless energy transfer between aequorin and *Aequorea* GFP does occur in solution, but only at very high protein concentrations (Cutler, 1995; Cutler and Ward, 1993; 1997). To eliminate trivial transfer at high-protein concentrations (10–20 mg mL^{-1}), experiments were performed in microcapillary tubes (200-μm diameter). Our calculations indicate that GFP concentration within *Aequorea* photocytes (cells where bioluminescence originates) is approximately 25 mg mL^{-1} (Cutler, 1995). Aequorin concentration appears to be several times higher. The conditions in these microcapillary tubes simulate intracellular protein concentrations and diameter of the photogenic mass such that aequorin and *Aequorea* GFP form a complex that emits green light upon Ca^{2+} addition. In such tubes, the pathlength of emitted light is too short for significant trivial transfer from aequorin to GFP. Present evidence suggests that the functional unit in *Aequorea* photocytes is a heterotetramer ($GFP_2 \cdot Aequorin_2$) (Cutler and Ward, 1997). Very high protein concentration (>10mg mL^{-1}), comparable to that found within the jellyfish photocytes, is required for formation of the heterotetramer (Cutler and Ward, 1997).

3.3 NATURAL SOURCES OF GFP

Green fluorescent proteins are found in a large number of bioluminescent coelenterates within the classes hydrozoa and anthozoa (Morin and Hastings, 1971; Cormier et al., 1974; Morin, 1974; Ward, 1979; Levine and Ward, 1982; Prasher, 1995). It is not clear, however, that they occur at all within the scyphozoans and there are no reports of GFPs in the closely related phylum ctenophora. All known naturally occurring GFPs, and most of the characterized mutants of wild-type recombinant *Aequorea* GFP (Heim et al., 1994, 1995; Heim and

Tsien, 1996; Ehrig et al., 1995; Cubitt et al., 1995; Delagrave et al., 1995; Cormack et al., 1996; Crameri et al., 1996) emit light with wavelength maxima in the 490–520-nm range; most are centered at 508–509 nm. However, the range of excitation (absorption) maxima of purified and partially characterized native GFPs is much greater (395–498 nm). Curiously, the only naturally occurring GFP molecule with an excitation maximum in the ultraviolet (UV) region (Morise et al., 1974) is the one found in the hydrozoan jellyfish *A. victoria* (λ_{max} = 395 nm), and this is the only GFP for which the gene has been cloned (Prasher et al., 1992; Chalfie et al., 1994). The second most blue-shifted excitation maximum of a natural GFP is seen in the GFP of *Halistaura* (*Mitrocoma*) *cellularia* (λ_{max} = 465 nm). Excitation and emission spectra of four naturally occurring GFPs are shown in Figure 3.1.

3.4 PHYSICAL CHARACTERISTICS OF GFP

Most of our knowledge of the physical characteristics of GFP comes from work on *R. reniformis* GFP and *A. victoria* GFP (Tables 3.1–3.4). Only minimal information is available about the 20 other naturally occurring GFPs that have been studied—usually with nothing more than the wavelength of peak emission being reported (Cormier et al., 1974; Ward, 1981; Levine and Ward, 1982; Prasher, 1995). All known GFPs are acidic, compact, globular molecules with monomer molecular weights (Fig. 3.2) of about 27 kDa (see Chapter 4 by Phillips for details of X-ray structure). With the exception of *Aequorea* GFP and recombinant forms thereof, which are monomers in dilute aqueous solution (Prendergast and Mann, 1978; W. Ward and T. Spires, unpublished), all other GFPs are stable, non-dissociable dimers (2×27 kDa) that remain dimeric (Fig. 3.3) unless denatured (Fig. 3.4) (Ward and Cormier, 1979; W. Ward, A. S. Sawyer, and T. Spires, unpublished results).

Aequorea and *Renilla* GFPs and many of the mutants of recombinant wild-type GFP are isolated, by ion exchange chromatography or isoelectric focusing, as a family of closely related isoforms (Cutler, 1995) with isoelectric points between 4.6 and 5.4 (for purification details see Protocol 1 by González and Ward in Protocol I, Section I.E). In the jellyfish or the sea pansy, a contributor to microheterogeneity is the presence of multiple GFP genes or alleles having several point mutations (Prasher et al., 1992). These mutations produce as many as five internal amino acid substitutions, some of which generate differences in the isoelectric points of resulting isoforms. Recombinant GFPs, products of single gene transfections, are not expected to have internal amino acid substitutions and should purify as a single isoform by ion exchange chromatography or isoelectric focusing. But this is not usually the case. Full length wild-type recombinant GFP is a protein containing 238 amino acids, the C-terminal segment of which has the sequence His-Gly-Met-Asp-Glu-Tyr-Lys (Prasher et al., 1992). Unlike the core of the protein, which is highly resistant to proteolysis (Roth, 1985), this C-terminal "tail" is quite susceptible to attack by carboxypeptidases and by nonspecific proteases such as proteinase K and pronase (Cutler, 1995). Because this tail contains two basic amino acid residues (His and Lys) and two acidic amino acid residues (Asp and Glu), partial proteolytic cleavage can gen-

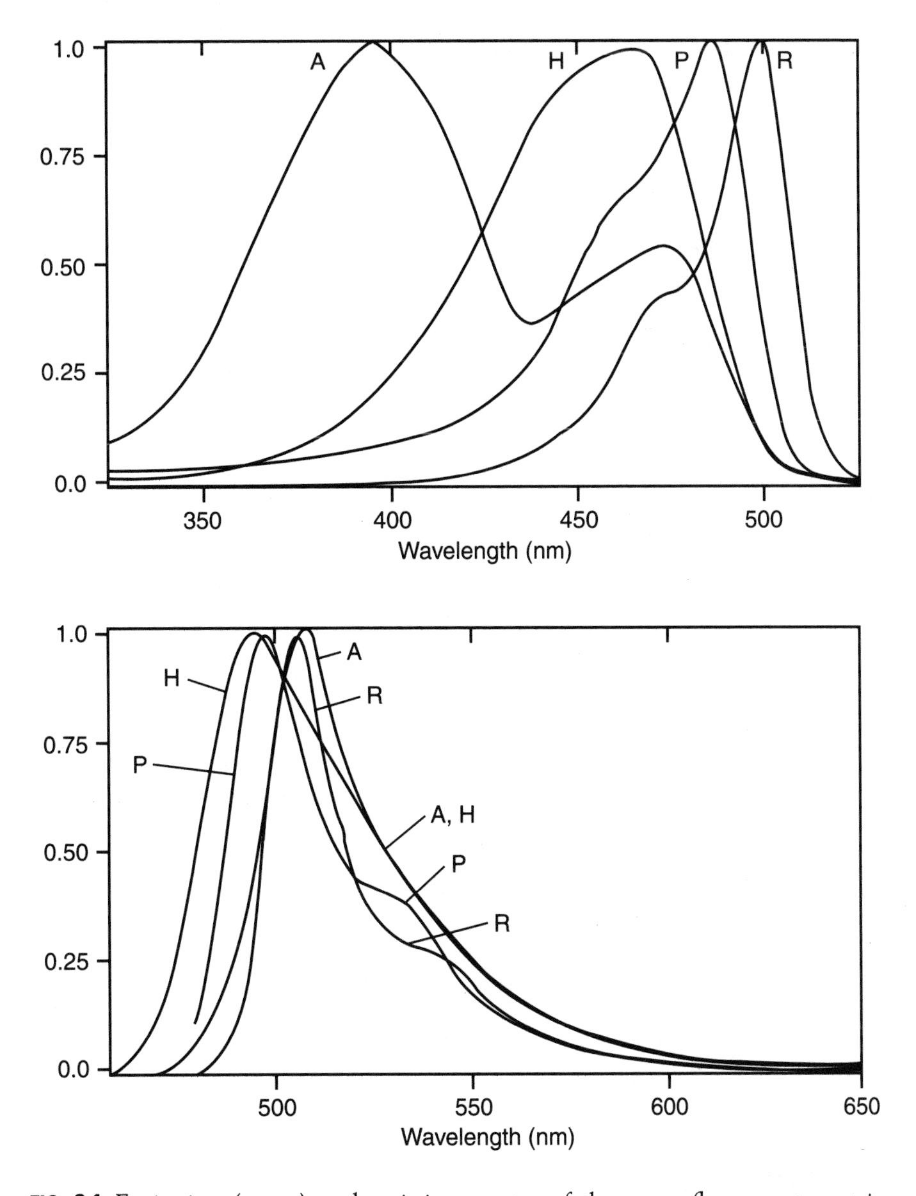

FIG. 3.1 Excitation (upper) and emission spectra of the green fluorescent proteins from *Aequorea* (A), *Halistaura (Mitrocoma)* (H), *Phialidium* (P), and *Renilla* (R). Excitation maxima are at 395, 465, 485, and 498 nm, respectively. Emission maxima are at 509, 497, 498, and 508 nm, respectively. Data were collected with a fully corrected Spex fluorometer at room temperature in 10 mM Tris-ethylenediamenetetracetic acid (EDTA) buffer at pH 8.0. Data were collected by Dr. Richard Ludescher in the Rutgers University Department of Food Science.

TABLE 3.1 *Aequorea* and *Renilla* GFP: Comparative Physical Data

	Aequorea	*Renilla*
Monomer molecular weight[a]	27 kDa	27 kDa[b]
	26.9 kDa[c]	
Isoelectric point(s) (pI)	4.6–5.1[d]	5.34 ± 0.07[b]
Fluorescence emission maximum	508[e]–509 nm	509 nm[b]
Fluorescence quantum yield	0.72–0.78[e]	0.80[b]
	0.80	
Molar extinction coefficient (monomer)		
$\varepsilon\lambda^{1M}$ (L mol^{-1} cm^{-1})		
$\lambda = 498$ nm	3,000	133,000[b]
$\lambda = 475$ nm	14,000	53,000[b]
$\lambda = 397$ nm	27,600	<1,000[b]
$\lambda = 280$ nm	22,000	22,000[b]
Absorption ratio (highest purity achieved)		
498 nm/280 nm		5.6[b]–6.0
397 nm/380 nm	1.25	

[a]At moderate protein concentration of *Aequorea* GFP (<0.5 mg mL^{-1}) the monomeric form predominates. At higher protein concentrations of *Aequorea* GFP (>2.0 mg mL^{-1}) the dimeric form predominates. Renilla GFP is dimeric (2 × 27 kDa) at all concentrations unless denatured. [b] From Ward and Cormier (1979). [c] From Prasher et al. (1992). Based upon sequence of cDNA. [d] From Cutler (1995). Nine isoforms have been characterized. [e] From Morise et al. (1974).

erate a rich array of isoforms including many that are separable by ion exchange chromatography, isoelectric focusing, and native gel electrophoresis. We have identified and sequenced nine isoforms of native *Aequorea* GFP having five internal substitutions and five different C-terminal truncations (Table 3.3). Fortunately, we see a much lower degree of microheterogeneity in recombinant forms of GFP, especially if we take precautions to reduce C-terminal proteolysis (by working at 0–4°C and by adding the serine protease inhibitor phenylmethylsulfonyl fluoride (PMSF).

3.4.A GFP Stability

One of the great advantages of GFP as a reporter of gene expression is its high level of stability. This stability appears to be the consequence of its unique three-dimensional structure (Ormö et al., 1996; Yang, et al., 1996a, 1997; Wu et al., 1997). Eleven beta strands surround and protect the chromophore that is positioned near the geometric center of a "beta-can" (see Chapter 4 by Phillips). Short loop regions and distorted alpha helices cap both ends of the "can." The protection is so complete that classical fluorescence quenching agents such as acrylamide, halides, and molecular oxygen have almost no effect on GFP fluorescence (Rao et al., 1980). The purified wild-type recombinant GFP derived from gfp10 (Chalfie et al., 1994) shows no significant differences in its stability properties when compared with natural GFP purified from A. *victoria*. In most respects,

TABLE 3.2 The amino acid compositions of *Renillas* and *Aequorea* GFP

Amino Acids	*Renilla* GFP Nearest Integer per 27,000 Da[a]	Aequorea GFP from cDNA Sequence gfp 10[b]
Lysine	19	20
Histidine	8	10
Arginine	7	6
Half-cystine	2[c]	2
Methionine	9	6
Aspartic acid	20 (Asp + Asn)	18
Asparagine		13
Glutamic acid	27 (Glu + Gln)	16
Glutamine		8
Threonine	17	15
Serine	15	10
Proline	11	10
Glycine	22	22
Alanine	14	8
Valine	18	17
Isoleucine	14	12
Tyrosine	11	11
Phenylalanine	13	13
Tryptophan	0[d]	1
Amino sugars	0[e]	0

[a] From Ward and Cormier (1979). Each value represents the average from hydrolyses of 24, 48, and 72 h unless otherwise indicated

[b] From Prasher et al. (1992).

[c] Determined as cysteic acid following performic acid oxidation.

[d] Determined by hydrolysis in the presence of thioglycolate.

[e] Determined by hydrolysis with *p*-toluenesulfonic acid.

purified *R. reniformis* GFP is even more stable than *A. victoria* GFP and its chromophore is less responsive to external perturbations (Ward et al., 1982). *Renilla* GFP, for example, has a broader pH stability profile than *Aequorea* GFP (Fig. 3.4) and much greater stability in protein denaturing solutions such as 8 M urea, 1% SDS, and 6 M guanidine hydrochloride (Ward and Cormier, 1979; Ward et al., 1982; W. Ward, A. S. Sawyer, and T. Spires, unpublished results).

3.4.B Denaturation and Renaturation

The T_m (temperature at which one-half of the endogenous fluorescence is lost) for *Aequorea* GFP is 76°C (Bokman and Ward, 1981; Ward, 1981). For *Renilla* GFP the value is 70°C (Ward, 1981) and for *Phialidium* GFP the T_m is 69°C (Levine and Ward, 1982). In both *Aequorea* and *Renilla* GFP, the far-UV circular dichroism (CD) signal (at 205, 207.5, and 252.2 nm) decays in parallel with the fluorescence as the temperature is raised [such CD changes are indicative of the loss of secondary and tertiary structure] that fluorescence depends on the intact secondary and tertiary structure of the protein (Bokman and Ward, 1981).

TABLE 3.3 Structural differences among native and recombinant GFP isoforms[a,b]

Structure Element	gfp2 gene	gfp10 cDNA	M1	M2	M3	L1	L2	L3	C1	C2[c]	rGFP
N-terminal amino acid	Met	Met	N-Acetylated serine								Ala
Residue 80	Gln	Gln	Gln	Gln	Gln	Gln	Gln	Gln	Gln	Gln	Arg
Residue 100	Tyr	Phe	Phe	Phe	Phe	Tyr	Tyr	Tyr	Tyr	Tyr	Phe
Residue 108	Ser	Thr	Thr	Thr	Thr	Ser	Ser	Ser	Ser	Ser	Thr
Residue 141	Met	Leu	Leu	Leu	Leu	Met	Met	Met	Met	Leu	Leu
Residue 172	Glu	Glu	Glu	Glu	Glu	Lys	Lys	Glu	Glu	Glu	Glu
Residue 219	Ile	Val	Val	Val	Val	Val	Ile	Ile	Ile	Ile	Val
Residue 230	Thr	Thr	Thr	Thr	Thr	Thr	Thr	Thr	Thr	Thr	Thr
Residue 231	His	His	His	His					His	His	His
Residue 232	Gly	Gly	Gly						Gly	Gly	Gly
Residue 233	Met	Met							Met	Met	Met
Residue 234	Asp	Asp							Asp	Asp	Asp
Residue 235	Glu	Glu								Glu	Glu
Residue 236	Leu	Leu									Leu
Residue 237	Tyr	Tyr									Tyr
Residue 238	Lys	Lys									Lys
Calculated MW (Da)			26,000	25,943	25,806	25,824	25,838	25,839	26,279	26,391	26,836
Experimental MW (Da)			26,009	25,953	25,816	25,830	25,841	25,841	26,279	26,393	26,845

[a] Shaded amino acids represent those that have been determined directly by cDNA sequencing of C-terminal protein sequencing.
[b] From Cutler (1995).
[c] An additional "c" isoform has been found (data not shown).

TABLE 3.4 Molar Extinction Coefficients of Native and Variant Green Fluorescent Proteins

GFP Source	Molar Extinction Coefficient[a] (L mol^{-1}/cm^{-1}) per Monomeric Unit			
	397 nm	475 nm	489 nm	498 nm
1. *Renilla* GFP[b]				
Dimer				133,000
2. *Aequorea* GFP				
Monomer[c]	27,600	14,000		
Monomer[d]	25,000	11,000		
Dimer	30,000	3,000		
3. wt Recombinant				
Monomer[e]	30,000	12,000		
Monomer[d]	28,000	11,000		
Dimer	34,000	2,500		
4. "Cycle 3" variant				
Monomer	30,000	8,000		
Monomer[d]	30,300	8,500		
5. S-65-T variant				
Monomer/dimer[f]			58,000	
Monomer/dimer[d]			56,000	
6. Mut 1 variant				
Monomer/dimer[f]			57,000	
Monomer/dimer[d]			55,000	

[a] Based on $E_4 4 = 44,000$ L $mol^{-1}cm^{-1}$ for the denatured protein in 0.1 M NaOH, unless otherwise indicated.
[b] From Ward and Cormier (1979); Renilla GFP has never been shown to exist as a fluorescent monomer
[c] From Ward (1981); λ_{max} for native *Aequorea* GFP at 395 and 470 nm.
[d] Based on measured extinction coefficients at 292 nm for Trp (3590 L mol^{-1} cm^{-1}) and Tyr (2340 L mol^{-1} cm^{-1}) in 0.1 NaOH. Each variant contains 1 Trp and 11 Tyr, $\therefore\ \varepsilon_{292} = 29,300$.
[e] Average of seven determinations.
[f] Average of five determinations.

Although heat-denatured GFP does not renature effectively, fully denatured GFP will recover most or all of its original fluorescence following other conditions of denaturation as described below.

The fluorescence of *Renilla* GFP is completely unperturbed over a wide pH range from about 5.5–12.6 (Ward, 1981) (Fig. 3.4). Between pH 4.5 and 5.5, the fluorescence is metastable and fluorescence bleaching increases over time, especially at elevated temperature. *Renilla* GFP is stable at its isoelectric point (pH 5.34) for several days at 4°C but is much less stable at room temperature. *Renilla* GFP will tolerate a pH of 12.6 for an hour or more, but pH values above 12.6 cause almost instantaneous (but reversible) loss of fluorescence as the protein denatures.

Native *Aequorea* GFP, the wild-type recombinant protein (Chalfie et al., 1994), and the "cycle 3" variant of GFP (Crameri et al., 1996) show a similar pattern of fluorescence as *Renilla* GFP at the acidic end of the pH spectrum, but

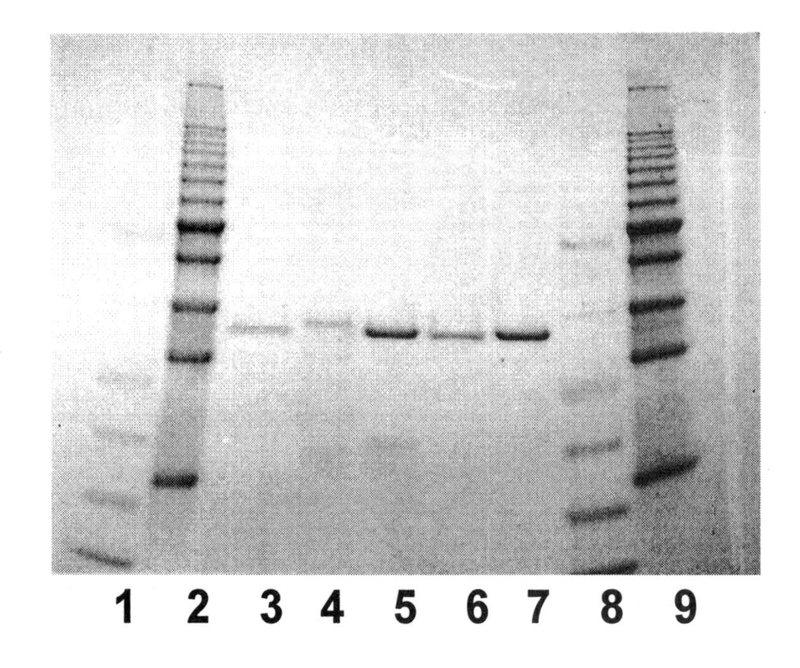

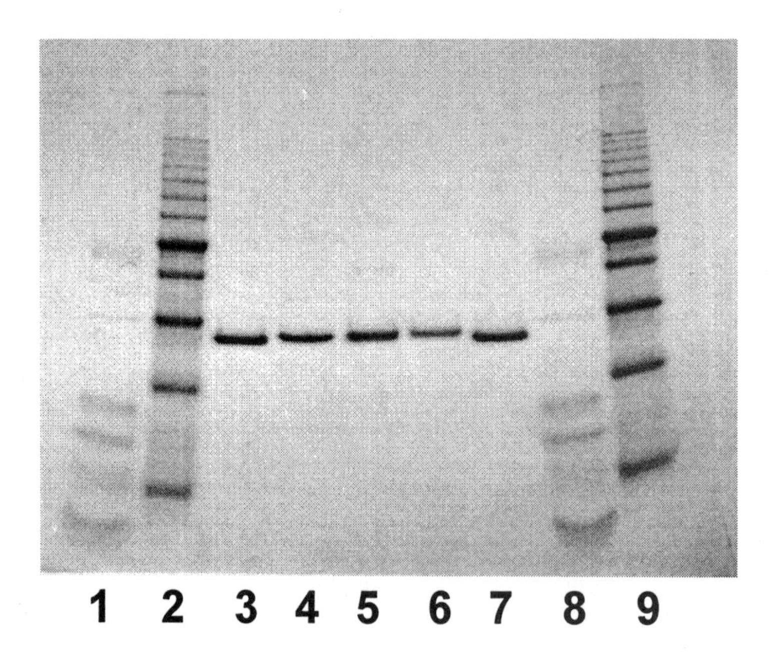

FIG. 3.2 Sodium dodecyl sulfate (SDS) gel electrophoresis of recombinant GFP samples. Samples were heated 10 min in 70 mM SDS solution containing 0.05 M dithiothreitol in bis–Tris buffer prior to application to a 1 mm, 4–20% Tris–Glycine polyacrylamide gel (Novex, San Diego, CA) with MOPS/SDS running buffer. Standards include molecular weight ladders (lanes 1 and 8) containing standards of 200, 120, 110, 100, 90, 80, 70, 60, 50, 40, 30, 20, and 10 kDa (Gibco, Gaithersburg, MD) and low molecular weight standards (lanes 2 and 9). The low

continued on next page

they are less stable than *Renilla* GFP at their isoelectric points (pH 4.7–5.1). These proteins will bleach almost completely during a 3-h long isoelectric focusing experiment at 15°C. At 10°C or below, they are stable for at least 3 h. These proteins are dramatically different from *Renilla* GFP under alkaline conditions (Ward et al., 1982). Beginning at about pH 10.0 and continuing to pH 12.2, native *Aequorea* GFP and the related recombinant forms (wild type and "cycle 3") undergo large shifts in their absorption (excitation) spectra (Fig. 3.4) (Ward et al., 1982; González et al., 1997). The familiar peak at 395 nm (Morise et al., 1974) drops in intensity and the shoulder at 475 nm increases three-fold (Fig. 3.5) (Ward et al., 1982). If, at pH 12, fluorescence is excited in the 475-nm region (e.g. FITC optics or 488-nm argon laser line), the intensity of fluorescence will appear to triple, making quantitation of fluorescence in the alkaline range more difficult. However, when, at pH 12, *Aequorea* GFP is excited in the UV (e.g., 365 nm), fluorescence will appear to drop by two-fold as compared with pH 8. The pH dependent spectral shifts in GFP absorption/excitation spectra in the extreme alkaline range may be the result of ionization of tyrosine at position 66 in the chromophore and/or deprotonation of arginine at position 96 that stabilizes the enol form of the chromophore imidazolone (Fig. 3.6).

Both *Renilla* and *Aequorea* GFPs will recover fluorescence after acid or base denaturation. With native *Aequorea* GFP, we have demonstrated up to 80% recovery of fluorescence following denaturation in acid (pH 1), base (pH 13), 6 M guanidine·HCl, and 8 M urea (Ward and Bokman, 1982); however, reproducibility has always been a problem and yields of renatured GFP vary greatly. Renaturation occurs rapidly (half-time in tens of seconds) upon adjustment to pH 8 (from acid or base) or following 12-fold dilution into aqueous buffer (from guanidine or urea). The percentage recovery is greater if the experiments are performed at 0°C. Denatured protein slowly oxidizes in air (half-time <3 days). After 7 days in air at 0–4°C, no recovery of fluorescence is seen. However, full fluorescence is restored by the addition of 2-mercaptoethanol (1 mM) (Surpin and Ward, 1989). The wild-type recombinant form of *Aequorea* GFP behaves in a very similar manner with respect to denaturation and renaturation. We have found, however, that the S65T variant recovers 100% from base denaturation (A. S. Sawyer, D. Gonzalez, and W. Ward, unpublished results). We have been unable to renature *R. reniformis* GFP from 6 M guanidine·HCl despite many attempts, but this protein will renature from acid or base as described above (Sawyer and Ward, unpublished). Better fluorescence recovery (up to 80%) is seen in renaturation from base than from acid (20–40%).

Fig. 3.2 *continued*

molecular weight standards include ovalbumin, carbonic anhydrase, ß-lactoglobulin,-lysozyme, bovine trypsin, and insulin α and β chains having molecular weights of 43, 29, 18.4, 14.3, 6.2, and 3 kDa, respectively. Sample lanes are as follows: *Upper gel:* lane 3-Mut 1, lane 4-S65T, lane 5-"cycle 3," lane 6-Y66H, lane 7-wild type recombinant. *Lower gel:* lane 3-wild-type recombinant, lane 4-native *Renilla mulleri*, lane 5-native *Renilla reniformis*, lane 6-native *Halistaura (Mitrocoma) cellularia*, lane 7- native *Phialidium gregarum*. All bands of native and recombinant GFP migrate as monomeric proteins of molecular weight 25–26 kDa in this system.

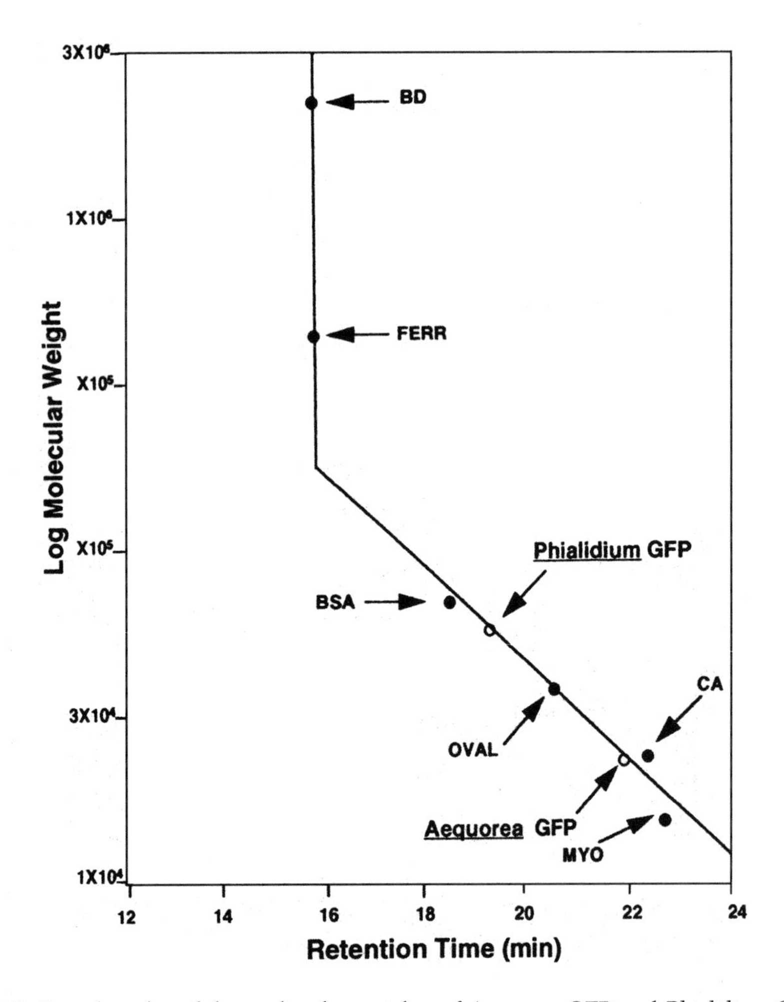

FIG. 3.3 Semi-log plot of the molecular weights of *Aequorea* GFP and *Phialidium* GFP vs elution time from a Phenomenex SEC S-2000 size exclusion high-performance liquid chromatography (HPLC) column. The six reference standards are Blue Dextran (BD), ferritin (FERR), bovine serum albumin (BSA), ovalbumin (OVAL), carbonic anhydrase (CA), and myoglobin (MYO) having molecular weights of 2000, 440, 67, 43, 29, and 17 kDa, respectively. Based on these standards, *Aequorea* GFP elutes at an apparent molecular weight of 28.5 kDa and *Phialidium* GFP elutes at an apparent molecular weight of 52 Kd. These apparent molecular weights are independent of GFP protein concentration for all detectable sample dilutions below 1 mg mL^{-1}. Sample concentrations as low as 1 μg/mL^{-1}, which can be detected with an on-line fluorimetric monitor, show no shifts in apparent molecular weight for any of the GFPs tested. The column was equilibrated with and eluted with an aqueous pH 6.5 buffer consisting of 50 mM sodium phosphate, 100 mM sodium chloride, and 0.02% w/v sodium azide. Flow was maintained at 0.5 ml/min and a sample volume of 80 μL was injected at time zero. Blue dextran and ferritin mark the column void volume (7.8 mL), which elutes at 15.6 min. All recombinant forms of GFP tested under these conditions (wt, S65T, "cycle 3," Mut 1, and Y66H) elute as monomers (26–30 kDa) while, with the exception of *Aequorea* GFP, all native forms of GFP [*R. reniformis*, *R. mulleri*, *R. kollikeri*, *Phialidium gregarum*, and *Halistaura* (*Mitrocoma*) *cellularia*] elute as dimers (50–54 kDa).

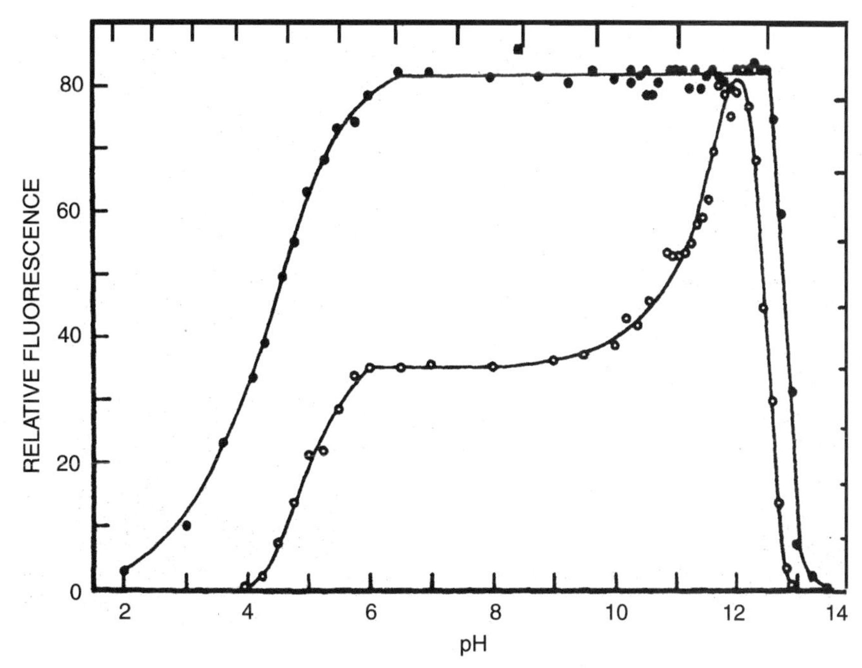

FIG. 3.4 GFP fluorescence versus pH. Solid circles for *Renilla* GFP, open circles for *Aequorea* GFP. Samples incubated 5 ± 0.5 min at 22± 2°C before fluorescence was measured on a Turner 110 fluorimeter with blue lamp, Ditric FITC excitation filter, and Corning 3–70 emission filter. Buffers were: 0.05 M each of glycine · HCl, sodium phosphate, sodium citrate (pH 2–11), 0.025 M Na_2HPO_4 (pH 11–11.9), 0.05 M KCl · NaOH buffer (pH 12–13), and 0.1–1.0 M NaOH (pH 13–14). From Ward (1981).

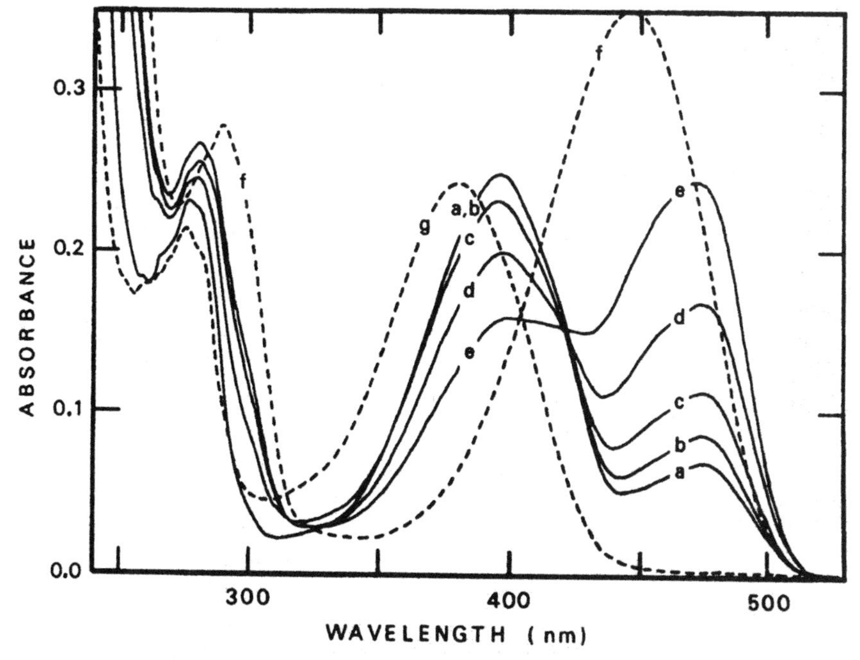

FIG. 3.5 *Aequorea* GFP absorption spectra at various pH values. Samples incubated 30 min at 22 ± 2°C at pH: 5.46 (*a*), 8.08 (*b*), 10.22 (*c*), 11.07 (*d*),11.55 (*e*), 13.0 (*f*) and 1.0 (*g*). For curves (*a–e*) the buffer contained 0.01 M each of sodium citrate, sodium phosphate and glycine. Sample (*f*) was in 0.1 M NaOH and sample (*g*) was in 0.1M HCl. Note isosbestic points at 422 and 405 nm. [From Ward 1981.]

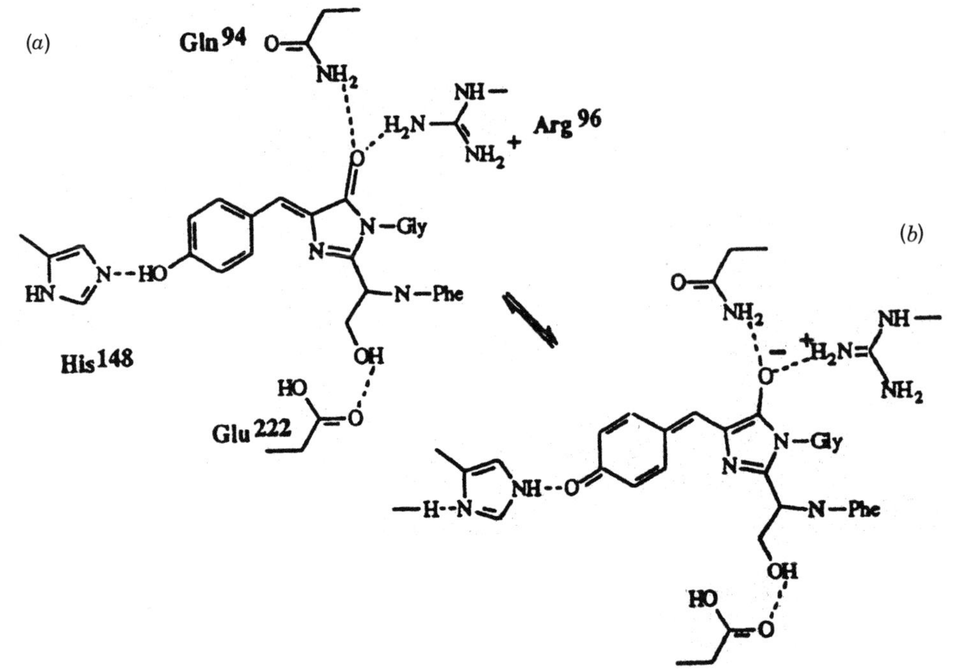

FIG. 3.6 Tautomeric forms of the GFP chromophore showing crucial interactions with four amino acid side chains. Form (*a*) is the protonated phenolic form of the chromophore with a keto oxygen on the imidazole ring. This form is thought to be responsible for the absorption band centered near 395 nm. Form (*b*) is the quinone–enol form of the chromophore with a full negative charge on the imidazolone oxygen. This form is thought to be responsible for the absorption band centered near 475 nm (wild type), 489 nm (S-65T, Mut 1), or 498 nm (*Renilla* GFP). [Diagram from Yang, et al., 1997 and Youvan and Michel-Beyerle, 1996.]

3.4.C Effect of Proteases

When used at moderate concentration (0.1 mg mL^{-1}), none of the common proteases has any effect on GFP fluorescence, even after 24 h incubation under optimum conditions for the protease. This protease resistance is true for *Renilla* GFP (subtilisin) and *Aequorea* GFP (trypsin, chymotrypsin, thermolysin, elastase, ficin, proteinase K, chymopapain, papain, subtilisin, pancreatin, bromelain, and pronase). At higher protease concentrations (43 mg mL^{-1}), following 60-h incubation with each of these twelve proteases, individually, only bromelain and pronase appear to have affected *Aequorea* GFP fluorescence significantly (Roth and Ward, 1983; Roth, 1985). Protease concentrations as high as 1 mg ml^{-1} (trypsin, chymotrypsin, papain, subtilisin, thermolysin, and pancreatin) have shown no effect on *Aequorea* GFP fluorescence (Bokman and Ward, 1981). In fact, GFP is so stable that it tolerates massive contamination with bacteria or fungi when, inadvertently, it is stored for months at room temperature in buffer or in buffered sucrose solutions without antimicrobial agents (W. Ward, unpublished results). Isolation of the chromophoric hexapeptide from

Aequorea GFP by papain digestion (Cody et al., 1993) required prior heat denaturation of the protein, as papain has no effect on undenatured GFP.

3.4.D Effect of Organic Solvents

A number of years ago, following purification of native *Aequorea* GFP by hydrophobic interaction chromatography on a phenyl sepharose column, in which we used 50% ethylene glycol as the eluting solvent, we measured a GFP recovery of 150% (Robart and Ward, 1990). The "150% recovery" was traced to a direct organic solvent effect on the fluorescence excitation spectrum of GFP (but not on the fluorescence emission spectrum). Since then, we have surveyed 25 different water-miscible organic solvents at 10% increments (from 10–90% v/v in 10 mM Tris–EDTA buffer at pH 8.0). Every organic solvent that we have surveyed, even glycerol, perturbs the absorbance/excitation spectrum of native and recombinant wild-type *Aequorea* GFP (González et al., 1997; A. S. Sawyer, E. Castriciones, and W. Ward, unpublished results). In each case, the solvent induces a spectral shift, qualitatively and quantitatively similar to the shift produced in the GFP absorption spectrum when pH is changed from 8.0–12.0 (Ward et al., 1982). The 395-nm absorption peak decreases in intensity while the 475-nm peak increases several-fold, suggesting a transition in the electronic form of the chromophore from that represented in Figure 3.6(*a*) to that represented in Fig. 3.6(*b*). The solvent-induced apparent increase in fluorescence, observed when exciting GFP with optics that favor the longer wavelength excitation band (Ditric FITC broad band interference filter, Ditric Optics) has, at times, exceeded four-fold. The most effective solvent in promoting this shift is acetonitrile, which induces large spectral shifts at concentrations as low as 5% v/v. Similar organic solvent experiments with native *R. reniformis* GFP and with *Aequorea* GFP variants S65T and Mut 1 (EGFP) show no changes at all in excitation or emission spectra, suggesting that in *Renilla* GFP, S65T, and Mut 1 the chromophore may always be in the electronic form represented in Figure 3.6(*b*). In most solvents, at very high organic solvent concentration (>60% v/v) both *Aequorea* and *Renilla* GFPs lose fluorescence intensity without undergoing further spectra shifts, consistent with solvent-induced denaturation.

3.4.E Effect of Detergents and Chaotropes

Surprisingly, none of the GFP types we have studied shows any significant change in the intensity or peak position of excitation or emission when observed within 60 min of treatment with a wide array of anionic, cationic, zwitterionic, and nonionic detergents (0.01, 0.10, 1.0% w/v) or chaotropic agents at room temperature (González et al., 1997; A. S. Sawyer, E. Castriciones, and W. Ward, unpublished results). Examples among the detergents include: SDS, CHAPS, CTAB, Tween 80, and Triton X-100. Among chaotropic agents, those that show little or no effect on GFP fluorescence include: 8 M urea, 4 M guanidine·HCl, 1 M guanidine·SCN, and 4 M KI. Over longer periods of time or at slightly higher temperature (40°C) both *Renilla* GFP and *Aequorea* GFP will lose fluorescence in detergents or chaotropes as the proteins slowly denature, but always *Aequorea* GFP fades more rapidly, suggesting, again, a more rigid tertiary conformation in the case of *Renilla* GFP.

3.4.F Effect of Fixatives and Preservatives

To the cellular and developmental biologists and microscopists who work with GFP, the most useful stability property of GFP may be its tolerance to fixatives such as formaldehyde and glutaraldehyde (Chalfie et al., 1994). Retention of GFP fluorescence in these traditional fixatives provides an opportunity to view and localize GFP in preserved tissue. We have, for example, been able to store solutions of GFP for several weeks in 3% buffered formaldehyde solutions (10% formalin) without appreciable loss of fluorescence. For long-term storage of preserved tissues, formaldehyde may be preferable to glutaraldehyde, as the latter tends to yellow with age and to develop its own bluish fluorescence. Thus opaque internal organs genetically transformed with GFP can be fixed with 10% formalin and prepared by frozen sectioning for microscopic localization of GFP. Water soluble imbedding, with substances such as carbowax, followed by conventional microtome sectioning could possibly be used instead of cryosectioning. Because GFP loses all fluorescence in absolute ethanol, it is very unlikely that GFP will tolerate complete dehydration, as required for paraffin or plastic imbedding.

3.5 SPECTROSCOPY AND CHROMOPHORE STRUCTURE

3.5.A Chromophore Structure

Shimomura (1979) was first to propose a structure for the *Aequorea* GFP chromophore, which he released from the protein by papain digestion. He also noted a similarity between this chromophore and coelenterate-type luciferin (coelenterazine). Later we showed (Cody et al., 1993) that the papain limit digest chromopeptide of *Aequorea* GFP is a cyclized hexapeptide derived from an internal portion of the GFP primary sequence (Phe64-Ser-Tyr-Gly-Val-Gln69). Mass spectroscopy of the isolated hexapeptide indicates the loss of 20 mass units, consistent with dehydration (–18) and dehydrogenation (–2) reactions. Peptide sequencing, two-dimensional nuclear magnetic resonance (2D NMR), and amino acid analysis following further proteolytic degradation of the hexapeptide with carboxypeptidase and pronase show the locations of posttranslational changes. The data demonstrate that cyclization involves condensation between the carboxyl carbon of serine and the amino nitrogen of glycine and the dehydrogenation of the tyrosine methylene bridge. The structure of this chromophore, shown in Figure 3.7(*a*) (Cody et al., 1993), has been confirmed by X-ray crystallography (see Chapter 4 by Phillips). The hexapeptide chromophore structure in *Renilla reniformis* GFP appears to be identical to that of *Aequorea* GFP except that Val68 and Gln69 are replaced by Asp and Arg, respectively (San Pietro et al., 1993).

Several chromophore variants of GFP that produce a fluorescent product have been isolated. Serine at position 65 has been replaced by Thr, Ala, Cys, Leu,

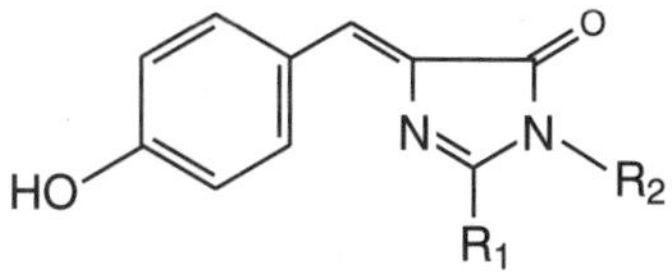

(*a*)

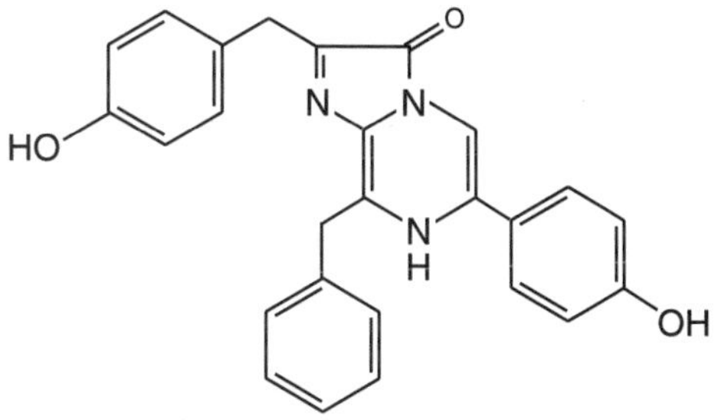

(*b*)

FIG. 3.7 (*a*) Structure of the chromophore of the GFP (*b*) Structure of coelenterate-type luciferin (coelenterazine).

and Gly, tyrosine at position 66 has been replaced by Phe, Trp, and His (Cubitt et al., 1995; Delagrave et al., 1995). To date, no functional GFP has been reported in which Gly67 is replaced by any other amino acid, suggesting either that changes at this position do not affect fluorescent properties of GFP or that there is an essential role for Gly67 in chromophore formation (Delagrave et al., 1995).

The chemical mechanism for chromophore formation is not completely understood; however, it is clear that molecular oxygen is required (Davis et al., 1994; Heim et al., 1994). Furthermore, the ability to express GFP in a cell-free translation system (Kolb et al., 1996) reinforces the belief that no other cofactors or enzymes are required for cyclization of the apoprotein to make a fluorescent chromophore. Heim et al. (1994) and Cubitt et al. (1995) proposed a plausible mechanism for GFP chromophore formation in which cyclization to form the imidazolone ring precedes tyrosine side chain oxidation by molecular oxygen, the latter reaction occurring with a time constant of about 4 h. Others have created or selected for GFP variants in *E. coli* with postinduction "greening" rate constants of less than 2 h. The "cycle 3" variant, for example (Crameri et al., 1996), reaches 50% of maximum fluorescence in *E. coli* in 95 min postinduction.

3.5.B The Question of Brightness

Most of the applications of GFP in cellular and developmental biology published in the years 1994–1996 have utilized the original wild-type clone of GFP that is based on the *gfp10* gene (Prasher et al., 1992; Chalfie et al., 1994). While some of the fluorescence images derived from the GFP product of this gene are indeed spectacular, such results are seen when the wild-type gene is controlled by an exceptionally strong promoter. Systems with weak promoters may give disappointing results, so there has been a demand for a "brighter" form of GFP. To the GFP applications expert, "brighter" may mean higher signal and improved contrast in fluorescence microscopy, confocal microscopy, or fluorescence activated cell sorting (FACS). Here, critical concerns include the intensity and spectral distribution of the exciting lamp, the selection of excitation and emission filters, and the relative photostability of the GFP in question. To the molecular biologist, a "brighter" GFP may be one that is more efficiently transcribed and translated because of optimum plasmid copy number, more efficient protein folding, elimination of cryptic introns, improved codon usage, and absence of GFP-containing inclusion bodies. To the biochemist and biophysicist, "brighter" means higher fluorescence quantum yield, higher molar extinction coefficient, and/or greater absorption cross-section.

The quantitation of any one of these parameters associated with "brightness," from photostability, to plasmid copy number, to molar extinction coefficient, is a nontrivial process. As a biochemist, I will focus on the "brightness" factors derived from quantum yield, molar extinction coefficient, and absorption cross-section.

3.5.C GFP Quantum Yield

Now, there seems to be general agreement that the fluorescence quantum yield for *R. reniformis* GFP (Ward and Cormier, 1979; Ward, 1979) and *A. victoria* GFP (Morise et al., 1974; Kurian et al., 1994) is 0.8. Earlier references, citing fluorescence quantum yields for *Renilla* GFP of 0.3 (Cormier et al., 1974; Wampler et al., 1971; Ward and Cormier, 1976), are in error.

Quantum yield measurements require the collection of fluorescence emission spectra from the unknown and from an accepted reference standard that closely approximates the unknown in spectral emission. Fluorescein satisfies these requirements for all forms of GFP except Y66H and related blue emission mutants. Both fluorophores are adjusted to the same low-absorbance value spectrophotometrically (generally 0.0100 absorbance units) at a wavelength common to both excitation spectra (470 nm can be used for all "non-blue" GFPs). In the case of the native GFP and the wild-type and cycle 3 recombinant GFPs, measurement of a concentrated protein stock solution followed by dilution to 0.0100 absorbance units may introduce a large error. In the 470-nm region, all three of these forms of GFP deviate greatly (by a factor of 4 or 5) from Beer's law upon dilution (Fig. 3.5), especially the 'cycle 3' variant (Morise et al., 1974; Ward et al., 1982; W. Ward, D. González, and A. S. Sawyer, unpublished results).

Fluorescence emission spectra are then recorded and the integrated spectra are compared mathematically to derive a quantum yield for the unknown. If the unknown and reference standard differ significantly in the shapes of their spectral emissions, then corrections for spectral sensitivity of the emission monochromator and photomultiplier tube must be employed. Further questions arise when dealing with unknowns having bimodal excitation spectra (native *Aequorea* GFP, wild-type recombinant, and the "cycle 3" variant), the peaks of which vary in relative intensity as a function of protein concentrations, pH, temperature, and ionic strength (Ward et al., 1982), solvent composition of the external environment (Robart and Ward, 1990; González et al., 1997), and exposure to intense light (Cubitt et al., 1995; Patterson et al., 1997).

Estimates of quantum yields, ranging from 0.21–0.77, for seven different GFP variants expressed in *E. coli*, including wild-type (0.77), have been reported (Heim and Tsien, 1996). However, raw data, calculations, and methodology are not reported in this publication, so the potential for independent interpretation is limited. Patterson et al. (1997) report quantum yields for wild-type, S65T, αGFP, EGFP, and EBFP of 0.79, 0.64, 0.79, 0.60, and 0.17, respectively. Methodology is clearly described.

In general, it would appear that enhanced "brightness" of GFP mutants cannot be attributed to improved fluorescence quantum yields. A value of 0.8 is very close to the theoretical limit of 1.0. Furthermore, all reported fluorescence quantum yields of so-called "brighter" mutants are, in fact, lower than values reported for the native GFPs of *Renilla* and *Aequorea*. If "brighter" variants exist, it must be for other reasons.

3.5.D Molar Extinction Coefficient

Accurate measurement of the molar extinction coefficient of GFP requires: (a) a very highly purified and correctly folded protein (e.g., >95%), with greater than 95% conversion of the chromogenic tripeptide (-Ser65 Tyr Gly-, or equivalent) into the mature fluorescent chromophore; (b) a reliable measurement of total protein concentration that is insensitive to changes in amino acid composition, (c) a measurement of absorbance at the wavelength of interest; and (d) a precise evaluation of the absorbance at that wavelength as a function of external conditions (pH, ionic strength, temperature, buffer composition, organic solvent strength, presence of buffer additives, etc.) as well as a function of protein concentration (e.g., is Beer's law obeyed?).

If it is assumed that an improperly folded protein (or one lacking the mature chromophore) can be separated from a properly folded, mature GFP by a combination of traditional methods of high-resolution protein purification (gel filtration, ion exchange, and hydrophobic interaction chromatography plus native gel electrophoresis and isoelectric focusing), then it should be possible to demonstrate empirically that a preparation of GFP is pure and properly formed. Furthermore, a GFP molecule with an improperly formed chromophore, one that has failed to cyclize, should be readily apparent by high-resolution mass spectrometry and, due to its different pI from that of mature GFP, should be separable from it by ion exchange chromatography or isoelectric focusing. In our many years of purifying and characterizing native and recombinant GFPs by traditional multiphase chromatographic and electrophoretic means, we have never seen any evidence of improperly folded or improperly cyclized GFP in our final product. If, however, one were to purify a recombinant GFP molecule containing a polyhistidine N- or C-terminal tail by immobilized metal affinity chromatography (IMAC) on a nickel column, for example, at the exclusion of other high-resolution methods (Inouye and Tsuji, 1994), it would be impossible to separate improperly folded or improperly cyclized GFP from properly folded GFP. Furthermore, IMAC purification of GFP having an N-terminal polyhistidine tag is likely to produce a mixed population of incomplete GFP molecules including dead-end products of translation, many of which would not be expected to fluoresce. Spectroscopic measurements to determine molar extinction coefficients of such a product could, in fact, be meaningless. The results of other attempts to physically characterize IMAC purified GFP could be equally meaningless.

Each of the four required conditions listed above (a–d) appears to have been met prior to the establishment of the molar extinction coefficient for *R. reniformis* GFP at its absorption wavelength maximum of 498 nm (Ward and Cormier, 1979; Ward et al., 1980; Ward, 1981). The value for the molar extinction coefficient of *Renilla* GFP is 133,000 L mol^{-1} cm^{-1} for the monomer and 266,000 L mol^{-1} cm for the dimer (Tables 3.1 and 3.4). We also established the molar extinction coefficient for fully denatured *R. reniformis* GFP under strongly alkaline conditions (0.1 M NaOH). Under these conditions (pH 13) *Renilla* GFP loses all of its fluorescence in 1–3 min at room temperature and the absorption spectrum shifts to a single broad band centered at 447 nm (Fig. 3.5, Curve *f*). The molar extinction coefficient for this "denatured chromophore" is 44,000 L mol^{-1} cm^{-1} (Ward et al., 1980; Ward, 1981). Then, because the absorption spectrum of denatured *A. victoria* GFP at pH 13 (0.1 M NaOH) is identical to that of denatured *R. reniformis* GFP at pH 13 (Ward et al., 1980), we were able to back calculate the extinction coefficient of native *Aequorea* GFP at 395 nm (in 1-mM sodium phosphate buffer at pH 7.2) and arrive at a value of 27,600 L mol^{-1} cm^{-1} (Ward and Bokman, 1982), a value that varies substantially with pH, temperature, ionic strength, and total GFP concentration (Ward, 1981). This comparative method, based upon the common absorption characteristics of denatured GFP, is suitable for calculating extinction coefficients (traceable back to the original measurements with *Renilla* GFP) for all GFPs, natural or recombinant, so long as the chromophore contains a dehydrotyrosine residue conjugated to the imidazolone group. Thus, this method works for the S65T mutant and for Mut 1 (Cormack et al., 1996; Yang, et al., 1996), but not for Y66H, Y66W, or Y66F. Table 3.4 summarizes the molar extinction coefficients we have determined. Values based on measurements of the "base-denatured chromophore" are in excellent agreement with those determined independently by measurements of known tyrosine and tryptophan content in the respective GFP variants. They also agree closely with those reported recently by Patterson et al. (1997). The spectrum of denatured GFP at pH 13 is stable for 30–60 min at room temperature, but at higher temperature (40°C) or after longer incubation time, the peak at 447 nm degrades irreversibly to a new spectral form peaking near 335 nm (Ward, 1981). When using this method for determining relative extinction coefficients, it is essential to follow the absorption spectrum of the denatured GFP over time to insure total denaturation without chromophore degradation. Determination of the molar extinction coefficients of Y66H and related Tyr^{66} variants requires alternate methods for measuring protein concentration. Summation of tryptophan and tyrosine absorbencies (Table 3.4, footnote *d*) or use of the BCA protein assay method (Pierce Chem. Co.) are suitable alternatives.

3.6 ABSORPTION/EXCITATION CROSS-SECTION

The true fluorescence efficiency of GFP is the product of its absorption cross-section and its quantum yield. The absorption/excitation cross-section of a molecule refers to the number of photons absorbed divided by the number of photons incident (Jagger, 1977). Thus, the more closely the exciting light (spectrum of the exciting light X spectral transmission of the exciting filter) matches the absorption spectrum of GFP, the higher the fluorescence efficiency. For a GFP molecule with a broad absorption spectrum but relatively low light extinction across the spectrum (e.g., wild-type recombinant GFP derived from *Aequorea* gfp10), it may be necessary to employ broad band exciting light (e.g., 350–490 nm) of moderately high intensity to generate sufficient fluorescence. GFP molecules with higher extinction coefficients (e.g., S65T, Mut 1, and other so-called "bright" mutants) may give satisfactory results with exciting lights that are narrower in spectral output (e.g., argon lasers with line spectrum emission at 488 nm). Fluorescence intensity will be proportional to the product of the GFP quantum yield and absorption cross-section, as described above, so long as the light intensity is below saturation levels (light levels that are so intense as to create a photostationary state where all molecules of GFP are always excited). When light intensity is high enough to maintain a photostationary state, as may occur in some cases in fluorescence microscopy, chromophores with very high extinction coefficients offer no particular advantage over those with lower extinction coefficients; for example, every chromophore is excited at all times. Under such conditions, all that matters is the number of chromophores in the illuminated field and the fluorescence quantum yield and relative photostability of the fluorophore.

3.6.A Monomer/Dimer Equilibrium

With one exception, all GFP molecules that we have studied are stable nondissociable dimers in dilute aqueous solution (Fig. 3.3). Relative to six other standard proteins of known molecular weight, they migrate as dimers (2 × 27 kDa) on a Phenomenex SEC S-2000 size exclusion HPLC column in an aqueous pH 6.5 buffer solution comprised of 50-mM sodium phosphate and 100-mM NaCl. The chromatographic positions, thus the apparent molecular weights, are unaffected by protein dilution, even at levels below 1 mg GFP/mL. The GFPs surveyed include those from *R. reniformis* (Ward and Cormier, 1979) *R. mulleri, R. kollikeri, Phialidium gregarium,* and *Halistaura (Mitrocoma) cellularia* (Fig. 3.3) (W. Ward, unpublished results). The only exception, so far, is the GFP from A. *victoria*. At moderate protein concentrations (<1 mg/mL), *Aequorea* GFP migrates as a 26–29 kDa monomer under the chromatographic conditions described above (Fig. 3.3). So do all of the recombinant forms of *Aequorea* GFP. In this regard, we have studied wild-type GFP (Chalfie et al., 1994), the DNA shuffled cycle 3 mutant (Crameri et al., 1996), S65T (Cubitt et al., 1995), Y66H (Cubitt et al., 1995), and Mut 1 (Cormack et al., 1996; Yang, et al., 1996b). All GFP variants, whether native or recombinant, have monomer molecular weights of 27 kDa as determined by SDS gel electrophoresis (Fig. 3.2).

Although they are monomers in dilute solution, native A. *victoria* GFP and all recombinant forms of GFP derived therefrom, including wild-type, S65T, Mut 1, cycle 3 variant, and Y66H, will form dimers at high-protein concentration. The concentration required to demonstrate dimer formation exceeds the loading capacity of the Phenomenex SEC S-2000 column described above. But dimer can be demonstrated clearly on a Bio-Rad P-100 BioGel column (fine particle size, 1 × 47 cm). This column is routinely run in our lab in 10-mM Tris–EDTA at pH 8.0 with 1 M $(NH_4)_2SO_4$ and 0.02% NaN_3. The high concentration of ammonium sulfate favors dimer formation by promoting hydrophobic interactions between GFP monomers. Gravity-generated flow rates of 2 mL/h are maintained and 200–500-μL samples of GFP (up to 100 mg ml^{-1}) are loaded. This protein

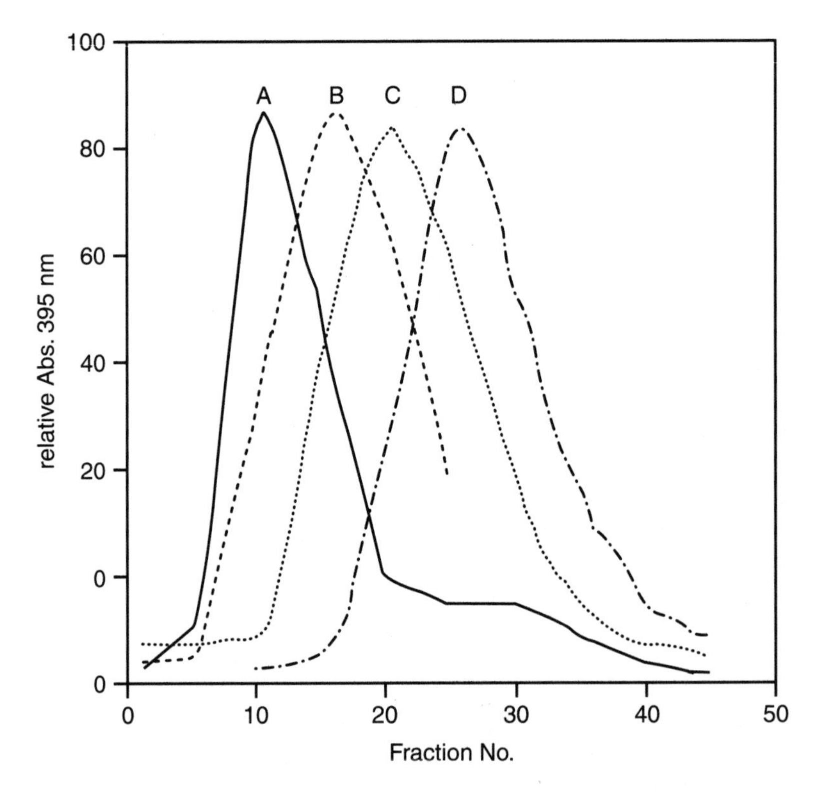

FIG. 3.8 Shift in apparent molecular weight of wild-type recombinant *Aequorea* GFP as a function of GFP concentration. The GFP samples are chromatographed at 24 ± 2°C on a BioRad P-100 BioGel column (225-ml total volume) equilibrated in 10-mM Tris–EDTA buffer (pH 8.0) containing 1 M $(NH_4)_2SO_4$ and 0.02% NaN_3. The concentrations of GFP samples loaded onto the column are 12 mg/mL (*a*), 4.2 mg/mL (*b*), 0.84 mg/mL (*c*), and 0.17 mg/mL (*d*). The elution behavior is consistent with there being nearly all dimer in sample (*a*), nearly all monomer in sample (*d*), and intermediate forms (in equilibrium) in samples (*b* and *c*).

concentration-dependent shift in apparent molecular weight of wild-type recombinant GFP is shown in Fig. 3.8.

As we have seen strong evidence in our laboratory (Cutler, 1995; Cutler and Ward, 1997) for hydrophobically driven dimer formation in native *Aequorea* GFP, in wild-type GFP, and in recombinant variants of GFP, we would expect to see a crystallographic unit cell that is also dimeric. Thus, when GFP crystals are grown in high-salt media, the unit cell is, in fact, a dimer (Perozzo et al., 1988; M. A. Perozzo, personal communication). But when the medium contains 22–26% polyethylene glycol, a solvent that does not favor hydrophobic interaction, monomers form the unit cell of the crystal (Ormö et al., 1996). If a nonpolar solvent (58% 2-methyl-2,4,-pentanediol) is used to promote crystallization of GFP, the unit cell is a dimer, but in this case most of the dimer interface contacts are polar (Yang, et al., 1996a, 1997). The dimer interface appears, by computer-generated molecular modeling of the Yang, F. et al. (1996a) coordinates, to contain both hydrophobic and hydrophilic contacts. A slight rotation around the long axis of the "beta can" can cause a switch from predominantly hydrophilic contacts at the dimer interface to predominantly hydrophobic contacts (M. A. Perozzo, personal communication). If such a rotation occurs as a function of crystallization solvent, it has little effect on the overall crystal structure, but it might be expected to affect the absorption spectrum of the crystalline protein.

If wild-type recombinant GFP is being used as a reporter of gene expression in living cells, it is likely to be of sufficient intracellular concentration (>5 mg mL^{-1}) as to be nearly all in the dimer form. In the photocytes of the jellyfish *A. victoria*, for example, GFP intracellular concentration (assuming uniform distribution of the protein throughout the cytoplasm) has been estimated to be 25 mg mL^{-1} (Cutler and Ward, 1997). Dimerization would not be of concern to those using GFP as a reporter if it were not for the fact that wild-type recombinant GFP and related mutants like the cycle 3 variant undergo large changes in absorption (excitation) spectra upon dimerization. The extinction coefficient of the peak at 395 nm increases about 15%, but the extinction coefficient of the shoulder near 475 nm (where most microscopists prefer to excite GFP) drops fourfold to fivefold (Fig. 3.5) from about 12,000 L mol^{-1} cm^{-1} to less than 3000 L mol^{-1} cm^{-1} (Ward et al., 1982). Thus, in the GFP concentration range from 0.2 to 10 mg mL^{-1}, the higher the concentration of intracellular GFP, the more poorly each chromophore absorbs and is excited by blue light.

The Ser65 mutants of GFP (e.g., S65T and Mut 1) do not show a concentration-dependent suppression of the blue absorption peak near 489 nm. They retain their relatively high extinction coefficients of 55,000–58,000 L $mol^{-1}cm^{-1}$ at 489 nm (Table 3.4), even when they dimerize at high protein concentration. Retention of a concentration-independent molar extinction coefficient is a distinct advantage in quantitating intracellular fluorescence signals. However, the Ser65 mutants are more sensitive to pH differences than wild-type GFP and show a steep pH-dependent loss of fluorescence below pH 7.0 (González et al., 1997; Patterson et al., 1997; A. S. Sawyer, D. González, and W. Ward, unpublished results). This loss of fluorescence intensity below pH 7.0 for S65T and Mut 1 may present a problem with quantitation and calibration of fluorescence signals within cells if the cytoplasmic or organellar pH is lower than 7.0 and/or variable.

3.7 A MOLECULAR MECHANISM FOR SUPPRESSION OF CHROMOPHORE ABSORPTION

For more than two decades, we have been aware that the 475-nm shoulder in the absorption spectrum of native A. *victoria* GFP is strongly suppressed at high-protein concentrations (Morise et al., 1974). As the protein concentration is raised from 0.112 to 18.6 mg mL^{-1}, the 475 nm absorption shoulder is suppressed four-fold (Fig. 3.5) while the main peak at 395 nm increases by 15% (Ward et al., 1982). At even higher protein concentrations, and particularly in the presence of antichaotropic salts, (e.g., ammonium sulfate) the shoulder suppression reaches five-fold (Cutler, 1995; Cutler and Ward, 1997) such that the molar extinction coefficient at 475 nm drops below 3000 L $mol^{-1}cm^{-1}$. The same behavior is seen with the wild-type recombinant GFP and the cycle 3 mutant (González et al., 1997; Patterson et al., 1997; W. Ward, unpublished results). It is also clear from calibrated gel filtration chromatography (Cutler, 1995; Cutler and Ward, 1997) that native *Aequorea* GFP has a strong tendency to form dimers at high-protein concentration and that ammonium sulfate (1 M) promotes dimerization at substantially lower GFP concentrations. High concentrations of antichaotropic salts, such as ammonium sulfate, are known to favor hydrophobic interactions, so we have proposed (Cutler, 1995) that GFP monomers interact at hydrophobic interfaces when they form dimers in aqueous solution, especially in the presence of antichaotropic salts. In organic solvents, such as the 58% 2-methyl-2,4-pentanediol used to induce crystal formation with wild-type recombinant GFP (Yang, et al., 1996a, 1997), hydrophobic interactions between monomers would be inhibited by the organic solvent. In fact, Yang, et al. (1996a) observe more hydrophilic contacts than hydrophobic contacts between GFP monomers in their crystals. This result is not at all surprising considering the crystallization solvent. However, just beyond these hydrophilic interfaces, on the surface of the GFP cylinder, there exists a region of strong surface hydrophobicity dominated by residues such as Leu221, Phe223, Pro211, and Tyr39 (Perozzo, personal communication). Amino acid residues 221 and 223 project into the hydrophobic interface while the bridge amino acid residue, Glu222, projects toward the interior of the beta can. Glu222 is hydrogen bonded to the hydroxyl residue of Ser65 in wild-type recombinant GFP (Yang, et al., 1996a).

Thus an unbroken chain exists, connecting surface hydrophobic residues to the chromophore via Glu222. This chain appears to "regulate" the distance between the keto oxygen of the chromophore imidazolone and the electron withdrawing groups Arg96 and Glu94 that are positioned above the imidazolone ring of the chromophore. A retracted chain (resulting from hydrophobic contacts between monomers at residues 221 and 223) "pulls" on the chromophore, separating the keto oxygen from the electron-withdrawing residues Arg96 and Glu94. An expanded chain (resulting from the relaxation of hydrophobic contacts at the dimer interface or direct elongation of the "chain") allows the chromophore to slip closer to Arg96 and Glu94, shortening the distance between the keto oxygen and residues 96 and 94. As the wild-type GFP dimer forms in aqueous solution (GFP dimerization is strongly favored in high ionic strength medium such as the cytoplasm of prokaryotic and eukaryotic cells where recombinant GFPs are being expressed], corresponding hydrophobic interfaces collapse upon each other. In this process, leucine221 and Phe223 are drawn into the interface and farther from the surface of the protein. This interfacial collapse exerts a "pull" on the chromophore, via Glu222, decreasing the electron-withdrawing influence of Arg96 and Glu94. Removed from the influence of these residues, the chromophore assumes the tautomeric form shown in Figure 3.6(*a*). This tautomeric form absorbs light at a wavelength maximum of 395 nm. The alternate tautomeric form, shown in Figure 3.6(*b*) is suppressed and so its corresponding spectral form (λ_{max} = 475 nm) is also suppressed.

In the case of S65T and Mut 1, the orientation of the 2° hydroxyl group (serine to threonine substitution) at position 65 may, in effect, lengthen the chain connecting the hydrophobic interface with the chromophore. This lengthening allows the chromophore to remain close to residues 96 and 94 at all times, favoring the quinone–enol form of the chromophore shown in Figure 3.6(*b*). As a consequence, S65T and Mut 1 display red-shifted absorption spectra (λ_{max} = 489 nm), which do not shift further upon protein dimerization. In the case of *Renilla* GFP, which is always dimeric, the interfacial contact between dimers may position the chromophore very close to electron-withdrawing groups (amino acid residues in *Renilla* GFP analogous to Arg96 and Glu94 in *Aequorea* GFP) at all times. Thus, the absorption peak is shifted to 498 nm and the protein fails to respond spectrally to a wide array of external perturbants.

Water-miscible organic solvents are capable of inducing large absorption/excitation spectral shifts (toward the red) in *Aequorea* GFP (Robart and Ward, 1990) and related recombinant forms (wild-type recombinant and the cycle 3 variant) (González et al., 1997). Such spectral shifts are observed by absorption spectroscopy at moderate protein concentration (0.1–1.0 mg mL^{-1}) and by fluorescence spectroscopy at much lower protein concentration (0.1–10 mg mL^{-1}), where *Aequorea* GFP is entirely monomeric, suggesting that these organic solvents must be affecting the monomer directly. No solvent-dependent shift in monomer–dimer equilibrium is involved. If the organic solvent interacts with surface residues Leu221 and Phe223, as might be expected, it may force these residues deeper into the protein interior. A retraction of these residues, communicated to the chromophore via Glu222, could allow the quinone–enol form of the chromophore to slip into the electron-withdrawing pocket formed by Arg96 and Glu94. Consistent with this model is the fact that all GFP forms with bimodal absorption/excitation spectra in the UV and visible regions of the spectrum (native, wild-type, and cycle 3 variants) undergo these large solvent-induced spectral shifts. Those that are already red-shifted (e.g., *Renilla* GFP, S65T, and Mut 1) show no further spectral shifts upon solvent addition.

3.8 FUTURE PROSPECTS

It is a daunting process to predict future directions of research, especially when one is so close to the field. No one, for example, within the bioluminescence community (even as late as 1990) predicted that GFP chromophore formation would be autocatalytic or that cDNA from A. *victoria* could ever be expressed in heterologous systems as a functional fluorescent protein. No one envisioned GFPs applications in developmental biology, in protein trafficking, in gene reporting, or in high throughput screening in cell-based assays. The GFP was just that other interesting protein in *Aequorea* that a handful of "aequorin researchers" stockpiled in their freezers and occasionally thawed out to study its magnificent physical properties.

When I started working in Milton Cormier's lab in 1973, GFP was almost unknown, except within a small circle of bioluminescence researchers. None of us ever expected to see GFP on the covers of *Science, Nature Biotechnology, Bioluminescence and Chemiluminescence, Biotechniques,* the CLONTECH catalog, or BioRad's "Explorer" series of GFP-based instructional kits for high schools.

So, having established my record as a prognosticator, here are my predictions. I believe that the next research frontier for GFP will be in the area of cell-based diagnostics and high throughput homogeneous screening of drugs, organic chemicals, toxicants, food additives, herbicides, pesticides, mutagens, carcinogens, and teratogens. Custom-designed cell lines in which GFP expression is controlled by specific promoters will be used to screen these sorts of compounds minutes after their robotic application to multiwell plates. The carefully designed specificity of each promoter-GFP response will establish directly and immediately the biochemical mode of action of the substance being tested. The information generated in such biochemically and pharmacologically specific screening assays will be of such high quality as to eliminate the need for many costly levels of secondary testing.

The search for *Aequorea* GFP variants with higher extinction coefficients, more rapid "greening" rates, and red-shifted excitation and emission spectra will continue to be fruitful, but the focus of the search will shift to other species of coelenterates that already display spectral biodiversity nearly as great as the range of *Aequorea* GFP variants. Incremental improvements in GFP will further stimulate development of instrumentation that can better handle the relatively high noise level of fluorometric screening methods.

I see GFP as one of several coelenterate bioluminescence reporters that will be used together, to an increasing extent, in cell-based diagnostics. *Renilla* luciferase and aequorin are already cloned and available for nonhomogeneous assays requiring luciferin (coelenterazine) addition. The need to add coelenterazine as an additional reagent, however, places the coelenterate system into the same category as other luciferin–luciferase diagnostic assays—only the luciferase has been cloned. But, there exists the hope that coelenterazine may soon be cloned as well, enabling *Renilla* luciferase, aequorin, and coelenterazine to be used together in a variety of homogeneous diagnostic assays. Development of rapid, sensitive, and homogeneous screening methods based on the coelenterate bioluminescence system could dominate this field over the next decade.

In the field of education, GFP will soon become the tool of choice to illustrate all aspects of biotechnology in the classroom laboratory from cell transformation to immobilized metal ion affinity chromatography. GFP based educational modules will be created for all educational levels, primary through postgraduate, "bringing to light" such diverse fields as general biology, environmental testing, microbiology, molecular biology, toxicology, biochemistry, and molecular medicine.

The best prediction for the future of GFP was written 2 years ago by Susan Hassler whose editorial in *Bio/Technology* (Hassler, 1995) says it all, "Green Fluorescent Protein: The Next Generation."

ACKNOWLEDGMENTS

This work was sponsored, in part, by a grant from the National Science Foundation–Advanced Technological Education (DUE-9602356) and research contracts from BioRad Corporation and Clontech Laboratories. The manuscript was critically reviewed by Daniel González. Figures and tables were prepared by Peter Anderson and Dr. Frank Petersen. Much of the previously unpublished work has been contributed by Dr. Mark Cutler, Amy Roth, Daniel González, Anita Sawyer, Tom Spires, and Elmer Castriciones.

REFERENCES

Bokman, S. H. and Ward, W. W. (1981). Renaturation of *Aequorea* green-fluorescent protein. *Biochem. Biophys. Res. Commun.* 101:1372–1380.

Chalfie, M., (1995). Green fluorescent protein. *Photochem. Photobiol.* 62:651–656.

Chalfie, M., Tu, Y. Euskirchen, G., Ward, W. W., and Prasher, D. C. (1994). Green-fluorescent protein as a marker for gene expression. *Science* 263:802–805.

Cody, C. W., Prasher, D.C., Westler, W. M., Prendergast, F. G. and Ward, W. W. (1993). Chemical structure of the hexapeptide chromophore of the *Aequorea* green-fluorescent protein. *Biochemistry* 32:1212–1218.

Cormack, B., Valdivia, R., and Falkow., S. (1996). FACS-optimized mutants of the green fluorescent protein (GFP). *Gene* 173:33–38.

Cormier, M. J., Hori, K. and Anderson, J. M., (1974). Bioluminescence in coelenterates. *Biochim. Biophys. Acta* 346:137–164.

Cormier, M. J., Hori, K., Karkhanis, Y. D., Anderson, J. M., Wampler, J. E., Morin, J. G. and Hastings, J. W. (1973). Evidence for similar biochemical requirements for bioluminescence among the coelenterates. *J. Cell. Physiol.* 81:291–297.

Crameri, A., Whitehorn, E. A., Tate, E., and Stemmer, W. P. C. (1996). Improved green fluorescent protein by molecular evolution using DNA shuffling. *Nature Biotech.* 14:315–319.

Cubitt, A. B., Heim, R., Adams, S. R., Boyd, A. E., Gross, L. A. and Tsien., R. Y. (1995). Understanding, improving and using green fluorescent proteins. *TIBS* 20:448–455.

Cutler, M. W. (1995). Characterization and energy transfer mechanism of the green-fluorescent protein from *Aequorea victoria*. Ph.D. Thesis, Rutgers University, New Brunswick, NJ.

Cutler, M. W. and Ward, W. W. (1993). Protein–protein interactions in *Aequorea* bioluminescence. Bioluminescence Symposium, Maui, HI, Nov. 5–10.

Cutler, M. W. and Ward, W. W. (1997). Spectral analysis and proposed model for GFP dimerization. *In Bioluminescence and Chemiluminescence; Molecular Reporting With Photons*. Hastings, J. W., Kricka, L. J., and Stanley, P. E. Eds., Wiley, New York, pp. 403–406.

Davis, D. F., Ward, W. W., and Cutler, M. W. (1994). Post-translational chromophore formation in recombinant GFP from *E. coli* requires oxygen. *In Bioluminescence and Chemiluminescence, Fundamentals and Applied Aspects*. Campbell, A. K., Kricka, L. J., and Stanley, P. E., Eds., Wiley, New York, pp. 596–599.

Delagrave, S., Hawtin, R. E. Silva, C. M., Yang, M. M., and Youvan, D. C. (1995). Red-shifted excitation mutants of the green fluorescent protein. *Bio/Technol.* 13:151–154.

Ehrig, T., O'Kane, D. J., and Prendergast, F. G. (1995). Green-fluorescent protein mutants with altered fluorescence excitation spectra. *FEBS Lett.* 367:163–166.

González, D.,Sawyer, A., and Ward, W. W. (1997). Spectral perturbations of mutants of recombinant *Aequorea victoria* green-fluorescent protein (GFP). *Photochem Photobiol.* 65:21S.

Hart, R. C., Matthews, J. C., Hori, K. and Cormier, M. J. (1979) *Renilla reniformis* bioluminescence: Luciferase-catalyzed production of nonradiating excited states from luciferin analogues and elucidation of the excited state species involved in energy transfer to *Renilla* green fluorescent protein. *Biochemistry* 18:2204–2210.

Hassler, S. (1995). Green fluorescent protein: the next generation. *Bio/Technol.* 13:103.

Heim, R., Cubitt, A. B., and Tsien, R. Y. (1995). Improved green fluorescence. *Nature (London)* 373:663–664.

Heim, R., Prasher, D. C., and Tsien, R. Y. (1994). Wavelength mutations and posttranslational autoxidation of green fluorescent protein. *Proc. Natl. Acad. Sci. USA* 91:12501–12504.

Heim, R. and Tsien, R. Y. (1996). Engineering green fluorescent protein for improved brightness, longer wavelengths, and fluorescence resonance energy transfer. *Curr. Biol.* 6:178–182.

Hori, K., Anderson, J. M., Ward, W. W., and Cormier, M. J. (1975). *Renilla* luciferin as the substrate for calcium induced photoprotein bioluminescence. Assignment of luciferin tautomers in aequorin and mnemiopsin. *Biochemistry* 14:2371–2376.

Hori, K., Charbonneau, H., Hart, R. C., and Cormier, M. J. (1977). Structure of native *Renilla* reniformis luciferin. *Proc. Nat. Acad. Sci. USA* 74:4285–4287.

Hori, K. and Cormier, M. J. (1973). Structure and chemical synthesis of a biologically active form of *Renilla* (sea pansy) luciferin. *Proc. Nat. Acad. Sci. USA* 70:120–123.

Hori, K., Wampler, J. E., Matthews, J. C., and Cormier, M. J. (1973). Identification of the product excited states during the chemiluminescent and bioluminescent oxidation of *Renilla* (sea pansy) luciferin and certain of its analogs. *Biochemistry* 12:4463–4469.

Inoue, S., Kakoi, H., Murata, M., Toto, T., and Shimomura, O. (1977a). Complete structure of *Renilla* luciferin and luciferyl sulfate. *Tetrahedron Lett.* 31:2685–2689.

Inoue, S., Okada, K., Kakoi, H., and Goto. T. (1977b). Fish bioluminescence I. Isolation of a luminescent substance from a myctophina fish, *Neoscopelus microchir*, and identification of it as *Oplophorus* luciferin. *Chem. Lett.* 1977:257–258.

Inouye, S. and Tsuji, F. I. (1994). Aequorea green fluorescent protein: expression of the gene and fluorescence characteristics of the recombinant protein. *FEBS Lett.* 341:277–280.

Jagger, J. (1977). Phototechnology and biological experimentation. *In The Science of Photobiology*. Smith, K. C., Ed., Plenum, New York, pp. 1–26.

Kain, S. R., Adams, M., Kondepudi, A., Yang, T.-T. Ward, W. W., and Kitts, P. (1995). Green fluorescent protein as a reporter of gene expression and protein localization. *BioTechniques* 19:650–655.

Kolb, V. A., Makeyev, E. V., Ward, W. W., and Spirin, A. S. (1996). Synthesis and maturation of green fluorescent protein in a cell-free translation system. *Biotech. Lett.* 18:1447–1452.

Kurian, E., Fisher, P. J., Ward, W. W., and Prendergast, F. G. (1994). Characterization of secondary and tertiary structure of the green fluorescent protein from A. *victoria. J. Biolum. Chemilum.* 9:333.

Levine, L. D. and Ward, W. W. (1982). Isolation and characterization of a photoprotein, "phialidin", and a spectrally unique green-fluorescent protein from the bioluminescent jellyfish *Phialidium gregarium. Comp. Biochem. Physiol.* 72B:77–85.

Morin, J. G. (1974). Coelenterate bioluminescence. *In Coelenterate Biology: Reviews and New Perspectives*. Muscatine, L. and Lenhoff, H., Eds., Academic, New York, pp. 397–438. .

Morin, J. G. and Hastings., J. W. (1971). Energy transfer in a bioluminescent system. *J. Cell. Physiol.* 77:313–318.

Morise, H., Shimomura, O., Johnson, F. H. and Winant, J. (1974). Intermolecular energy transfer in the bioluminescent system of *Aequorea. Biochemistry* 13:2656–2662.

Ormö, M., Cubitt, A. B., Kallio, K., Gross, L. A., Tsien, R. Y., and Remington, S. J. (1996). Crystal structure of the *Aequorea victoria* green fluorescent protein. *Science* 273:1392–1395.

Patterson, G. H., Knobel, S. M., Sharif, W. D., Kain, S. R., and Piston, D. W. (1997). Use of the green fluorescent protein and its mutants in quantitative fluorescence microscopy. *Biophys. J.* 73:2782–2790.

Perozzo, M. A., Ward, K. B., Thompson, R. B., and Ward, W. W. (1988). X-ray diffraction and time-resolved fluorescence analyses of *Aequorea* green fluorescent protein crystals. *J. Biol. Chem.* 263:7713–7716.

Prasher, D.C. (1995). Using GFP to see the light. *Trends Genet.* 11:320–323.

Prasher, D. C., Eckenrode, V. K., Ward, W. W., Prendergast, F. G. and Cormier, M. J. (1992). Primary structure of the *Aequorea victoria* green-fluorescent protein. *Gene* 111:229–233.

Prendergast, F. G. and Mann, K. G. (1978). Chemical and physical properties of aequorin and the green-fluorescent protein isolated from *Aequorea forskalea. Biochemistry* 17:3448–3453.

Rao, B., Kemple, M., and Prendergast, F. (1980). Proton nuclear magnetic resonance and fluorescence spectroscopic studies of segmental mobility in aequorin and a green fluorescent protein from *Aequorea forskalea. Biophys. J.* 32:630–632.

Robart, F. D. and Ward, W. W. (1990). Solvent perturbations of *Aequorea* green-fluorescent protein. *Photochem. Photobiol.* 51:92s.

Roth, A. (1985). Purification and protease susceptibility of the green-fluorescent protein of *Aequorea aequorea* with a note on *Halistaura*. M.S. Thesis. Rutgers University, New Brunswick, NJ.

Roth, A. F. and Ward, W. W. (1983). Conformational stability after protease treatment in *Aequorea* GFP. *Photochem. Photobiol.* 37S:S71.

SanPietro, R. M., Prendergast, F. G., and Ward, W. W. (1993). Sequence of the chromogenic hexapeptide of *Renilla* green-fluorescent protein. *Photochem. Photobiol.* 57:63s.

Shimomura, O. (1979). Structure of the chromophore of *Aequorea* green fluorescent protein. *FEBS Lett.* 104:220–222.

Shimomura, O., Inoue, S., Johnson, F. H. and Haneda, Y. (1980). Widespread occurrence of coelenterazine in marine bioluminescence. *Comp. Biochem. Physiol.* 65B:435–437.

Shimomura, O. and Johnson, F. H. (1979). Comparison of the amounts of key components in the bioluminescence systems of various coelenterates. *Comp. Biochem. Physiol.* 64B:105–107.

Stearns, T. (1995). Green fluorescent protein. The green revolution. *Curr. Biol.* 5:262–264.

Surpin, M. A. and Ward, W. W. (1989). Reversible denaturation of *Aequorea* green-fluorescent protein–thiol requirement. *Photochem. Photobiol.* 49:62S.

Wampler, J. E., Hori, K., Lee, J., and Cormier, M. J. (1971). Structured bioluminescence. Two emitters during both the *in vitro* and the *in vivo* bioluminescence of *Renilla*. *Biochemistry* 10:2903–2910.

Wampler, J. E., Karkhanis, Y. D., Morin, J. G., and Cormier, M. J. (1973). Similarities in the bioluminescence from the *Pennantulacea*. *Biochim. Biophys. Acta* 314:104–109.

Ward, W. W. (1979). Energy transfer processes in bioluminescence. *In Photochemical and Photobiological Reviews*, Vol. 4, Smith, K., Ed., Plenum, New York, pp. 1–57.

Ward, W. W. (1981). Properties of the coelenterate green-fluorescent proteins. *In Bioluminescence and Chemiluminescence: Basic Chemistry and Analytical Applications*. DeLuca, M., and McElroy, D, W., Eds., Academic, New York, pp. 235–242.

Ward, W. W. and Bokman, S. H., (1982). Reversible denaturation of *Aequorea* green-fluorescent protein: Physical separation and characterization of the renatured protein. *Biochemistry* 21:4535–4540.

Ward, W .W., Cody, C., Hart, R. C., and Cormier, M. J. (1980). Spectrophotometric identity of the energy transfer chromophores in *Renilla* and *Aequorea* green-fluorescent proteins. *Photochem. Photobiol* 31:611–615.

Ward, W. W. and Cormier, M. J. (1975). Extraction of *Renilla*-type luciferin from the calcium-activated photoproteins aequorin, mnemiopsin and berovin. *Proc. Natl. Acad. Sci.* 72:2530-2534.

Ward, W. W. and Cormier, M. J. (1976). *In vitro* energy transfer in *Renilla* bioluminescence. Michael Kasha Symposium—Electronic Processes and Energy Transfer in Organic, Inorganic and Biological Systems (special edition, *J. Phys. Chem.* 80:2289–2291).

Ward, W. W. and Cormier, M. J. (1978a). Energy transfer via protein–protein interaction in *Renilla* bioluminescence. *Photochem. Photobiol.* 27:389–396.

Ward, W. W. and Cormier, M. J. (1978b). Protein–protein interactions as measured by bioluminescence energy transfer. *Methods Enzymol.*, 62:257–267.

Ward, W. W. and Cormier, M. J. (1979). An energy transfer protein in coelenterate bioluminescence: Characterization of the *Renilla* green-fluorescent protein (GFP). *J. Biol. Chem.* 254:781–788.

Ward, W. W., Prentice, H. J., Roth, A. F., Cody, C. W., and Reeves, S. C. (1982). Spectral perturbations of the *Aequorea* green-fluorescent protein. *Photochem. Photobiol.* 35:803–808.

Wu, C.-K., Liu, Z.-J., Rose, J. P., Inouye, S., Tsuji, F., Tsien, R. Y., Remington, S. J., and Wang, B.-C. (1997). The three-dimensional structure of green fluorescent protein resembles a lantern. *In Bioluminescence and Chemiluminescence, Molecular Reporting With Photons*. Hastings, J. W., Kricka, L. J. and Stanley, P. E. Eds., Wiley, New York, pp. 399–402.

Yang, F., Moss, L. G., and Phillips, G. N., Jr. (1996a). The molecular structure of green fluorescent protein. *Nature Biotech.* 14:1246-1251.

Yang, F., Moss, L. G., and Phillips, G. N., Jr. (1997). The three-dimensional structure of green fluorescent protein. *In Bioluminescence and Chemiluminescence, Molecular Reporting With*

Photons. Hastings, J. W., Kricka, L. J., and Stanley, P. E., Eds., Wiley, New York, pp. 375–382.

Yang, T. T., Cheng, L., and Kain, S.R. (1996b). Optimized codon usage and chromophore mutations provide enhanced sensitivity with the green fluorescent protein. *Nucleic Acids Res*. 24(22):4592–4593.

Youvan, D. C. and Michel-Beyerle, M. E. (1996). Structure and fluorescence mechanism of GFP. *Nature Biotech*. 14:1219–1220.

4

The Three-Dimensional Structure of Green Fluorescent Protein and Its Implications for Function and Design

GEORGE N. PHILLIPS, JR.
Department of Biochemistry and Cell Biology, Rice University, Houston, TX

4.1 INTRODUCTION

Green fluorescent protein (GFP) is a naturally fluorescent protein isolated from the jellyfish *Aequorea victoria* and other marine organisms. It converts the blue chemiluminescence of other proteins, aequorin or luciferase, into green fluorescent light (Morin and Hastings, 1971; Ward, 1979), presumably to reduce scattering, and hence improve penetration of the light over longer distances. The molecular cloning of GFP cDNA from the Pacific jellyfish, *A. victoria* (Prasher et al., 1992) and the demonstration by Chalfie et al. (1994) that this GFP can be functionally expressed in bacteria and nematodes have opened exciting new avenues of investigation in cell, developmental, and molecular biology, as pointed out in prior reviews (Chalfie, 1995; Heim and Tsien, 1996) and by the many techniques and applications described in this volume. As a consequence of this interest, knowledge of the three-dimensional (3D) structure of GFP has become highly desirable for engineering modified GFPs for various purposes, and at least six laboratories are currently carrying out structural studies on GFP and its mutations.

Green Fluorescent Protein: Properties, Applications, and Protocols, Edited by Martin Chalfie and Steven Kain
ISBN 0-471-17839-X

Green fluorescent protein from the jellyfish, A. *victoria*, has several isotypes (Cutler, 1995); the cDNA of GFP that was first cloned and expressed, called TU No. 58 (Chalfie et al., 1994) is comprised of 238 amino acids. As first expressed in recombinant form, the N-terminal methionine was replaced by methionine–alanine, and there was also an inadvertent substitution of arginine for glutamine at position 80, probably due to a polymerase chain reaction (PCR) error in the original cloning. For the purposes of this chapter, I refer to this form as "wild type", since the key amino acids are identical to the one of the native isotypes, and its spectral properties are like those from the native protein.

4.2 SPECTRAL AND PHYSICAL PROPERTIES OF GREEN FLUORESCENT PROTEIN

The wild-type fluorescence absorbance/excitation peak is at 395 nm with a minor peak at 475 nm (molar absorbances of roughly 30,000 and 7000 M^{-1} cm^{-1}, respectively (Ward, 1979; Ward and Cormier, 1979; Kahana and Silver, 1996). The normal emission peak is at 508 nm. Interestingly, continued excitation at wavelengths that excite the major 395-nm peak lead to a decrease over time of the 395-nm excitation peak and a reciprocal increase in the 475-nm excitation band (Chalfie et al, 1994; Cubitt et al., 1995). This interconversion effect is especially evident with irradiation of GFP by ultraviolet (UV) light.

Recently, femtosecond time-resolved spectroscopic studies have revealed that the two states corresponding to two major absorption bands can interconvert quickly in the excited state (Chattoraj et al., 1996; Lossau et al., 1996). Both Chattoraij et al. (1996) and Lossau et al. (1997) have shown that excited-state deprotonation transfer is a key process in GFP photochemistry (see also Youvan and Michel-Beyerle, 1996). Both groups demonstrated a slowing of kinetic features upon deuteration in the picosecond decay of the excited protonated species and the concomitant rise of the green fluorescence of the deprontonated fluorophore. While deuterium effects are consistent with proton movement, they are not in themselves proof. In the initial work, both groups simply incubated the protein in deuterated buffer therefore, neither the extent nor the position of deuteration can be determined. However, Lossau et al. (1996) and Brejc et al. (1997) further substantiated a deprotonation mechanism through the comparative spectroscopic study of wild-type, GFP containing the $Phe^{64}Leu$ mutation plus $Tyr^{66}His$ (blue emitter) or $Ser^{65}Thr$ (ref-shifted excitation), or GFP with only the $Ser^{65}Thr$ mutant. The $Tyr^{66}His$ mutant has a fluorophore that does not deprotonate in the excited state and the $Ser^{65}Thr$ mutant have fluorophores that are deprotoneted in the electronic ground state, respectively.

The site of the deprotonation is controversial. The carboxylic acid side chain of Glu222 has been postulated to be highly sensitive to the details of its environment, and hence a likely candidate (Brejc et al., 1997). A hydrogen-bonding network that involves Ser^{205} and a bound water molecule provides a putative pathway for proton movement. The results from analysis of spectral and structural differences between wild-type and $Ser^{65}Thr$ mutants are nicely explained. However, the use of semiempirical quantum mechanical modeling has also been used to study the relative stability; proton affinities, geometrical structures, and absorption spectra of wild-type and several mutant GFPs suggest an alternative site for deprotonation (Voityuk et al., 1997). Agreement between the computed and observed absorption spectra of the fluorophore in aqueous solution at acidic and basic pH suggests that the structure of important states of the fluorophore in the native and mutant proteins can be identified: The absorption maximum of GFP at 477 nm is assigned to a zwitterion where the phenolic oxygen of Tyr^{66} is deprotonated and the nitrogen of the heterocyclic ring is protonated. Furthermore, the high-energy absorption peak at 397 nm is assigned to the excitation of the protonated form of the nitrogen in the heterocycle. This conclusion is further corroborated by the agreement between the calculated absorption energies at 355, 433, and 387 nm in the $Tyr^{66}Phe$, $Tyr^{66}Trp$, and $Tyr^{66}His$ mutants as compared to the experimental values of 360, 436, and 382 nm, respectively. In summary, these computations suggest that the protonated nitrogen constitutes a crucial factor in the function of GFP.

Physical and chemical studies of purified GFP also identified several important characteristics that relate directly to its structure. GFP it is very resistant to denaturation, requiring treatment with 6 M guanidine hydrochloride at 90°C or pH of less than 4.0 or greater than 12.0. Partial to near total renaturation occurs within minutes following reversal of denaturing conditions by dialysis or neutralization (Ward and Bokman, 1982). Over a nondenaturing range of pH, increasing pH leads to a reduction in fluorescence by 395-nm excitation and an increased absorption at 475 nm. (Ward et al., 1982).

Because GFP *in crystallum* exhibits nearly identical fluorescence spectra and excited-state lifetimes to that for GFP in aqueous solution (Perozzo et al., 1988) and fluorescence is not an inherent property of the isolated fluorophore, the elucidation of its 3D structure of GFP (Ormö et al., 1996; Yang et al., 1996) helped provide an explanation for the generation of fluorescence in the mature protein, as well as the mechanism of autocatalytic fluorophore formation. Furthermore, the development of fluorescent proteins with varied emission and excitation or other characteristics based on the predicted changes in the structure would dramatically expand biological applications of GFP and its variants.

4.3 THE β-CAN STRUCTURE OF GFP

The structure of the wild-type protein was solved by Yang et al. (1996) and that of the $Ser^{65}Thr$ mutant by Ormö et al. (1996). The density maps of both determinations of GFP were very clear, revealing quite regular β-barrels with 11 strands on the outside of cylinders (Fig. 4.1). These cylinders have a diameter of about 30 Å and a length of about 40 Å. Inspection of the density within the cylinders revealed the fluorescent center of the molecule, a modified tyrosine side chain and cyclized protein backbone as a part of an irregular α-helical segment. Small sections of α helices and loops also form caps on the ends of the cylinders. This motif, with a single α helix inside a very uniform cylinder of β-sheet structure, represents a new protein class, which we have named the β-can (Yang et al, 1996).

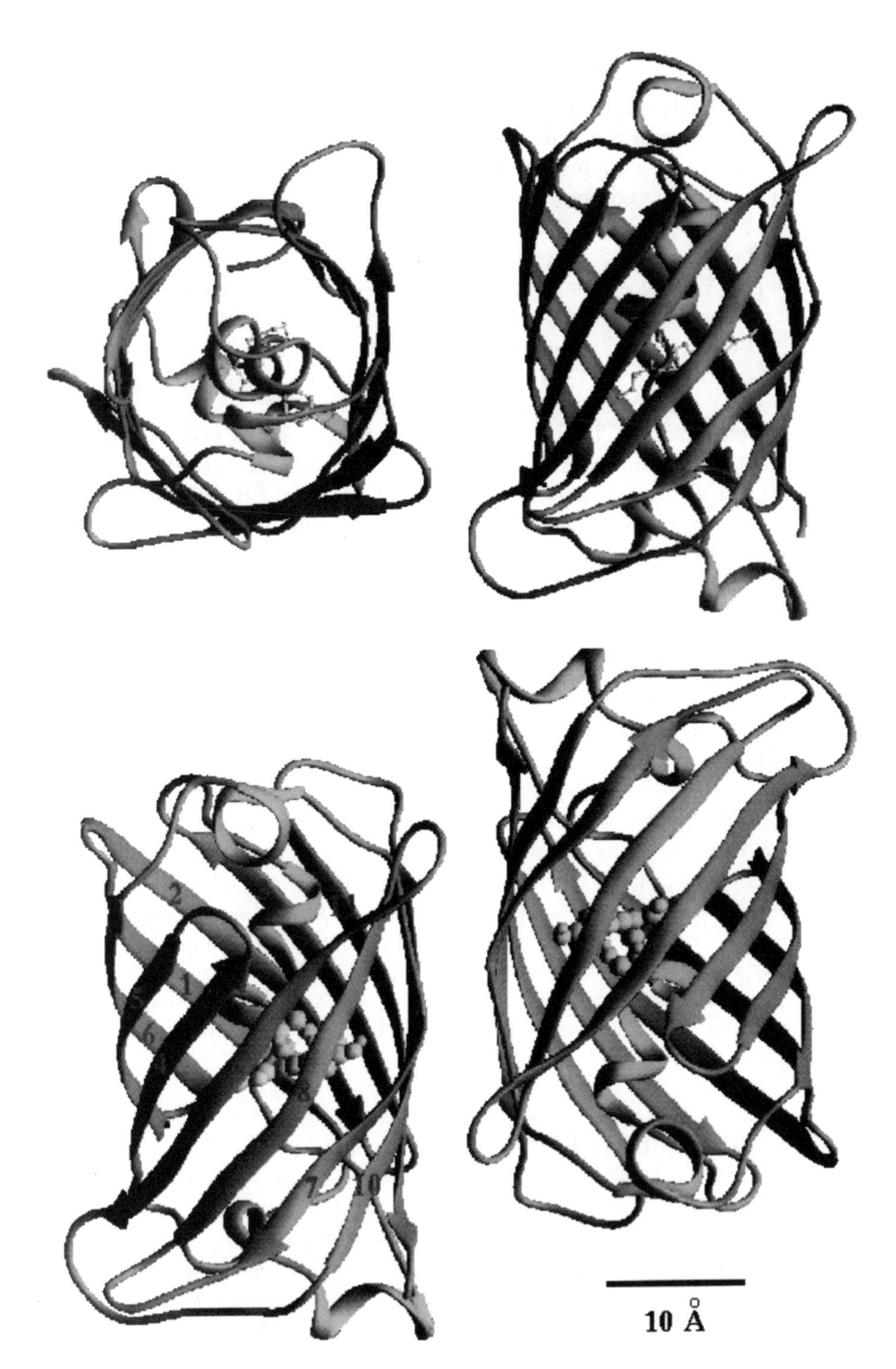

FIG. 4.1 End-on (top left) and side (top right) views of the cylindrical β-can structure of GFP. Eleven strands of β-sheet form an antiparallel barrel with short helices forming lids on each end. The fluorophore is inside the can, as a part of a distorted α helix, which runs along the axis of the cylinder. The GFP usually forms dimers in the crystal, aligned largely along the sides of the cylinders. Drawing by Ribbons (Carson, 1987), coordinates (Protein Data Bank entry 1GFL). [From Yang et al., 1996.] Reprinted with permission from Nature Biotechnology.

The regularity of the β-can of GFP is quite remarkable. The 11 strands of the sheet form an almost seamless symmetrical structure, the only irregularities being between two of the strands. In fact, the structure is so regular, that water molecules on the outside of the can also form "stripes" around the surface of the cylinder. A surprising number of water molecules are also found inside the can. The tightly constructed β barrel would appear to serve the role of protecting the fluorophore well, providing overall stability and resistance to unfolding by heat and denaturants.

The known proteins that most closely resemble the β-can fold of GFP are porin, which has not 11, but 16 antiparallel strands and has no "lids" at the ends of the barrel (Kreusch et al., 1994) and strepavidin, which is a smaller, eight-stranded antiparallel β barrel (Fig. 4.2) (Hendrickson et al., 1989). Unlike streptavidin, both GFP and porin have water molecules inside the barrel, however, as well as small segments of polypeptide chain inside. In the case of porin, whose function is to allow passage of small molecules through its center, its design needs to be open, whereas GFPs function is better served with a closed structure that can restrict access of quenchers to the fluorophore and perhaps also contain damaging free radical photoproducts. Because of the smaller number of strands in streptavidin and hence smaller inside diameter, its center consists simply of side chains originating from the staves of the barrel.

4.4 THE FLUOROPHORE AND ITS ENVIRONMENT

Analysis of a hexapeptide derived by proteolysis of purified GFP led to the prediction that the fluorophore originates from an internal Ser-Tyr-Gly sequence, which is posttranslationally modified to a 4-(*p*-hydroxybenzylidene)-imidazolidin-5-one structure (Cody et al., 1993). Studies of recombinant GFP expression in *Escherichia coil* led to a proposed sequential mechanism initiated by a rapid cyclization between Ser^{65} and Gly^{67} to form a imidazolin-5-one intermediate followed by a much slower (hours) rate-limiting oxygenation of the Tyr^{66} side chain by O_2 (Heim et al., 1994). Extensive combinatorial mutagenesis (Delagrave et al., 1995) and other mutational studies (Heim et al., 1995; Cormack et al., 1996) suggests that the Gly^{67} is required in the functional fluorophore. While no known cofactors or enzymatic components are required for this apparently autocatalytic process, it is rather thermosensitive with the yield of fluorescently active to total GFP protein decreasing at temperatures greater than 30°C (Lim et al., 1995). However, once the functional form of GFP is produced, it is quite thermostable.

The critical fluorophore-forming sequence, Ser-Tyr-Gly, occurs many times in proteins. How is it that cyclization occurs in GFP, but not in these other proteins? Two factors appear to be required for fluorophore formation: close proximity of the backbone atoms of amino acids 65 and 67 and acid–base chemistry to catalyze the cyclization. Close proximity is achieved by the removal of steric hindrance by a side chain at position 67, where a glycine is present. As described above, despite many mutations at position 67, no functionally fluorescent GFPs have been found with anything but glycine at position 67. Among the 3D structures in the protein data bank with a Ser-Tyr-Gly sequence, 10 of

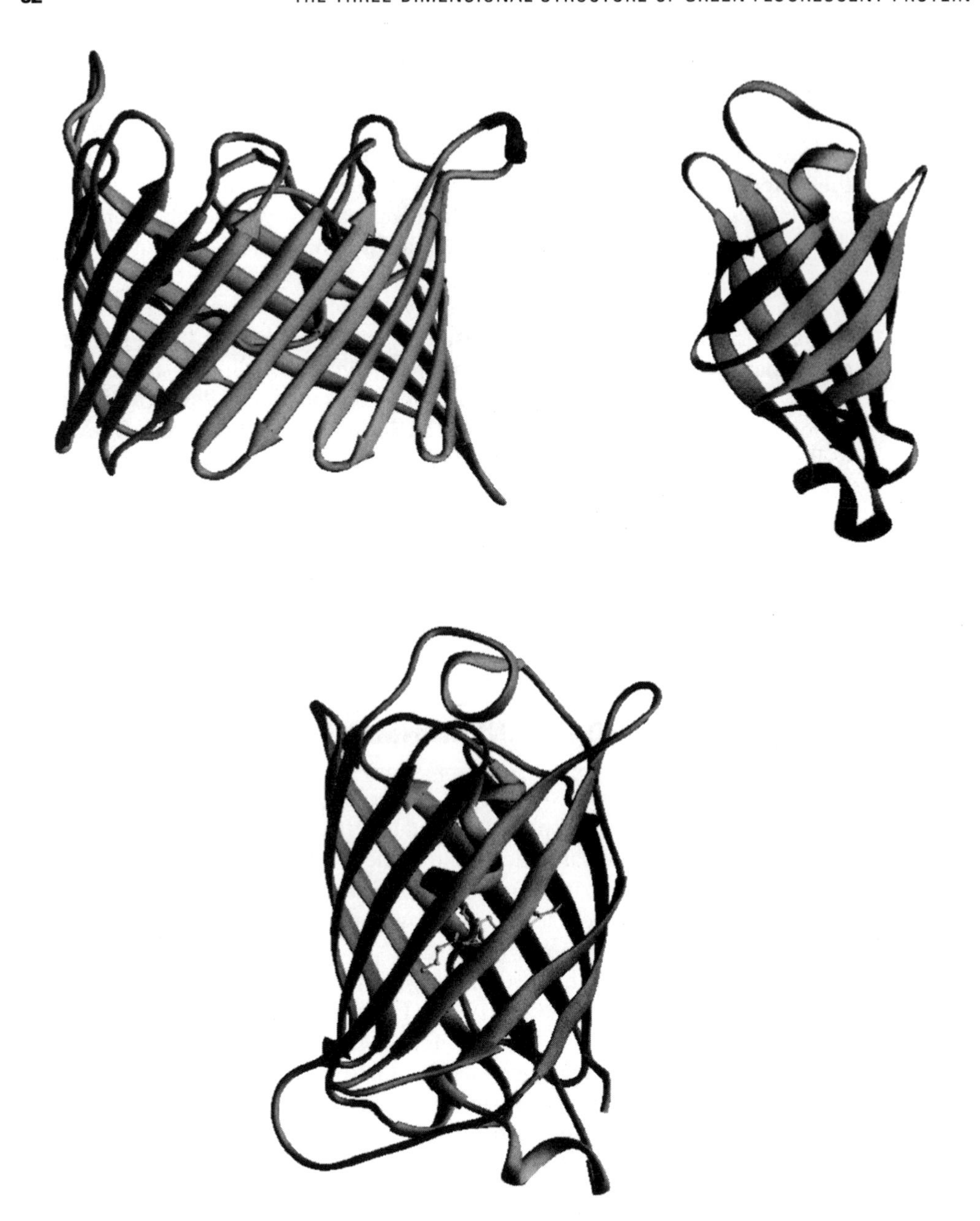

FIG. 4.2 Tertiary structures of porin (a, top left), GFP (b, bottom), and strepavidin (c, top right) showing three sizes of antiparallel β-barrel proteins. Porin is open on both ends with a water channel through the middle, GFP has water and protein on the inside, but is sealed on both ends, and streptavidin is too small to have anything in the core except side chains from the strands of the barrel.

the 206 found have the required proximity of backbone atoms (Zimmer and Branchini, 1997), so steric factors seem to be necessary but not sufficient for cyclization. Arginine at position 96 is close by an could act as a base, withdrawing electrons by hydrogen bonding with carbonyl oxygen of Ser^{65} and activating the carbonyl carbon for nucleophilic attack by the amide nitrogen of Gly^{67}. Aspects of this scheme have been supported by *ab initio calculations and by database* searches of similar compounds and protein sequences (Branchini et al., 1997).

Since model compounds identical to the hydroxyphenyl imidizolidinone core of the fluorophore have been synthesized and shown NOT to be significantly fluorescent in solution (Niwa et al., 1997), the protein and its strategically placed acids and bases at the edges of the fluorophore are implicated in providing key resonance stabilization. The remarkable cylindrical fold of the protein seems ideally suited for the function of the protein—it provides a scaffold that surrounds the fluorophore by 360° and provides a wide range of possible protein side-chain interactions. Together with the short α helices and loops on the ends, the barrel structure forms a single compact domain and does not have obvious clefts for easy access of diffusable ligands to the fluorophore. The fluorophore is protected from collisional quenching by oxygen ($K_{\mathrm{bm}} < 0.004M^{-1}s^{-1}$) (Rao et al., 1980), and hence reduction of the quantum yield. For comparison, the bimolecular quenching rate for free tryptophan by oxygen is about $10^{10}M^{-1}\ \mathrm{s}^{-1}$ and for tryptophan within small proteins is on the order of $10^{9}M^{-1}\ \mathrm{s}^{-1}$ (Lakowicz and Weber, 1973). Perhaps more seriously, photochemical damage by the formation of singlet oxygen from the collision of oxygen with the fluorophore in the excited state is reduced by the restricted access of the structure. Such collisions could result in reactive oxygen species. Thus, the β can structure may serve both to protect the fluorophore from inactivation and also to contain any reactive products that form inadvertently.

Two aspects of the structure may explain the lack of quenching of the fluorophore by molecular oxygen. The β-can structure could provide significant barriers to the penetration by oxygen and/or the fluorophore is tightly held by local interactions. The observation that the fluorophore is near the geometric center of the can supports the notion that GFP structure either protects the fluorophore from diffusional penetrative quenching or protects the organism from eventual and unavoidable free radical reactions begun by photochemical processes at the fluorophore by contained self-destruction, or both. The crystallographic "temperature factors", are indeed lowest in the center of the can, implying more rigidity, but it is often the case in protein crystal structures that centers have low mobility, and thus this cannot be taken as proof of a special design. The quenching experiments like those described above on mutants of GFP could help to resolve this question. If fluorescence quenching by oxygen is dramatically increased by mutations that affect the integrity of the can but are not directly associated with the fluorophore, then the barrel structure is a barrier. If quenching is not affected, then it is the local interactions around the fluorophore that provide protection of the photoexcited states from collisional quenching. GFP from *Renilla reniformis* (a sea pansy) may have even more specialized structures for maintaining the rigidity of the fluorophore, as its absorption and emission spectra are essentially mirror images (Ward, 1979). This relationship is a hallmark of highly immobilized fluorescent molecules, since it implies that the environment of the fluorophore is exactly the same during the absorbing and emitting states.

The fluorophore is located on the central helix within a couple of angstroms of the geometric center of the cylinder. The pocket containing the fluorophore has a surprising number of charged residues in the immediate environment (Fig. 4.3). The environment around the fluorophore includes both apolar and polar amino acid side chains and immobilized water molecules. Both Phe^{64} and Phe^{46} are near the fluorophore and separate the single tryptophan, Trp^{63} from direct contact with fluorophore (closest distance of 13 Å).

Most of the other polar residues in the pocket form an extensive hydrogen-bonding network on the side of Tyr^{66} that requires abstraction of protons in the oxidation process. It is tempting to speculate that these residues help abstract the protons. Atoms in the side chains of Thr^{203}, Glu^{222}, and Ile^{167} are in van der Waals contact with Tyr^{66}, so their mutation would have direct steric effects on the fluorophore and would also change its electrostatic environment if the charge were changed, as suggested previously (Ehrig et al., 1995). Additional testing may show that mutation of other residues near the fluorophore also affect the absorption and/or emission spectra.

4.5 STRUCTURES OF AEQUOREA GREEN FLUORESCENT PROTEIN MUTANTS AND OTHER GREEN FLUORESCENT PROTEINS

The location of certain amino acid side chains in the vicinity of the fluorophore also begins to explain the fluorescence and the behavior of certain mutants of the protein. At least two resonant forms of the fluorophore can be drawn: one with a

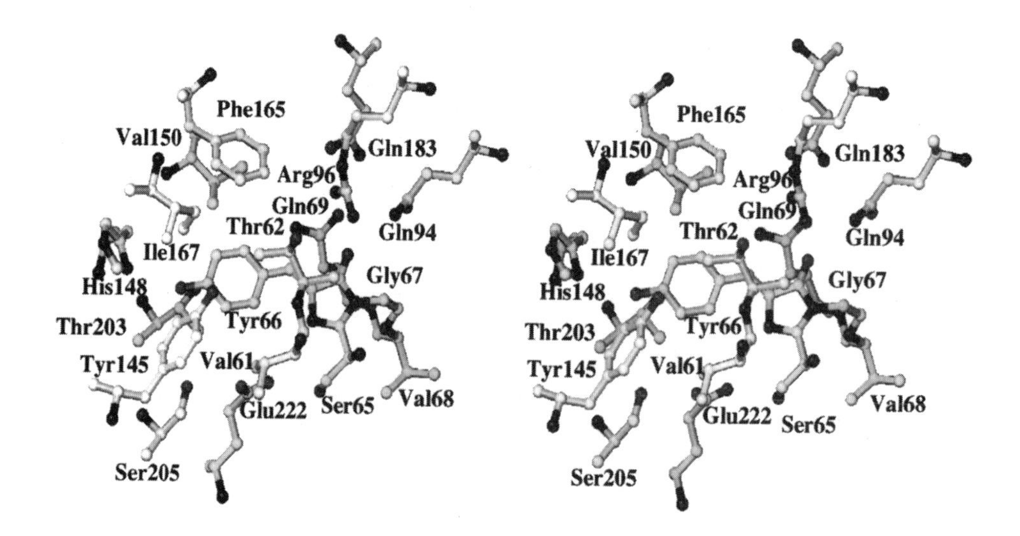

FIG. 4.3 Stereoview of the fluorophore and its environment. His^{148}, Gln^{94}, Arg^{96}, and Glu^{222} can be seen on opposite ends of the fluorophore and probably stabilize anionic resonant forms. Water molecules, charged, polar, and nonpolar side chains all contact the fluorophore in various ways.

partial negative charge on the benzyl oxygen of Tyr^{66}, and one with the charge on the carbonyl oxygen of the imidizolidone ring (Fig. 4.4). Interestingly, basic residues appear to form hydrogen bonds with each of these oxygen atoms, His^{48} with Tyr^{66} and Gln^{94} and Arg^{96} with the imidizolidone. These bases presumably act to stabilize and possibly further delocalize the charge on the fluorophore. Recent kinetic studies have suggested that proton transfers may play an important role in the excitation–fluorescence process (Chattoraj et al., 1996; Youvan and Michel-Beyerle, personal communication). Based on this structure, the carbonyl oxygen of the imidizolidone, arising originally from the backbone of Tyr^{66}, is a likely candidate for the acceptor, with Arg^{96} as the donor. Another possibility is the Tyr^{66} distal oxygen-His^{148} donor–acceptor pair or the Glu^{222}-Ser^{65} interactions. Perturbation of the interaction by mutation to His at position 66 result in only the high-energy absorption peak and a blue-shifted emission band, whereas disruption of the Ser^{65} hydroxyl-Glu^{222} interactions result in a red-shifted absorption maximum and an unchanged emission spectrum. The Tyr^{66} to Phe mutation has been reported to have dramatically reduced fluorescence (Heim et al., 1994), presumably due to poorer charge delocalization.

The structure of the GFP mutant, Ser^{65}Thr (Ormö et al., 1996) shows only minor differences in the β-can structure when compared to the wild-type protein.

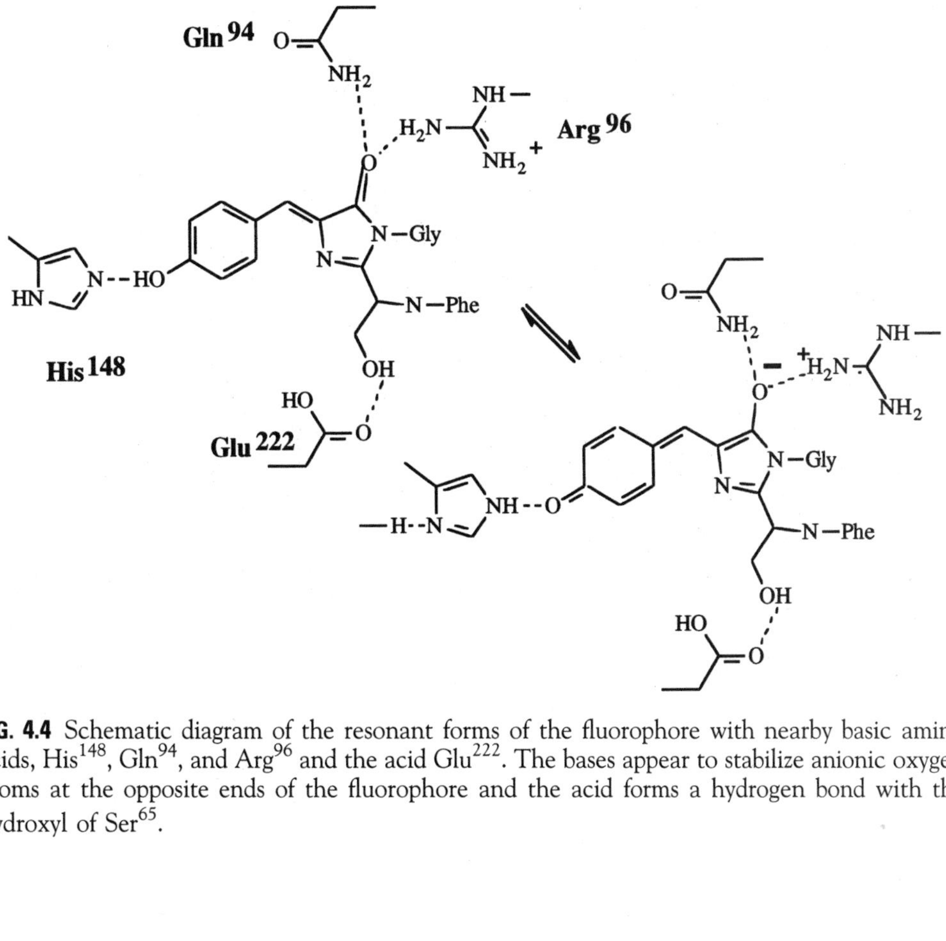

FIG. 4.4 Schematic diagram of the resonant forms of the fluorophore with nearby basic amino acids, His^{148}, Gln^{94}, and Arg^{96} and the acid Glu^{222}. The bases appear to stabilize anionic oxygen atoms at the opposite ends of the fluorophore and the acid forms a hydrogen bond with the hydroxyl of Ser^{65}.

The different absorption properties are consistent with the general trend that disruption of the interaction between the Ser^{65} hydroxyl and the Glu^{222} carboxylate eliminate the 395-nm absorption peak. Inspection of the structure of the Ser^{65}Thr GFP confirms that this interaction has been changed in this mutant as well, via a rotation of the hydroxyl of Thr^{65} by about 120° about the Cα–Cβ bond relative to the wild-type protein (Fig. 4.5). Glutamic acid at position 222 has corresponding adjustments. The net result is a longer hydrogen bond in the Ser^{65}Thr structure. Ormö et al. (1996) postulated that the pattern of hydrogen bonding is different in the Thr^{65} mutation and that the net result is that an anionic Glu^{222} cannot coexist with an anionic fluorophore without serine at position 65.

The availability of *E. coli* clones expressing GFP has led to extensive mutational analysis of GFP, with more crystal structures are now starting to appear that test ideas about the designed rearrangements of side chains resulting from mutagenesis (Table 4.1). Screens of random and directed point mutations for changes in fluorescent behavior have uncovered a number of informative amino acid substitutions. Mutation of Ser^{65} to Thr, Ala, Cys, or Leu causes a loss of the 395-nm excitation peak with a major increase in blue excitation (Delagrave et al., 1995; Heim et al., 1995). When combined with Ser^{65} mutants, mutations at other sites near the fluorophore such as Val^{68} Leu and Ser^{72} Ala can further enhance the intensity of green fluorescence produced by excitation at 488 nm (Delagrave et al., 1995; Cormack et al., 1996). However, amino acid substitutions significantly outside this region also affect the protein's spectral character. For example, Ser^{202} Phe and Thr^{203} Ile both cause the loss of excitation in the 475-nm region with preservation of 395-nm excitation (Heim et al., 1994; Ehrig et al., 1995). The residue Ile^{167} Thr change results in reversed ratio of 395–475-nm sensitivity (Cubitt et al., 1995), while Glu^{222} Gly is associated with the elimination of only the 395-nm excitation (Ehrig et al., 1995). The pH dependence of the excitation bands at 395 and 475 nm (Ward et al., 1982) is almost certainly due to His^{148}, whose Nδ atom is 3.3 Å from the Tyr^{66} hydroxyl oxygen atom of the fluorophore, although NMR pK_a measurements or mutagenesis studies would be needed for confirmation.

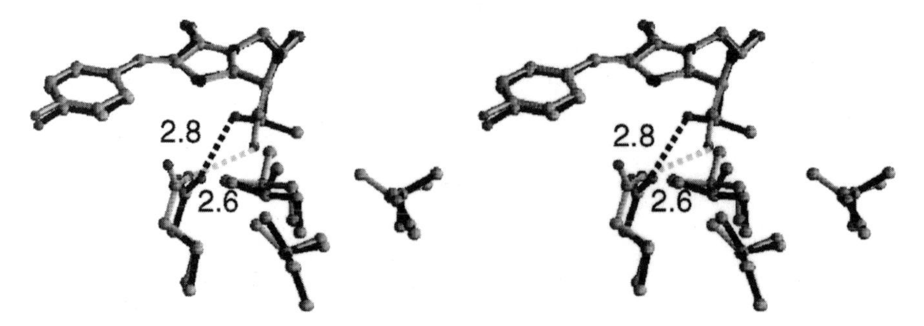

FIG. 4.5 Comparison of the wild-type and Ser^{65} Thr structures in the vicinity of the fluorophore. Consistent with the pattern that disruption of the Glu^{222}-Ser^{65} interactions lead to elimination or severe reduction of the absorption at 395 nm, there are different hydrogen-bonding arrangements in the Ser^{65} Thr mutant. [Drawing by Ribbons, coordinates from entry 1GFL and 1EMA from the protein data bank.]

TABLE 4.1

Variant	Space Group	Cell size [A. (deg)]	Res (A)		Condition	Reference	Footnotes
Wild type	*P41212*	87 × 120	1.9	Dimer	MPD, pH 6.8	Yang et al., 1996	*a*
Native	C2	93 × 66 × 46 (108)	2.1	Dimer	2M Phos, pH 7	M. Perozzo, unpublished results	*b*
Wild-type His tag	*P6122*	77 × 77 × 330	<2.5	Dimer	?	Wu et al., 1997	c
Wild type	*P212121*	52 × 63 × 69	2.1	Monomer	?	Brejc et al., 1997 and Sixma, T.	
	P6122(?)	93 × 93 × 188	2.8	?		(personal communication)	
S65T	*P212121*	52 × 63 × 71	1.9	Monomer	PEG pH 8	Ormö et al., 1996	*a*
F64L	*P6122*	90 × 90 × 130	2.3	Dimer		Palm et al., 1997	*a*
F64L, I167T, K238N	*P6122*	90 × 90 × 130	2.3	Dimer	2 M AS pH 8.5	Palm et al., 1997	*a*
	P21	93 × 52 × 103 (101)	2.3	Monomer	MPD pH 8		*a, d*
	I4132	176	2.5	Dimer	2M AS pH 8.5		*a*
F64L, S65C, I167T, K238N	*P6122*	90 × 90 × 130	2.1	Dimer	(see above)	Palm et al., 1997	*a*
F64L, Y66H	*P6122, I4132*	90 × 90 × 130	2.3	Dimer	(see above)	Palm et al., 1997	*a*
			2.1	Dimer			
F64L, Y66H, V163A	*P6122*	90 × 90 × 130	2.5	Dimer	(see above)	Palm et al., 1997	*a*
	I4132		2.4	Dimer			

[a] Includes the inadvertent Q80R mutation, and an additional A at the N-terminus.
[b] Has the following sequence differences from the cloned isoform: R80G, F100Y, T108S, L141M, E172L.
[c] Has a 37 residue His-tag sequence at the N-terminal (only 0, and –1 positions seen in the e–d maps)
[d] Four monomers per asymmetric unit, but not the commonly seen dimer.

4.6 STRUCTURE-BASED ENGINEERING OF GREEN FLUORESCENT PROTEIN

Mutations in regions of the sequence adjacent to the fluorophore, that is, in the range of positions 65–67, have been systematically explored (Delagrave et al., 1995), some having significant wavelength shifts and most suffer a loss of fluorescence intensity. For example, mutation of the central Tyr to Phe or His shifts the excitation bands but there is an overall loss of intensity. Secondary mutations to compensate for the deleterious intensity effects may also now be possible. The Ser^{65} Thr mutant is particularly interesting because of its reported increase in fluorescence intensity (Heim et al., 1994, 1995). These authors suggested that this effect is through improved conversion of the tyrosine to dehydrotyrosine. This is not likely, however, as we see essentially fully cyclized structure in the wild-type protein (Yang et al., 1996). An alternative possibility for increased fluorescence may be reduced collisional quenching, as the additional methyl group may make for better packing in the interior of the protein. The report of improvements in the brightness of cells producing by DNA shuffling of GFP constructs, comprising mutations Phe^{99} Ser, Met^{153} Thr and Val^{163} Ala, as numbered in the TU#58 system, are difficult to explain simply based on the structure (Crameri et al., 1996). Positions 153 and 163 are on the surface of the protein and may exert their effects through improved solubility and/or reduced aggregation. The Phe-Ser mutation at first glance would appear to destabilize the core of the protein and at present we have no good ideas how it would improve the fluorescence properties of GFP.

4.7 CONTROL OF ACTIVATION OF THE FLUOROPHORE

The mechanism of activation of the fluorophore from ordinary protein structure is consistent with a nonenzymatic cyclization mechanism like that of Asn-Gly deamidation (Wright, 1991) followed by oxidation of the tyrosine to dehydrotyrosine, as previously suggested. The role of molecular oxygen in this mechanism and in GFP fluorescence is paradoxical, however. Molecular oxygen is proposed to be needed for oxidation of tyrosine to form an extended aromatic system, but oxygen must also be excluded from regular interactions with the fluorophore or else collisional quenching of the fluorescence or damaging photochemistry will occur. The low bimolecular quenching rate suggests that the protein's design sacrifices efficient fluorophore formation for stability and higher quantum yields once fully formed.

Once formed, the dehydrotyrosine can be chemically reduced by sodium dithionite to produce a mixture of D and L amino acid geometries at the carbon alpha of tyrosine 66 (Yang, F. personnel communication). Because most enzymatic reactions are stereospecific, this result implies that the catalysis of the original dehydrogenation may, in fact, be nonenzymatic or may involve amino acid side chains that are only transiently available in the oxidation, and are not present (or needed) in the reduction.

4.8 GFP TRUNCATION AND FUSION CONSTRUCTS

Truncation of more than seven amino acids from the C-terminus or more than the N-terminal Met lead to total loss of fluorescence (Dopf and Horiagan, 1996). These N- and C-termini truncation studies and the fluorescent fusion products are now understandable, given the structure of the protein. Since the C-terminus loops back outside the cylinder and the last seven or so amino acids are disordered it should not be critical to have them present and further addition would seem to be easily tolerated. These residues do not form a stave of the barrel. The role of the N-terminus is a little less clear, as the first strand in the barrel does not begin until amino acid 10 or 11. Thus barrel formation does not require the N-terminal region. The N-terminal segment, is however, an integral part of the "cap" on one end of the protein, and may be essential in folding events or in protecting the fluorophore. Again, extensions at the N-terminus would not disrupt the motif structure of the protein. In fact, the GFP crystal structure solved by Wu et al. (1997) has a 37 residue His tag on the N-terminus, of which only two amino acids are ordered and hence visible in the electron density map.

4.9 CONTROL OF DIMERIZATION OF GREEN FLUORESCENT PROTEIN

Green fluorescent protein can form homodimers in solution and in crystals. The equilibrium dissociation constant, K_d is approximately 100 μM as measured by analytical ultracentrifugation (F. Yang and G. N. Phillips, unpublished results). This weak association is consistent with the observation that most, but not all crystal forms exist as dimers and correlation microscopy of purified GFP, which revealed a diffusion coefficient consistent with monomers at low-protein concentrations (Terry et al., 1996). The fluorescence of GFP also changes on dimerization (Ward et al., 1982).

In both the wild-type and native crystal structures, the crystallographic contacts are all rather tenuous, consisting of a few amino acid side chains for each. In contrast, the dimer symmetry is maintained by extensive contacts (Fig. 4.6 and Table 4.2), is similar in at least three structures (Yang et al., 1996; Wu et al., 1997) and (M. Perozzo, W. Ward, and K. Ward, personnel communication), and thus is likely to be the source of the dimerization seen in solution studies (Table 4.2). The dimer contacts are fairly tight and consist of a core of hydrophobic side chains from each of the two monomers and a wealth of hydrophilic contacts. The smaller hydrophobic patch could conceivably be involved in physiological interactions with aequorin, as there would be a natural advantage to close proximity for efficient energy transfer. There are fluorescence changes on dimerization (Cutler, 1995), and subtle rearrangements of the fluorophore may also occur upon dimerixation (Wu et al., 1997).

Control of the dimerization will be important for fluorescence resonance energy transfer (FRET) studies of protein–protein interactions using GFP (Heim and Tsien, 1996; Mitra et al., 1996), as one would not want to induce association and hence resonance energy transfer between the differently colored GFP proteins by mechanisms other than of the target protein interactions.

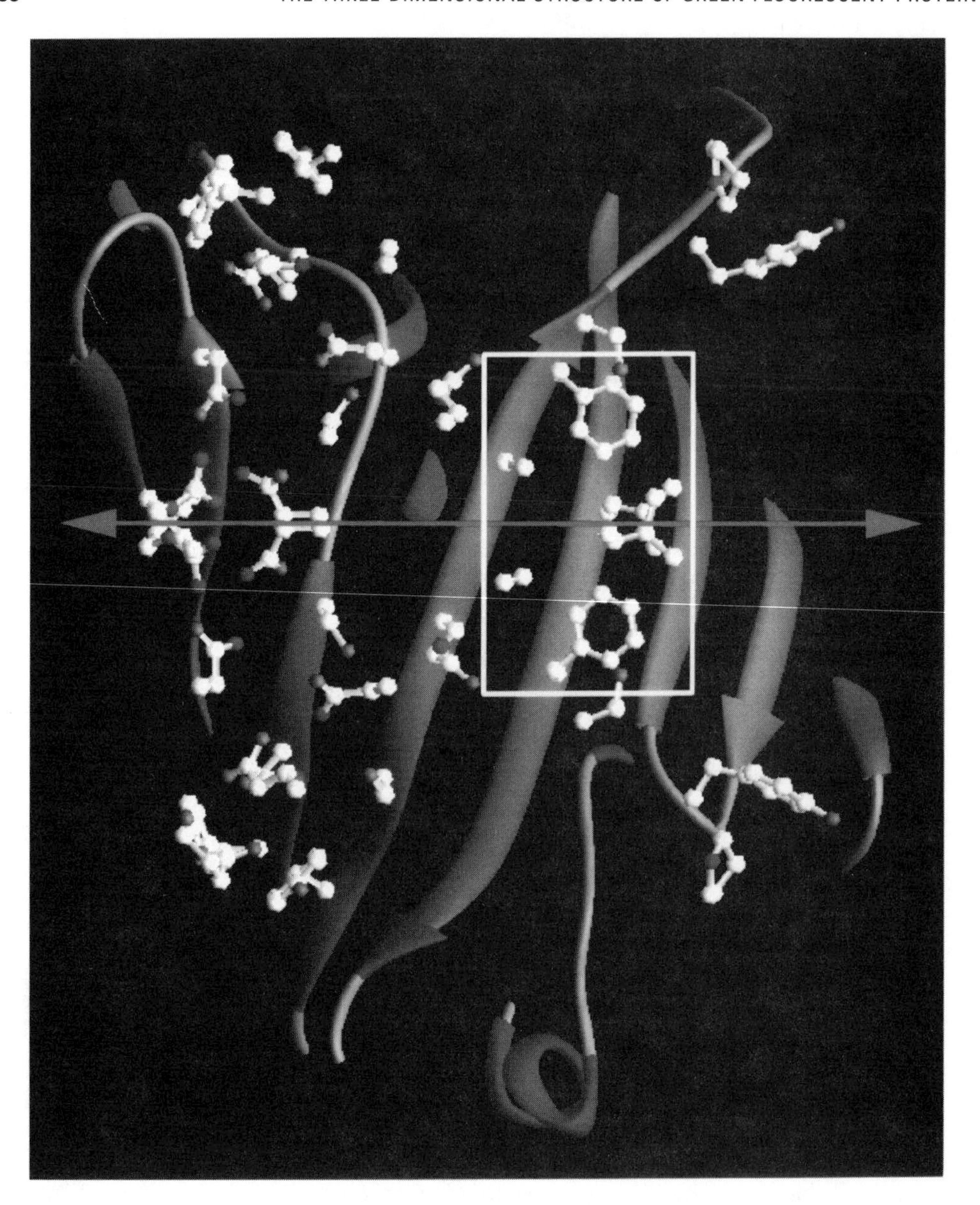

FIG. 4.6 The dimer contact region. The two polypeptide chains associate over a broad area, with a small hydrophobic patch (in the box) and numerous hydrophilic contacts. The twofold dimer axis is in the plane of the page (arrow). The side chains from each of the two monomers are shown as dark balls or light colored balls. A list of all the contacts is given in Table 4.2. Figure also appears in color section.

TABLE 4.2 Dimer interactions[a]

Monomer 1		Monomer 2		Distance (Å)
Residue	Atom	Residue	Atom	
Ala206	CB	Phe223	CD1	3.5
Leu221	CD2	Phe223	CZ	4.4
Leu207	O	Phe223	CB	4.7
Tyr145	O	Ser147	OG	2.8
Gln204	OE1	Leu207	O	2.8
Tyr39	OH	Ser208	OG	2.9
Glu142	OE2	Asn149	OD1	3.1
Asn144	CB	Ser147	OG	3.2
Asn144	CA	Gln204	NE2	3.3
Asn146	OD1	Ser147	N	3.3
Tyr145	N	Gln204	NE2	3.4
Tyr39	CZ	Lys209	O	3.5
Tyr39	OH	Val219	CG1	3.5
Gln204	NE2	Ala206	CA	3.5
Tyr143	O	Gln204	OE1	3.6
Asn146	OD1	Arg168	NH1	3.6
Tyr39	CD1	Pro211	CD	3.7
Ser208	CB	Phe223	CD2	3.7
Asn146	CB	Asn146	OD1	3.7
Thr38	O	Pro211	CG	4.0
Asn170	ND2	Ser147	O	4.0
Tyr39	CE1	Asp210	CA	4.1
Arg73	NH1	Pro211	CB	4.1
Arg168	NH1	Asn170	ND2	4.2
Asn144	ND2	Ser202	OG	4.4
Gln204	NE2	Ser206	O	4.4
Tyr145	O	Asn146	C	4.6
Tyr143	O	Ser202	OG	4.8
Asn144	ND2	Asn149	ND2	4.9

[a] The first three lines are the residues of the hydrophobic core.

Dimerization of GFP and the concommitent spectral changes have also been exploited in the construction of novel reporters of calcium and calmodulin activities (Romoser et al., 1997; Persechini et al., 1997). A calmodulin target sequence was inserted as a linker between a red-shifted and a blue-shifted GFP sequence to yield a dimer GFP that exhibits FRET changes when calcium-calmodulin binds (Romoser et al., 1997). These authors went on to fuse calmodulin to the molecule to yield a complete calcium indicator that is tunable by changing the particular sequences of the calmodulin target (Persechini et al., 1997). Target ability of similar GFP-based Ca^{++} indicators for the cytosol, nucleus, and endoplasmic reticulum has also been achieved by adding appropriate localization signals (Miyawaki et al., 1997)

4.10 IMPLICATIONS OF THE STRUCTURE ON FOLDING MECHANISMS

The pattern of connections of the antiparallel β strands (Figs 4.7 and 4.8) suggests that the initial stages of folding involve the coalescence of two- or three-stranded segments of contiguous polypeptide chain, followed by assembly of the segments into a complete can around the central helix. Many variants of GFP fold with different rates or extents relative to wild-type protein, producing better markers in that less protein winds up in inclusion bodies. One change, Val^{163} Arg, increases the temperature tolerance for functional GFP expression (Kahana and Silver, 1996). This effect may also be the basis of several known

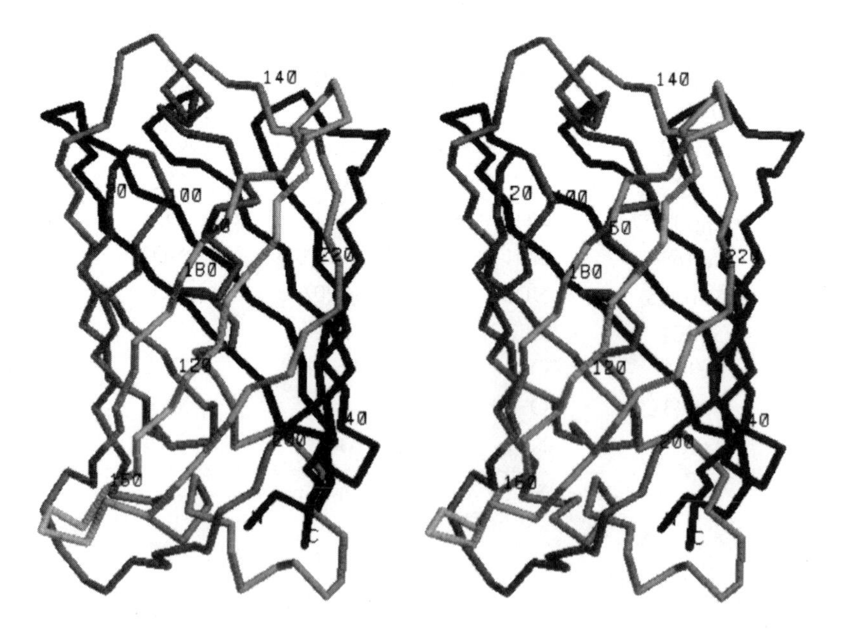

FIG. 4.7 Stereoview of the polypeptide trace of GFP, showing the N- and C-termini and the central helix. Note that not much of either of the termini can be cleaved without losing a stave in the barrel structure. [Figure produced with Rasmol (Sayle and Milner-White, 1995).

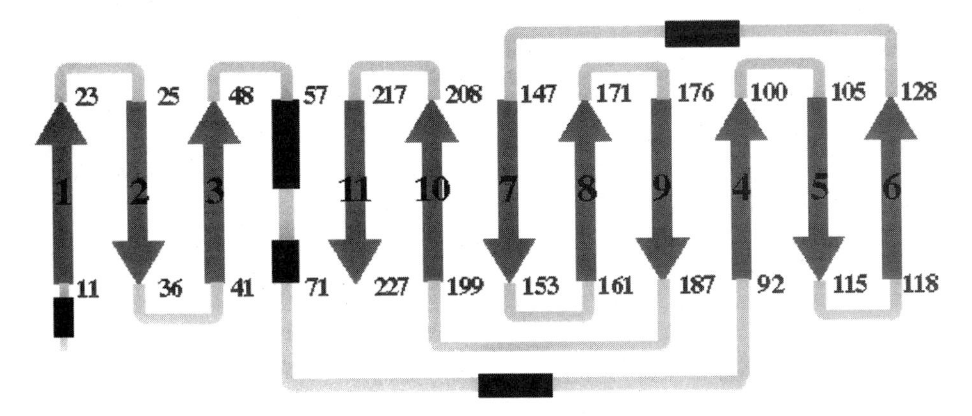

FIG. 4.8 Schematic diagram of secondary structure components and their connections. The protein appears to be built of sets of two or three adjacent β strands that come together to form the can structure.

improvements in total fluorescence by GFP (Cormack et al., 1996) and further studies in this area are certainly warranted. Folding kinetics are usually complicated, and it is probably difficult to predict results of mutants in advance, but GFP could be used as a convenient reporter of folding in biophysical studies of β-barrel formation that might lead to better understanding of folding phenomena.

4.11 CONCLUSIONS AND FUTURE PROSPECTS

New insights into the mechanism of GFP photochemistry should be of great interest to cell biologists, especially if such new mechanisms suggest improved procedures for imaging. In the complex problems of multicolor analyses and quantitation of FRET, careful attention to the biophysical details will be essential. Furthermore, future studies may include imaging devices that have the capability of acquiring full excitation, emission, and absorption spectra for every pixel in a scene.

Clearly for biological purposes, it may also sometimes be desirable to sacrifice long-term stability for "real-time" color development. In fact, fast photodestruction could be desirable to follow increases and then decreases in reporter gene applications. Another potential use of GFP is in electron microscopy. If the photochemistry can be controlled, the free radical reactions could in principle be used to produce deposits of metal stains for high-resolution localization experiments.

The 3D structures of GFP have provided a physicochemical basis of many observed features of the protein, including its stability, protection of its fluorophore, behavior of mutants, dependence of the spectra on pH, and dimerization properties. The structures will also allow directed mutation studies to complement random mutagenesis and also improve combinatorial approaches.

ACKNOWLEDGMENTS

I thank Fan Yang for help with preparing the figures, Doug Youvan for helpful discussions and the following organizations for financial support: the Robert A. Welch Foundation, the W. M. Keck Center for Computational Biology, and the NIH (AR40252 and AR32764).

REFERENCES

Branchini, B. R., Lusins, J. O., and Zimmer, M. (1997). A molecular mechanics and database analysis of the structural preorganization and activation of the chromophore-containing hexapeptide fragment in green fluorescent protein. *J. Biomol. Struct. Funct.* 14:1–8.

Brejc, K., Sixma, T. K., Kitts, P. A., Kain, S. R., Tsien, R. Y., Ormö, M., and Remington, S. J. (1997). Structural basis for dual excitation and photoisomerization of the *Aequorea victoria* green fluorescent protein. *Proc. Natl. Acad. Sci. USA* 94:2306–2311.

Carson, M. (1987). Ribbon models of macromolecules. *J. Mol. Graphics* 5:103–106.

Chalfie, M. (1995). Green fluorescent protein. *Photochem. Photobiol.* 62:651–656.

Chalfie, M., Tu., Y., Euskirchen, G., Ward, W. W., and Prasher, D. C. (1994). Green fluorescent protein as a marker for gene expression. *Science* 263:802–805.

Chattoraj, M., King, B. A., Bublitz, G. U., and Boxer, S. G.. (1996). Ultra-fast excited state dynamics in green fluorescent protein: Multiple states and proton transfer. *Proc. Natl. Acad. Sci. USA* 93:8362–8367.

Cody, C. W., Prasher, D. C., Westler, W. M., Prendergast, F. G., and Ward, W. W. (1993). Chemical structure of the hexapeptide chromophore of the Aequorea green-fluorescent protein. *Biochemistry* 32:1212–1218.

Cormack, B. P., Valdivia, R. H., and Falkow, S. (1996). FACS-optimized mutants of the green fluorescent protein (GFP). *Gene* 173:33–38.

Crameri, A. E., Whitehorn, E. A., Tate, E., and Stemmer, W. P. C. (1996). Improved green fluorescent protein by molecular evolution using DNA shuffling. *Nature Biotech.* 14:315–319.

Cubitt, A. B., Heim, R., Adams, S. R., Boyd, A. E., Gross, L. A., and Tsien, R. Y. (1995). Understanding, improving and using green fluorescent proteins. *TIBS* 20:448–455.

Cutler, M. W. (1995). Characterization and energy transfer mechanism of the green-fluorescent protein. Rutgers, the State University of New Jersey, New Brunswick, NJ.

Delagrave, S., Hawtin, R. E., Silva, C. M., Yang, M. M., and Yoiuvan, D. C. (1995). Red-shifted excitation mutants of the green fluorescent protein, *Biotechnology* 13:151–154.

Dopf, J. and Horiagan, T. M. (1996). Deletion mapping of the *Aequorea victoria* green fluorescent protein. *Gene* 173:39–44.

Ehrig, T., O'Kane, D. J., and Prendergast, F. G. (1995). Green-fluorescent protein mutants with altered fluorescence excitation spectra. *FEBS Lett.* 367:163–166.

Heim, R., Cubitt, A. B., and Tsien, R. Y. (1995). Improved green fluorescence. *Nature (London)* 373:663–664.

Heim, R., Prasher, D. C., and Tsien, R. Y. (1994). Wavelength mutations and posttranslational autoxidation of green fluorescent protein. *Proc. Natl. Acad. Sci. USA* 91:12501–12504.

Heim, R. and Tsien, R., (1996). Engineering green fluorescent protein for improved brightness, longer wavelengths and fluorescence resonance energy transfer. *Curr. Biol.* 6:178–182.

Hendrickson, W. A., Pahler, A., Smith, J. L., Satow, R., Meritt, E. A., and Phizackerley, R. D. (1989). Crystal structure of core strepavadin determined from multiwavelength anomalous diffraction of synchrotron radiation. *Proc. Natl. Acad. Sci. USA* 86:2190–2194.

Kahana, J. and Silver, P. A. (1996). *Current Protocols in Molecular Biology*. F. Ausabel, R. Brent, R. Kingston Moore, D. D., Seidman, J. G., Smith, J. A., and Struhl, K. Green and Wiley, New York, Suppl. 34: 9.7.22–9.7-28.

Kreusch, A., Newbueser, A., Schlitz, E., Weckessser, J., and Schultz, G. E. (1994). The structure of the membrane channel porin from Rhodopseudomonas blastica at 2.0 Angstrsom resolution. *Protein Sci.* 3:58.

Lakowicz, J. and Weber, G. (1973). Quenching of fluorescence by oxygen. A probe for structural fluctuation in macromolecules. *Biochemistry* 12:4161–4170.

Lim, C. R., Kimata, K., Oka, M., Nomaguchi, K., and Kohno, K. (1995). Thermosensitivity of a green fluorescent protein utilized to reveal novel nuclear-like compartments. *J. Biochem. (Tokyo)* 118:13–17.

Lossau, H., Kumer, A., Heinecke, R., Poellinger-Dammer, F., Kompa, C., Bieser, G., Jonsson, T., Silva, C. M., Yang, M. M., Youvan, D. C., and Michel-Beyerle, ME. (1996). Time-resolved spectroscopy of wild type and mutant green fluorescent proteins reveals excited state deprotonation consistent with fluorophore–protein interactions. *Chem. Phys.* 213:1–16.

Mitra, R. D., Silva, C. M., and Youvan, D. C. (1996). Fluorescence resonance energy transfer between blue-emitting and red-shifted excitation derivatives of the green fluorescent protein. *Gene* 173:13–17.

Miyawaki, A., Llopis, J., Heim, R., McCaffery, J. M., Adams, J. A., Ikung, M., Tsien, R. Y. (1997) Fluorescent indicators for the Ca^{2+} based on green fluorescent proteins and calmodulin. *Nature* 388:822–887.

Morin, J. G., and Hastings, J. W. (1971). Energy transfer in a bioluminescent system. *J. Cell Physiol.* 77:313–318.

Niwa, H., Matsuno, T., Kojima, T., Kubota, M., Hirano, T., Ohashi, M., Inouye, T., Ohmiya, Y., and Tsuji, F. I., (1977). *Aequorea* green fluorescent protein: structural elucidation of the chormophore. *In Bioluminescence and Chemoluminescence*, Hastings, J. W., Kricka, L. J., and Stanley, P. E., Eds., Wiley: Chichester, New York, pp. 2395–2398.

Ormö, M., Cubitt, A., Kallio, K., Gross, L., Tsien, R., and Remington, S. (1996). Crystal structure of the Aequorea victoria green fluorescent protein. *Science* 273:1392–1395.

Palm G. J., Zdanov, A., Gaitanaris, G. A., Stauber, R., Pavlakis, G. N., and Wlodawer, A. (1997). The structural basis for spectral variations in green fluorescent protein. *Nature Struct. Biol.* 4:361–365.

Perozzo, M. A., Ward, K. B., Thompson, R. B., and Ward, W. W. (1988). X-ray diffraction and time-resolved fluorescence analyses of Aequorea green fluorescent protein crystals. *J. Biol. Chem.* 263:7713–7716.

Persechini, A., Lynch, J. A. and Romoser, V. A. (1977). Novel fluorescent indicator proteins for monitoring free intracellular Ca^{2+}. *Cell Calcium* 22:209–216.

Prasher, D. C., Eckenrode, V. K., Ward, W. W., Prendergast, F. G., and Cormier, M. J. (1992). Primary structure of the Aequorea victoria green-fluorescent protein. *Gene* 111:229–233.

Rao, B. D. N., Kemple, M. D., and Prendergast, F. G. (1980). Proton nuclear magnetic resonance and fluorescence spectroscopic studies of segmental mobility in aequorin and a green fluorescent protein from aequorea forskalea. *Biophys. J.* 32:630–632.

Romoser, V. A., Hinkle, P. M., and Persechini, A. (1997). Detection on living cells of Ca^{2+}-dependent changes in the fluorescence emission of an indicator composed of two green fluorescent protein variants linked by a calmodulin binding sequence. *J. Biol. Chem.* 272:13270–13274.

Sayle, R. and E. Milner-White (1995) RasMol: Biomolecular graphics for all. *TIBS* 20: 374–375.

Terry, B. R., Matthews, E. K., and Haseloff, J. (1996). Molecular characterisation of recombinant green fluorescent protein by fluorescence correlation microscopy. *Biochem. Biophys. Res. Commun.* 217:21–27.

Voityuk, A. A., Michel-Beyerle, M. E., Rosch, N. (1997). Protonation effects on the chromophore of green fluorescent protein – quantum chemical study of the absorption spectrum. *Chemical Physics Letters* 272:162–167.

Ward, W. W. (1979). Energy transfer processes in bioluminescence. *In Photochemical and Photobiological Reviews*. Smith, K., Ed., Plenum, New York, vol. 4, pp. 1–57.

Ward, W. W. and Bokman, S. H. (1982). Reversible denaturation of Aequorea green-fluorescent protein: physical separation and characterization of the renatured protein. *Biochemistry* 21:4535–4540.

Ward, W. W. and Cormier, M. J. (1979). An energy transfer protein in coelenterate bioluminescence: characterization of the Renilla green-fluorescent protein. *J. Biol. Chem.* 254:781–788.

Ward, W. W., Prentice, H., Roth, A., Cody, C., and Reeves, S. (1982). Spectral perturbations of the *Aequorea* green fluorescent protein. *Photochem. Photobiol.* 35:803–808.

Wright, H. T. (1991). Nonenzymatic deamidation of asparaginyl and glutaminyl residues in proteins. *Crit. Rev. Biochem. Mol. Biol.* 26:1–52.

Wu, C.-K., Liu, Z.-J., Rose, J. P., Inouye, S., Tsuji, F., Tsien, R. Y., Remington, S. J., and Wang, B.-C. (1997). The three-dimensional structure of green fluorescent protein resembles a lantern. *In Bioluminescence and Chemoluminescence* Hastings, J. W., Kricka, L. J., and Stanley, P. E., Eds., Wiley, Chichester, UK, pp. 399–402.

Yang, F., Moss, L. G., and Phillips, G. N., Jr. (1996). The molecular structure of green fluorescent protein. *Nature Biotechnol.* 14:1246–1251.

Youvan, D. C. and Michel-Beyerle, M. E. (1996). Structure and fluorescence mechanism of GFP. *Nature Biotechnol.* 14:1219–1220.

Zimmer, M. and Branchini, B. (1997). A computational and database analysis of the structural preorganization and activation involved in chromophore formation if green fluorescent protein. *In Bioluminescence and Chemoluminescence*. Hastings, J. W., Kricka, L. J., and Stanley, P. E., Eds., Wiley, Chichester, UK, pp. 407–410.

5

Molecular Biology and Mutation of Green Fluorescent Protein

ROGER TSIEN
Department of Pharmacology, University of California, San Diego, La Jolla, CA
DOUGLAS PRASHER
USDA APHIS, Otis ANGB, MA

> "Some things, indeed, are not seen in daylight, though they produce sensation in the dark: as for example the things of fiery and glittering appearance for which there is no distinguishing name. . . . But in no one of these cases is the proper color seen.'
>
> (translation of Aristotle from Harvey, 1957)

5.1 INTRODUCTION

Probably not since the discovery by McElroy (1947) of the involvement of adenosine triphosphate (ATP) in the reaction catalyzed by firefly luciferase has there been as much interest in bioluminescence as currently exists. This interest is due, in large part, to the work of Chalfie et al. (1994) first showing the usefulness of the *Aequorea victoria* GFP. For the first time, there exists a genetically encoded reporter molecule that is detectable in the absence of an enzymatic substrate or cofactor in a variety of cell types. The fluorescent properties of GFP make it especially useful in living tissue. This report has brought a somewhat esoteric area of biology into the main stream of science.

This work built on the efforts of the research groups led by John Blinks, Milt Cormier, Woody Hastings, Frank Johnson, Frank Prendergast, Osamu Shimomura, and Bill Ward that led to the cloning of the GFP cDNA by Prasher et al. (1992). Prasher began the cloning while still working as part of

Green Fluorescent Protein: Properties, Applications, and Protocols, Edited by Martin Chalfie and Steven Kain
ISBN 0-471-17839-X

Cormier's group at the University of Georgia. At that time, Prasher fully realized GFPs potential but was cautious of potential technical difficulties associated with the formation of the chromophore in a recombinant form of GFP. The first *gfp* cDNA isolated, pGFP1, originated from the same cDNA library from which Prasher isolated aequorin cDNAs (Prasher et al., 1985). Partial amino acid sequence provided by Ward and Prendergast permitted the design of oligonucleotide hybridization probes toward *gfp* cDNAs. Unfortunately, the cDNA was a partial one encoding 168 amino acids and lacked both 5′- and 3′-sequences of the coding region (Prasher et al., 1992). However, it contained the amino acid sequence of the chromophore, which explained much of the chemical analysis of a chromophore-containing hexapepetide derived from native GFP (Cody et al., 1993). After Prasher collected new *Aequorea* tissue (frozen in liquid nitrogen at the University of Washington Friday Harbor Labs), a new *Aequorea* cDNA library of 1.4×10^6 recombinants was constructed in the λgt10 vector (Prasher et al., 1992). Using the cDNA insert from pGFP1 as a hybridization probe, four GFP recombinants, later shown to be siblings, were isolated from the amplified lambda library because no signals were observed when the primary library was probed. Thus, the library contained only a single *gfp* cDNA and, fortunately, it contained the entire coding sequence. This relative abundance (0.00007%) was unusually low if one assumes the mRNAs of GFP and aequorin exist in the same ratio in *Aequorea* as the proteins which is 1:5, respectively (Ward, personal communication). Since the apoaequorin mRNA represents 0.3% of the mRNA (Prasher et al., 1985), it can be expected the GFP mRNA might represent 0.06% of the total. Thus, the frequency of GFP recombinants in the cDNA library was nearly 1000-fold lower than predicted!

The protein sequence, derived from the cDNA nucleotide sequence, contains 238 amino acid residues and enabled determination of the chromophore structure (Shimomura, 1979; Cody et al., 1993). The p-hydroxybenzylideneimidazolinone chromophore is formed by the cyclization of Ser^{65}, Tyr^{66}, and Gly^{67} and dehydrogenation of the tyrosine. A mechanism of the chromophore formation has been proposed (Heim et al., 1994; Cubitt et al., 1995). The chromophore is strongly fluorescent only in the intact protein (Ward and Bokman, 1982).

Nucleotide sequences derived from *Aequorea* indicate at least five variants of GFP (Table 5.1). Four of the DNA sequences are encoded by cDNAs, while the fifth is encoded in the exons of a *gfp* gene. The variants differ generally by conservative amino acid replacements, suggesting they would have nearly identical physical properties. At least nine isoforms of native GFP have been purified from a large collection of jellyfish collected at Friday Harbor, Washington (Cutler, 1995). One of the *gfp* genomic clones contains three exons that can be matched to the cDNA, while a fourth exon must exist to account for the 5′ end of the cDNA (Prasher et al., 1992). The tripeptide encoding the chromophore is located near the 3′-end of exon II (Prasher et al., 1992).

Chalfie et al. (1994) showed the chromophore forms when GFP is expressed from its cDNA in a prokaryote (*Escherichia coli*) or a eukaryote (*Caenorhabditis elegans*). The spectral properties of the recombinant GFP are identical to native (*Aequorea*-derived) GFP. Thus, no *Aequorea*-specific factor is required to form the chromophore. Subsequently, Heim et al. (1994) reported that extensive

TABLE 5.1 Heterogeneous amino acid residues derived from *gfp* nucleotide sequences

	Locus[a]				
Residue Position	AEVGFPA[b]	AEVGFP[c]	AVGFP1[d]	AVGFP2[e]	AEVGFPB[f]
14	Ile	Ile	Val	Ile	Ile
25	His	Gln	Gln	Gln	His
30	Ser	Ser	Ser	Arg	Ser
45	Lys	Lys	Asn	Lys	Lys
84	Phe	Phe	Phe	Leu	Phe
100	Phe	Tyr	Tyr	Tyr	Phe
108	Thr	Thr	Thr	Thr	Ser
141	Leu	Met	Met	Met	Met
154	Ala	Gly	Gly	Gly	Ala
157	Gln	Pro	Pro	Pro	Gln
172	Glu	Lys	Lys	Lys	Glu
209	Lys	Lys	Lys	Gln	Lys
212	Asn	Asn	Asn	His	Asn
213	Glu	Glu	Glu	Gly	Glu
219	Val	Ile	Ile	Val	Ile
226	Ala	Ala	Ala	Ser	Ala
228	Gly	Gly	Arg	Gly	Gly

[a] As assigned by GenBank.
[b] Derived from the *gfp10* cDNA reported by Prasher et al. (1992), Accession No. M62653.
[c] Derived from the *gfp* cDNA reported by Inouye and Tsuiji (1994), Accession No. L29345.
[d] Derived from a cDNA submitted to GenBank by Watkins and Campbell, Accession No. X83959.
[e] Derived from a cDNA submitted to GenBank by Watkins and Campbell, Accession No. X83960.
[f] Derived from the exons of the gene encoded by *gfp2* (Prasher et al., 1992), Accession No. M62654.

dilution of nonfluorescent recombinant GFP does not affect the rate of aerobic fluorescence development, suggesting that no cellular factor is required and the final formation of the fluorophore is spontaneous.

A flood of activity describing the use of the wild-type GFP and its modification followed the report of Chalfie et al. (1994). For reviews, see Cubitt et al. (1995), Marshall et al. (1995), Stearns (1995), Coxon and Bestor (1995), Prasher (1995), Simon (1996), and application chapters in this book. The modifications of wild-type GFP have been attempts to improve on its limitations. (1) The excitation spectra is not optimal for use with commonly used filter sets. Excitation near the major peak (i.e., 395 nm) causes photoisomerization (Cubitt et al., 1995) and significant autofluorescence when used *in vivo*. (2) The 470-nm excitation peak is subject to less photoisomerization and autofluorescence, but its much lower extinction coefficient causes the fluorescence to be less bright. (3) The chromophore in the wild-type GFP forms rather slowly, potentially limiting its use to detect fast transcriptional events (Heim et al., 1994, 1995). (4) Additional colors would be valuable to permit multicolor comparison of two or more protein localizations or gene expression levels, as well as to enable fluorescence resonance energy transfer (Heim et al., 1994; Cubitt et al., 1995; Heim and Tsien, 1996; Rizzuto et al., 1996; Yang et al., 1996b). A variety of GFP mutants have been described to overcome these deficiencies. We first describe the methods used to obtain the mutants. We then discuss the mutants in order of increasing distance from the chromophore and decreasing effect on the wavelengths of excitation and emission. A useful shorthand for mutations is the one-letter code for the wild-type amino acid, its position number, and the one letter code(s) for the replacement amino acids. For example, V163A denotes valine 163 mutated to alanine.

5.2 MUTATIONAL STRATEGIES

Many strategies have been used to generate mutations in GFP. In order of increasingly precise location of the mutation sites, they include (1) random mutagenesis by chemical mutagens or error-prone polymerase chain reaction (PCR) over the entire or at least a major portion of the coding sequence (Heim et al., 1994; Ehrig et al., 1995; Heim and Tsien, 1996); (2) DNA "shuffling" (Crameri et al., 1996); (3) randomization (possibly with codon biases) of a predetermined, limited stretch of residues (Delagrave et al., 1995; Cormack et al., 1996); (4) deliberate site-directed mutations to test specific hypotheses (Heim et al., 1995; Ormö et al., 1996), eventually aided by the crystal structure. All four methods have yielded valuable results. The first three create massive genetic diversity and must be coupled with a high-throughput screen for the desired properties, such as (a) visual inspection of Petri plates, (b) video imaging, and (c) fluorescence-activated cell sorting (FACS). These mutation and screening methods are briefly compared below

5.2.A Mutation Methods

5.2.A.1 Random Mutagenesis by Chemical Mutagens or Error-Prone PCR

The cDNA vector encoding GFP can be treated with mutagens such as hydroxylamine or nitrous acid before transfection into a host (Sikorski and Boeke, 1991). Alternatively, the cDNA can be amplified by PCR under conditions when polymerase fidelity is impaired, for example, by replacing Mg^{2+} by Mn^{2+} in the reaction mixture and restricting the concentration of one nucleoside triphosphate at a time (Muhlrad et al., 1992). Error-prone PCR has the advantages of confining the mutations only to the desired region of the gene between the two primers and in producing a greater variety of mutations at the nucleic acid level. Nevertheless, most mutations are changes of just one base in the relevant codon, so that amino acids whose codons differ in two or three positions from the starting codon are much less likely to be accessed. The roles of Y66, I167, T203, and E222 (Heim et al., 1994; Ehrig et al., 1995) were discovered this way, as well as some of the folding mutations such as Y145F, M153T, and V163A (Heim and Tsien, 1996).

5.2.A.2. DNA Shuffling or Sexual PCR

This method recombines an existing set of mutations spread throughout the gene of interest. For example, it would be an ideal method to shuffle the individual folding mutations listed above and produce composite proteins containing multiple mutations that might synergize with each other. The collection of cDNAs, each of which contains only one or a few mutations, is subjected to limited digestion with DNAse. The fragments are annealed back together in a PCR-like reaction with nucleoside triphosphates but without primers. This processes allows fragments with different mutations to reassort with each other, heals the breaks, and introduces additional point mutations. The resulting mixture is finally tidied and amplified by conventional PCR with primers, spliced into an expression vector, and screened. Obviously, this cycle can be repeated as many times as is profitable. Crameri et al. (1996) thereby achieved a 45-fold increase in brightness over unoptimized wild-type GFP. Inspection of the actual mutations (F99S, M153T, V163A) shows that the latter two have also been found by more traditional random mutagenesis. So the individual mutants produced by DNA shuffling are not particularly unique, but the potential for synergistically recombining them is the main attraction. It should also be noted that Crameri et al. (1996) screened by exciting bacteria with 365-nm and looked for the brightest cells expressing large amounts of GFP, where the main problem was the avoidance of misfolding and inclusion bodies. Their use of 365-nm excitation would discard the large class of mutants that are excitable only at 470–490 nm. Although their GFP clearly has a much higher probability of folding correctly at 37°C, there is no evidence that it is intrinsically brighter than properly matured wild-type GFP. Also, it is still strongly photoisomerizable, that is, ultraviolet (UV) irradiation depresses the 395-nm excitation peak and enhances the 475-nm peak (Yokoe and Meyer, 1996).

5.2.A.3 Randomization of a Predetermined Stretch of Amino Acids

Delagrave et al. (1995) randomized positions 64–65 and 67–69 with the random codon NNK, where N is any base and K is G or T. They found (F64M, S65G, Q69L) as their favored mutant, code-named "RSGFP4." Later Cormack et al. (1996) randomized positions 55–74 with a 10% probability of substituting NNK for each wild-type codon. This procedure produced three different mutants (F64L, S65T), (S65A, V68L, S72A), and (S65G, S72A). All of these have excitation spectra shifted to 470–490 nm because of their replacement of Ser^{65}, plus one or more mutations that improve folding efficiency. Evidently positions 55–63 and 73–74 are not particularly helpful. In the future, regions to be subjected to randomization will probably be selected on the basis of the crystal structure (Ormö et al., 1996; Yang et al., 1996a), that is, proximity to the chromophore in three-dimensional (3D) space rather than primary sequence.

5.2.A.4 Deliberate Site-Directed Mutants

The S65 mutations (Heim et al., 1995) were first found this way, even though the hypothesis being tested, that Ser^{65} was uniquely capable of undergoing dehydration to a dehydroSer, proved incorrect. The most rational and recent example is the replacement of Thr^{203} by aromatic residues such as Tyr as in Table 5.2 (Ormö et al., 1996).

5.2.B Screening Methods

5.2.B.1 Visual screening

Visual screening requires the least expensive equipment. The minimum is a source of excitation light, either 365 nm from a 'black-light' illuminator or about 480 nm from a xenon lamp and interference filter or monochromator, that can illuminate Petri dishes. Observation is through a long-pass filter. In the case of UV excitation, an external UV-blocking filter is advisable for health reasons. For blue excitation, one can tape pieces of yellow or orange gelatin filters (e.g., Kodak Wratten filters) over disposable laboratory safety glasses. Alternatively, one can scan the dish with a fluorescence microscope with a low-power objective, typically 4X, using the normal filter cube to select excitation and emission wavelengths. The main advantage of visual screening is low capital cost; however, it is both laborious and insensitive to small changes in wavelength below the capacity of human color vision to recognize easily. Individual mutations with such small effects are still worth collecting because when compounded, their effects might sum into a larger and more valuable alteration. Although such additivity of small perturbations has not been specifically demonstrated for GFP, it seems to be the mechanism by which rhodopsin is tuned to different wavelengths (Neitz et al., 1991; Merbs and Nathans, 1992).

5.2.B.2 Video Imaging

Video imaging with a frame grabber and computerized false-color display of spectral features has been entitled *Digital Imaging Spectroscopy* (Youvan et al., 1995). This method is similar to the excitation or emission ratioing now common in imaging physiological indicators inside cells, except that the optics

TABLE 5.2 Effects of different aromatic amino acids at position 203 of GFP[a]

Mutant (code name)	λ_{exc}	ε	λ_{em}	Comments and References
T203H/S65T (5B, 9B)	512	19,400	524	(Ormö et al., 1996)
T203H/S65G/S72A	508		518	Cubitt (unpublished)
T203F/S65G/S72A	512		522	(Dickson et al., 1997)
T203Y/S65T (6C)	513	14,500	525	(Ormö et al., 1996)
T203Y/F64L/S65G/S72A (10B)	513	30,800	525	(Ormö et al., 1996)
T203Y/S65G/V68L/S72A (10C)	513	36,500	527	(Ormö et al., 1996); Quantum yield 0.64
T203W/S65G/S72A (11)	502	33,000	512	(Ormö et al., 1996)

[a] Abbreviations: λ_{exc}, excitation maximum in nanometers; ε, extinction coefficient in $M^{-1}cm^{-1}$ at λ_{exc}; λ_{em}, emission maximum in nanometers.

illuminate and observe a macroscopic rather than microscopic field of view. Digital imaging is more objective and sensitive than human vision to small changes in spectra, although the combined optical and computer setup is relatively expensive and not widely available. Also, it may take some practice to coordinate the manual removal of a given colony with its image on a computer screen. Several mutants in which the 395-nm excitation peak was suppressed in favor of the 475-nm peak were found by this method (Delagrave et al., 1995).

5.2.B.3 Fluorescence-Activated Cell Sorting (FACS)

This method would seem the obvious high-throughput screen, especially if the sorting criterion is an unusual ratio of emissions at two wavelength bands. However, early attempts with wild-type GFP (unpublished results of Chused and Tsien, 1994) were unsuccessful because the signals from individual bacteria were weak and noisy, necessitating high laser power, and the sorted bacteria did not retain the phenotype for which they had been selected. In retrospect, the UV laser probably altered the protein spectra by photoisomerization and damaged the DNA. Cormack et al. (1996) were much more successful, perhaps because they were simply seeking increased brightness at 488-nm excitation. Ultimately, FACS analysis should be the optimal screening technique for spectral properties that can be ascertained in one pass through the instrument. The equipment is relatively ubiquitous, has high throughput ($>10^3$ cells^{-1} s), is not restricted to bacteria, does not require growing entire colonies, and obviates manual isolation of the desired cells.

5.3 MUTANTS THAT COVALENTLY ALTER THE FLUOROPHORE: Y66F,H, OR W

The most fundamental mutations that do not destroy fluorescence entirely are those that alter the covalent core structure of the chromophore. Such mutations are inevitably at position 66, because the crucial π-electron conjugated framework is derived from Tyr^{66} plus the invariant carbonyl carbon and amino nitrogen of residues 65 and 67, respectively. All the aromatic amino acids allowable by the genetic code, namely, phenylalanine, histidine, and tryptophan, have been tested and found to generate fluorescent proteins of drastically different wavelengths (Heim et al., 1994; Heim and Tsien, 1996). The structures of the chromophores expected to be formed from these amino acids and the observed excitation and emission wavelengths are compared with wild-type GFP in Figure 5.1. The rank order of shortest-to-longest excitation wavelengths is as expected from the electronic properties of the side chains. The benzene ring of phenylalanine has no electron-donor group to conjugate to the electron-withdrawing carbonyl at the other end of the chromophore, so Phe gives the shortest wavelengths. The imidazole of His has a moderately electron-donating HN< group, but this is weakened by the electron-withdrawing =N– on the same ring. Next after imidazole is the nonionized phenol of tyrosine, which is responsible for the 395-nm excitation maximum of wild-type GFP. Yet, more electron rich is the >NH in the indole of tryptophan, which also has the largest conjugated system.

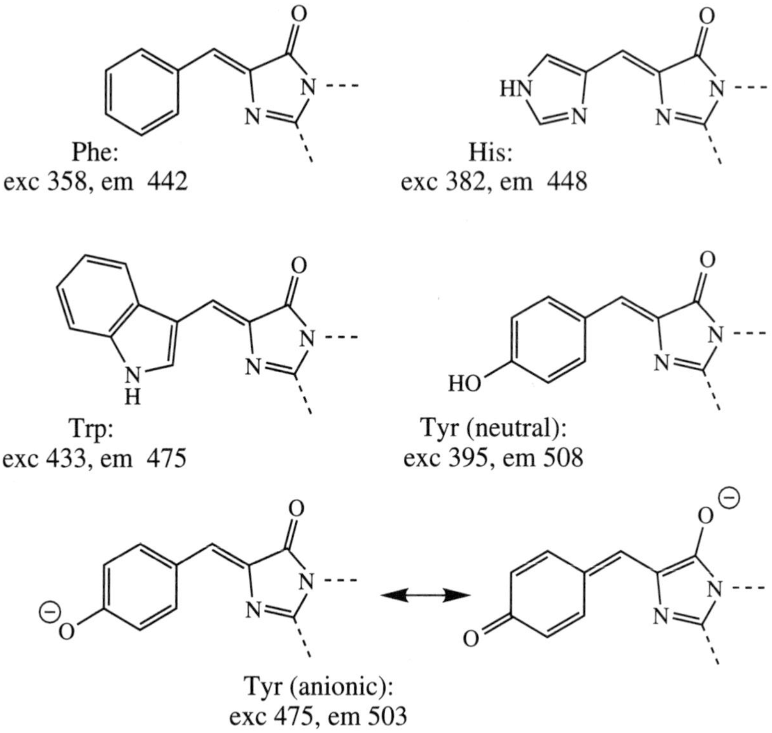

FIG. 5.1 Structures and fluorescence wavelengths of fluorophores resulting from different aromatic amino acids at position 66 of GFP. Each structure is labeled by the amino acid occupying position 66 before chromophore formation. Excitation (exc) and emission (em) peak wavelengths are given in nanometers. The native amino acid, Tyr, gives two peaks depending on whether the phenolic hydroxyl is neutral or ionized. For the latter, two of the possible resonance structures are drawn. Spectral data from (Heim and Tsien, 1996) and (Cubitt et al., 1995).

The most electron-donating group is the phenolate anion of tyrosine, which is responsible for the subsidiary 470 nm excitation maximum of wild-type GFP. The chromophore with an ionized phenolate has about twice the extinction coefficient as that with a neutral phenol (Chattoraj et al., 1996). Therefore wild-type GFP, whose 395-nm peak is typically about three times as high as its 470-nm peak, is approximately a 6:1 mixture of neutral and anionic chromophores. This ratio has recently been confirmed by X-ray crystallography of wild-type GFP, in which two internal isomers corresponding to neutral and anionic chromophores in 85:15 ratio can be resolved (Brejc et al., 1997).

Although it was surprising that some fluorescence could be retained after replacing Tyr^{66}, one would not expect that such mutants would be optimally folded, especially in view of the compact and highly organized chromophore environment subsequently revealed in the crystal structures (Ormö et al., 1996; Yang et al., 1996a; Brejc et al., 1997). Indeed, all practical applications of position 66 mutants have required additional amino acid substitutions to increase folding efficiency and fluorescence brightness (Heim and Tsien, 1996;

Mitra et al., 1996; Siemering et al., 1996). Such folding mutations are discussed in Section 5.6. The visual appearance of bacteria expressing first-generation improved versions of Y66H (blue) and Y66W (cyan) are shown in color Figure 5.2; normalized excitation and emission spectra are in color Figure 5.3. Crystal structures of mutants containing Y66H have now been solved (Palm et al., 1997; Wachter et al., 1997).

5.4 MUTANTS THAT CHANGE THE RATIO BETWEEN THE TWO WT EXCITATION PEAKS: I167T, T203I, E222G, S65GACTVL, Q69L

The next most fundamental mutations are those that alter the ionization of Tyr^{66} and change the ratio between the excitation peaks due to the neutral (~395 nm) and anionic (~470 nm) species. Mutation of Thr^{203} to Ile (T203I) suppresses the anionic peak (Heim et al., 1994; Ehrig et al., 1995). This result is now readily understandable from the crystal structure (Ormö et al., 1996; Yang et al., 1996a), because the hydroxyl of Thr^{203} donates a hydrogen bond to the chromophore phenolate. In the neutral form of the chromophore, the hydroxyl of Thr^{203} rotates away from the protonated phenol (Brejc et al., 1997). Ile cannot form a hydrogen bond, so the anionic phenolate would be destabilized relative to the neutral phenol. However, once the neutral chromophore absorbs light and reaches the excited state, the phenol ionizes (Chattoraj et al., 1996; Brejc et al., 1997) despite the relatively poor solvation, so the subsequent emission still peaks at 511 nm.

Most workers have been more interested in shifting the excitation to longer (visible) rather than shorter (UV) wavelengths, because UV is potentially more injurious to cells, excites more cellular autofluorescence, and generally requires more expensive optics and instrumentation. Shifts toward longer wavelengths require favoring the anionic peak. The first such mutants were I167T and I167V, which gave anionic excitation peaks about twice the amplitude of the neutral peaks (Heim et al., 1994). Given the ratio of extinction coefficients, these mutants probably contain about a 1:1 mol ratio of the two species. The crystal structure of GFP (Ormö et al., 1996; Yang et al., 1996a) can rationalize why Thr at position 167 gives a moderately but not overwhelmingly higher ratio of anion to neutral than wild-type Ile^{167}. Thr should point its OH group toward the phenolate without coming close enough to form a direct hydrogen bond. The dipole moment of the OH should therefore favor the phenolate. This rationalization, however, does not explain why Val at position 167, which lacks the OH, produces much the same effect as Thr. Perhaps the smaller steric bulk of both Val and Thr relative to Ile is more important.

A much more effective way of suppressing the neutral species was found to be mutation of Ser^{65} to any of a variety of small uncharged amino acids including Gly, Ala, Cys, Thr, or Leu (Heim et al., 1995). Such mutations cause essentially complete ionization of the chromophore even in the ground state and simplify the excitation spectrum to a single peak between 470 and 490 nm, whose amplitude is about six times higher than that of the 475-nm subsidiary peak of the wild-type protein. Thus at such blue excitation wavelengths, the mutants are about six times brighter per molecule than wild type. The improved brightness of

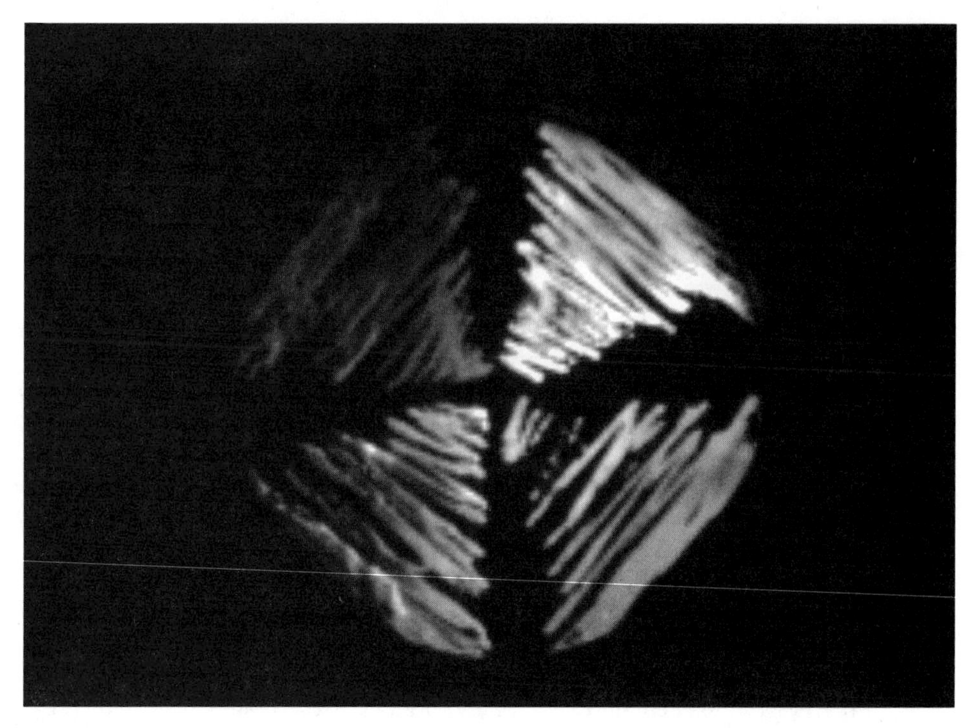

FIG. 5.2 Visual appearance of *E. coli* expressing four differently colored mutants of GFP. Clockwise from upper right: blue mutant P4-3 (=Y66H, Y145F) (Heim and Tsien, 1996); cyan mutant W7 (=Y66W, N146I, M153T, V163A, N212K) (Heim and Tsien, 1996); green mutant S65T (Heim et al., 1995); yellow mutant 10C (=S65G, V68L, S72A, T203Y) (Ormö et al., 1996). In each of these lists of mutants, the mutation most responsible for the spectral alterations is underlined, while the other substitutions improve folding or brightness. The bacteria were streaked onto nitrocellulose, illuminated with a Spectroline B-100 mercury lamp (Spectronics Corp., Westbury, NY) emitting mainly at 365 nm, and photographed with Ektachrome 400 slide film through a low-fluorescence 400-nm and a 455-nm colored glass long-pass filter in series. The relative brightnesses of the bacteria in this image are not a good guide to the true brightnesses of the GFP mutants. Expression levels were not normalized, the 365 nm excites the blue and cyan mutants much more efficiently than the green and yellow mutants, but the blue emission is significantly filtered by the 455-nm filter required to block violet haze. Figure also appears in color section.

such Ser^{65} mutants when excited in this blue region is understandable, because the wild-type protein was only about one-sixth ionized. The exact positions of the excitation and emission maxima depend on which amino acid replaces Ser^{65}, as shown in Table 5.3. Alanine gives the shortest wavelengths, Thr or Gly give the longest. Very polar or bulky residues such as Arg, Asn, Asp, Phe, or Trp at position 65 do not seem to be tolerated (Heim et al., 1995).

The mutant with Thr at position 65, or S65T, was chosen by Heim et al. (1995) for further analysis and use because it combined the longest peak wavelengths (489-nm excitation, 511-nm emission) with the most conservative substitution. The visual appearance of bacteria expressing S65T is included in Figure 5.2, and its excitation and emission spectra are shown in Figure 5.3. The S65T

TABLE 5.3 Effects of different amino acids at position 65 of GFP[a]

Amino Acid	λ_{exc}	ε	λ_{em}	Q	Comments and References
Ser (wt)	395	21,000	508	0.77	At room temp. (Heim et al., 1995); the 395- and 475- nm peaks probably represent excitation of the neutral and anionic fluorophore respectively
	475	7,150	503		
Ser (wt)	404		504		At 77 K in 1:1 glycerol–water to slow down proton transfers (Chattoraj et al., 1996)
	471		482		
Ala	471		503		(Heim et al., 1995)
Ala	481		507		Measured in combination with V68L, S72A (Cormack et al., 1996)
Cys	479	47,400	507		(Heim et al., 1995)
Leu	484		510		R. Heim (unpublished data)
Thr	489	39,200	511	0.68	(Heim et al., 1995)
Thr		52,900			R. Heim (unpublished redetermination)
Thr	488		507		Measured in combination with F64L (Cormack et al., 1996)
Gly	490		505		Measured in combination with F64M, Q69L (Delagrave et al., 1995)
Gly	501		511		Measured in combination with S72A (Cormack et al., 1996)
Arg,Asn,Asp, Phe,Trp					Weak or no fluorescence (Heim et al., 1995)

[a] Abbreviations: λ_{exc}, excitation maximum in nanometers; ε, extinction coefficient in $M^{-1}cm^{-1}$ at λ_{exc}; λ_{em}, emission maximum in nanometers; Q, fluorescence quantum yield. Discrepancies of 2–3 nm between different laboratories in estimating λ_{exc} and λ_{cm} are probably not significant.

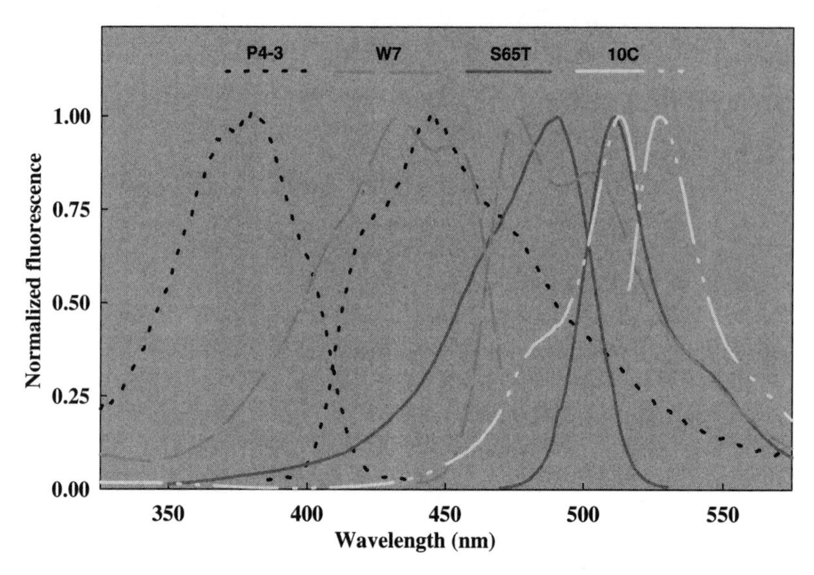

FIG. 5.3 Normalized excitation and emission spectra for mutants P4-3, W7, S65T, and 10C (see Fig. 5.2 for amino acid substitutions and references). All spectra were normalized to a maximal value of 1 to facilitate wavelength comparison. Each pair of excitation and emission spectra is depicted in a different line pattern and color as listed above the spectra. Within each pair the emission spectra is always the one at longer wavelengths. Figure also appears in color section.

mutant also proved to have at least two other advantages over wild type: a fourfold increase in the rate of the final oxygen-dependent step in the fluorophore formation, and a considerable increase in photostability (Heim et al., 1995). The other possible substitutions at position 65 have not been explicitly tested for these properties. Also, the time required for the entire conversion of nascent polypeptide to mature fluorescent protein was not measured, only the final oxidative step. Although the former value may be what matters in most applications, it is relatively difficult to measure. For example, one would have to compare GFP fluorescence versus quantitative immunoreactivity over the gradual time course of translation. It is also likely to be highly dependent on expression conditions such as the presence of chaperones and the level of expression, higher GFP levels tending to promote nonproductive aggregation or precipitation. By contrast, the final oxidative step is much easier to isolate and start synchronously, merely by suddenly exposing anaerobically expressed protein to air and observing the exponential time constant for fluorescence development. Also, because the final oxidative step is dilution independent, its rate is more likely to be an intrinsic property of the protein itself.

The increased photostability of S65T is manifested in two ways. Wild-type GFP when irradiated at wavelengths short enough to excite the neutral species tends to photoisomerize toward the anionic species (Cubitt et al., 1995; Chattoraj et al., 1996). Therefore excitation of the 395-nm peak rapidly diminishes its amplitude and boosts that of the 470-nm peak (Chalfie et al., 1994). This conversion probably involves transfer of a proton from the neutral chromophore to Glu^{222} (Brejc et al., 1997) and can partially reverse upon standing in the dark over the time scale of hours to days (Chattoraj et al., 1996). Since the fluorophore of S65T is already entirely anionic, there is no way for photoisomerization to push the ionization any further. Aside from the wild-type photoisomerization, both wild-type and mutant GFPs irreversibly photobleach. Comparison of determinations from separate groups (Cubitt et al., 1995) indicates that blue light photobleaches S65T about 3.5 times more slowly than wild type, though these results should be redetermined in side-by-side comparison. Considering that S65T has a much higher extinction coefficient, the ratio of photobleaching quantum yields is even more in favor of S65T (Cubitt et al., 1995).

The crystal structures of S65T (Ormö et al., 1996) and wild-type (Yang et al., 1996a; Brejc et al., 1997) GFPs have now clarified why position 65 can control whether the protein contains mostly neutral (with Ser at 65) or wholly anionic (with any other small amino acid including Thr at 65) fluorophores. The side-chain OH contributed by Ser^{65} is uniquely able to solvate the side-chain carboxylate anion of Glu^{222} by donating a hydrogen bond to it. The negative charge on Glu^{222} prevents a negative charge on the immediately adjacent fluorophore, because two like charges buried deep inside the rest of the protein would be energetically unfavorable (Ormö et al., 1996; Brejc et al., 1997). Removal of the side-chain OH, as in Gly, Ala, or Val at position 65, prevents solvation of Glu^{222} and forces it to remain neutral. The fluorophore is then free to ionize, aided by the extensive solvation of its phenolate and carbonyl oxygens. Even a Thr at position 65 cannot solvate the anion of Glu^{222} because the bulk of the extra methyl on the Thr side chain prevents the OH from rotating into position to donate a hydrogen bond to Glu^{222}. Instead the Thr OH donates a hydrogen bond to the backbone carbonyl of Val^{61} and receives a hydrogen bond from the uncharged carboxylate of Glu^{222}. As mentioned in Section 5.3, careful X-ray structural analysis shows that about 15% of the molecules with Ser^{65} have a different internal conformation in which Thr^{203} rotates to solvate the fluorophore phenolate (Brejc et al., 1997). This fraction agrees with the fraction of wild-type protein estimated spectroscopically to contain anionic fluorophores in the ground (i.e., nonexcited) state at normal pH (Chattoraj et al., 1996). At pH values approaching 12, just below those required to denature the protein, the neutral fluorophores are progressively ionized to anions, as signaled by diminution of the 395-nm excitation peak and amplification of the 470-nm peak (Ward et al., 1982). Presumably, as the protein begins to unfold, the proton on the neutral chromophore can escape to the external alkaline solution. However, in the intact protein at less extreme pH values, the proton can only be internally transferred to the buried carboxylate Glu^{222}, so the ratio of neutral to anionic chromophores is independent of the pH outside (Ward et al., 1982; Brejc et al., 1997).

In the initial description of the S65T crystal structure (Ormö et al., 1996), extra electron density was seen next to the imidazolidinone moiety and was interpreted to mean that 30% of the molecules had either failed to form their chromophore or had left the C=N double bond hydrated as a carbinolamine. This assignment seemed to be confirmed by electrospray mass spectra, that indicated that the polyhistidine-tagged fusion protein had a mass 6 or 7 Da higher than predicted, consistent with 30% mol fraction of species that should have 20 or 18 extra daltons, respectively. However, both these results are now believed to be artifacts (unpublished data of M. Ormö, S. J. Remington, and L. A. Gross). The extra electron density in the crystal structure is only seen in the selenomethionine derivative, whose folding was more difficult than that of the S65T protein with normal methionines. The extra mass in electrospray mass spectra disappears when the N-terminal fusion tag is removed by proteolysis, indicating that the extra mass is probably partial oxidation of one of the many methionines in the disordered fusion tag rather than in the GFP itself. The upshot is that fluorophore formation is fully efficient in S65T.

Delagrave et al. (1995) combinatorially mutagenized positions 64–69 and screened colonies by imaging spectroscopy. They found six mutants, which they described as "red shifted" because their excitation spectra peaked near 490 nm. These mutations are now interpretable as emphasizing the anionic fluorophore at the expense of the neutral form. Many biologists have misunderstood the term "red shifted" to imply that the fluorescence is actually red, but in fact the emission spectra are not significantly changed from the normal green emission of wild-type protein, peaking at 505–510 nm. Five out of six of these mutants had Gly, Ala, Cys, or Leu at position 65, which are undoubtedly the key substitutions responsible for the shift in excitation spectra.

The reciprocal interaction between Glu^{222} and the fluorophore predicts that if Glu^{222} were replaced by a nonionizable group, the fluorophore would be permanently anionic. Indeed, Ehrig et al. (1995) found E222G by random mutagenesis and visual screening and shown that its only excitation peak is at 481 nm, consistent with full ionization of the fluorophore.

The only other mutant with an increased ratio of anionic to neutral excitation peaks that cannot be explained by the above substitutions of Ser^{65} and Glu^{222} is "RSGFP1" of Delagrave et al. (1995), which is F64G, V68L, Q69L. Mutations of Phe^{64} and Val^{68} are known to affect folding efficiency (see below) but not wavelengths, so Q69L is most likely responsible for the increased ionization of the fluorophore. The crystal structure (Ormö et al., 1996) shows that Q69 anchors a cluster of water molecules that also participates in solvating the Glu^{222} carboxylate. Replacement by Leu would disrupt this hydrogen-bonding network and destabilize the carboxylate anion more than the neutral carboxyl of Glu^{222}, thereby indirectly promoting fluorophore ionization. The actual spectra of RSGFP1 have not been reported, so the extent of ionization is not known.

5.5 MUTANTS THAT MORE SUBTLY MODIFY THE ENERGY LEVELS

A few mutants shift both the excitation and emission spectra along the wavelength axis. Obviously the major effort has been directed toward obtaining mutants with longer rather than shorter wavelengths. It is this class of mutants that can most accurately be termed red shifted, unlike the previously discussed mutations that promote ionization of the phenol and thereby alter the ratio of preexisting excitation peaks, but that have relatively little effect on the emission spectrum. The best understood mutant is T203Y, which was designed when the crystal structure of GFP (Ormö et al., 1996) revealed the close proximity of

Thr^{203} to the fluorophore. Replacement of that aliphatic residue by aromatic residues next to the fluorophore was intended to increase the local polarizability. Polarizability indicates the ability of the medium to instantaneously shift its own electron densities to solvate and stabilize any change in dipole moment in the chromophore or fluorophore, and is correlated with refractive index. Pi-electron clouds as in aromatic rings are much more polarizable than aliphatic groups. Because transitions in both directions from the ground to the excited state and vice versa involve changes in dipole moments, increased polarizability is a fairly safe way to lower those transition energies, albeit only modestly (Lakowicz, 1983). Indeed, starting from S65 mutants that were already fully ionized, T203Y increased the excitation and emission maxima by 24 and 16 nm, respectively, to 513 and 527 nm, respectively (Ormö et al., 1996). These wavelengths, which are the longest so far published for a GFP mutant, permit detection through at least some standard filter sets for rhodamines, for example, 510–560-nm excitation, 565-nm dichroic, 572–647-nm emission. Although pure 527 nm is itself still green, the long tail of emission at yet longer wavelengths makes the emission yellowish to the eye and distinguishable from S65T in side-by-side comparisons. Figure 5.2 shows such a comparison captured on color film. The hue difference is somewhat more impressive in real life than in this picture. Also T203Y mutants are readily distinguished from S65T in flow cytometry when the GFPs are excited at 488 nm and the 505–535 and 535–585-nm emission bands are ratioed (Knapp, Heim, and Miyawaki, personal communication). Excitation and emission spectra for an T203Y mutant (with several additional substitutions to improve folding) are shown in Figure 5.3. The T203F and T203H mutants were almost as effective (Table 5.2), whereas T203W produced less of a shift, perhaps because its steric bulk was inconveniently excessive or because dipole moments of the new substituents are playing contributory roles. The T203Y mutant would have been extremely difficult to find by screening random mutants, because all three bases of the native Thr^{203} codon (ACA) would have had to change to encode Tyr (TAT or TAC). Normal techniques for introducing random mutations would probably put many incapacitating substitutions elsewhere long before all three bases of a single codon were mutated, so this is a case where the crystal structure was invaluable in guiding mutagenesis.

Another mutant of even smaller and less easily explained effect is M153A. This mutant increases the excitation and emission wavelengths of S65T by 15 and 3 nm, respectively (Heim and Tsien, 1996).

5.6 MUTANTS THAT DECREASE AGGREGATION OR IMPROVE FOLDING, ESPECIALLY AT HIGHER TEMPERATURES

A large class of mutations improve the percentage yield of GFP molecules that fold correctly and become fluorescent without seeming to affect the spectral properties of those properly matured proteins. The yields of fluorescent protein expressed from the original jellyfish gene fall steeply as the temperature increases above about 15–20°C, which is not surprising given the low temperature of Puget Sound in which *A. victoria* lives. Also, high-level expression in *E. coli* tends to give extensive deposition of nonfluorescent protein in inclusion bodies, in which the chromophore has not even formed (Siemering et al., 1996). Both of these problems can be greatly ameliorated by suitable amino acid substitutions, which have mostly been found by random mutagenesis and visual or flow-cytometric screening for brighter bacteria at temperatures up to 37°C. Table 5.4 gives a list of these mutations. Many have been arrived at independently by several groups (Cormack et al., 1996; Crameri et al., 1996; Heim and Tsien, 1996; Kahana and Silver, 1996; Siemering et al., 1996), which may imply that existing mutational strategies are approaching saturation. In our opinion, all new GFP constructs should incorporate a selection of such mutations that improve folding, because they do no harm and can often produce large increases in brightness. The only exception would be experiments when one desires to shut off the formation of newly fluorescent GFP molecules by raising the temperature (Kaether and Gerdes, 1995; Lim et al., 1995). Note that even wild-type GFP is fairly heat stable once properly folded and matured (Lim et al., 1995), becoming denatured only above 65°C (Ward and Bokman, 1982); only during the folding process is it highly temperature sensitive (Siemering et al., 1996).

Although folding mutations are quite valuable, one should also not expect indefinite further improvements in GFP brightness. The ultimate brightness of any fluorophore is limited at the molecular level by the product of extinction coefficient and fluorescence quantum efficiency. Folding mutations do not significantly improve the brightness of those molecules that are correctly folded (A.B. Cubitt, personal communication; Siemering et al, 1996), although unfortunately few laboratories characterizing GFP mutants have measured extinction coefficients or quantum yields. The fact that the brightness of bacteria expressing very high levels of GFP can be raised by a factor of 20–50 (Cormack et al., 1996; Crameri et al., 1996; Siemering et al., 1996) is more indicative of the wretched folding efficiency of the wild-type protein at high temperature and concentrations; at lower temperatures or expression levels, that is, when one wants to be able to detect as few GFP molecules as possible, the improvement due to folding mutations is not as impressive. Furthermore, the improvement factors due to individual mutations do not multiply: two mutations that separately improve folding efficiency by factors of x and y do not give a improvement of xy when combined (A. B. Cubitt, personal communication). Folding efficiency can asymptotically approach but can never exceed 100%.

The existing crystal structures (Ormö et al., 1996; Yang et al., 1996a) do not clearly explain how most of the folding mutations exert their favorable effects. Many of the mutations may only make a difference while the protein is still relatively disordered and in an unfavorable environment, whereas the crystals are grown from mature protein produced under conditions chosen to maximize successful expression. In a few cases, such as F99S and M153T, the mutations do reduce obvious patches of surface hydrophobicity and could inhibit aggregation. Indeed, the triple mutant (F99S, M153T, V163A) (Crameri et al., 1996) has a diffusion coefficient inside mammalian cells one order of magnitude higher than that of wild-type GFP, implying a corresponding reduction in binding to other macromolecules (Yokoe and Meyer, 1996). Because Val^{163} points into the interior of the protein (Ormö et al., 1996), whereas F99 Phe^{99} and Met^{153} face outward, the latter two are most likely the culprits in wild-type GFP. However, the triple mutation did not alter the overall speed of fluorescence development at

TABLE 5.4 Mutations that primarily improve folding of GFP

Mutation	Side-chain location	Phenotype and references. The ε denotes molar extinction coefficient.
F64L	Buried	Improves bacterial expression of S65T (Cormack et al., 1996; Youvan and Michel-Beyerle, 1996) and ε of T203Y (Ormö et al., 1996)
V68L	Buried	Improves bacterial expression of S65A (Cormack et al., 1996) and ε of T203Y
S72A	Buried	Improves bacterial expression (Cormack et al., 1996) of S65G and S65A and ε of T203Y
F99S	Surface	Improves bacterial expression of WT (Crameri et al., 1996)
Y145F	Buried	Improves ε and brightness of Y66H (Heim and Tsien, 1996)
N146I	Surface	Improves ε and brightness of Y66W (Heim and Tsien, 1996)
M153T	Surface	Improves ε and brightness of Y66W (Heim and Tsien, 1996); improves bacterial expression of wild type (Crameri et al., 1996)
V163A	Buried	Improves ε and brightness of Y66W (Heim and Tsien, 1996); improves expression and thermotolerance of nascent WT (Crameri et al., 1996; Kahana and Silver, 1996), S65T (Kahana and Silver, 1996), and Y66H (Siemering et al., 1996)
I167T	Buried	Together with V163A and S175G, improves brightness of S65T (Siemering et al., 1996)
S175G	Surface	Improves thermotolerance of nascent WT, Y66H (Siemering et al., 1996)
N212K	Surface	Improves ε and brightness of Y66W (Heim and Tsien, 1996)

37°C compared to wild type (Crameri et al., 1996). Mutations V163A and S175G together actually slow the final aerobic development of fluorescence (Siemering et al., 1996), even though they greatly improve the yield of properly matured protein. Molecular understanding of how other folding mutations work, especially those affecting buried side chains, may require detailed comparison of the resulting final structures with the baseline structures already on hand, as well as investigations of the folding intermediates and dynamics.

5.7 TRUNCATIONS OF GFP

One of the most frequently asked questions about GFP is whether it can be significantly reduced in size. In the most systematic investigation of this possibility, Dopf and Horiagan (1996) produced a family of genetic truncations from the N- and C-termini. The N-terminal methionine could be replaced by a polyhistidine tag, but deletion of residues 2–8 prevented fluorescence or chromophore development. The C-terminus was slightly more tolerant, in that 6 but not 13 residues could be eliminated. These narrow limits are in good agreement with the crystal structure (Ormö et al., 1996), in which Met^1 and residues 230–238 are too disordered to be located. There are no major regions of internal sequence that appear dispensible. The core domain of GFP, residues 2–229, seems to be a monolithic entity required to shield the fluorophore from solvent. A few local loops might conceivably be shortened somewhat, but the slight net reduction in size would hardly be worth the effort. Fortunately, it seems possible to target full-length GFP to essentially any compartment in the cell, so its size has not yet severely restricted its versatility.

5.8 SILENT AND LOSS-OF-FUNCTION MUTATIONS

During any random or semirandom mutagenesis screen, the great majority of colonies typically are either indistinguishable from the starting phenotype or of significantly reduced brightness. In principle, these mutants could be sequenced to provide a list of neutral or deleterious substitutions, but such a list would be laborious to collect and of negligible interest to those wishing to improve GFP or obtain novel properties. Perhaps this list deserves compilation, because it might increase understanding of how GFP folds and builds its chromophore. A few mutations are known to be neutral, such as the ubiquitous Q80R, which may have arisen from a PCR error in the initial cDNA clone (Chalfie et al., 1994). The substitutions within the natural isoforms of Table 5.1 are presumably all permissive for fluorescence, though other properties may well be altered. As mentioned above, mutations of Ser^{65} to bulky or highly polar residues, mutations of Tyr^{66} to any nonaromatic amino acid, and mutations of Gly^{67} to anything else are probably not tolerated.

5.9 NUCLEIC ACID CHANGES THAT DO NOT CHANGE THE PREDICTED AMINO ACID SEQUENCE

Green fluorescent protein expression levels can often be increased by redesigning the nucleic acid sequence in ways that should have no significant effect on the final protein sequence. For example, the codon usage in the jellyfish gene is not optimal for mammalian cells, so the gene has been resynthesized with mammalian-preferred codons (Crameri et al., 1996; Levy et al., 1996; Zolotukhin et al., 1996). Translation in eukaryotes can be optimized by inclusion of an optimal translation–initiation sequence (Kozak, 1989). This redesign sometimes involves inserting a new codon that begins with G immediately after the start (AUG) codon. This introduces an extra amino acid such as Ala or Val, which adds one to the numbering of all amino acids from 2 upward (Crameri et al., 1996). Fortunately, the N-terminus is tolerant of such additions. For ease of comparison of mutants, this chapter numbers residues according to their position in the original *gfp* gene. In plant cells, mRNA derived from the original *gfp* gene undergoes undesired splicing, which can be eliminated by codon changes (Haseloff and Amos, 1995); see also the chapters by Haseloff (Chapter 10) and Fire (Chapter 8).

5.10 TANDEM CONCATENATIONS OF TWO GFPS

The availability of GFP mutants of different colors, UV-excited blue emitters and blue-excited green emitters, enables fluorescence resonance energy transfer (FRET) from one to the other. Fluorescence resonance energy transfer is strongly dependent on the angular orientation and distance of the fluorophores from one another, falling off steeply as the distance exceeds the Förster distance R_0 at which FRET is 50% efficient (Lakowicz, 1983; Tsien et al., 1993). For the blue emitter P4-3, containing the point mutations Y66H and Y145F, donating energy to either S65T or S65C, R_0 is calculated to be about 40 Å (Heim and Tsien, 1996) assuming that the mutual orientation of the fluorophores is random or freely tumbling. Pairs of mutants that would give larger values of R_0 would be welcome, because GFP is a cylinder of about 12-Å radius and 42-Å length (Ormö et al., 1996), so much of R_0 is used up simply within the two GFPs. Other pairs have R_0 values exceeding 50 Å due to higher quantum yields of the donor and better overlap between its emission spectrum and the excitation spectrum of the acceptor (A. B. Cubitt and R. Y. Tsien, unpublished data). Another pair (Mitra et al., 1996) is BFP5, consisting of F64M/Y66H/V68I, donating to RSGFP4, F64M/S65G/Q69L, though here the quantum efficiencies and extinction coefficients have not been published so R_0 cannot be calculated. Concatenation of the genes encoding S65C or S65T and P4-3 yields a tandemly fused protein with a 25 residue linker containing a trypsin clearage site connecting the two GFP derived domains (Heim and Tsien, 1996). Likewise BFP5 and RSGFP4 have been fused with a 20 residue linker sensitive to Factor X_a (Mitra et al., 1996). In either case, before protease cleavage, UV excitation gives rise to some blue emission but also substantial green emission due to FRET from the blue to the green emitting domain. After protease cleavage to separate the two domains, FRET is abolished, the blue emission is increased (i.e., dequenched) and the green emission is nearly abolished. For the S65C::P4-3 construct, the ratio between blue and green emission intensities increased by a factor of 4.6 upon cleavage (Heim and Tsien, 1996), while the RSGFP4::BFP5 fusion showed about a 1.9-fold increase in emission ratio (Mitra et al., 1996). At least with the S65C::P4-3 construct, the large change in ratio between blue and green emissions was shown to result from separation of the two fluorophores rather than an effect on either one separately, because control experiments with the two separate proteins showed no spectral sensitivity to protease under matching conditions. Previous applications of FRET to monitor protease activity have always used artificially synthesized fluorophores conjugated to small synthetic peptides (Krafft and Wang, 1994; Knight, 1995); the big advantage of the protease substrates based on GFP is that they are directly encoded by gene sequences and can thus be expressed in situ inside cells or transgenic organisms and targeted to appropriate locations.

ACKNOWLEDGMENTS

Milt Cormier, former Professor at The University of Georgia, for giving Prasher the opportunity and encouragement to initiate the cloning efforts; Dennis Willows, Director of the University of Washington Friday Harbor Labs, for making it possible for numerous research groups to collect *Aequorea*; Roger Heim and Andrew Cubitt, for helpful discussion and permission to cite unpublished data.

REFERENCES

Brejc, K., Sixma, T. K. Kitts, P. A., Kain, S. R., Tsien, R. Y., Ormö, M., and Remington, S. J. (1997). Structural basis for dual excitation and photoisomerization of the *Aequorea victoria* green fluorescent protein. *Proc. Natl. Acad. Sci. USA* 94:2306–2311.

Chalfie, M., Tu, Y., Euskirchen, G., Ward, W. W., and Prasher, D. C. (1994). Green fluorescent protein as a marker for gene expression. *Science* 263:802–805.

Chattoraj, M., King, B. A., Bublitz, G. U., and Boxer, S. G. (1996). Ultra-fast excited state dynamics in green fluorescent protein: multiple states and proton transfer. *Proc. Natl. Acad. Sci. USA* 93:8362–8367.

Cody, C. W., Prasher, D. C., Westler, W. M., Prendergast, F. G., and Ward, W. W. (1993). Chemical structure of the hexapeptide chromophore of the Aequorea green-fluorescent protein. *Biochemistry* 32:1212–1218.

Cormack, B. P., Valdivia, R. H., and Falkow, S. (1996). FACS-optimized mutants of the green fluorescent protein (GFP). *Gene* 173: 33–38.

Coxon, A., and Bestor, T. H. (1995). Proteins that glow in green and blue. *Chem. Biol.* 2: 119–121.

Crameri, A., Whitehorn, E. A., Tate, E., and Stemmer, W. P. C. (1996). Improved green fluorescent protein by molecular evolution using DNA shuffling. *Nature Biotechnol.* 14: 315–319.

Cubitt, A.B., Heim, R., Adams, S. R., Boyd, A. E., Gross, L. A., and Tsien, R. Y. (1995). Understanding, using and improving green fluorescent protein. *Trends Biochem. Sci.* 20: 448–455.

Cutler, M.W. (1995). Characterization and Energy Transfer Mechanism of the Green Fluorescent Protein from *Aequorea victoria*. Rutgers University, New Jersey, Ph.D. Thesis.

Delagrave, S., Hawtin, R. E., Silva, C. M., Yang, M. M., and Youvan, D. C. (1995). Red-shifted excitation mutants of the green fluorescent protein. *BioTechnol.* 13:151–154.

Dickson, R. M., Cubitt, A. B., Tsien, R. Y., and Moerner, W. E. (1997). On/off blinking and switching behaviour of single green fluorescent protein molecules. *Nature* 388:355–358.

Dopf, J., and Horiagon, T. (1996). Deletion mapping of *Aequorea victoria* green fluorescent protein. *Gene* 173:39–44.

Ehrig, T., O'Kane, D. J., and Prendergast, F. G. (1995). Green fluorescent protein mutant with altered fluorescence excitation spectra. *FEBS Lett.* 367:163–166.

Harvey, E. N. (1957). A History of Luminescence from the Earliest Times Until 1900, American Philosophical Society, Philadelphia.

Haseloff, J., and Amos, B. (1995). GFP in plants. *Trends Genet.* 11:328–329.

Heim, R., Prasher, D. C., and Tsien, R. Y. (1994). Wavelength mutations and post-translational autooxidation of green fluorescent protein. *Proc. Natl. Acad. Sci. USA* 91: 12501–12504.

Heim, R., Cubitt, A. B., and Tsien, R. Y. (1995). Improved green fluorescence. *Nature (London)* 373:663–664.

Heim, R., and Tsien, R. Y. (1996). Engineering green fluorescent protein for improved brightness, longer wavelengths and fluorescence energy transfer. *Curr. Biol.* 6:178–182.

Inouye, S., and Tsuji, F. I. (1994). *Aequorea* green fluorescent protein. Expression of the gene and fluorescence characteristics of the recombinant protein. *FEBS Lett.* 341:277–280.

Kaether, C. and Gerdes, H. H. (1995). Visualization of protein transport along the secretory pathway using green fluorescent protein. *FEBS Lett.* 369:267–271

Kahana, J., and Silver, P. (1996). Use of the *A. Victoria* green fluorescent protein to study protein dynamics *in vivo. In Current Protocols in Molecular Biology*. Ausabel, F. M., Brent, R. Kingston, R. E., Moore, D. D., Seidman, J. G., Smith, J. A., and Struhl, K. Eds., Wiley, New York, pp. 9.7.22–9.7.28.

Knight, C. G. (1995). Fluorimetric assays of proteolytic enzymes. *Methods Enzymol.* 248:18–34.

Kozak, M. (1989). The scanning model for translation: an update. *J. Cell Biol.* 108:229–241.

Krafft, G. A., and Wang, G. T. (1994). Synthetic approaches to continuous assays of retroviral proteases. *Methods Enzymol.* 241:70–86.

Lakowicz, J. R. (1983). Principles of Fluorescence Spectroscopy, Plenum, New York.

Levy, J. P., Muldoon, R. R., Zolotukhin, S., and Link, C. J. J. (1996). Retroviral transfer and expression of the humanized, red-shifted green fluorescent protein into human tumor cells. *Nature Biotechnol.* 14:610–614.

Lim, C. R., Kimata, Y., Oka, M., Nomaguchi, K., and Kohno, K. (1995). Thermosensitivity of green fluorescent protein fluorescence utilized to reveal novel nuclear-like compartments in a mutant nucleoporin NSP1. *J. Biochem.* 118:13–17.

Marshall, J., Molloy, R., Moss, G. W. J., Howe, J. R., and Hughes, T. E. (1995). The jellyfish green fluorescent protein: a new tool for studying ion channel expression and function. *Neuron* 14:211–215.

McElroy, W. D. (1947). The energy source for bioluminescence in an isolated system. *Proc. Natl. Acad. Sci. USA* 33:342–345.

Merbs, S. L., and Nathans, J. (1992). Absorption spectra of the hybrid pigments responsible for anomalous color vision. *Science* 258:464–466.

Mitra, R. D., Silva, C. M., and Youvan, D.C. (1996). Fluorescence resonance energy transfer between blue-emitting and red-shifted excitation derivatives of the green fluorescent protein. *Gene* 173:13–17.

Muhlrad, D., Hunter, R., and Parker, R. (1992). A rapid method for localized mutagenesis of yeast genes. *Yeast* 8:79–82.

Neitz, M., Neitz, J., and Jacobs, G. H. (1991). Spectral tuning of pigment underlying red-green color vision. *Science* 252:971–974.

Ormö, M., Cubitt, A. B., Kallio, K., Gross, L. A., Tsien, R.Y., and Remington, S. J. (1996). Crystal structure of the *Aequorea victoria* green fluorescent protein. *Science* 273: 1392–1395.

Palm, G. J., Zdanov, A., Gaitanaris, G. A., Stauber, R., Pavlakis, G. N., and Wlodawer, A. (1997). The structural basis for spectral variations in green fluorescent protein. *Nature Struct. Biol.* 4:361–365.

Prasher, D. C., McCann, R. O., and Cormier, M. J. (1985). Cloning and expression of the cDNA coding for *Aequorin*, a bioluminescent calcium-binding protein. *Biochem. and Biophys. Res. Comm.* 126:1259–1268.

Prasher, D. C., Eckenrode, V. K., Ward, W. W., Prendergast, F. G., and Cormier, M. J. (1992). Primary structure of the *Aequorea victoria* green-fluorescent protein. *Gene* 111: 229–233.

Prasher, D. C. (1995). Using GFP to see the light. *Trends Genet.* 11:320–323.

Rizzuto, R., Brini, M., DeGiorgi, F., Rossi, R., Heim, R., Tsien, R.Y., and Pozzan, T. (1996). Double labelling of subcellular structures with organelle-targeted GFP mutants *in vivo*. *Curr. Biol.* 6:183–188.

Shimomura, O. (1979). Structure of the chromophore of *aequorea* green fluorescent Protein. *FEBS Lett.* 104:220–222.

Siemering, K. R., Golbik, R., Sever, R., and Haseloff, J. (1996). Mutations that suppress the thermosensitivity of green fluorescent protein. *Curr. Biol.* 6:1653–1663.

Sikorski, R. S., and Boeke, J. D. (1991). *In vitro* mutagenesis and plasmid shuffling: from cloned gene to mutant yeast. *Methods Enzymol.* 194:302–318.

Simon, S. M. (1996). Cellular probes on the move. *Nature Biotechnol.* 14:1221

Stearns, T. (1995). Green fluorescent protein—the green revolution. *Curr. Biol.* 5:262–264.

Tsien, R. Y., Bacskai, B. J., and Adams, S.R. (1993). FRET for studying intracellular signalling. *Trends Cell Biol.* 3:242–245.

Wachter, R. M., King, B. A., Heim, R., Kallio, K., Tsien, R. Y., Boxer, S. G., and Remington, S. J. (1997). Crystal structure and photodynamic behavior of the blue-emission variant Y66H/Y145F of green fluorescent protein. *Biochemistry* 36:9759-9765.

Ward, W. W., Prentice, H. J., Roth, A. F., Cody, C. W., and Reeves, S. C. (1982). Spectral perturbations of the *Aequorea* green-fluorescent protein. *Photochem. Photobiol.* 35: 803–808.

Ward, W. W., and Bokman, S. H. (1982). Reversible denaturation of *Aequorea* green-fluorescent protein: physical separation and characterization of the renatured protein. *Biochemistry* 21:4535–4540.

Yang, F., Moss, L. G., and Phillips, G. N. Jr. (1996a). The molecular structure of green fluorescent protein. *Nature Biotechnol.* 14:1246–1251.

Yang, T.-T., Kain, S.R., Kitts, P., Kondepudi, A., Yang, M.M., and Youvan, D. C. (1996b). Dual color microscopic imagery of cells expressing the green fluorescent protein and a red-shifted variant. *Gene* 173:19–23.

Yokoe, H., and Meyer, T. (1996). Spatial dynamics of GFP-tagged proteins investigated by local fluorescence enhancement. *Nature Biotechnol.* 14:1252–1256.

Youvan, D. C., Goldman, E. R., Delagrave, S., and Yang, M. M. (1995). Digital imaging spectroscopy for massively parallel screening of mutants. *Methods Enzymol.* 246:732–748.

Youvan, D. C., Michel-Beyerle, M. E. (1996). Structure and fluorescence mechanism of GFP. *Nature Biotechnol.* 14:1219–1220.

Zolotukhin, S., Potter, M., Hauswirth, W., Guy, J., and Muzyczka, N. (1996). A "humanized" green fluorescent protein cDNA adapted for high levels of expression in mammalian cells. *J. Virol.* 70:4646–4654.

PART THREE

Applications of Green Fluorescent Protein

6

The Uses of Green Fluorescent Protein in Prokaryotes

RAPHAEL H. VALDIVIA AND BRENDAN P. CORMACK
Department of Microbiology and Immunology, Stanford University School of Medicine, Stanford, CA

STANLEY FALKOW
Rocky Mountain Laboratories, National Institute of Allergy and Infectious Diseases, Hamilton, MT and Department of Microbiology and Immunology, Stanford University, School of Medicine, Stanford, C.A.

6.1 INTRODUCTION

The green fluorescent protein (GFP) of *Aequorea victoria* is a unique tool that permits the monitoring of gene expression and protein localization in living cells. Green fluorescent protein is stable, does not require cofactors for activity (Chalfie et al., 1994; Inouye and Tsuji, 1994a), and can be functionally expressed in different bacterial species. Because of GFPs unique properties, it can be used as a reporter of gene expression, dynamic processes during bacterial development, and the behavior of single bacteria in complex environments. Several properties of wild-type GFP, however, limit its applications in prokaryotes: (1) GFP tends to precipitate in the cytoplasm as insoluble, nonfluorescent inclusion bodies; (2) the posttranslational formation of a functional chromophore occurs approximately 2 h after synthesis of GFP (Heim et al., 1995); and (3) the magnitude of the fluorescence signal obtained is low com-

Green Fluorescent Protein: Properties, Applications, and Protocols, Edited by Martin Chalfie and Steven Kain
ISBN 0-471-17839-X

pared to that of other reporter proteins [e.g., β-galactosidase (LacZ), chloramphenicol acetyl transferase (CAT), luciferase (Lux)]. Despite these concerns, GFP has already had a significant impact in the fields of bacterial development, pathogenesis, and ecology. Moreover, the advent of new generations of GFP variants with increased sensitivity and solubility should widen the spectrum of uses to which GFP is applied.

Green fluorescent protein cDNA has been expressed in a variety of both gram-positive and gram-negative bacteria. The bacterial DNA GC content does not seem to pose a barrier to expression, since *gfp* is expressed in GC rich organisms such as *Mycobacteria* sp. (Dhandayuthapani et al., 1995; Kremer et al., 1995; Valdivia et al., 1996) and in AT rich bacteria such as *Helicobacter pylori* (Covacci, personal communication). Codon optimization has not been required to achieve detectable levels of the fluorescent protein. However, in several eukaryotic systems (see Chapters 10 and 12 and Protocol I.B.3) *gfp* codon optimization has resulted in increased fluorescence, and it would not be surprising to find that species-tailored optimizations will lead to enhanced fluorescence in prokaryotic systems as well. Table 6.1 shows a summary of the different bacterial species in which GFP has been successfully expressed.

TABLE 6.1 Examples of bacterial species in which GFP has been successfully expressed

Bacterial Species	References
Anabaena	Buikema and Haselkorn (1996)
Bacillus subtilis	Arigoni et al., 1995; Barak, et al., 1996: Lewis and Errington, 1996; Resnekov et al., 1996; Sharpe and Errington, 1996; Webb et al., 1995
Bartonella henselae	Lee and Falkow (unpublished results)
Caulobacter crescentus	Skerker and Shapiro (unpublished results)
Escherichia coli	Chalfie et al., 1994; Cormack et al., 1996; Delagrave et al., 1995; Leff and Leff, 1996; Heim et al., 1994; Ma et al., 1996
Helicobacter pylori	Covacci, 1996
Legionella pneumophila	Martin and Tompkins (unpublished results)
Mycobacterium bovis BCG	Dhandayuthapani et al., 1995; Kremer et al., 1995; Via et al., 1996
Mycobacterium marinum	Valdivia et al., 1996
Mycobacterium smegmatis	Dhandayuthapani et al., 1995; Kremer et al., 1995; Valdivia et al., 1996
Myxococcus xanthus	Licking and Kaiser (unpublished results)
Pseudomonas putida	Burlage et al., 1996; Christensen et al., 1996
Pseudomonas fluorescens	Tombolini et al., 1996
Rhizobium meloliti	Gage et al., 1996
Salmonella typhimurium	Kain et al., 1995; Valdivia and Falkow, 1996; Valdivia et al., 1996
Yersinia pseudotuberculosis	Valdivia et al., 1996
Yersinia pestis	Hinnebusch et al., 1996

While GFP does not require any additional cofactors for fluorescence, the amount of fluorescence is not always proportional to the total pool of GFP. The assembly of the GFP chromophore requires molecular oxygen; therefore GFP fluorescence often decreases under anaerobic or strong reducing environments (Heim et al., 1994; Inouye and Tsuji, 1994b). Temperature can also affect GFP fluorescence because the tendency of GFP to precipitate into nonfluorescent inclusion bodies increases with temperature (Ogawa et al., 1995; Cormack et al., 1996). The fluorescence signal from GFP can be enhanced by growing bacteria aerobically at low temperatures (25–30°C) (Heim et al., 1994; Webb et al., 1995). This requirement is less important when using some of the enhanced GFP mutants discussed in this chapter (also references in Chapter 5).

Unlike conventional bacterial gene reporters such as *lacZ*, *lux*, or *cat*, GFP is not an enzyme and thus there is no signal amplification derived from multiple substrate cleavage by one molecule of reporter protein. The fluorescence signal in a particular cell, therefore, depends largely on two broad parameters: the rate of synthesis of functional protein and, because GFP is very stable, the rate of dilution as the cell divides. Most reports of GFP expression in bacteria have been performed with *gfp* on multicopy copy plasmids (Chalfie et al., 1994; Dhandayuthapani et al., 1995; Buikema and Haselkorn, 1996; Gage et al., 1996; Hinnebusch et al., 1996; Ma et al., 1996; Valdivia et al., 1996). But single copy fusions with strong promoters in *Pseudomonas* sp. have also yielded fluorescent bacteria (Burlage et al., 1996; Christensen et al., 1996; Tombolini et al., 1996). In *B. subtilis*, single copy gene constructs driving GFP–protein fusions have been routinely imaged, even though the promoters are only of moderate strength (Arigoni et al., 1995; Webb et al., 1995; Lewis and Errington, 1996; Resnekov et al., 1996; Sharpe and Errington, 1996). The rate of dilution is probably lower here than in other systems because *B. subtilis* ceases to divide after sporulation and translational activity from the GFP protein fusion is essentially cumulative. In contrast, the pool of posttranslationally modified GFP is rapidly diluted in a fast-growing organism. For example, Tombolini et al. (1996) followed the fluorescence of GFP-tagged *P. fluorescens* by flow cytometry during different stages of growth. Individual bacteria were virtually nonfluorescent during the exponential phase of growth and did not achieve maximal fluorescence until stationary phase.

Several recently isolated GFP mutants provide greater sensitivity, faster kinetics of formation, and greater protein solubility (Delagrave et al., 1995; Heim et al., 1995; Crameri et al., 1996; Cormack et al., 1996). These variants improve sensitivity in most if not all systems, and some should be sensitive enough as to allow for the imaging of gene fusions driven from moderate strength single copy constructs, especially with advanced fluorescence imaging systems (CCD camera, laser scanning confocal microscopes; see discussion in Protocol A-III). Indeed, in *Salmonella typhimurium* we have routinely imaged single copy *gfp* fusions with the aid of enhanced GFPs (unpublished observations). A review of these mutants is covered in detail elsewhere (see Chapter 5). The use of these GFP variants have been crucial to some of the applications described below. Where appropriate, we will point out which GFP variant was used in each particular example described.

6.2 BACTERIAL DEVELOPMENT AND CELL BIOLOGY

Green fluorescent protein has been successfully used as a tool to study cell differentiation in two models of bacterial development, the cyanobacterium *Anabaena* and *B. subtilis*, where cell specific gene expression has been observed with the use of *gfp* transcriptional and translational fusions. In addition, GFP has been used to visualize bacterial subcellular structures such as the cytoskeletal division apparatus during *E. coli* division and to observe localization of proteins important in *B. subtilis* spore formation. The following examples illustrate how GFP is uniquely suited for the study of bacterial development.

6.2.A Spore Formation in *B. subtilis*

Under certain environmental conditions, *B. subtilis* undergoes asymmetrical cell division leading to the formation of a forespore and a mother cell (reviewed in Errington, 1993; Stragier and Losick, 1996). A cascade of sigma factors controls the transcription of forespore and mother cell specific genes that lead to different developmental outcomes (Losick and Stragier, 1992). Understanding cell-specific gene expression during sporulation has been key to the dissection of this developmental program. Cell-specific gene expression has been analyzed either by immunofluorescence or with the use of fluorogenic enzyme substrates (Lewis et al., 1994; Harry et al., 1995; Pogliano et al., 1995). GFP is a powerful addition to these methods. Transcriptional and translational GFP fusions can be used to study cell-specific gene expression and the localization of proteins important in bacterial development, obviating the need to raise antibodies to specific cellular components. Most importantly, GFP allows analyses to be performed in living cells.

The potential of GFP for the study of *B. subtilis* development has been demonstrated in two separate investigations. Webb et al. (1995) visualized cell-specific gene expression in the forespore and the mother cell at different stages of sporulation: Fusions of GFP to either *sspE-2G* or *csfB* resulted in forespore specific fluorescence. By contrast, translational GerE-GFP or transcriptional *cotE-gfp* fusions showed mother-cell specific fluorescence (see color Fig. 6.1). A GFP fusion to the spore coat protein CotE, localized to the region surrounding the forespore and appeared as uneven green halos on mature spores.

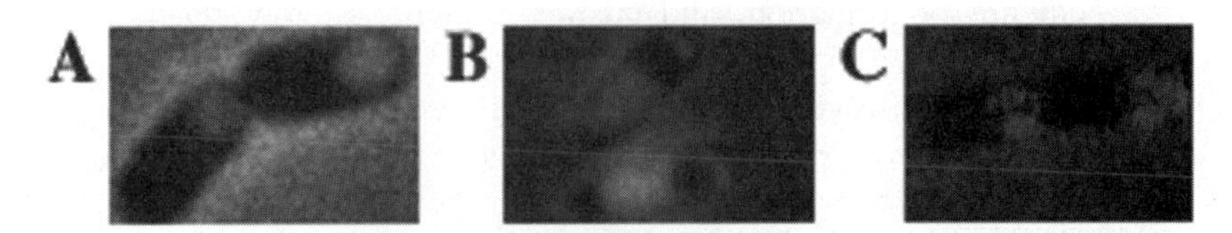

FIG. 6.1 Fluorescence images of sporulating *B. subtilis* cells expressing transcriptional and translational GFP fusions. Two sporangia are shown per panel (*a*) Forespore-specific expression of a σF-dependent SspE2G-GFP fusion. (*b*) Mother cell-specific expression of a σE-dependent *cotE-gfp* fusion. (*c*) Localization of a SpoIVFB-GFP translational fusion (note localized fluorescence seen as a shell at one end of each sporangium). Courtesy of O. Resnekov and C. Webb, Harvard University. Figure also appears in color section.

Lewis and Errington (1996), using the S65T variant of GFP (Heim et al., 1995), showed that a DacF-GFP fusion localizes to the forespore. By contrast, a SpoIVA-GFP fusion localized to the mother cell only if a wild-type copy of *spoIVA* was coexpressed. Full-length SpoIVA fused to GFP did not retain its native function, and the strain was rendered sporulation deficient. This result points out that caution must be exercised when interpreting results from GFP fusions. Optimally, to avoid artifactual results, one would like to verify that any translational GFP fusion can functionally replace the native protein. Another note of caution is indicated by the paradoxical difference in fluorescence intensity between CsfB-GFP and GerE-GFP translation fusions. Analogous LacZ fusions to these two proteins show much higher expression of GerE than CsfB, while the reverse seems to be true for the GFP fusions (Webb et al., 1995). Not withstanding these notes of caution, these experiments show that GFP fusions (transcriptional and translational) behave largely as predicted by other methods.

In other work, GFP has played a role in understanding the molecular functions of these proteins. For example, Arigoni et al. (1995) and Barak et al. (1996) independently showed that SpoIIE, a key protein in triggering the developmental fate of progeny after cell division, localizes to a sharp zone close to the pole of sporangia prior to septum formation. Resnekov et al. (1996) using an enhanced GFP mutant (F64S) showed that SpoIVFB, which is proposed to be a proteolytic activator of a mother-cell-specific sigma factor (σ^k), localizes initially to the sporulation septum and subsequently to the forespore. The use of SpoIVFB-GFP fusions was central to this work since raising and purifying adequate anti-SpoIVB antibodies for immunofluorescece proved to be exceedingly difficult. The localization of SpoIVFB to the mother cell membrane surrounding the sporangium lends support to a proposed model in which a sporangia protein (SpoIVB) couples development in the forespore to mother cell transcription by activating the proteolytic conversion of pro-σ^k to σ^k, thereby activating σ^k-directed gene expression.

Sharpe and Herrington (1996) used GFP (S65T) to analyze DNA transfer. One of the major events during *B. subtilis* sporulation is the translocation of one of the daughter chromosomes into the forespore. Bacteria deficient in SpoIIIE are unable to translocate the entire chromosome and are arrested after 30% transfer (Wu et al., 1995). The same region of the chromosome is always transferred, suggesting an interaction between a centrosome-like segment of the *B. subtilis* chromosome and one or more bacterial targeting proteins (Wu and Errington, 1994). Wu and Errington used a forespore-specific *dacF::gfp* fusion to show that Soj and Spo0J, homologs of proteins needed for efficient plasmid partitioning in other bacterial systems, are important in the specificity of chromosomal translocation. The *dacF* mutant is located in a region of the chromosome that is not translocated to the forespore in an *spoIIIE* mutant. Forespore fluorescence from *dacF-gfp* fusions could not be detected in *spoIIIE* mutants but could be detected in double mutants of *spoIIIE* and a deletion of *soj-spo0J*.

6.2.B Heterocyst Formation in *Anabaena*

The cyanobacterium *Anabaena* grows as long filaments of photosynthetic cells that, under conditions of fixed-nitrogen starvation, undergo genetic changes leading to the development of specialized nitrogen-fixing cells known as heterocysts (reviewed in Haselkorn, 1992). W. J. Buikema and R. Haselkorn (personal communication) showed that fusions of *gfp* and the developmentally regulated gene *hetR* were expressed only in the cells destined to become heterocysts (see color Fig. 6.2). GFP fluorescence was observed only during early stages of heterocyst formation. This downregulation of fluorescence is likely not the result of a change in *hetR* transcription, but rather the result of the anaerobic environment present within the nitrogen-fixing heterocyst, and the resulting failure to form a functional GFP chromophore. Indeed, in a strain deficient for the deposition of glycolipids important for maintaining the anaerobic environment in heterocysts (HglK) (Black et al., 1995), the *hetR::gfp* fusion was fluorescent throughout heterocyst development.

6.2.C Cell Division in *E. coli*

During bacterial division, a complex network of proteins localize to the midpoint between daughter chromosomes in the diving cell. Ma et al. (1996) used GFP fusions to localize two of these proteins, FtsZ and FtsA, in living bacteria. To visualize the formation of the division ring in actively dividing cells, these authors utilized a fast chromophore assembly GFP mutant (S65A/V68L/S782A) (Cormack et al., 1996). Both FtsA-GFP an FtsZ-GFP fusions localize to the

FIG. 6.2 Heterocyst-specific gene expression in the cyanobacterium *Anabaena*. Fluorescence from nitrogen-starved *Anabena* cells bearing a plasmid with a *hetR-gfp* fusion. Green fluorescence is observed preferentially in heterocyst where *hetR* is exclusively expressed. Courtesy of W. Buikema, University of Chicago. Figure also appears in color section.

division site and support the concept of a multiprotein septator complex. A polymeric form of FtsZ-GFP could be seen as a ring structure in actively dividing cells (Fig. 6.3) and as localized aggregates and spiral tubules when the fusion protein was overexpressed. Deletion analysis of the FtsZ-GFP fusions defined the NH_2 terminus of FtsZ as necessary for polymerization. The behavior of the FtsZ-GFP fusion was consistent with the noted similarities between FtsZ and tubulin (Erickson, 1995). For example, FtzZ-GFP polymerizes to form long tubules in *E. coli* cells, the C-terminus domain of FtsZ is not essential for this polymerization, and furthermore, C-terminal truncations produce tubules that are unusually large and stable, suggesting a lack of dynamic instability as seen in C-terminal truncations of tubulin. FtsA-GFP appeared to localize to the membrane and to ring structures in the division plane at both early and late stages of septum formation. FtsA-GFP also colocalizes with FtsZ into tubular structures when FtsZ is overproduced, suggesting that FtsA directly interacts with FtsZ. The use of these fusions will allow for the simple identification of mutants that alter the pattern of FtsZ and FtsA localization. More importantly, prokaryotic cell division components can now be followed dynamically during cell division.

6.3 BACTERIA IN COMPLEX ENVIRONMENTS

Green fluorescent protein expression is uniquely suited to track live bacteria within complex environments and can be conveniently assayed within fractionated samples. Bacterial localization, association, and multiplication, as monitored by fluorescence, can be followed temporally and spatially. These features have tremendous implications for the study of bacterial behavior in natural habitats such as soil and biofilms and in the study of host colonization by bacterial pathogens and symbionts. Thus, GFP should facilitate not only the localization of individual bacteria in complex microbial environments, but also the analysis of gene regulation in those environments.

6.3.A Bacterial–Host Interactions

6.3.A.1 Bacteria in an Animal Host

Green fluorescent protein can be used to study the interactions between pathogenic bacteria and their mammalian hosts (Dhandayuthapani et al., 1995;

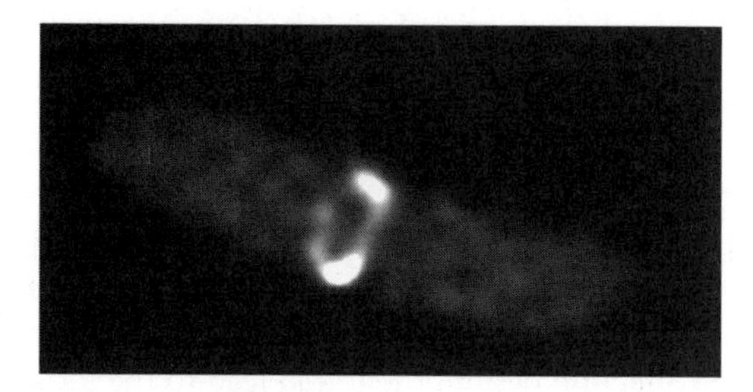

FIG. 6.3 Localization of FtsZ-GFP during septum formation in live *E. coli*.

Kremer et al., 1995; Valdivia et al., 1996). GFP has been expressed in *S. typhimurium*, *Y. pseudotuberculosis*, and *Mycobacteria sp.* with no adverse effect on the ability of these pathogenic organisms to interact with their hosts or cause disease (Valdivia et al., 1996). Bacteria–host interactions can be followed microscopically by epifluorescence, and can be quantitated with a spectrofluorimeter or a flow cytometer. This quantitation is very sensitive; for *Y. pseudotuberculosis*, the bacterial load present within single infected mammalian cells can be determined by flow cytometric analysis (Valdivia et al., 1996).

GFP simplifies the analysis of the cell biology of bacterial infections since endogenous labeling of the microorganism guarantees a constant level of fluorescence signal and obviates the need to raise antibodies against the pathogen for immunolabeling. Furthermore, since GFP is intracellular, there is a reduced potential of interfering with surface contact between the bacteria and host cells. The GFP–labeled organisms have been used to visualize host cell actin rearrangements incuded by *S. typhimurium* invasion (Kain et al., 1995), the interaction of *Y. pseudotuberculosis* with cytoskeletal components in murine macrophages (Hromockyj, Amieva, and Falkow, unpublished results), and to purify *M. bovis* BCG containing vesicles after homogenization of infected macrophages (Dhandayuthapani et al., 1995).

GFP tagged bacteria can also be visualized in the tissues of experimentally infected animals. For example, *M. marinum* expressing *gfp* has been imaged in cryosections of chronically infected frog spleens up to 5-weeks postinfection (Valdivia et al., 1996) and from lungs sections of *M. bovis* BCG infected mice (Kremer et al., 1995). In addition, Hinnebusch et al. (1996) used GFP tagged plague bacillus, *Y. pestis*, to visualize the colonization and blockage of the flea midgut. The GFP tagged *Y pestis* bearing a deletion in the hemin storage locus *(hms)* were unable to block the foregut of a colonized flea. This blockage is important in the transmission of plague since it profoundly affects the feeding behavior of the insect. A flea colonized with Hms+ *Y. pestis* is unable to feed and thus aggressively attempts to take a blood meal. Eventually, a successful feeding occurs in which the bacterial mass that blocked the foregut is regurgitated into the host's bloodstream.

In addition to histology, flow cytometry can also identify specific classes of infected cells within an organ. For example, we have sorted infected cells from the spleens of mice infected with GFP labeled *S. typhimurium* (R. H. Valdivia, D., Monack, and S. Falkow, unpublished results). This technique is particularly helpful in the study of bacterial pathogenesis, because it can potentially identify subsets of cells specifically targeted by intracellular pathogens during acute and chronic infections. Furthermore, *gfp* gene fusions allow one to measure bacterial gene expression in infected animal tissues.

6.3.A.2 Bacteria in a Plant Host

GFP labeled bacteria can also be used to analyze bacterial–plant interactions. The clearest illustration of this is the interaction between *Rhizobium* sp. and their plant host. The symbiotic relationship between a plant host and *Rhizobium* begins with the infection of a root hair and lead to the formation of a plant nodule that is colonized by nitrogen-fixing *Rhizobium* (reviewed in Fisher and Long, 1992).

Gage et al. (1996) recently examined *R. meloliti* growth and behavior during the early stages of nodule formation using bacteria expressing *gfp* (S65T) (see color Fig. 6.4). From time lapse observations of infected root hairs, the investigators determined that bacterial growth occurs only from bacteria in the tip region of the infection thread. Interestingly, bacterial fluorescence was found at all stages of colonization, including the nodules. This result is surprising since the nodule environment is thought to be oxygen-free in order to permit efficient nitrogen fixation. This finding suggests that the S65T GFP mutants might be less dependent than wild-type GFP on free-oxygen for chromophore assembly.

Tombolini et al. (1996) also examined the adherence of GFP tagged *P. fluorescens* to the roots of Japanese lotus plants. In this particular work, *P. fluorescens* was chromosomally tagged with *gfp* (I167T) fused to a strong constitutive promoter that was present in a Tn5 delivery vector.

6.3.B Bacterial Ecology and Behavior

GFP labeled bacteria have been used in the fields of bioremediation (the use of biological agents to remove toxic contaminants from soil and water, for example) and bacterial ecology. One obvious application is the tracking of bacteria in soil,

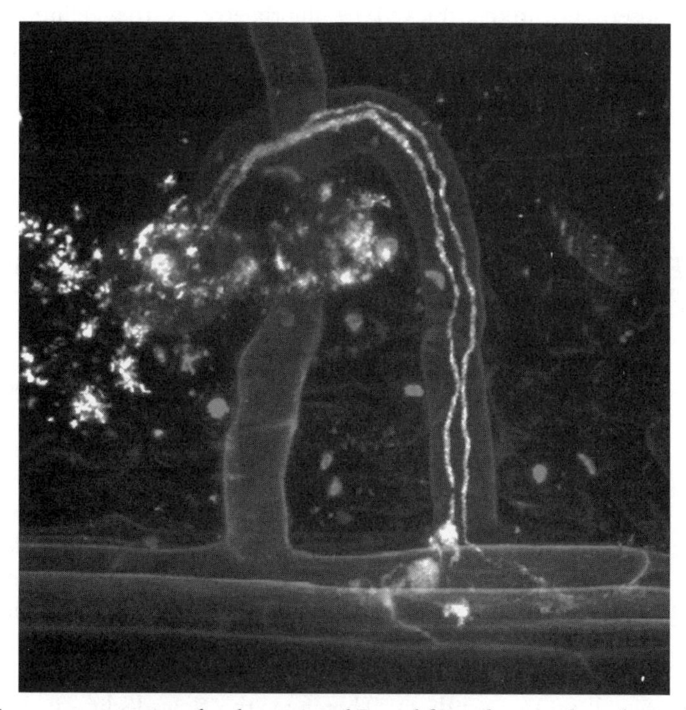

FIG. 6.4 Laser scanning confocal images of *R. meloliti* infection threads in plant root hairs. The *R. meloliti* bearing a plasmid with a *trp-gfp* fusion was used to infect alfalfa plants. Infection threads can be seen within individual root hair as they extend toward the main root body (stained red with propidium iodide). Figure also appears in color section.

biofilms, and complex microbial communities. Another is the monitoring of gene transfer in bacterial populations.

Christensen et al. (1996) monitored conjugation-mediated DNA transfer in situ by following the expression of *gfp* in recipient cells of *P. putida*. Donor cells had a (toluene degradation) TOL plasmid containing *gfp* under the control of the *Pϕ10* promoter of bacteriophage T7. Recipient cells constitutively expressed the T7 RNA polymerase. Transfer of the TOL plasmid to recipient cells was monitored by following the T7-driven synthesis of GFP. In addition, donor cells expressed luciferase, facilitating identification of donor and recipient cells. By monitoring the DNA flux within a mixed community over time, the authors provided new insights into the dynamics of horizontal gene transfer. Specifically, they found that conjugative transfer of DNA occurs very rapidly upon initial bacterial contact and is limited under poor growth conditions. While these conclusions took into consideration the lag in posttranslational chromophore oxidation, the analysis might be further refined by the use of fast-chromophore assembly GFP mutants, such as those described previously (Heim et al., 1994; Cormack et al., 1996).

GFP also makes a suitable marker to follow genetically engineered microorganisms as they move through porous materials or in aquatic environments (Leff and Leff, 1996). Burlage et al. (1996) showed that either *E. coli* expressing *gfp* from a plasmid or *P. putida* expressing *gfp* from a Tn5*gfp* chromosomal insertion can be tracked by fluorimetry as they elute from sand columns. Tombolini et al. (1996) showed that GFP tagged *P. fluorescens* can be visualized in soil samples. Bacterial fluorescence was easily detected, even after prolonged carbon starvation conditions, suggesting that GFP tags will be useful in monitoring bacteria growing under energy limiting conditions (i.e., in soil and water samples).

6.4 GREEN FLUORESCENT PROTEIN AS A GENETIC TOOL

GFP synthesis is easily assayed as green fluorescence. This fluorescence can be visualized directly on culture plates upon illumination with either blue- or long-wave ultraviolet (UV) light (Chalfie et al., 1994). For some applications, a qualitative comparison of fluorescence intensity between two bacterial colonies bearing different *gfp* fusion is sufficient to determine gross differences in levels of gene expression. However, spectrofluorimetry provides a more accurate, quantitative measurement of GFP fluorescence. Spectrofluorimetric measurements from *gfp*-expressing bacteria are simple, and, because it does not require cell lysis or the addition of exogenous substrates, can be monitored in the same sample over time. Kremer et al. (1995) used the spectrofluorimetric measurement of GFP fluorescence to assess bacterial sensitivity to antibiotics. While GFP is not a vital marker, the levels of GFP synthesis roughly correlate with the levels of overall protein synthesis. Therefore, the slow decline in bacterial fluorescence during drug exposure is an indirect measure of the antibiotic's adverse effects on bacterial metabolism. Fluorimetry has also been used to compare the relative strength of different mycobacterial promoters in M. *smegmatis* (Dhandayuthapani et al., 1995). Dhandayuthapani et al examined expression of the *ahp*C gene, which encodes alkyl hydroperoxide reductase whose levels have been linked to resistance to the antimycobacterial drug isoniazid. The investigators compared the levels of fluorescence expressed by *ahpC-gfp* fusions to demonstrate that the *ahpC* promoter region from *Mycobacterium tuberculosis* had substantially lower transcriptional activity than its *Mycobacterium leprae* counterpart, potentially explaining why M. *tuberculosis*, unlike other mycobacteria, is exquisitely sensitive to isoniazid.

6.4.A Flow Cytometry and Bacterial Genetics

The use of fluorescence as a reporter of gene expression permits the use of fluorescence-based technologies that not only quantitify fluorescence but also physically separate microorganisms on the basis of their relative fluorescence intensities (reviewed in Parks et al., 1989; Shapiro, 1995). A fluorescence-activated flow cytometer reads the fluorescence intensity of every particle that passes through the laser sensing area. This useful feature allows one to examine a large population of cells and determine the levels of gene expression for each bacterium in the sample. *Escherichia coli, S. typhimurium, Y. pseudotuberculosis, P. fluorescens*, and *Mycobacteria* sp. have been successfully analyzed by flow cytometry (Dhandayuthapani et al., 1995; Kremer et al., 1995; Cormack et al., 1996; Tombolini et al., 1996; Valdivia et al., 1996). Unlike spectrofluorimetry, which gives an average fluorescence for a sample, flow cytometry provides a multiparameter record of all cells sampled (Parks et al., 1989). This information can be displayed as histograms or contour plots that provide the frequency of bacteria with particular light scatter and fluorescence characteristics (see Protocol III.B.4.c). For example, Dhandayuthapani et al. (1995) have used flow cytometry to characterize four different mycobacterial promoters fused to *gfp*. The per bacterium fluorescence intensity mimicked that found by spectrofluorimetry. Furthermore, Dhandayuthapani et al. (1995) could separate a mixed population of mycobacteria bearing a transcriptionally weak (*mtrA::gfp*) and strong (*hsp60::gfp*) gene fusions using a fluorescence activated cell sorter (FACS). In another example, demonstrating the power of this approach, Cormack et al. (1996) used FACS to isolate highly fluorescent GFP mutants from a library of over 4×10^6 bacteria carrying different GFP chromophore mutations (See chapter 5).

These genetic approaches to isolating productive *gfp* fusions have also been used to identify genes induced under complex or poorly defined conditions. In conventional bacterial genetics inducible genes are often isolated by screening the expression of fusions with a measurable reporter gene. For example, bacteria bearing random *lacZ* gene fusions can be scored on nutrient agar plates, in the presence of an inducing stimulus, for the synthesis of β-galactosidase. Positive fusions are then assayed for the synthesis of β-galactosidase in the absence of the inducer and thus inducible (and repressible) gene fusions can be identified. However, if the induction conditions are hard to replicate in solid media or are detrimental to bacterial growth, such genes are often difficult to isolate. Some of these problems can be overcome by using FACS to isolate bacteria bearing *gfp* gene fusions. The transient expression of *gfp* is a phenotype that is easily scored by the cell sorter. The physical separation of individual bacteria on the basis of fluorescence is analogous to the manual screening of colonies on agar plates, but the processivity of the FACS machine (2–3 thousand bacteria per second; Parks et al., 1989), makes this screening process similar in power to genetic selection. Furthermore, FACS can discriminate among different fluorescence intensities. Theoretically, individual bacteria bearing inducible gene fusions with any absolute fluorescence level can be specifically isolated (Parks et al., 1989). We will use two examples to illustrate how these flow cytometric based gene selection and enrichment strategies, termed differential fluorescence induction (DFI) (Valdivia and Falkow, 1996), have been successfully applied to the isolation of inducible genes.

6.4.A.1 Isolation of Acid-inducible Genes in *S. typhimurium*

The response of bacteria to acidic conditions has been difficult to study because bacterial growth is hindered at low pH. We developed a DFI enrichment cycle to identify bacterial genes that are induced by transient exposure to highly acidic conditions (pH 4.5) (Valdivia and Falkow, 1996). A library of *S. typhimurium* bearing random DNA fragments fused to *gfp* (S65G/S72A) (Cormack et al., 1996) in a plasmid vector was exposed to media at pH 4.5 for 2 h and all fluorescent bacteria in the pool were sorted. The sorted sample was expanded and exposed to media at neutral pH. Since acid-inducible fusions will not be expressed at neutral pH, bacteria bearing non-fluorescent fusions at pH 7 were collected. This nonfluorescent bacterial population was exposed to acidic pH again, and all fluorescent organisms sorted. The final collected pool was highly enriched ($\sim$30–50%) for bacteria bearing gene fusions whose activity was upregulated under acidic conditions. The DNA sequence analysis of eight of these acid-inducible gene fusions revealed high homology to promoter regions from genes involved in stress response and multidrug resistance; and genes with previously reported pH regulated activity (Valdivia and Falkow, 1996).

FACS can also be used to identify loci that regulate the activity of a gene of interest. For example, we have used an acid-inducible *aas::gfp* fusion (Valdivia and Falkow, 1996) to isolate miniTn5 insertions that abolish this fusion's induction at acidic pH (R. H. Valdivia, M. Rathman, and S. Falkow, unpublished data). After a generalized insertional mutagenesis, over 10^7 *S. typhimurium* bearing an *aas::gfp* fusion were exposed to pH 4.5 and sampled by FACS. Nonfluorescent organisms were present at a frequency of 0.01%. These bacteria were collected and the miniTn5 insertions present within this sorted population were transduced into a nonmutagenized *S. typhimurium* background. The inability to induce *aas::gfp* under acidic conditions cotransduced with the insertion elements. Several insertions mapped to the *ompR/envZ* locus. OmR/EnvZ is a two component system known to regulate the expression of several genes in response to changes in osmolarity (reviewed in Pratt and Silhavy, 1995). The finding that this locus is also involved in the acid response is not surprising since low pH regulates the expression of at least one porin gene, under the control of OmpR (Foster et al., 1994).

6.4.A.2 Isolation of Macrophage-Inducible Genes in *S. typhimurium*

S. typhimurium is a facultative intracellular pathogen that modifies the macrophage's phagocytic vacuole to permit its own survival (reviewed in Garcia and Finlay, 1995) Several bacterial genes are preferentially expressed in the intracellular environment (Alpuche-Aranda et al., 1992; Garcia et al., 1992; Fierer et al., 1993), but the mechanisms that allow *S. typhimurium* to survive in this adverse environment are poorly understood. To explore the genetic basis of intracellular survival, we have applied DFI selections to identify genes from pathogenic bacteria that are induced within the host cell (R. H. Valdivia and S. Falkow, unpublished results). We infected macrophages with *S. typhimurium* bearing random *gfp* gene fusions (see above) and sampled them by FACS. Macrophages containing GFP-fluorescent bacteria were collected, lysed, and the released bacteria grown in the absence of host cells. Bacteria that did not fluoresce under these conditions were collected by FACS and used for a second round of macrophage infection. Bacteria recovered from fluorescent macrophages contained *gfp* fusions that were upregulated in the host cell's intracellular environment. Bacteria bearing individual gene fusions were tested by fluorescence microscopy to demonstrate upregulation of *gfp* expression in the intracellular environment. In a nonsaturating screen, we identified 18 macrophage-inducible loci, including two previously identified acid-inducible genes (Valdivia et al., 1996) (see color Fig. 6.5). Some of these genes are known to be upregulated intracellularly or to be important for in vivo survival. The large majority of the identified genes, however, have not been previously described. The role of these genes in *Salmonella* pathogenesis is currently under investigation.

GFP is particularly well suited for the in vivo imaging of bacterial gene expression within host cells. Via et al. (1996) showed both microscopically

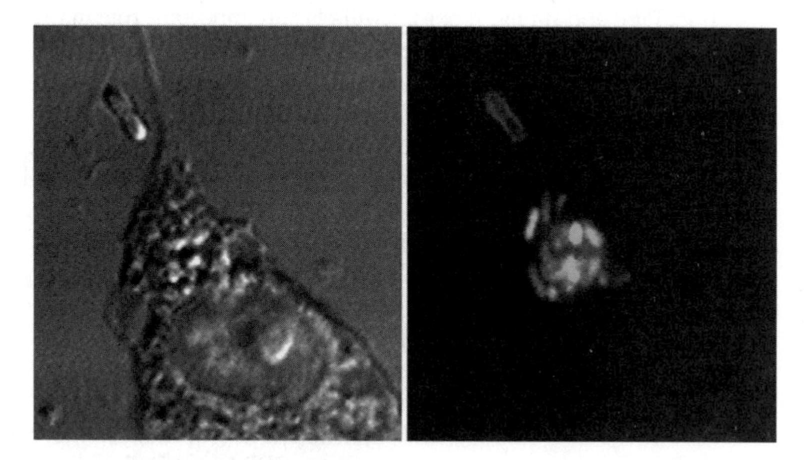

FIG. 6.5 Visualization of *S. typhimurium* intracellular-specific gene expression by fluorescence microscopy. *S. typhimurium* bearing a *pagA::gfp* fusion shows gene induction inside an infected mammalian cell but not in the extracellular medium. The corresponding DIC images show the relative topology of bacteria with respect to the infected cell. Figure also appears in color section.

and by flow cytometry that an *mtrA::gfp* is upregulated in M. *bovis* BCG during its residence in macrophages. A direct quantitation of *gfp* expression by intracellular organisms was determined by the flow cytometric analysis of bacteria released after mechanical disruption of infected cells. The MtrA protein is homologous to response regulators of many bacterial two-component systems. The finding that the transcription of *mtrA* is upregulated intracellularly makes this gene a putative candidate as a mycobacterial virulence factor. We used similar flow cytometric quantitation to examine the kinetics of *S. typhimurium* gene induction with phagocytic cells, and showed that there are at least two classes of macrophage inducible genes: the first class induced within 1 h of cell entry, and the second induced only after 4 h (Valdivia and Falkow 1996), suggesting that the bacterial response to macrophage internalization is transcriptionally complex. This result emphasizes the advances that GFP has brought to the analysis of gene regulation, and hints at the promise it holds.

6.5 PERSPECTIVES

Microbes exist in nature in complex interactive communities. Appreciation of this fact has increased the interest in following particular bacterial cells and their gene expression in mixed populations. GFP is probably the single-best experimental tool available for this purpose, and it enters the repertoire of research tools just as microbiologists begin to shift from the study of microorganisms grown in the laboratory to their study in the "wild." With the use of GFP and other available reporting molecules, like LacZ and Lux, we will undoubtedly see an increased focus on microbial interactions in the "real world." In the field of host–parasite relationships, particularly among the medically important microbes, the study of the initial interaction of the microbe with the innate elements of the immune system is now seen as key to understanding the pathogenesis of infection. GFP and similar reporters will play an important role in revealing the details of the interaction between host and microbe that occurs at the earliest times after exposure to infectious agents. Such studies will undoubtedly aid in the design of novel antiinfective agents and in the development of vaccines.

GFP also adds to the family of new experimental approaches (Mahan et al., 1993) that identify genes expressed only in the natural microbial habitat. Recently, a negative selection method using signature-tagged transposition (Hensel et al., 1995) has proved successful for identifying genes essential in one environment (e.g., pathogen genes essential inside the animal host), but dispensible under another set of growth conditions. Detection of GFP expression provides another powerful approach to the identification of genetic sequences expressed under unique and complex environmental conditions. Since GFP is extremely stable, it has the advantage of permitting even transient gene expression to be detected by flow cytometry. In our hands, the combined use of signature-tagged transposition and the GFP-based DFI provides complementary information about genetic sequences essential for *S. typhimurium* growth within the murine host. We believe that this and similar GFP-based methods that exploit the features of contemporary cell sorting technology have the potential to become highly useful tools for the identification of novel genetic sequences important for growth under conditions too complex to be duplicated easily in the laboratory.

One can expect the complete nucleotide sequence of the most medically and commercially important microorganisms to be widely available in the next decade. These sequences will be extensively analyzed by computer algorithms to identify homologous sequences with known biochemical motifs. The fact remains that understanding the roles of these genetic sequences in the biology of the microbe will necessarily include understanding where, when, and how particular genes are expressed. The use of GFP can provide insight that sequences cannot into the coordinate transcription and assembly of products underlying such complex activities as cell division or the contact-dependent translocation of proteins into mammalian cells. The use of GFP and its derivatives will permit the exploration of many functional facets of the biology of microbes.

ACKNOWLEDGMENTS

We wish to thank W. Buikema, W. Margolin, D. Gage, F. Bruijn, O. Resnekov, and C. Webb for contributing unpublished data, figures, and helpful discussions.

REFERENCES

Alpuche-Aranda, C. M., Swanson, J. A., Loomis, W. P., and Miller, S. I. (1992). *Salmonella typimurium* activates virulence gene transcription within acidified macrophage phagosomes. *Proc. Natl. Acad. Sci. USA* 89:10079–10083.

Arigoni, F., Pogliano, K., Webb, C. D., Stragier, P., and Losick, R. (1995). Localization of protein implicated in establishment of cell type to sites of asymmetric division. *Science* 270:637–640.

Barak, I., Behary, J., Olmedo, G., Guzman, P., Brown, D. P., Castro, E., Walker, D., Westpheling, J., and Youngman, P. (1996). Structure and function of the *Bacillus* SpoIIE protein and its localization in sites of sporulation septum assembly. *Mol. Microbiol.* 19:1047–1060.

Black, K., Buikema, W. J., and Haselkorn, R. (1995). The *hglK* gene is required for the localization of heterocyst-specific glycolipids in the cyanobacterium *Anabaena* strain PCC 7120. *J. Bacteriol.* 172:6440–6448.

Burlage, R. S., Yang, Z. K., and Mehlhorn, T. (1996). A transposon for green fluorescent protein transcriptional fusions: application for bacterial transport experiments. *Gene* 173: 53–58.

Chalfie, M., Tu, Y., Euskirchen, G., Ward, W. W., and Prasher, D. C. (1994). Green Fluorescent Protein as marker of gene expression. *Science* 263:802–805.

Christensen, B. B., Sternberg, C., and Molin, S. (1996). Bacterial plasmid conjugation on semi-solid surfaces monitored with the green fluorescent protein (GFP) from *Aequorea victoria* as a marker. *Gene* 173:59–65.

Cormack, B. P., Valdivia, R. H., and Falkow, S. (1996). FACS-optimized mutants of the green fluorescent protein (GFP). *Gene* 173:33–38.

Crameri, A., Whitehorn, E. A., Tate, E., and Stemmer, P. C. (1996). Improved green fluorescent protein by molecular evolution using DNA shuffling. *Nature Biotech.* 14:315–319.

Delagrave, S., Hawtin, R. E., Silva, C. M., Yang, M. M., and Youvan, D. C. (1995). Red-shifted excitation mutants of the green fluorescent protein. *Bio/technology* 13:151–155.

Dhandayuthapani, S., Via, L. E., Thomas, C. A., Horowitz, P. M., Deretic, D., and Deretic, V. (1995). Green fluorescent protein as a marker for gene expression and cell biology of mycobacterial interactions with macrophages. *Mol. Microbiol.* 17:901–912.

Driks, A., Roels, S., Beall, B., Jr., Moran, C. P., and Losick, R. (1994). Subcellular localization of proteins involved in the assembly of the spore coat of *Bacillus subtilis. Genes Dev.* 8:234–244.

Errington, J. (1993). *Bacillus subtilis* sporulation: regulation of gene expression and control of morphogenesis. *Microbiol. Rev.* 57:1–33.

Erickson, H. P. (1995). FtsZ, a prokaryotic homolog of tubulin? *Cell* 80:367–370.

Fierer, J., Eckmann, L., Fang, F., Pfeifer, C., Finlay, B. B., and Guiney, D. (1993). Expression of the *Salmonella* virulence plasmid gene *spvB* in cultured macrophages and nonphagocytic cells. *Infect. Immunol.* 61:5231–5236.

Fisher, R. F. and Long, S. R. (1992). *Rhizobium*-plant signal exchange. *Nature (London)* 356:655–660.

Foster, J. W., Park, Y. K., Bang, I. S., Karem, K., Betts, H., Hall, H. K., and Shaw, E. (1994). Regulatory circuits involved with pH-regulated gene expression in *Salmonella typhimurium. Microbiology* 140:341–352.

Gage, D. J., Bobo, T., and Long, S. R. (1996). Use of green fluorescent protein to visualize the early events of symbiosis between *Rhizobium meloliti* and alfalfa, *Medicago sativa. J. Bacteriol* 178:7159–7166.

Garcia del Portillo, F. and Finlay, B. B. (1995). The varied lifestyles of intracellular pathogens within eukaryotic vacuolar compartments. *Trends Microbiol.* 3:373–380.

Garcia del Portillo, E., Foster, J. W., Maguire, M. E. and Finlay, B. B. (1992). Characterization of the micro-environment of *Salmonella typhimurium*-containing vacuoles within MDCK epithelial cells. *Mol. Microbiol.* 6:3289–3297.

Harry, E. J., Pogliano, K., and Losick, R. (1995). Use of immunofluorescence to visualize cell-specific gene expression during sporulation in *Bacillus subtilis. J. Bacteriol.* 177:3386–3393.

Haselkorn, R. (1992). Developmentally regulated gene rearrangements in prokaryotes. *Ann. Rev. Genet.* 26:113–130.

Heim, R., Cubitt, A. B., and Tsien, R. Y. (1995). Improved green fluorescence. *Nature (London)* 373:663–664.

Heim, R., Prasher, D. C., and Tsien, R. Y. (1994). Wavelength mutations and post-translational autoxidation of green fluorescent protein. *Proc. Natl. Acad. Sci. USA* 91:12501–12504.

Hensel, M., Shea, J., Gleeson, C., Jones, M., Dalton, E., and Holden, D. (1995). Simultaneous identification of bacterial virulence genes by negative selection. *Science* 269:400–403.

Hinnebusch, B. J., Perry, R. D., and Schwan, T. G. (1996). Role of *Yersinia pestis* hemin storage (*hms*) locus in the transmission of plague by fleas. *Science* 273:367–370.

Inouye, S. and Tsuji, F. I. (1994a). *Aequorea* green fluorescent protein: Expression of the gene and fluorescent characteristics of the recombinant protein. *FEBS Lett.* 341:277–280.

Inouye, S. and Tsuji, F. I. (1994b). Evidence for redox forms of the *Aequorea* green fluorescent protein. *FEBS Lett.* 351:211–214.

Kain, S. R., Adams, M., Kondepudi, A., Yang, T.-T., Ward, W. W., and Kitts, P. (1995). Green fluorescent protein as a reporter of gene expression and protein localization. *BioTech.* 19:640–655.

Kremer, L., Baulard, A., Estaquier, J., Poulain-Godefroy, O., and Locht, C. (1995). Green fluorescent protein as a new expression marker in mycobacteria. *Mol. Microbiol.* 17:913–922.

Leff, L. G. and Leff, A. A. (1996). The use of green fluorescent protein to monitor survival of genetically engineered bacteria in aquatic environments. *App. Environ. Microbiol.* 62:3486–3488.

Lewis, P. J. and Errington, J. (1996). Use of green fluorescent protein for detection of cell-specific gene expression and subcellular protein localization during sporulation in *Bacillus subtilis*. *Microbiology* 142:733–740.

Lewis, P. J., Nwoguh, C. E., Barer, M. R., Harwood, C. J., and Errington, J. (1994). Use of digitized video microscopy with a fluorogenic enzyme substrate to demonstrate cell- and compartment-specific gene expression in *Salmonella enteriditis* and *Bacillus subtilis*. *Mol. Microbiol.* 13:655–662.

Losick, R. and Stragier, P. (1992). Crisscross regulation of cell-type specific gene expression during development in *B. subtilis*. *Nature (London)* 355:601–604.

Ma, X., Ehrhardt, D. W., and Margolin, W. (1996). Co-localization of cell division proteins FtsZ and FtsA to cytoskeletal structures in living *Escherichia coli* cells using green fluorescent protein. *Proc. Natl. Acad. Sci. USA* 93:12998–13003.

Mahan, M. J., Slauch, J. M. and Mekalanos, J. J. (1993). Selection of bacterial virulence genes that are specifically induced in host tissues. *Science* 259:686–688.

Ogawa, H., Inouye, S., Tsuji, F. I., Yasuda, K., and Umesono, K. (1995). Localization, trafficking and temperature-dependence of the *Aequorea* green fluorescent protein in cultured vertebrate cells. *Proc. Natl. Acad. Sci. USA* 92:11899–11903.

Parks, D. R., Herzenberg, L. A., and Herzenberg, L. A. (1989). Flow cytometry and Fluorescence-Activated Cell Sorting. *In Fundamental Immunology*. Paul, W. E., Ed. Raven, New York; pp. 781–802.

Prasher, D. C., Eckenrode, V. K., Ward, W. W., Prendergast, F. G., and Cormier, M. J. (1992). Primary structure of the *Aequorea victoria* green fluorescent protein. *Gene* 111:229–233.

Pratt, L. and Silhavy, T. J. (1995). Porin regulon of *Escherichia coli*. *In Two-Component Signal Transduction*. Hoch, J. A., and Silhavy, T. J., Eds., The American Society of Microbiology, Washington, DC, pp. 105–127.

Pogliano, K., Harry, E. J. and Losick, R. (1995). Visualizing the subcellular localization of sporulation proteins in Bacillus subtilis using immunofluorescence microscopy. *Mol. Microbiol.* 18:459–470.

Resnekov, O., Alpera, S., and Losick, R. (1996). Subcellular localization of proteins governing the proteolytic activation of a developmental transcription factor in *Bacillus subtilis*. *Genes Cells* 1:529–542.

Shapiro, H. M. (1995). *Practical Flow Cytometry*, 3rd ed. Wiley-Liss, New York.

Sharpe, M. E. and Errington, J. (1996). *The Bacillus subtilis soj-spo0J* locus is required for a centromere-like function involved in prespore chromosome partitioning. *Mol. Microbiol.* 21:501–509.

Stragier, P. and Losick, R. (1996) Molecular genetics of sporulation in *Bacillus subtilis*. *Ann. Rev. Genet.* 30:297–241.

Tombolini, R., Unge, A. Davey, M. E., de Bruijn, F. J., and Jansson, J. K. (1997). Flow cytometric and microscopic analysis of GFP-tagged *Psedumonas fluorescens* bacteria. *FEMS Microbiol Ecol.* 22:17–28.

Valdivia, R. H. and Falkow, S. (1996). Bacterial Genetics by Flow Cytometry: Rapid Isolation of *Salmonella typhimurium* acid-inducible promoters by Differential Fluorescence Induction. *Mol. Microbiol.* 22:367–378.

Valdivia, R. H., Hromockyj, A. E., Monack, D., Ramakrishnan, L., and Falkow, S. (1996). Applications for the green fluorescent protein (GFP) in the study of host-pathogen interactions. *Gene* 173:47–52.

Via, L. E., Curcic, R., Mudd, M. H., Dhandayuthapani, S., Ulmer, R. J., and Deretic, V. (1996). Elements of signal transduction in *Mycobacterium tuberculosis*. In vitro phosphorylation and in vivo expression of the response regulator MtrA. *J. Bacteriol.* 178:3314–3321.

Webb, C. D., Decatur, A., Teleman, A., and Losick, R. (1995). Use of Green Fluorescent Protein for visualization of cell-specific gene expression and subcellular protein localization during sporulation in *Bacillus subtilis*. *J. Bacteriol.* 177:5906–5911.

Wu, L. J. and Errington, J. (1994). *Bacillus subtilis* SpoIIE protein required for DNA segregation during asymmetric cell division. *Science* 264:572–575.

Wu, L. J., Lewis, P. J., Allmansberger, R., Hauser, P. M., and Errington, J. (1995). A conjugation-like mechanism for prespore chromosome partitioning during sporulation in *Bacillus subtilis*. *Genes Dev.* 9:1316–1326.1

7

The Uses of Green Fluorescent Protein in Yeasts

JASON A. KAHANA AND PAMELA A. SILVER
Department of Biological Chemistry and Molecular Pharmacology, Harvard Medical School and Dana Farber Cancer Institute, Boston, MA

7.1 INTRODUCTION

The budding yeast *Saccharomyces cerevisiae* and the fission yeast *Schizosaccharomyces pombe* serve as excellent model systems for studying a variety of cytological phenomena. Because such basic biological processes as cell cycle progression, intracellular protein transport, gene expression, and metabolism are similar throughout all eukaryotic organisms, these yeasts are a convenient means to study these processes. In particular, several characteristics of yeast make it a useful model system: Both budding and fission yeasts are easy to culture and grow extremely quickly. Yeast can be manipulated genetically, and the entire genome for budding yeast has been sequenced. Moreover, the availability of a variety of gene expression systems including inducible promoters, minichromosomes, and conditional-lethal alleles greatly facilitate the study of protein functions.

However, yeasts have several shortcomings as model organisms. Most notably the small size (5–8 μm diameter) of individual cells makes cytology difficult. The presence of a cell wall makes microinjection of fluorescent markers impossible. Furthermore, in contrast to the relatively flat cells of mammalian tissue culture, the significant thickness (~60% of diameter) of yeast cells prevents the visualization of larger organelles (such as the nucleus) by standard light micro-

Green Fluorescent Protein: Properties, Applications, and Protocols, Edited by Martin Chalfie and Steven Kain
ISBN 0-471-17839-X

scopy techniques. Thus a convenient method for studying cytology would make yeast a more "complete" system.

The recently-cloned (Prasher et al., 1992) *Aequorea victoria* green fluorescent protein (GFP) now provides such a method. Because GFP is a naturally fluorescent protein and requires no cofactors for its spectral characteristics (Chalfie et al., 1994), it has the potential to be an extremely useful cytological tag in a variety of cell types. In particular, the use of this "vital fluorophore" has a variety of advantages over traditional methods of protein localization: In contrast to indirect immunofluorescence, the use of GFP fluorescence is relatively inexpensive (because antibodies are not required), free of fixation artifacts, and has the potential for performing dynamic, *in vivo* studies of cellular phenomena.

7.2 EXPRESSION OF GFP IN YEASTS

The budding and fission yeasts are particularly good organisms in which to utilize GFP. A wide variety of yeast gene expression systems enable the researcher to express native GFP and fusion proteins at levels which are easily visible and physiologically relevant. In this section, we will review several of the systems which can be used to express recombinant (e.g., GFP-fusion) genes in yeasts.

7.2.A *S. cerevisiae* Expression Systems

In contrast to the expression systems used with higher eukaryotes, several budding yeast recombinant gene expression systems can closely mimic the expression of endogenous genes. Most notably, autonomously-replicating, centromeric (ARS/CEN) plasmids mimic the behavior of single chromosomes in yeast. Furthermore, *S. cerevisiae* genes have relatively simple promoter and terminator sequences that regulate their levels of expression. With few exceptions, centromeric constructs that include about 300–600 bp of sequences both upstream and downstream of an open reading frame allows expression of genes at similar levels to the chromosomal locus. When GFP fusion genes are introduced into yeast on ARS/CEN plasmids with the endogenous promoter and terminator sequences (of the original gene being fused) they tend to be expressed at approximately physiological levels, levels that are unlikely to cause toxic side effects. For instance, we have constructed an ARS/CEN plasmid which includes 275 bp of the *NUF2* 5′ upstream sequence, a fusion of a GFP cDNA to the 3′ end of the *NUF2* ORF, and 300 bp of the *NUF2* 3′ UTR (Fig. 7.1). Using anti-Nuf2p antibodies in immunoblots, we have noticed the level of Nuf2-GFP is virtually indistinguishable from that of Nuf2p.

Regulated expression of recombinant genes in *S. cerevisiae* is extremely convenient as well. First and foremost, several inducible promoters can drive expression of exogenous genes in *S. cerevisiae*. The most common inducible promoter, the GAL1-10 promoter (Yocum et al., 1984) is activated in the presence of the sugar galactose and strongly repressed in the presence of glucose. The *MET3* promoter is active in the absence of the amino acid methionine. Addition of methionine to the medium results in strong repression of the *MET3* promoter (Cherest et al., 1987). The *CUP1* promoter is activated in the presence of soluble

Cu^{2+} ion (Karin et al., 1984). In each of these examples, a given open reading frame can be expressed at relatively high levels (up to ~0.5–1% total cellular protein) under conditions which promote otherwise normal cell metabolism and growth. Thus, for genes whose level of expression is too low to permit visualization using GFP fluorescence, use of a high-level inducible promoter may be a good solution. Furthermore, an inducible promoter is often the most convenient system for expressing fusions of non-yeast genes (or any ORF without a yeast promoter) to GFP.

In contrast to the use of inducible promoters, the use of high-copy (2μ) plasmids allows for the overexpression of genes under the control of their own promoters. Some genes have toxic effects when expressed during inappropriate periods of the cell cycle or at the levels of induced promoters. In these cases, it is not feasable to use such promoters. For genes that are temporally expressed during the cell cycle, 2μ constructs allow overexpression without the non-physiological effects of atemporal gene expression. In general, 2μ plasmids are maintained in 20–40 copies per cell; this increased "dosage effect" typically leads to a proper temporal overexpression of genes on high-copy constructs.

7.2.B *S. pombe* Expression Systems

In contrast to budding yeast, fission yeast can maintain plasmids which lack S. *pombe* origins of replication. For instance, the bacterial plasmid pBR322 can replicate to high copy number in S. *pombe*. Addition of the S. cerevsiae 2μ or S. *pombe* $ars1^+$ sequences to such plasmids tend to lower the copy number and increase transformation efficiency. However, such plasmids are mitotically unstable. Addition of the S. *pombe* stb^+ element tend to render these constructs mitotically and meiotically more stable to the point that they are roughly 10% as stable as S. cerevisiae ARS/CEN plasmids (Moreno, et al., 1991). Use of a plasmid containing $ars1^+$, stb^+, a selectable marker (e.g. $ura4^+$ for uracil prototrophy), and a S. *pombe* cDNA flanked by its endogenous 5′ and 3′ untranslated sequences should have similar, levels of expression to the chromosomally-encoded gene.

The most common promoters used to express exogenous genes in fission yeast are the viral SV40 early promoter and the S. *pombe* $nmt1^+$ (no message in thiamine) and $adh1^+$ promoters. While the SV40 and $adh1^+$ promoters drive moderate- to high-level constitutive expression of genes, the high-level expression driven by the $nmt1^+$ promoter can be repressed by addition of thiamine to the medium.

7.2.C Function and Proper Localization

In the event that the use of a GFP-fusion is the only means of localizing a protein, it is necessary to determine whether the fusion localizes properly. Perhaps the most direct means of assaying this is to determine if the GFP-fusion protein is functional. Because conditional-lethal (e.g., temperature-sensitive) alleles of essential yeast genes are commonly available, it is possible to test the ability of a GFP fusion construct to permit growth under otherwise nonpermissive conditions. In cases where conditional lethal alleles are not available, a "plasmid-swap experiment" may be possible (Rothstein, 1991). For example, when the only source of the essential gene is a vector with the *URA3* (ura4$^+$ in *S. pombe*) selectable marker, cells are highly-sensitive to the *URA3*-specific poison 5-fluoro-orotic acid (5-FOA, Boeke et al., 1987). However, when the GFP fusion gene is introduced on a plasmid with a different marker (e.g., *TRP1* or leu1$^+$), cells can then lose the original plasmid if and only if the fusion protein is expressed and functional. Cells which have "swapped" plasmids will exhibit the ability to grow in the presence of 5-FOA. Using either of these functional assays, it is often possible to ascertain whether a) the fusion gene is being expressed and b) the fusion correctly localizes to the proper subcellular compartment (as assayed by its functionality). For instance, our Nuf2-GFP construct (see Fig. 7.1) rescues temperature-sensitive alleles of *NUF2* and can be swapped for a *NUF2; URA3* plasmid in a *nuf2*Δ strain. From this, we can conclude that the fusion protein

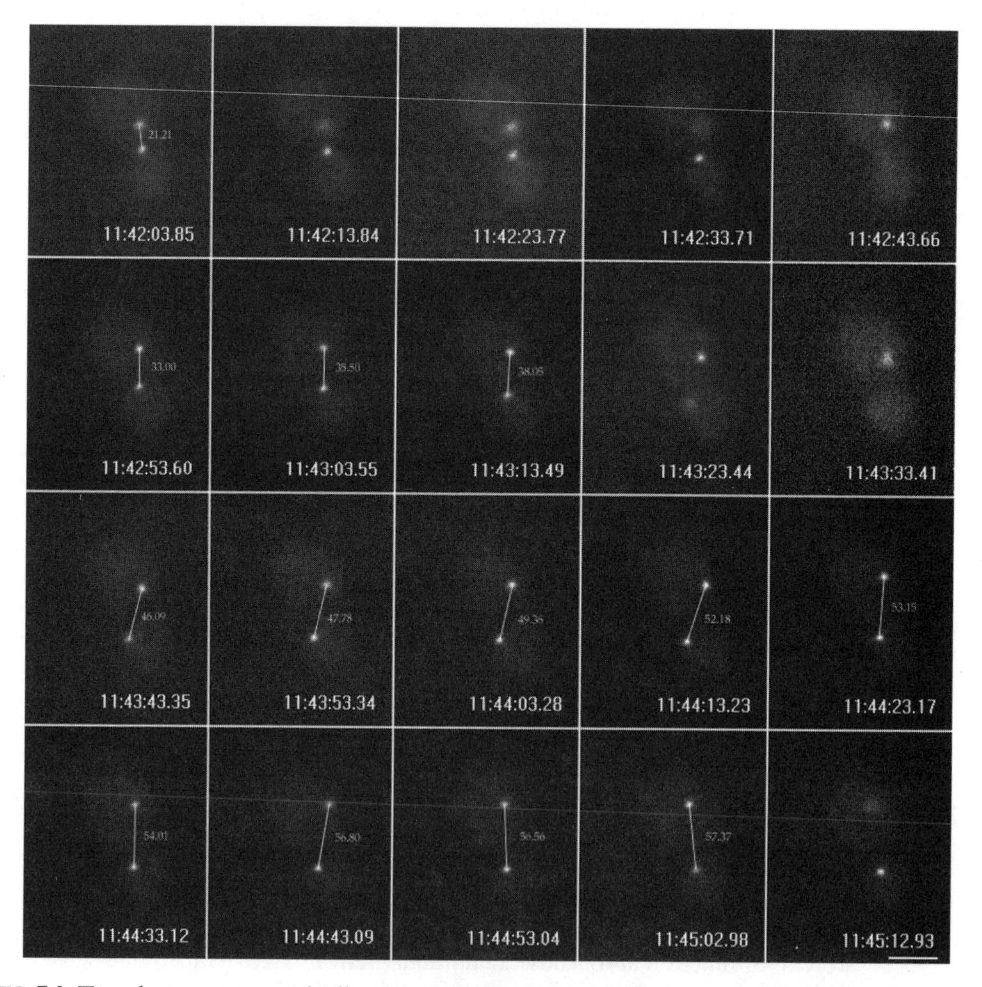

FIG. 7.1 Time-lapse montage of cells expressing the centrosomal antigen, Nuf2-GFP. In frames in which the centrosomes (white dots) are parfocal, the distance between them is measured in CCD pixels. (14.7 pixels = 1μm). A time stamp is included at the bottom of each frame.

is expressed and functional. Moreover, it must localize to the correct compartment (although additional localization is possible). Thus, in cases in which a GFP fusion is the only means of protein localization, a functional assay provides a useful control for proper localization. In contrast, using tissue culture systems, it is often difficult to assess the functionality of a GFP fusion *in vivo*.

7.3 UTILIZATION OF GFP- AND GFP-FUSION PROTEINS IN YEAST

Several groups have utilized GFP to study a variety of cytological phenomena in yeast. In this section, we review the use of GFP or GFP fusions to study organelle behavior, protein localization, gene expression, and protein folding in *S. cerevisiae* and *S. pombe*.

Several groups have studied the phenomenon of nuclear localization signal (NLS)-dependent protein transport in budding yeast by using GFP fusions. Flach et al. (1994) showed that a fusion of the histone H2B protein to GFP (H2B-GFP) localizes exclusively to the nuclei of wild-type yeast cells. However, in cells which harbor a mutant allele of a gene required for the accurate transport of nuclear NLS-containing proteins into the nucleus (*npl3-1*), H2B-GFP is cytoplasmically localized. Sidorova et al. (1995) have shown that a fusion between the N-terminal one third of Swi6p to GFP displays a cell cycle-dependent nuclear localization similar to that of unfused Swi6p. An NLS-GFP fusion has recently been used to monitor the rate of nuclear transport in vivo (Shugula, 1996; Roberts, 1998). In yeast cells depleted of ATP by treatment with azide and 2-deoxy-D-glucose, the NLS-GFP diffuses out of the nucleus. Re-addition of glucose results in a temperature-dependent re-import of the NLS-GFP, which can be visually quantitated. Finally, GFP-fusion proteins have also been effectively used to study the active export of nuclear proteins (Lee et al., 1996) and the localization and movements of the nuclear transport factors themselves in relationship to the nuclear pore complex (State et al., 1997; Seedorf and Silver, 1997). Thus, the use of GFP-fusions has a number of advantages over immunofluorescence in studying nuclear protein transport: (1) It is free of fixation artifacts (i.e., washing out of cytoplasmic protein after cell permeabilization), (2) It does not require an antibody to the nuclear protein or NLS being studied, and (3) It allows for dynamic, in vivo studies of protein transport using time-lapse fluorescence microscopy. (4) line cell applications

The dynamic behavior of the mitotic spindle in yeast has also been studied by the use of GFP fusion proteins. Using a fusion of the budding yeast centrosomal antigen Nuf2p to the N-terminus of GFP and fluorescence microscopy, we determined the rate and polarity of mitotic spindle growth in vivo (see Fig. 7.2 Kahana, et al., 1995). Similarly, Carminati and Stearns (1997) have expressed a fusion of the *S. cerevisiae* α-tubulin subunit to both the N- and C- termini of GFP. The fluorescent fusion proteins localize to mitotic spindle fibers *in vivo* and are being utilized to determine the role of astral microtubule positioning in spindle and nuclear migration during mitosis. Nabeshima et al. (1995) used a GFP fusion to localize p93dis1, a protein required for sister chromatid separation, in *S. pombe*. Time-lapse videos of dividing cells revealed that Dis1-GFP moves from cytoplasmic microtubules during interphase to centromeres during metaphase.

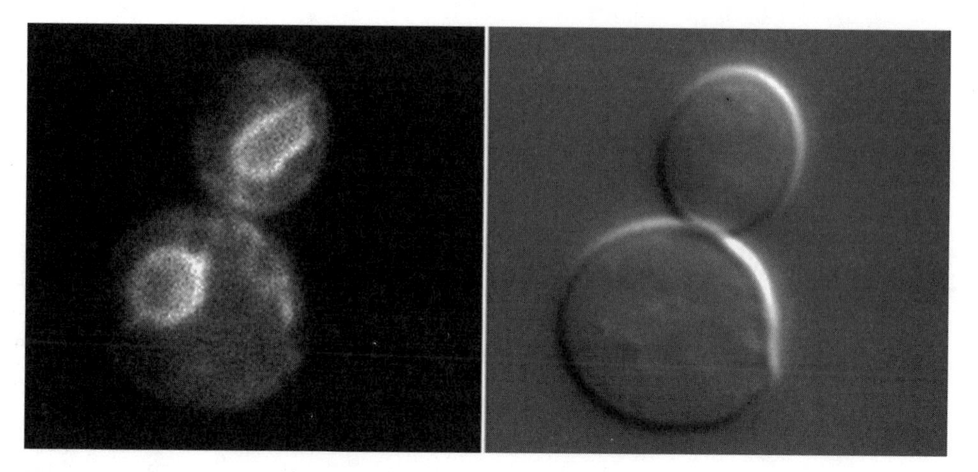

FIG. 7.2 Expression of Sec63-GFP highlights the nuclear rim of living yeast cells. A fluorescence and differential interference contrast image of yeast cells constitutively expressing Sec63-GFP from the *Sec63* promoter.

Then, during anaphase, Dis1-GFP moves to the spindle microtubules. Based on these observations, the authors have proposed that Dis1p elicits chromatid separation through its action on the mitotic spindle rather than upon the chromatids themselves. Hiraoka and co-workers (1996) fused GFP to α-tubulin and observed the relationship between DNA movement and microtubule polymerization during meiotic prophase in *S. pombe*. By using time-lapse fluorescence microscopy, they determined that nuclear movement in fission yeast is mediated by the dynamic instability of microtubules. Recently, Robinett et al., 1996 extended this type of analysis to monitor chromatin dynamics by inserting a 256 *lac* operator repeat into budding yeast chromosomes. Co-expression of a GFP-lac repressor protein fusion that can bind to the operator repeats allowed extended in vivo observations of the operator-tagged chromosomal DNA. In principle, all of these systems can be used to determine directly the precise effects of a myriad of particular proteins upon spindle and chromosome behavior *in vivo* by performing these assays in strains expressing mutant forms of such proteins. Prior to the availibility of such vital GFP tags for spindle antigens, these dynamic studies were impossible in yeast.

The budding yeast actin cytoskeleton has also been visualized using GFP fusion proteins. Doyle and Botstein (1996) fused of GFP to the C-terminus of actin (Act1p) and to the N-termini of the actin-binding proteins Abp1p and Sac6p to study the dynamics of yeast microfilaments in vivo. Similarly, Waddle et al. (1996), fused the actin capping protein (Cap2p) to GFP to observe the movement of cortical actin patches and microfilament polarization during the cell cycle. Both groups used time-lapse fluorescence microscopy to determine the velocity and direction of actin patch movement during the cell cycle. For instance, Waddle et al. showed that actin patch movement is arrested in the presence of metabolic inhibitors such as sodium azide. Thus, using GFP to perform dynamic experiments, the authors showed that patch motion is not due to simple diffusion, but rather occurs in an energy-dependent fashion.

To study endoplasmic reticulum (ER) protein expression, Hampton et al. (1996) constructed a C-terminal fusion between the *S. cerevisiae* hydroxymethylglutaryl CoA (HMG CoA) reductase protein and GFP. The fusion localized to the ER lumen and was used to track ER position and the formation of large stacks of ER and nuclear membranes, termed karmellae, which are visualized as bright patches of the ER. Hampton et al. showed that, as with unfused HMG CoA reductase, HMG CoA reductase-GFP is degraded with a concomitant loss of fluorescence, in response to flux in the mevalonate synthesis pathway. Thus, GFP can also be used as a reporter for the specific degradation of proteins (Hampton et al., 1996). We have constructed a fusion of the ER-membrane protein Sec63p to the N-terminus of GFP. As with HMG CoA reductase-GFP, the endoplasmic reticulum is visible by fluorescence *in vivo* (Fig. 7.2). Since the ER is coincident with the yeast nuclear membrane, and the yeast nucleus is not readily visible by transmitted-light microscopy, these fusions are also useful as nuclear markers in vivo. Thus, either of these GFP-fusions may be used to determine the dynamics of nuclear movement during the cell cycle by time-lapse microscopy. Similarly, the dynamics of the nuclear pore complexes themselves have been followed in yeast (Bucci, 1997; Belgareh, 1997) by monitoring the movements of GFP fused to the nucleoporins Nup49p and Nup133p. These studies suggested for the first time that nuclear pore complexes can move laterally through the double nuclear membrane.

To study the function of cytoplasmically inherited genetic elements, Patino et al. (1996) constructed a fusion of the N-terminal prion-determining domain of SUP35p, to GFP (NPD-GFP). Under conditions when the prion is in its "non-pathogenic" form, the GFP fusion is distributed evenly throughout the cell. However, when the conditions are altered to "activate" the pathogenic form of the protein, the GFP fusion is found in discrete "condensates" within the cell. This localization data corresponds to biochemical data which had shown that normal SUP35p is soluble, while the "activated" prion form is insoluble. Thus, GFP can be utilized to determine visually the solubility state of certain proteins to which it has been fused and to observe, *in vivo*, the transition between conformational states of the prion.

GFP has also been used to study the function of peroxisomes in yeast. By fusing the first 16 amino acids of thiolase to the N-terminus of GFP, Huang and Lazarow (1996) have directed the fluorescent fusion protein into the peroxisome. Furthermore, the authors have constructed a fusion of the first 54 amino acids of the peroxisomal membrane protein Pas3p to GFP, which anchors the fusion protein in the peroxisomal membrane with its GFP moiety in the cytosol. Using the Pas3-GFP protein, the authors detected "ghost" peroxisomal structures in yeast mutants in which no physiological perixosomes had not been seen previously. Furthermore, they observed peroxisome inheritance into buds during cell division in vivo using time-lapse fluorescence microscopy on dividing yeast cells.

Halme et al. (1996) have used GFP to study bud-site selection in *S. cerevisiae*. A fusion of GFP to the C-terminus of Bud10p localized the fusion protein to the bud site during cell division. Because the fusion protein expressed under the control to the *BUD10* promoter functionally complements the phenotype of a *bud10Δ* mutant, the observed localization of the GFP fusion likely reflects the localization of the endogenous unfused protein. The authors further showed that the localization of Bud10-GFP was dependent upon the function of another bud-site selection protein, Bud3p. Whereas in wild-type cells, Bud10-GFP is found in punctate "ring" structures at the bud neck during budding, the signal is considerably more diffuse in a *bud3*Δ strain. Thus, visualization of Bud10-GFP in living cells can be used as a convenient assay for the effects of other proteins involved in bud-site selection.

In addition to protein localization, the use of GFP fusions has been used to determine levels of protein expression. Atkins and Izant (1995) showed that the fission yeast, *Schizosaccharomyces pombe*, expressing GFP can be isolated using fluorescence-activated cell sorting (FACS). The FACS method is sufficiently sensitive to detect GFP-expressing cells compromising as little as 1% of a total population. Quantitative analysis of GFP-expressing cells comprising as little as 5% of the total population was possible as well. This high-throughput, quantitative system should be valuable for determining the effects of both mutations and environmental factors (e.g. drugs) upon gene expression in yeasts.

7.4 METHODS

7.4.1 Expression Systems

Ross-Macdonald et al. (1997) have developed an innovative system for fusing GFP to virtually any cloned *S. cerevisiae* gene using transposon-mediated insertion. A construct composed of the GFP cDNA, *URA3*, and *tet* genes flanked by transposon-3 (Tn3) 38 bp terminal repeats is used to create in-frame fusions of GFP to a given gene. This plasmid is grown in a bacterial strain that constitutively expresses Tn3 transposase. A construct with the yeast gene to be studied is transformed into a bacterial strain that constitutively expresses Tn3 resolvase. When the two bacterial strains are mated (by transduction), the Tn3-flanked GFP-*URA3-tet* fragment transposes randomly into sites in the yeast gene. (The transposition event is selected in bacteria by multiple antibiotic resistance of transductants.) Roughly one-sixth (due to the six possible reading frames) of transpositions should produce in-frame fusions of the GFP to the yeast gene. The new "transposed" fusion plasmid is isolated from the bacteria, and the fusion gene is excised by restriction digestion. The fusion fragment is then transformed into diploid yeast using the *URA3* selectable marker for homologous integration. For transformants which receive an in-frame fusion gene, the expression level and localization of the protein being studied can be assayed by fluoresence microscopy. For example, various Bdf1-GFP fusions prepared using this method localize properly to the nuclei of living yeast cells. In summary, this method provides an easily-extensible means of tagging yeast genes with the GFP cDNA sequence.

Recently, a similar methodology has been reported for the fission yeast *S. pombe* (Sawin and Nurse, 1996). *S. pombe* were transformed with a gene library in which *S. pombe* genomic sequences were fused to GFP and intracellular localizations determined by rapid fluorescence screening in vivo. Several novel genes whose products are found in certain nuclear sub-regions, including chromatin, the nucleolus and the mitotic spindle were identified by this approach.

We have designed a series of yeast plasmid vectors for the construction of GFP fusion proteins in *S. cerevisiae*. The pCGF (C-terminal GFP Fusion) series of high-copy vectors are designed for inducible expression of fusions of yeast genes to the 3′ end of the wild-type GFP cDNA (Kahana and Silver, 1996). When an ORF is ligated (in any of three frames) to the 3′ end of the P_{Gal}-GFP sequence, high-level fusion gene expression can be induced by addition of galactose to the medium. Use of the *Gal1-10* promoter can be extremely useful for expression of non-yeast cDNAs or yeast genes normally expressed at levels too low to be seen with GFP.

While this system may be useful for some proteins, the atemporal and typically high-level expression may cause several problems. First and foremost, such expression may, depending on the gene fused, have toxic side effects. Secondly, overexpression of a GFP-fusion gene may cause it to mislocalize. For instance, if a Nuf2-GFP fusion is expressed from a pCGF vector in the presence of 2% galactose, fluorescence is observed throughout the cell. However, normal localization of the Nuf2 protein is limited to the spindle pole body when it is expressed constitutively at low levels from the *NUF2* promoter. Some of these problems can be ameliorated by repressing expression by replacing of the galactose with glucose. Thus, "pulse-chase" expression can be achieved with this system. We have observed that expression of Nuf2-GFP from a pCGF vector using a one-hour galactose pulse followed by a six-hour glucose chase leads to accurate localization of the fusion protein (at ~1000–5000 molecules Nuf2-GFP/centrosome). Presumably this effect is attributable to dilution of protein level due to turnover and cell division (Kahana et al., 1995). Moreover, protein levels can be further modulated by adding mixtures of galactose and glucose to give low- and intermediate-level expression.

Accurate localization of most GFP proteins likely requires expression levels similar to the endogenous yeast genes. To achieve this, we developed the pNGF vector, an ARS/CEN vector with a spectrally altered mutant of GFP (S65T, V163A; Heim et al., 1995; Kahana and Silver, 1996) and the *NUF2 3′* UTR sequence (Fig. 7.3). When a gene fragment containing a 5′ promoter sequence coupled to an ORF is ligated into pNGF, the resulting construct should express the fusion at a level similar to that of the endogenous gene. Because the fusion is not overexpressed, toxic side effects are not likely, and functionality can be readily tested (see above). In our hands, this is the most convenient and accurate means of expressing most GFP fusions. Fusions of GFP (in the pNGF vector) to the C-terminus of Nuf2p, Cbf2p (Jiang et al., 1993), and Cdc23p (Sikorski et al., 1990) have all produced functional, fluorescent products.

Several approaches can assess GFP fusion expression. The first, and most straightforward, is to look for fluorescence by microscopy. While this method will show that the GFP moiety is being accurately expressed, it does not prove that the entire fusion protein is being made. For instance, if an ORF is ligated "out-of-frame" into a pCGF vector, cells will fluoresce in the presence of galactose due to expression of unfused GFP. Furthermore, if a protein is rapidly degraded within the cell, the GFP apoprotein may not "mature" into its fluorescent form in time to be detected.

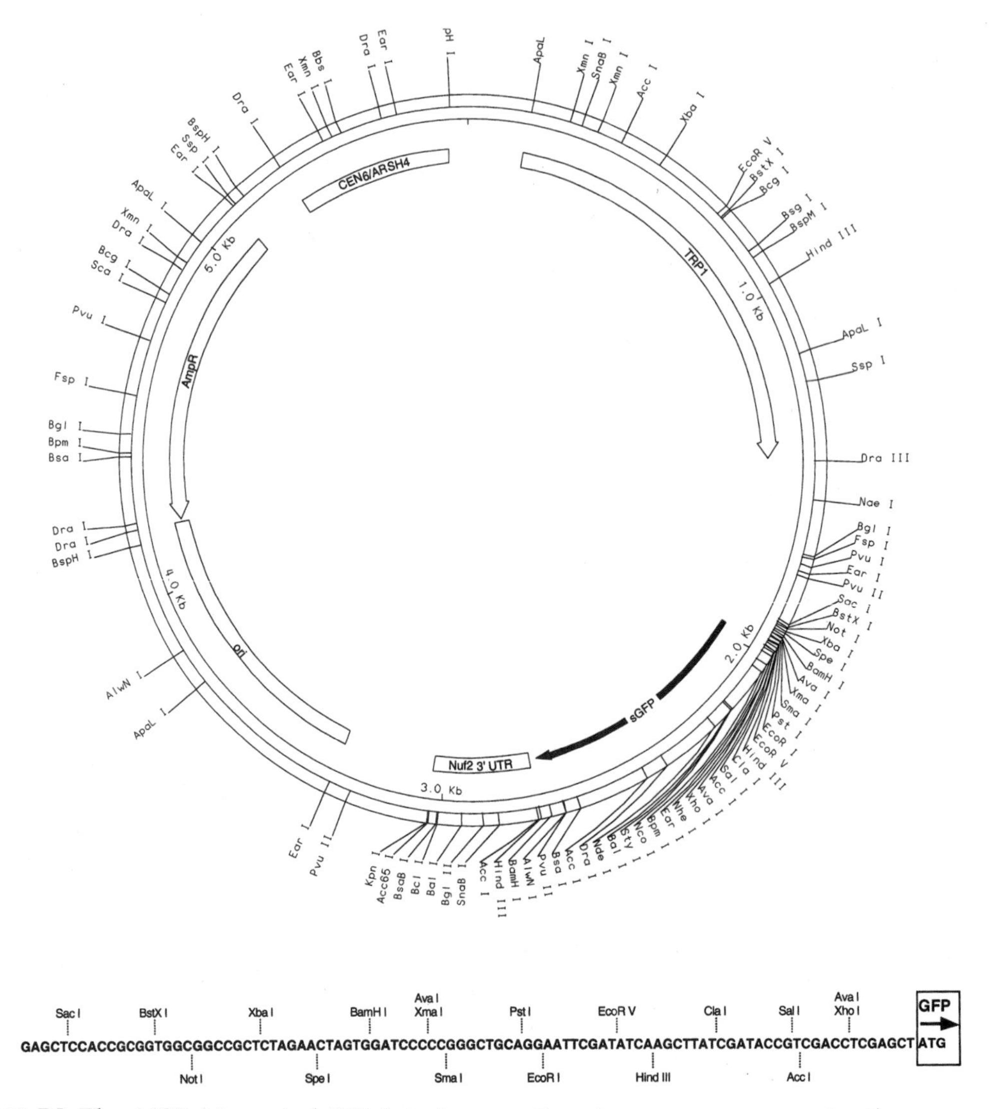

FIG. 7.3 The pNGF (N-terminal-GFP fusion) vector. Gene fragments comprising the 5′ promoter and open reading frame regions are ligated to GFP at the XhoI site. The polylinker sequence is included.

Hence, the most accurate method of assessing expression is the immunoblot. If antibodies against the protein being fused are available, they should recognize the fusion protein (which should run ~28 kDa larger than the unfused protein). Furthermore, such antibodies can be used to assess the relative levels of the fused to the unfused protein by the relative intensities of the signals on the blots. Anti-GFP monoclonal and polyclonal antibodies are commercially available from CLONTECH laboratories. However, we have observed that these antibodies recognize a variety of non-GFP bands in yeast lysates. Thus, when using these antibodies, a negative control experiment (e.g. yeast lysate without GFP) must be performed.

7.4.2 Microscopy

Because yeast are among the smallest eukaryotes (~5–10 μm in diameter), a high level of magnification must be used in fluorescence microscopy. For visual observations, we typically use a 60 or 100X objective lens and a 10X eyepiece lens. For maximal brightness and resolution, we always use objectives with a numerical aperture of 1.4 (the highest commercially available).

For recording images with a digital camera, we typically set the magnification in such a manner as to project 1μm of the specimen onto 8–15 pixels of the detector. Using a CCD camera with 6.8 × 6.8-μm^2 pixels, we need only a total magnification of 60X to achieve about 9 pixels/μm sample (or ~0.1 μm resolution). To achieve this, we use a 60X 1.4 N.A. objective lens without an intermediate eyepiece or projection lens. A detailed protocol for time-lapse digital acquisition of images is included in Protocol III.B.4.d.

Both yeast and many types of yeast media exhibit yellow autofluorescence when excited with ultraviolet or blue light. Thus, the use of a fluorescence filter set which maximizes GFP detection while minimizing autofluorescence must be use. We have found that the use of a standard "barrier pass" FITC filter set (Excitation 460–500 nm, Dichroic 505 nm; Barrier 510–560 nm; Chroma Technology No. #41001 or equivalent) with the S65T isolate of GFP (ex. 488 nm, em. 520 nm) gives the highest signal/noise ratio for detection. Furthermore, the use of low-fluorescence media is often advantageous. Rich media such as YEPD (Rose et al., 1990) tends to have high-background fluorescence. Less rich media such as "synthetic complete" (Rose et al., 1990) generally fluoresce much less brightly. Furthermore, media which lack tryptophan tend to have the lowest levels of autofluorescence. The recipe for one such type of media is included in the protocols section.

Budding yeast that have mutations in the *ADE1* or *ADE2* gene tend to accumulate a metabolic intermediate which interferes with the observation of GFP. Under normal room lighting, colonies of *ade1* or *ade2* cells appear pink while individual cells appear bright yellow or green in FITC filter sets when observed by epifluorescence. While it has been reported that addition of supplemental adenine to yeast media diminishes the pinkness of colonies, we have observed that this method does not completely ablate the autofluorescence observed by fluorescence microscopy. Thus, use of *ade1* and *ade2* strains should be avoided with GFP.

Green fluorescent protein is not affected by the presence of formaldehyde or other chemicals typically used in immunofluorescence protocols (Chalfie et al., 1994). Thus a GFP fusion can be colocalized with a protein being detected by immunofluorescence. Furthermore, in the aforementioned case that a fusion protein is degraded too rapidly for the GFP chromophore to properly oxidatively cyclize, formaldehyde fixation may help. By fixing cells in formaldehyde, the protein degradation machinery is ostensibly arrested. Afterward, if the cells are washed and stored in water (formaldehyde is a reducing agent), the GFP will be able to mature into the fluorescent form. While the use of this method precludes the possibility of dynamic observations, it is perhaps the most efficacious means for using GFP to localize proteins which are rapidly turned over.

7.5 CONCLUSIONS

To summarize, GFP is an extremely useful tool for studying biological processes in the yeast. Conversely, the use of yeast as a model system offers many advantages for expression of GFP. By utilizing such recently developed cytological methods in the genetically tractable yeasts, new types of studies of gene expression and gene action can be undertaken. For instance, the effects of gene disruption on dynamic phenomena (e.g., actin patch movement or mitotic spindle growth) can now be studied. Thus, the use of the "genetic fluorophore" will certainly augment the already valuable yeast model system.

REFERENCES

Atkins, D. and Izant, J. G. (1995). Expression and analysis of the green fluorescent protein gene in the fission yeast *Schizosaccharomyces pombe. Curr. Genet.* 28:585–588.

Belgareh, N. and Doye, V. (1997) Dynamics of nuclear pore distribution in nucleoporin mutant yeast cells. *J. Cell Biol.* 136:747–759.

Boeke, J. D., Trueheart, J., Natsoulis, G., and Fink, G. R. (1987). 5-Fluoroorotic acid as a selective agent in yeast molecular genetics. *Methods Enzymol.* 154:164–175.

Bucci, M. and Wente, S. R. (1997). In vivo dynamics of nuclear pore complexes in yeast. *J. Cell Biol.* 136:1185–1199.

Carminati, J. L. and Stearns, T. (1997). Microtubules orient the mitotic spindle in yeast through dynein-dependent interactions with the cell cortex. *J. Cell. Biol.* 138:629–641.

Chalfie, M., Tu, Y., Euskirchen, G., Ward, W. W., and Prasher, D. C. (1994). Green fluorescent protein as a marker for gene expression. *Science* 263:802–805.

Cherest, H., Kerjan, P., and Surdin-Kerjan, Y. (1987). The *Saccharomyces cerevisiae MET3* gene: nucleotide sequence and relationship of the 5′ non-coding region to that of *MET25. Mol. Gen. Genet.* 210:307–313.

Ding, D.-Q., Chikashige, Y., Yamamoto, A., Haraguchi, T, and Hiraoka, Y. (1996). Microtubule-mediated nuclear movement in meiotic prophase in fission yeast. Abstract for the American Society of Cell Biology meeting, 1996. *Mol. Biol. Cell* 7S:51a.

Doyle, T. and Botstein, D. (1996) Movement of yeast cortical actin cytoskeleton visualized *in vivo. Proc. Natl. Acad. Sci. USA* 93:3886–3891.

Flach, J., Bossie, M., Vogel, J., Corbett, A. H., Jinks, T., Willins, D. A., and Silver, P. A. (1994). A yeast RNA-binding protein shuttles between the nucleus and the cytoplasm. *Mol. Cell. Biol.* 14:8399–8407.

Hampton, R. Y., Koning, A., Wright, R., and Rine, J. (1996). *In vivo* examination of membrane protein localization and degradation with the green fluorescent protein. *Proc. Natl. Acad. Sci. USA* 93:828–833.

Halme, A., Michelitch, M., Mitchell, E. L., and Chant, J. (1996). Bud10p directs axial cell polarization in budding yeast and resembles a transmembrane receptor. *Curr. Biol.* 5:570–579.

Heim, R., Cubitt, A. B., and Tsien, R. Y. (1995). Improved green fluorescence. *Nature (London)* 373:663–664.

Hiraoka, Y. and Haraguchi, T. (1996). Analysis of chromosome organization and dynamics by three-dimensional fluorescence microscopy. *Tanpakushitsu Kakusan Koso* 41:2165–2173.

Huang, K. and Lazarow, P. B. (1996). Targeting of green fluorescent protein to peroxisomes and peroxisome membranes in *S. cerevisiae*. Abstract for the American Society of Cell Biology meeting, 1996. *Mol. Biol. Cell* 7S:494a.

Jiang, W., Lechner, J., and Carbon, J. (1993). Isolation and characterization of a gene (CBF2) specifying a protein component of the budding yeast kinetochore. *J. Cell Biol.* 121:513–519.

Kahana, J. A., Schnapp, B. J., and Silver, P. A. (1995). Kinetics of spindle pole body separation in budding yeast. *Proc. Natl. Acad. Sci. USA* 92:9707–9711.

Kahana, J. A. and Silver, P. A. (1996). Use of the *A. victoria* green fluorescent protein to study protein dynamics *in vivo*. *In Current Protocols in Molecular Biology*, Ausubel, F. M., Brent, R., Kingston, R. E., Moore, D. E., Seidman, J. G. Smith, J. A., and Struhl, K., Eds. Wiley, New York, pp. 9.6.13–9.6.19. .

Karin, M., Najarian, R., Haslinger, A., Valenzuela, P., Welch, J., and Fogel, S. (1984). Primary structure and transcription of an amplified genetic locus: the CUP1 locus of yeast. *Proc. Natl. Acad. Sci. USA* 81:337–341.

Lee, M. S., Henry, M., and Silver, P. A. (1996). A protein that shuttles between the nucleus and the cytoplasm is an important mediator of RNA export. *Genes Dev.* 10:1233–1246.

Marschall, L. G., Jeng, R. L., Mulholland, J., and Stearns, T. (1996). Analysis of Tub4p, a yeast γ-tubulin-like protein: Implications for microtubule-organizing center function. *J. Cell Biol.* 134:443–454.

Moreno, S., Klar, A., and Nurse, P. (1991). Molecular genetic analysis of fission yeast *Schizosaccharomyces pombe*. *Methods Enzymol.* 194:795–823.

Nabeshima, K, Kurooka, H., Takeuchi, M., Kinoshita, K., Nakaseko, Y., and Yanagida, M. (1995). p93^{dis1}, which is required for sister chromatid separation, is a novel microtubule and spindle pole body-associating protein phoshphorylated at the Cdc2 target sites. *Genes Dev.* 9:1572–1585.

Patino, M. M., Liu, J.-J., Glover, J. R., and Lindquist, S. (1996). Support for the prion hypothesis for inheritance of a phenotypic trait in yeast. *Science* 273:622–626.

Prasher, D. C., Eckenrode, V. K., Ward, W. W., Pendergast, F. G. and Cormier, M. J. (1992). Primary structure of the *Aequorea victoria* green-fluorescent protein. *Gene* 111:229–233.

Roberts, P. M. and Goldfarb, D. S. (1998). In vivo nuclear transport kinetics in Saccharomyces cerevisiae. *Methods Cell Biol.* 53: 545–557.

Robinett, C. C., Straight, A., Li, G., Willhelm, C., Sudlow, G., Murray, A., and Belmont, A. S. (1996). In vivo localization of DNA sequences and visualization of large-scale chromatin organization using lac operator/repressor recognition. *J. Cell Biol.* 125:1685–1700.

Rose, M. D., Winston, F., and Hieter, P. (1990). Methods in Yeast Genetics: A Laboratory Manual. Cold Spring Harbor Laboratory Press, Cold Spring Harbor, NY.

Ross-Macdonald, P., Sheehan, A., Roeder, G. S., and Snyder, M. (1997). A multipurpose transposon system for analyzing protein production, localization, and function in *Saccharomyces cerevisiae*. *Proc. Natl. Acad. Sci. USA* 94:190–195.

Rothstein, R. (1991) Targeting, disruption, replacement, and allele rescue: Integrative DNA transformation in yeast. *Methods Enzymol.* 194:281–301.

Sawin, K. E. and Nurse, P. (1996). Identification of fission yeast nuclear markers using random polypeptide fusions with green fluorescent protein. *Proc. Natl. Acad. Sci. USA* 93:15146–15151.

Seedorf, M. and Silver, P. A. (1997). Importin/karyopherin protein family members required for mRNA export from the nucleus. *Proc. Natl. Acad. Sci. USA* 96:8590–8595.

Shulga, N., Roberts, P., Gu, Z., Spitz, L., Tabb, M. M., Nomura, M. and Goldfarb, D. S. (1996). In vivo nuclear transport kinetics in Saccharomyces cerevisiae: a role for heat shock protein 70 during targeting and translocation. *J. Cell. Biol.* 135:329–339.

Sidorova, J. M., Mikesell, G. E., and Breeden, L. L. (1995). Cell cycle-regulated prosphorlyation of Swi6 controls its nuclear localization. *Mol. Biol. Cell.* 6:1641–1658.

Sikorski, R. S., Boguski, M. S., Goebl, M., and Hieter, P. (1990). A repeating amino acid motif in CDC23 defines a family of proteins and a new relationship among genes required for mitosis and RNA synthesis. *Cell* 60:307–317.

Stade, K., Ford, C. S., Guthrie, C. and Weiss, K. (1997). Exportin 1 (Crm1p) is an essential nuclear export factor. *Cell* 90:1041–1050.

Waddle, J. A., Karpova, T. S., Waterston, R. H., and Cooper, J. A. (1996). Movement of cortical actin patches in yeast. *J. Cell Biol.* 132:861–870.

Ward, W. W., Cody, C. W., Hart, R. C. and Cormier, M. J. (1980). Spectrophotometric identity of the energy transfer chromophores in *Renilla* and *Aquorea* green-fluorescent proteins. *Photochem. Photobiol.* 31:611-615.

Yocum, R. R., Hanley, S., West, R., and Ptashne, M. (1984). Use of LacZ fusions to delimit regulatory elements of the inducible divergent *GAL1-GAL10* promoter in *Saccharomyces cerevisiae*. *Mol. Cell. Biol.* 4:1985–1998.

8

The Uses of Green Fluorescent Protein in *Caenorhabditis elegans*

ANDREW FIRE, WILLIAM G. KELLY, MEI HSU AND SI-QUN XU
Carnegie Institution of Washington, Department of Embryology, Baltimore, MD

JOOHONG AHNN
Carnegie Institution of Washington, Department of Embryology, Baltimore, MD
Kwangju Institute of Science and Technology, Kwangju, Korea

BRIAN D. HARFE, STEPHEN A. KOSTAS, AND JENNY HSIEH
Carnegie Institution of Washington, Department of Embryology, Baltimore, MD
Biology Graduate Program, John Hopkins University, Baltimore, MD

8.1 INTRODUCTION

Over the last three decades, the nematode *Caenorhabditis elegans* has been the subject of intensive developmental, ultrastructural, genetic, and molecular characterization (Wood et al., 1988; Riddle et al., 1997). These analyses yielded (among other outcomes) a description of the complete lineage history of every cell in the animal (Sulston and Horvitz, 1977; Kimble and Hirsh, 1979; Sulston et al., 1983), the complete neurological wiring diagram of the 302 cell nervous system (White et al., 1986), an extensive library of lethal and visible mutations, and a detailed genetic map (Brenner, 1974; for updates see Hodgkin et al., 1988, 1997). Ongoing "genomics" efforts have progressed to the point where the majority of the genome sequence will be available before the publication of

Green Fluorescent Protein: Properties, Applications, and Protocols, Edited by Martin Chalfie and Steven Kain
ISBN 0-471-17839-X

this book (Waterston and Sulston, 1995). Experimental manipulations of the system can be carried out using a variety of techniques. These include chemical mutagenesis, isolation or ablation of specific cells, and DNA mediated transformation (see Epstein and Shakes, 1995 for a description of many of the methods employed for C. *elegans*).

The availability of detailed catalogs for many different aspects of C. *elegans* biology sets the stage to address a wide variety of cellular and developmental questions. As with the other experimental organisms described in this book, functional analysis would be greatly augmented by the ability to follow specific molecules and track specific cell populations in a minimally invasive manner in the whole organism. Several features make C. *elegans* ideal for applying fluorescence techniques to a whole animal system. First, the organism has a rapid life cycle (12 h for embryogenesis followed by 48 h of postembryonic development to form a fertile adult). Second, the organism is virtually transparent throughout its life cycle. Third, the organism exhibits only modest levels of autofluorescence.

Caenorhabditis elegans was the first system for which applications of *gfp* were described. Chalfie et al. (1994) expressed the *gfp* coding region in a subpopulation of neuronal cells in the animal, obtaining transgenic lines in which these cells were specifically labeled. Since then, *gfp*-based constructs have been used for a variety of applications and experimental approaches in C. *elegans*. Nonetheless, the field is just beginning to realize the breadth of questions to which *gfp* technology can be applied.

8.2 TECHNICAL ASPECTS OF *gfp*-BASED EXPERIMENTS IN *C. elegans*

8.2.A Optical properties of *C. elegans*

Caenorhabditis elegans is virtually clear when viewed by brightfield illumination (Fig 8.1). The lack of absorbance is so striking that virtually all published photographs are prepared with some type of contrast enhancement (usually interference contrast for high magnification and illumination diffusion for low magnification). The optical clarity of the animal is ideal for experiments in which a fluorescent molecule is to be tracked in a whole organism.

A subset of C. *elegans* tissues exhibit moderate autofluorescence. It should be stressed that the characteristic localization and color for autofluorescence usually allows facile identification of exogenous *gfp* fluorescence.

1. Starting in mid-stage embryogenesis and extending through the full life cycle, the gut is moderately fluorescent (Clokey and Jacobson, 1986). This fluorescence has a broad spectrum for both excitation and emission (Davis et al., 1982), and is thus visible with all microscope filter combinations tested to date. The different emission spectra of *gfp* and gut autofluorescence allow some distinction between the two signals. Nevertheless, faint *gfp* expression in or around the gut may be obscured. Gut autofluorescence varies somewhat during molt cycle and between

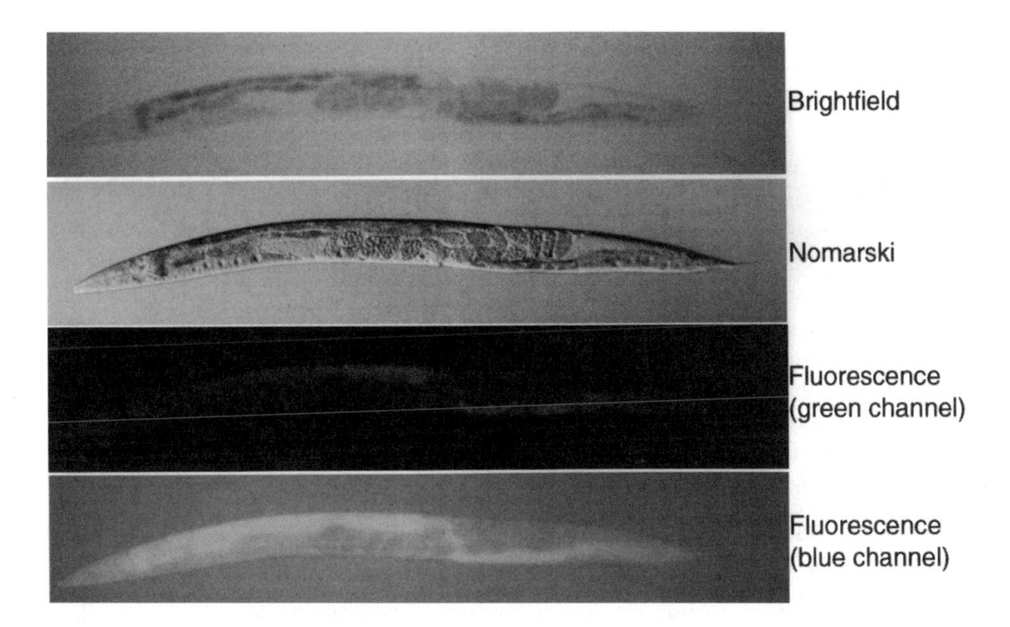

FIG. 8.1 Wild-type (nontransgenic) C. *elegans* adult viewed under several different optical conditions: Brightfield observation. The animal is virtually transparent; thus distinctions between tissues are barely visible under brightfield observation. Nomarski differential interference contrast microscopy. Optically enhanced microscopy produces the familiar image of C. *elegans* with well defined internal structures. Fluorescence observation with blue illumination [standard blue-excitation (fluorescein) relatively broad band filter set (Olympus BH2-DMB)]: predominant staining is seen in the gut. Faint staining is seen in other areas. Fluorescence observation with violet illumination [Standard UV/Violet (DAPI) relatively broad band filter set (Olympus BH2-DMUV)]: predominant staining is seen in the gut. Scale: The animal is 1 mm long. Figure also appears in color section.

strains; mutants with altered gut fluorescence have been isolated (Babu, 1974), but none completely lacks this signal.

2. Nucleoli of hypodermal cells are weakly autofluorescent (yellow) when viewed under blue illumination.
3. Certain cuticular structures (particularly male tail and vulva) have a yellow fluorescence.
4. Extremely faint autofluorescence is seen in various contractile or motile tissues (e.g., gonad surface, muscle, sperm). This autofluorescence is generally not problematic unless fluorescence observation is assayed at very high illumination intensity (or electronically amplified).

8.2.B Creation of Transgenic Animals

Transformation in C. *elegans* is carried out by injection of deoxyribonucleic acid (DNA) into the large syncytial germline in adult hermaphrodites (Stinchcomb et al., 1985; see Mello and Fire, 1995 for an extensive review of DNA transformation in C. *elegans*). Many (up to 50%) of the next generation show expression of the injected construct. The initial expression is frequently mosaic. In particular, only a fraction (5–20%) of the F1 transformed animals transmit the DNA to their progeny. Transmission to the second generation generally indicates that the DNA has been incorporated into some type of heritable structure. Large extrachromosomal arrays formed from the injected DNA are the most common type of heritable structure. These are heritable indefinitely and have transmission frequencies of 5–95%. Nonhomologous integration of the injected DNA is observed at a much lower frequency; integration of existing extrachromosomal elements can also be deliberately induced by treatments that transiently break endogenous chromosomes (see Mello and Fire, 1995). A very limited number of extremely rare homologous recombination events have been observed in C. *elegans* (Broverman et al., 1993; Mello and Fire, 1995).

8.2.C Properties of the Wild-Type *gfp* Coding Region and "Improved" Variants in *C. elegans*.

Initial efforts to use *gfp* as a reporter in C. *elegans* made use of the wild-type version of the protein. In experiments with a strong transcriptional promoter and in a stable differentiated tissue, an acceptable level of fluorescent signal could be obtained with this reporter (Chalfie et al., 1994; Sengupta et al., 1996; unpublished results from our laboratory). For many experiments analyzing small cells or rapidly changing developmental processes, however, the wild-type *gfp* was nonoptimal. First, the time required for wild-type *gfp* to fold and carry out the chemical reactions needed to form a fluorescent chromophore (2–4 h; Heim et al., 1994) was longer than the expected half life of many labile (but nonetheless interesting) proteins in C. *elegans*. In particular, cell cycles in the cleaving embryo are much shorter than this 2–4 h fluorescence-acquisition time (Sulston et al., 1983). Second, the level of fluorescence under standard fluorescent microscope illumination was rather dim in all but the most highly expressed constructs. Third, the wild-type protein exhibits photobleaching over several minutes with strong illumination; this complicates the tasks of photography and long-term observation.

One particularly striking observation using the original wild-type *gfp* concerned expression in the early embryo. We found that we could express high amounts of the *gfp* polypeptide in the early embryo. Although the protein could easily be detected by antibody staining or fusion to a second reporter protein (*Escherichia coli* β-galactosidase), we saw no fluorescence in these early embryos. Comparable levels of *gfp* in terminally differentiated tissues were easily visible by fluorescence.

These observations put us at a temporary roadblock in attempts to make full use of *gfp* reporters in the early embryo. Fortunately, this obstacle appeared at the same time that bacterial genetics was providing an answer to the problem. By using selection for strong fluorescence in bacteria, Heim et al. (1995; also see Chapter 5 in this volume) has been able to select "improved" variants. In particular, mutations of the Ser65 residue resulted in three desirable changes in the fluorescent properties of the protein (Heim et al., 1995):

1. Fluorescence yield using standard microscopic illumination systems (e.g., for fluorescein) was increased by a factor of approximately 7.
2. The time ($t - \frac{1}{2}$) required for acquisition of fluorescence was decreased by at least four-fold (from ~2 h to $\frac{1}{2}$ h).
3. The rate of fluorescence photoconversion was greatly reduced.

We found that introduction of these mutations was sufficient to allow detection of *gfp* in the early embryo.

To maximize the fluorescent signal in transgenic animals expressing *gfp* fusions, we also modified the *gfp* coding region to introduce canonical intron sequences at intervals of several hundred base pairs. We had used similar multiple-intron insertion to improve activity of *C. elegans lacZ*; methods for insertion, and mechanism of stimulation will be presented elsewhere.

The combination of the activating mutations in the *gfp* protein and the introduction of intervening sequences in the coding region has resulted in a set of vectors, which should be applicable in *C. elegans* to a wide variety of cell types and experimental questions (Fig. 8.2).

Since the original report by Heim et al. (1994), several groups have described *gfp* variants with additional alterations in the protein sequence, each of which is likewise more active than the original *gfp* (Cormack et al., 1996; Yang et al., 1996b). In many cases, the published data on new mutants does not allow a direct comparison with the S65C and S65T versions described by Heim and Tsien. We have now tried a large number of *gfp* variants in addition to the original. There is no simple or uniform means of quantitatively comparing the different activated forms of *gfp* in *C. elegans*. Nonetheless, several qualitative observations from our comparisons may be of use (note, of course, that many of these observations simply recapitulate the observations made for the original bacterially expressed *gfp* mutants).

Below is a listing of "red-shifted" *gfp* variants tested so far in *C. elegans*:

Wild type (Chalfie et al., 1994)

S65T (Heim et al., 1995)

S65C (Heim et al., 1995)

EGFP (F64L+S65T) (Cormack et al., 1996)

mut2 (S65A+V68L+S72A) (Cormack et al., 1996)

mut3 (S65G+S72A) (Cormack et al., 1996)

1. With all of the modified versions tested, we see marked improvement over wild-type *gfp* fluorescence observed in *C. elegans* transgenics under standard [e.g., fluoresce in isothiocyanate FITC] filter sets).
2. Each of the *gfp* modifications noted above causes an excitation red shift, greatly reducing fluorescence with near ultraviolet (UV) (e.g., DAPI, Hoechst filter sets), and yielding a surprisingly strong red signal using rhodamine filter sets (green illumination).
3. For a cytosolic expressed construct viewed under standard conditions for fluorescence microscopy (using Fluorescein filter sets from

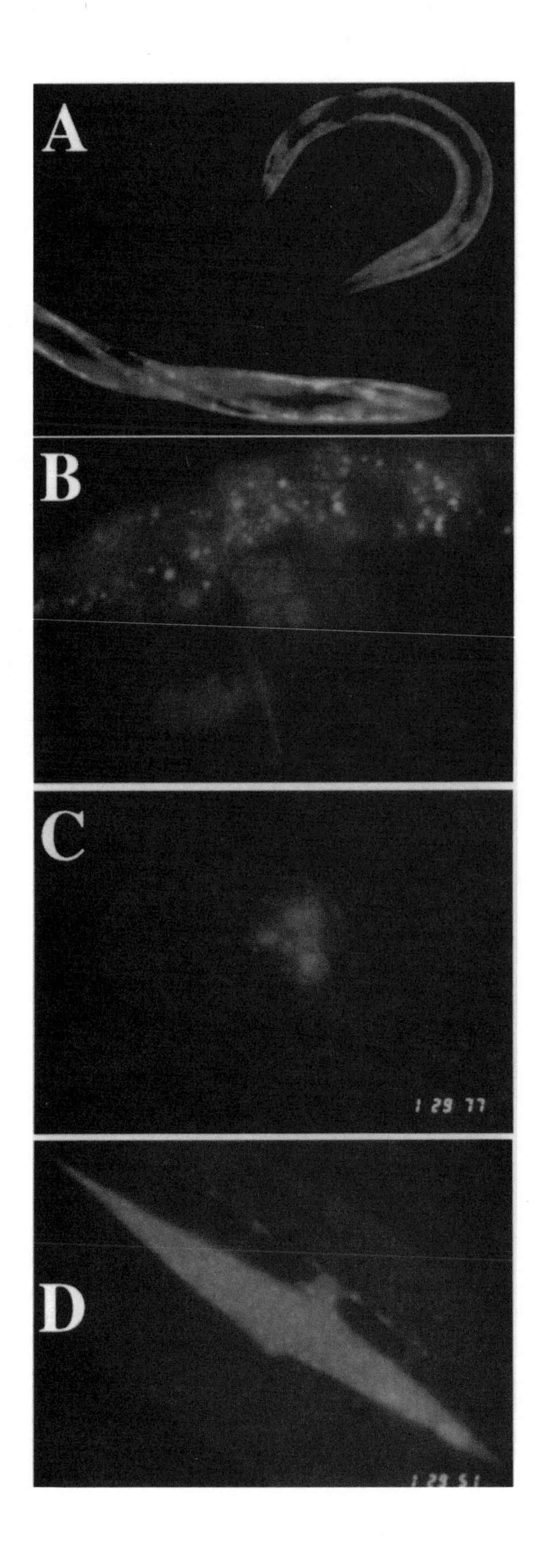
A
B
C
1 29 77
D
1 29 51

Olympus), we saw comparable initial activity levels from S65T, S65C, EGFP, mut2, and mut3.

4. The S65C variant was substantially more photostable than any of the other variants tested (these where compared when illuminated with the strong blue light of a standard fluorescein filter set). The mut3 variant was markedly less stable to photobleaching than the others.
5. We found that secreted proteins can be tagged with *gfp* (see Section 8.2D); of the two originally reported mutations, S65T and S65C, the former seems more active in secreted constructs. We have in general used S65T or derivatives in tagging experiments for which the extracellular domain is labeled.

8.2.D Double-Labeling Experiments Using GFP Variants with Different Chromatic Properties.

In addition to the above red-shifted variants, we recently produced animals expressing the blue-shifted double mutant Y66H+Y145F. This mutant absorbs in the violet range and emits a distinctly blue fluorescence (Heim et al., 1996). The blue fluorescence of this variant can indeed be distinguished visibly from the green fluorescence of the wild-type GFP (or the six red-shifted GFP variants described above): body wall muscle cells that express a combination of nucleus-localized blue GFP(Y66H+Y145F) and cytoplasm-localized green GFP(S65C) are shown in Figure 8.3. We hope that the blue GFP will provide a useful tool for experiments in which multiple labels are required. Two features of the Y66H+Y145F variant may require special experimental setup or analysis. First, the fluorescence of this GFP (Y66H+Y145F) fades rapidly (on the order of seconds) under standard Hoechst or DAPI channel illumination. Second, the gut autofluorescence is much stronger under near-UV illumination conditions used for observing the blue fluorescence of GFP (Y66H Y145F). It is likely that these two problems can be circumvented with highly selective optical filter combinations and/or digitally enhanced image acquisition.

A variety of additional GFP emission variants, or "colors", as they are developed by the community (e.g., Heim et al., 1996) will be very useful. In

---

FIG. 8.2 (opposite) Examples of GFP expression in C. *elegans*. All photos show transgenic lines carrying specific promoters driving S65C version of GFP. Panel (*a*) *myo-3* promoter (Okkema et al., 1993) is used to drive expression of partially nuclear-targeted GFP in body-wall muscles. Panel (*b*): pes-10 promoter (Seydoux and Fire, 1994) is used to drive expression of mitochondrially targeted GFP in early somatic blastomeres. Top (granular) signal is yellowish gut fluorescence of parent animal. Embryonic fluorescence (lower) is due to *gfp* construct. The *gfp* signal is present in punctate structures that are presumed to be mitochondria. More diffuse general staining in these cells may represent a combination of out-of-focus fluorescence, and *gfp*, which has not entered mitochondria. Panel (*c*): *hlh-1* promoter (Krause et al., 1994) is used to drive *gfp* expression in myogenic precursor cells of the midstage C. *elegans* embryo. Panel (*d*): High-magnification photo of mitochondrial-targeted *gfp* expressed in a single body-wall muscle cell. Faint projections from the cell are presumably the muscle arms that join the cell to the adjacent nerve chord. Figure also appears in color section.

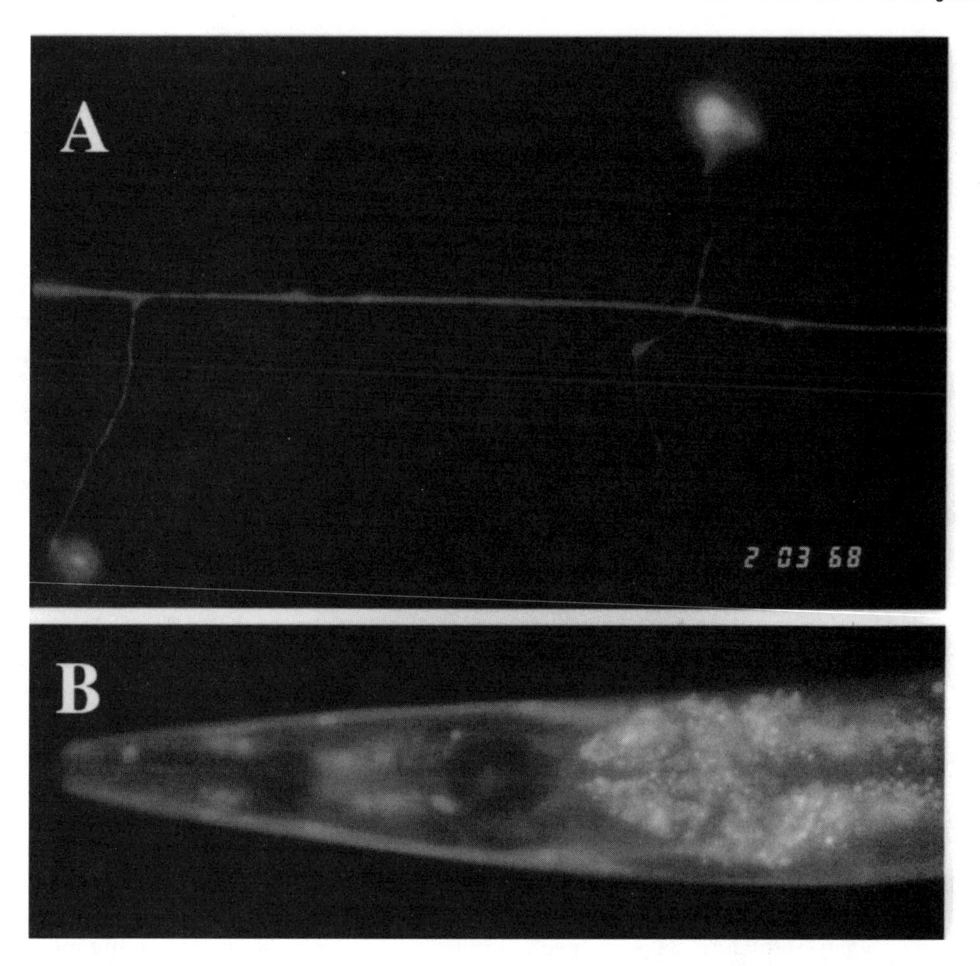

FIG. 8.3 Examples of two GFP applications in C. *elegans*. Panel (*a*): Localization of unaltered GFP throughout the cytosol of touch neurons (see Chalfie et al., 1994). The cell bodies (slightly out of focus) are uniformly stained. Axon tracks for the two cells are clearly visible in the plane of focus. The GFP was expressed from the *mec-7* promoter using similar constructs to those of Chalfie et al. (1994), with modifications to increase activity (F64L S65T, and three internal introns) as described in the text. Panel (*b*): A transgenic animal expressing two different *gfp* constructs, each driven from the body-muscle-specific *unc-54* promoter. This single-exposure simple photograph of an animal is illuminated using standard "DAPI" (Violet excitation) filter set. The two *gfp* constructs used are the *gfp* (S65C) expressed throughout the cytosol and the "Blue" *gfp* (Y66H Y145F) targeted to the nucleus by insertion into a construct with the SV40 nuclear localization signal appended to *lacZ* (Fire et al., 1990). At the left in the photo (anterior in the animal) are muscle cells with blue nucleus and green cytoplasm; strong blue fluorescence on the right side of the photo derives from autofluorescent granules in the gut. Figure also appears in color section.

particular, it might be hoped that structural biologists will be able to apply the recently available crystal structure of *gfp* (Yang et al., 1996a) and an activated variant (Ormö et al., 1996) to provide tailor-made *gfp* coding regions for the types of cell-marking and protein-tagging studies for which *gfp* has been applied in C. *elegans*.

8.2.E Intentional Targeting of GFP to Distinct Cellular Compartments

For various applications, it has been valuable to manipulate the localization of *gfp* reporters. We find that native *gfp* localizes primarily to the cytoplasm but is small enough to diffuse also into the nucleus. The standard reporter protein (S65C) appears to be unanchored and can rapidly diffuse throughout the cell (as observed using fluorescence photobleaching assays). In cells with irregular shapes, complete filling of cytosolic spaces is seen; this includes filling of extended neuronal processes (Chalfie et al., 1994; Sengupta et al., 1996) and muscle arms (unpublished results from our laboratory).

Targeting of GFP to the nucleus can be partially achieved by appending a nuclear localization signal (e.g., from SV40 large T antigen). The nuclear localized protein is apparently only transiently retained by the nucleus, so that the steady-state distribution is only slightly more concentrated in the nucleus. Increasing the molecular weight of the protein apparently improves the retention by the nucleus. We have used a *gfp-lacZ* fusion protein to obtain well-defined nuclear localization; in this case the protein is virtually completely nuclear (with exclusion from the nucleolus).

Mitochondrial localization of *gfp* can be engineered using a targeting signal derived from an enzyme from the mitochondrial lumen (chicken aspartate amino transferase; Fire and Xu, unpublished results). The mitochondrial-localized *gfp* is highly fluorescent and may have a longer half-life in some cells than cytosolic or nuclear *gfp*.

A synthetic secretion signal sequence has been used to produce *gfp*, which is destined for extracellular compartments. We find that the resulting *gfp* is efficiently secreted, although in some cases a signal can be seen in the secretory apparatus of the producing cell. We observed secreted *gfp* from a variety of cellular sources (body muscles, neurons, and gut) accumulating in the coelomic fluid and can be subsequently taken up by the six coelomocyte cells. These cells may have a scavenger function of removing or cycling molecules from the coelomic fluid.

For some purposes (e.g., tracing of neural connectivity), it is most useful to have an outline of the *gfp* expressing cell. Harald Hutter has targeted GFP to the plasma membrane by using the membrane targeting signal from let-60 RAS. Similar schemes have been used in mammalian systems (Moriyoshi et al., 1996). It is not yet clear whether this protein has any direct or indirect effect on the RAS signaling pathway that might interfere with its use as a neutral reporter.

In some cases, it is useful to have two populations of cells differentially labeled. An example of such an application in cell lineage studies is the use of a fixed set of "reference cells" (e.g., muscles) while following a specific cell of interest (e.g., a migrating myoblast). An elegant way to do this would be to use *gfp* reporters with different fluorescent properties in the two different cell types. This double labeling can be done with the original (or blue shifted) and absorption-red-shifted *gfp* reporters and should prove extremely useful. As an alternative, two different cell populations can easily be labeled with differently localized *gfp* reporters. We have made use of the combination of mitochondrial targeted *gfp* and nuclear targeted *gfp-lacZ* fusions for this purpose.

8.2.F Fluorescent Labeling of Specific Cell Populations: Promoter Fusions with *gfp*

Transgenic strains that express a reporter construct fused to a tissue or cell-type specific promoter can be used to label a specific cell population during development. The prerequisite for this experiment is the availability of tissue specific promoter segments with characterized activity patterns. For many (but not all) genes, upstream sequences can drive a reporter construct in an appropriate pattern. Many earlier studies of *C. elegans* promoters were carried out using *E. coli lacZ* as a reporter segment (e.g., Fire et al., 1990). The *lacZ* and *gfp* vectors in general use for *C. elegans* have essentially identical structures with the exception of the different reporter insertions (Fire et al., 1990; Chalfie et al., 1994; and more recent vector kits available from our laboratory at ftp://www.ciwemb.edu/pub/FireLabVectors). This modularity allows reporter coding regions to be easily exchanged once a single fusion construct with either *gfp* or *lacZ* is available.

As is evident from recent *C. elegans* meetings (and should soon be evident from the more general scientific literature) many different populations of *C. elegans* cells have now been labeled by generation of *gfp* constructs with tissue-, stage-, or cell-specific promoters. Neural cells are among the most striking: the production of *gfp* lines expressed in subsets of neurons has the remarkable utility of providing a population of animals in which a specific set of neural processes (or a whole section of the nervous system wiring pattern) can be mapped out both in wild type and as a function of experimental or genetic manipulation (Chalfie et al., 1994; Sengupta et al., 1996).

8.2.G Fluorescent Labeling of Molecular Populations: *gfp*-Tagging of Specific Genes.

The GFP forms a short and easily portable coding region that can be used to "tag" specific proteins so that they can be followed by their fluorescence. In many cases, the resulting proteins also can retain their normal function. There is a remarkable degree of flexibility in designing *gfp* tagged constructs, since *gfp* appears active in virtually any position within the coding region. We have analyzed fusion constructs in which *gfp* is appended to the N- or C-terminus, as well as several constructs in which *gfp* has been inserted in-frame into an internal part of the coding region. The latter constructs are particularly straightforward to assemble by insertion of *gfp* coding sequence in-frame at internal unique restriction sites (*gfp* coding regions in the proper reading frame can either be obtained from existing vectors or by the polymerase chain reaction (PCR) with appropriately designed adapter primers).

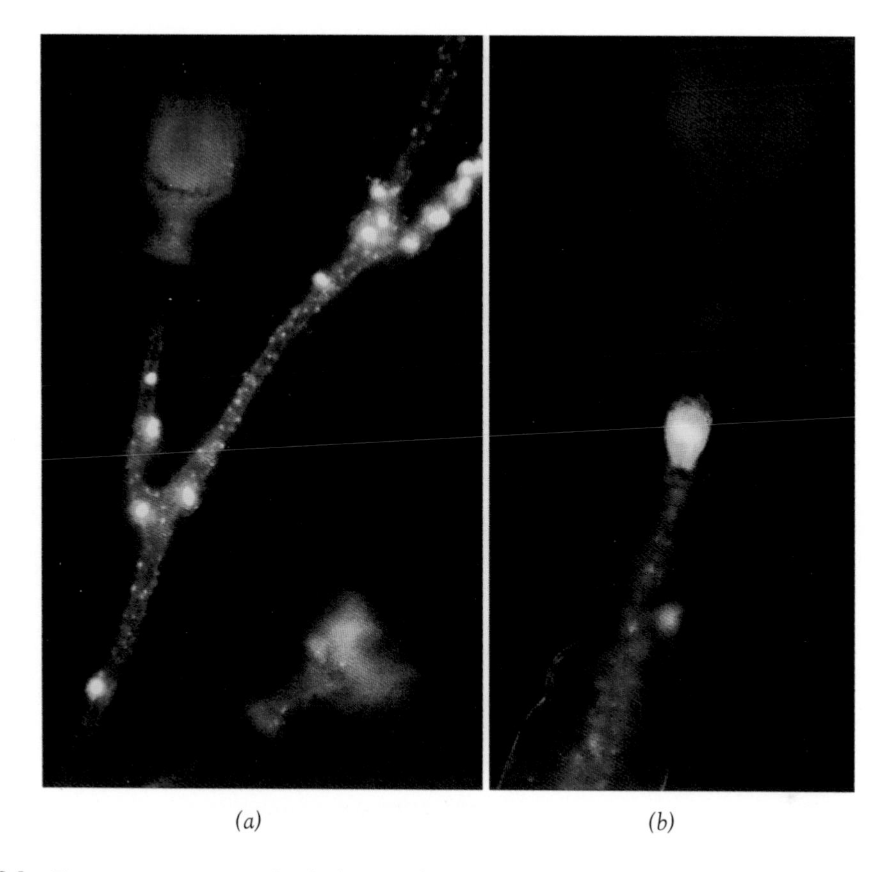

(*a*) (*b*)

FIG. 2.4 Fluorescence micrograph of a living colony of *Obelia* species showing photocytes visualized by GFP. Height of field shown is about 3 mm in (a) and 1.5 mm in (*b*). (*a*) Dispersed photocytes (bright green spots) in an upright of O. *geniculata* (two polyps also shown at lower right and upper left). (*b*) Concentrated photocytes at the tip of a pedicel below the base of a hydranth of O. *bidentata* (=bicusp-idata).

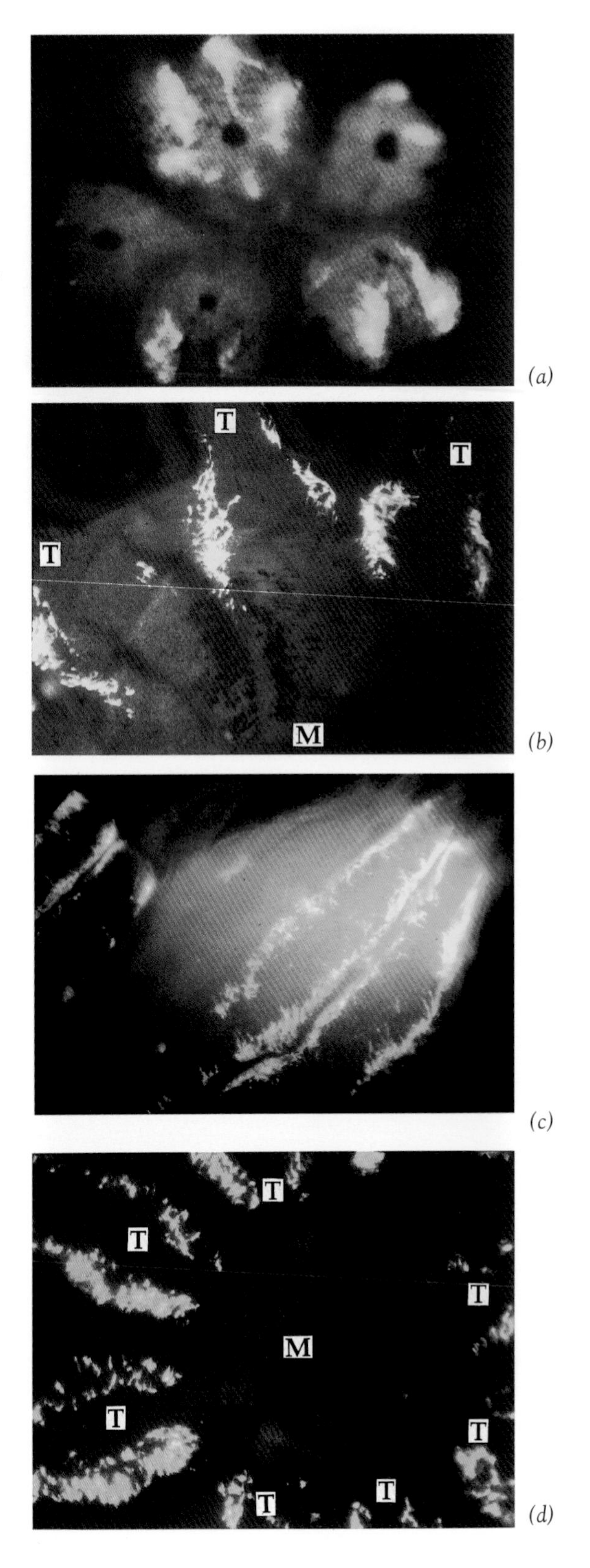

FIG. 2.6 Fluorescence micrographs of photocytes visualized by GFP in living pennatulacean (sea pen) colonies. Width of field shown is about 0.8 mm in *(a)* and 1.6 mm in *(b-d)*. *(a)* Photocytes in a cluster of five siphonozooides (water pumping polyps) of *Renilla kollikeri*. *(b)* Photocytes clustered in the lateral-axial region of the tentacles and oral disk of an autozooid (feeding polyp) of *Renilla kollikeri* (mouth [M] and base of three of eight tentacles [T] shown). *(c)* photocytes clustered in only the two outer (of the eight) chambers within the calyx of the column (and not the tentacles) of an autozooid of *Acanthoptilum gracile*. *(d)* photocytes clustered laterally along the length of each of the eight tentacles (T) of an autozooid of *Ptilosarcus guerneyi* (M = mouth).

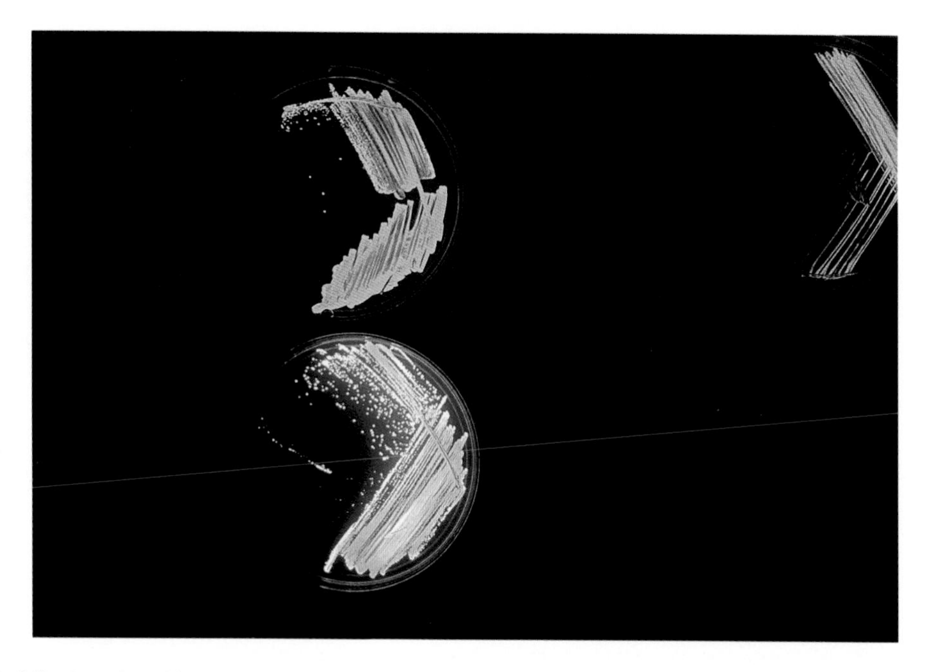

FIG. 2.7 Streaks of luminescent bacteria photographed by their own light, showing two strains of *Photobacterium fischeri*, one of which emits yellow light by virtue of having YFP (yellow fluorescent protein). The other lacks YFP, emitting only blue light.

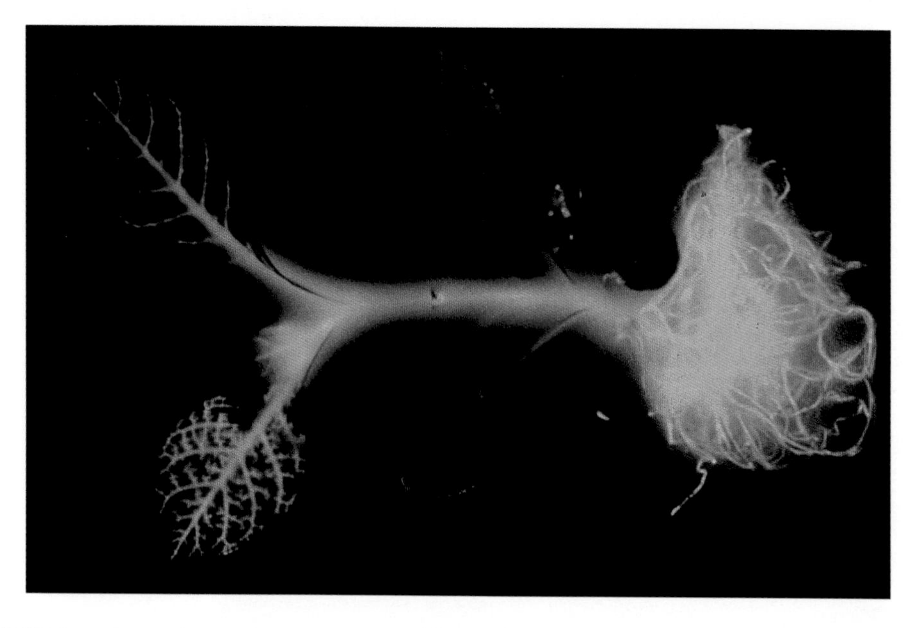

FIG. 2.10 Transgenic tobacco plant carrying the firefly luciferase gene photographed by its own light. The continuous luminescence occurs following the uptake of luciferin by the roots. [From Ow et al., 1986].

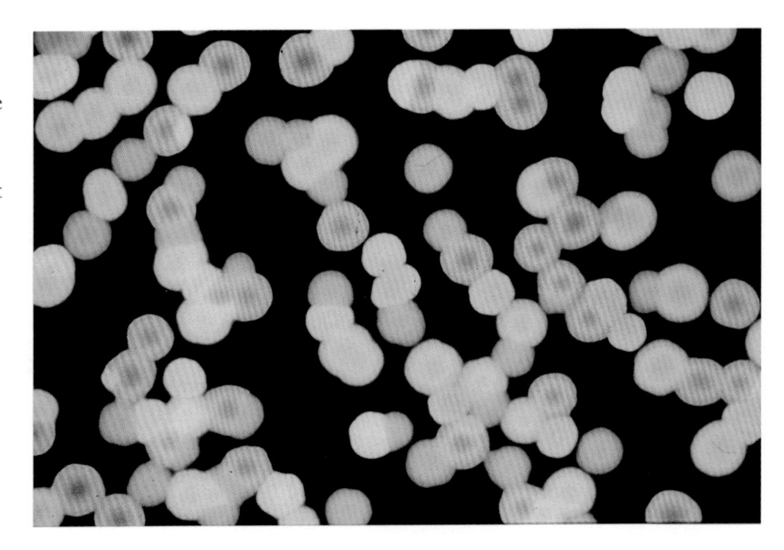

FIG. 2.11 Bacterial colonies carrying four different beetle luciferase genes cloned from the ventral organ, distinguished by their different luminescence colors: green, yellow-green, yellow and orange. [Wood et al., 1989.]

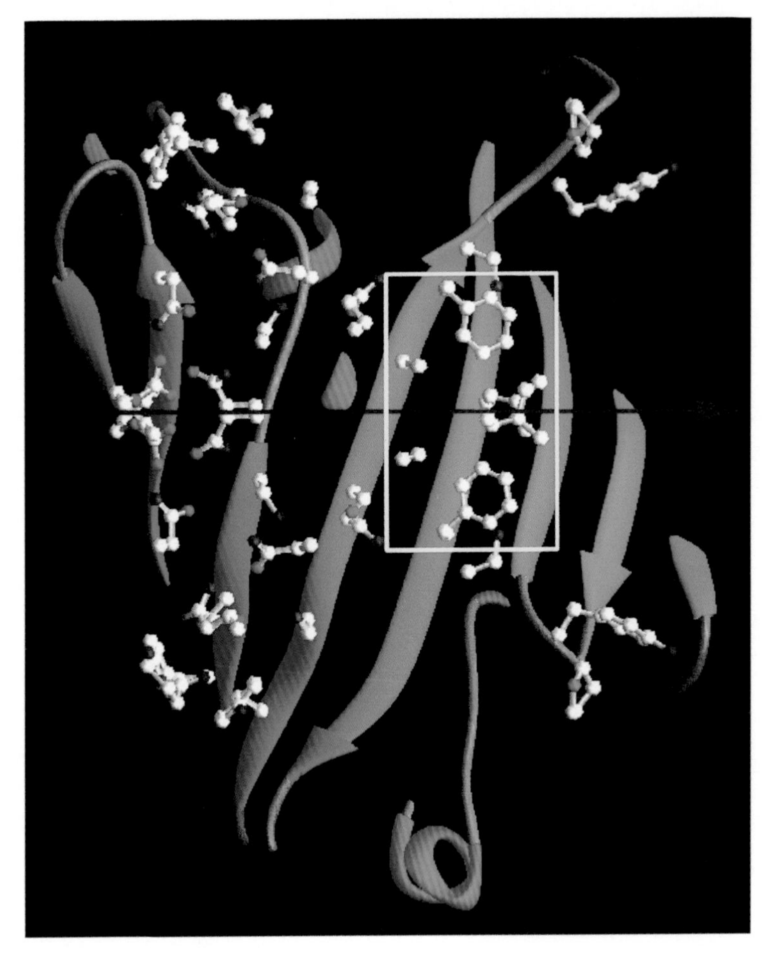

FIG. 4.6 The dimer contact region. The two polypeptide chains associate over a broad area, with a small hydrophobic patch (in the box) and numerous hydrophilic contacts. The twofold dimer axis is in the plane of the page (arrow). The side chains from each of the two monomers are shown as dark balls or light colored balls. A list of all the contacts is given in Table 4.2.

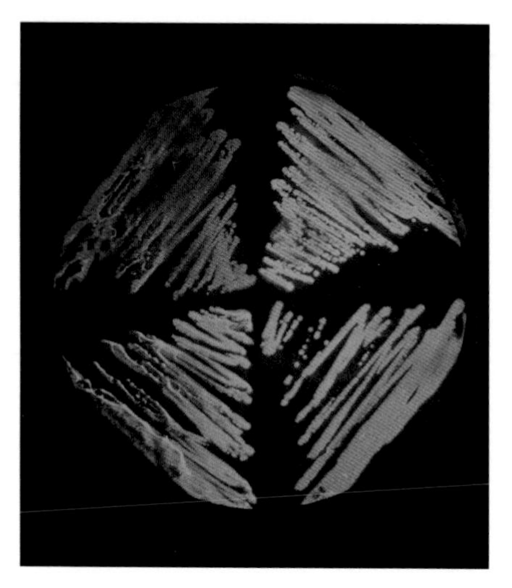

FIG. 5.2 Visual appearance of *E. coli* expressing four differently colored mutants of GFP. Clockwise from upper right: blue mutant P4-3 (=Y66H, Y145F) (Heim and Tsien, 1996); cyan mutant W7 (=Y66W, N146I, M153T, V163A, N212K) (Heim and Tsien, 1996); green mutant S65T (Heim et al., 1995); yellow mutant 10C (=S65G, V68L, S72A, T203Y) (Ormö et al., 1996). In each of these lists of mutants, the mutation most responsible for the special alterations is underlined, while the other substitutions improve folding or brightness. The bacteria were streaked onto nitrocellulose, illuminated with a Spectroline B-100 mercury lamp (Specronics Corp., Westbury, NY) emitting mainly at 365 nm, and photographed with Ektachrome 400 slide film through a low-fluorescence 400 nm and a 455-nm colored glass long-pass filter in series. The relative brightnesses of the bacteria in this image are not a good guide to the true brightnesses of the GFP mutants. Expression levels are not normalized, the 365 nm excites the blue and cyan mutants much more efficiently than the green and yellow mutants, but the blue emission is significantly filtered by the 455-nm filter required to block violet haze.

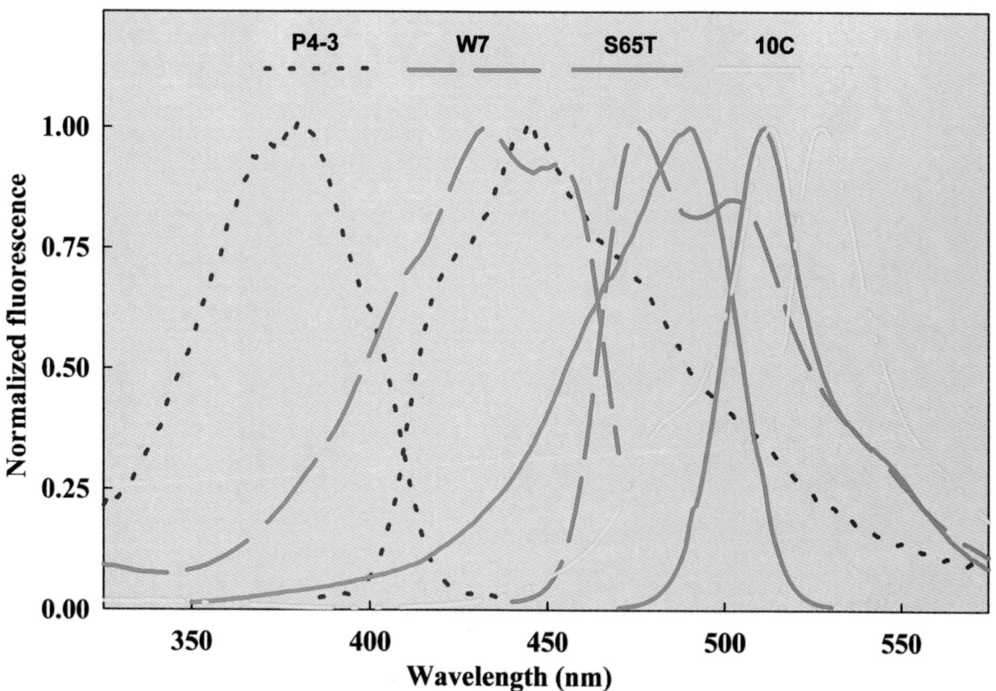

FIG. 5.3 Normalized excitation and emission spectra for mutants P4-3, W7, S65T, and 10C (see Fig. 5.2 for amino acid substitutions and references). All spectra were normalized to a maximal value of 1 to facilitate wavelength comparison. Each pair of excitation and emission spectra is depicted in a different line pattern and color as listed above the spectra. Within each pair the emission spectra is always the one at longer wavelengths.

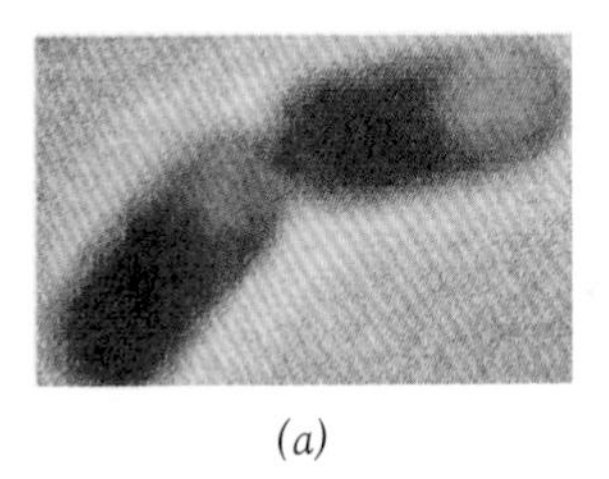

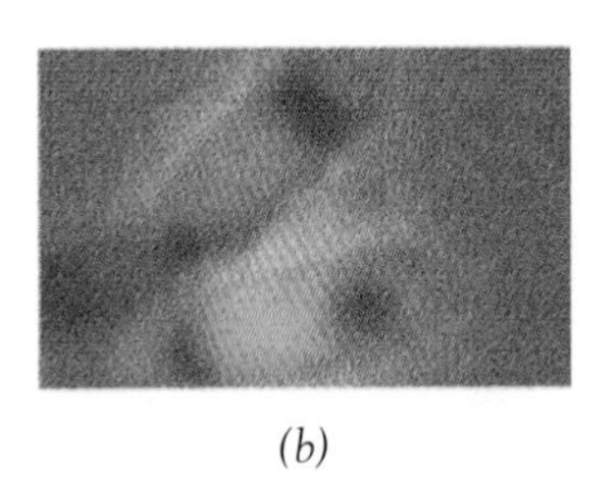

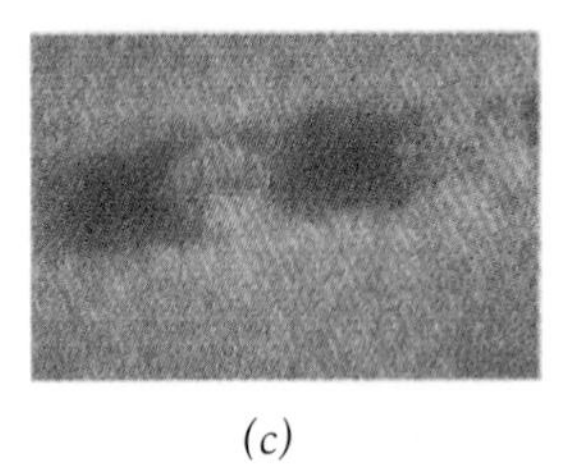

(*a*) (*b*) (*c*)

FIG 6.1 Fluorescence images of sporulating *B. subtilis* cells expressing transscriptional and translational GFP fusions. Two sporangia are shown per panel (*a*) Forespore-specific expression of a σF-dependent SspE2G-GFP fusion. (*b*) Mother cell-specific expression of a σE-dependent *cotE-gfp* fusion. (*c*) Localization of a SpoIVFB-GFP translational fusion (note localized fluorescence seen as a shell at one end of each sporangium). Courtesy of O. Resnekov and C. Webb, Harvard University.

FIG. 6.2 Heterocyst-specific gene expression in the cyanobacterium *Ababaena*. Fluorescence from nitrogen-starved *Anabena* cells bearing a plasmid with a *hetR-gfp* fusion. Green fluorescence is observed preferentially in heterocyst where *hetR* is exclusively expressed. Courtesy of W. Buikema, University of Chicago.

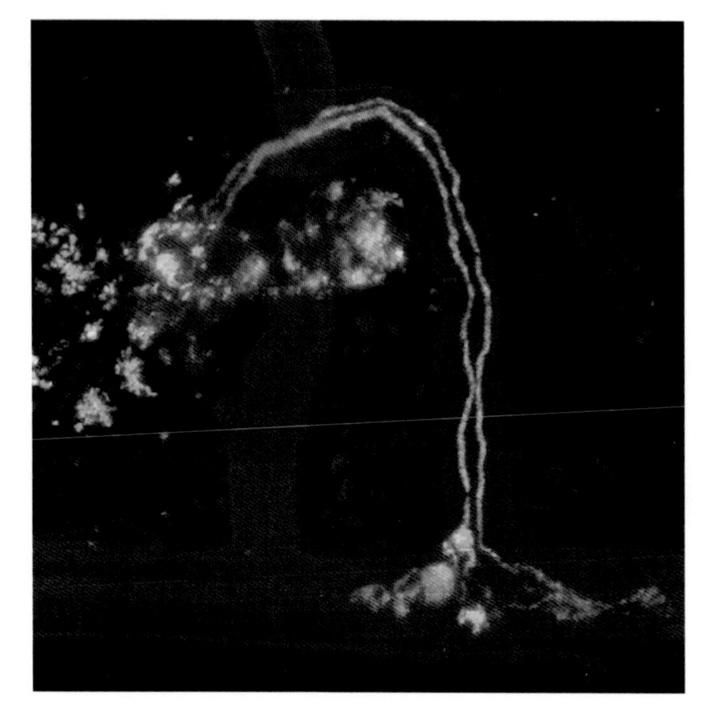

FIG. 6.4 Laser scanning confocal images of *R. meloliti* infection threads in plant root hairs. The *R. meloliti* bearing a plasmid with a *trp-gfp* fusion was used to infect alfalfa plants. Infection threads (green) can be seen within individual root hair (red) as they extend toward the main root body (stained red with propidium iodide).

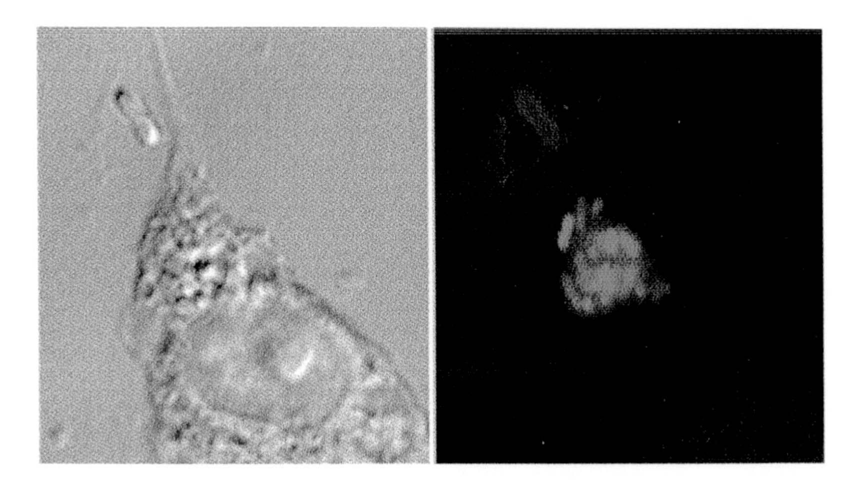

FIG. 6.5 Visualization of *S. typhimurium* intracellular-specific gene expression by fluorescence microscopy. *S. typhimurium* bearing a *pagA::gfp* fusion shows gene induction (green) inside an infected mammalian cell but not in the extracellular medium (immunostained red). The corresponding DIC images show the relative topology of bacteria with respect to the infected cell.

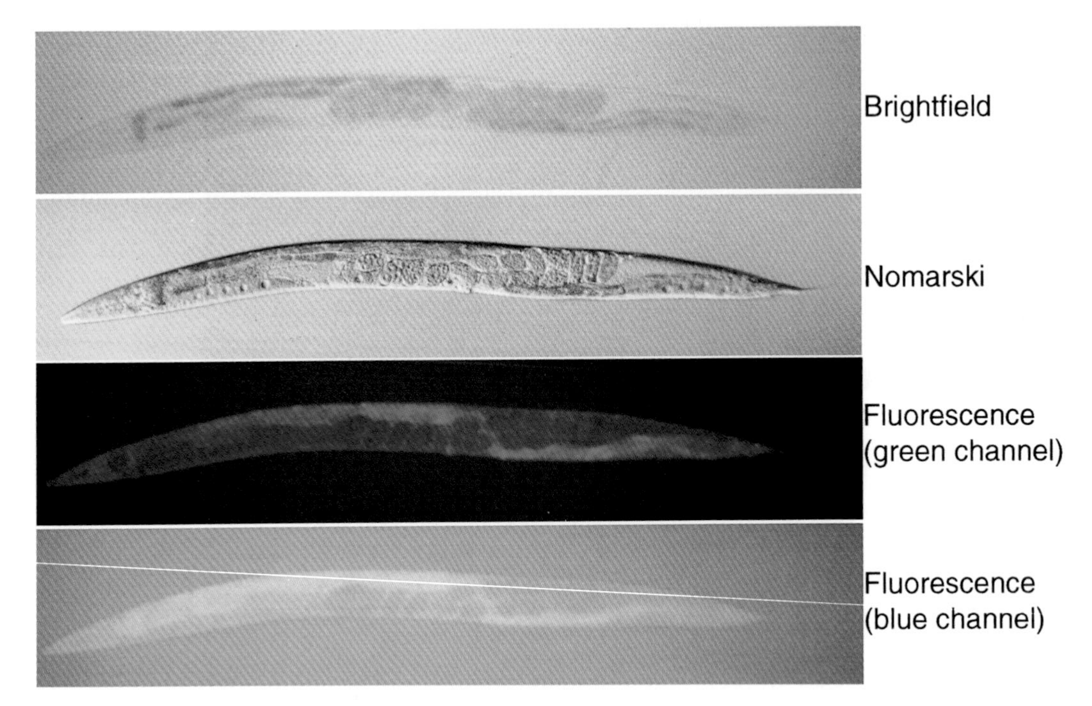

FIG. 8.1 Wild-type (nontransgenic) C. *elegans* adult viewed under several different optical conditions: Brightfield observation. The animal is virtually transparent; thus distinctions between tissues are barely visible under brightfield observation. Nomarski differential interference contrast microscopy. Optically enhanced microscopy produces the familiar image of C. *elegans* with well defined internal structures. Fluorescence observation with blue illumination [standard blue-excitation (fluorescein) relatively broad band filter set (Olympus BH2-DMB)]: predominant staining is seen in the gut. Faint staining is seen in other areas. Fluorescence observation with violet illumination [Standard UV/Violet (DAPI) relatively broad band filter set (Olympus BH2-DMUV)]: predominant staining is seen in the gut. Scale: The animal is 1 mm long.

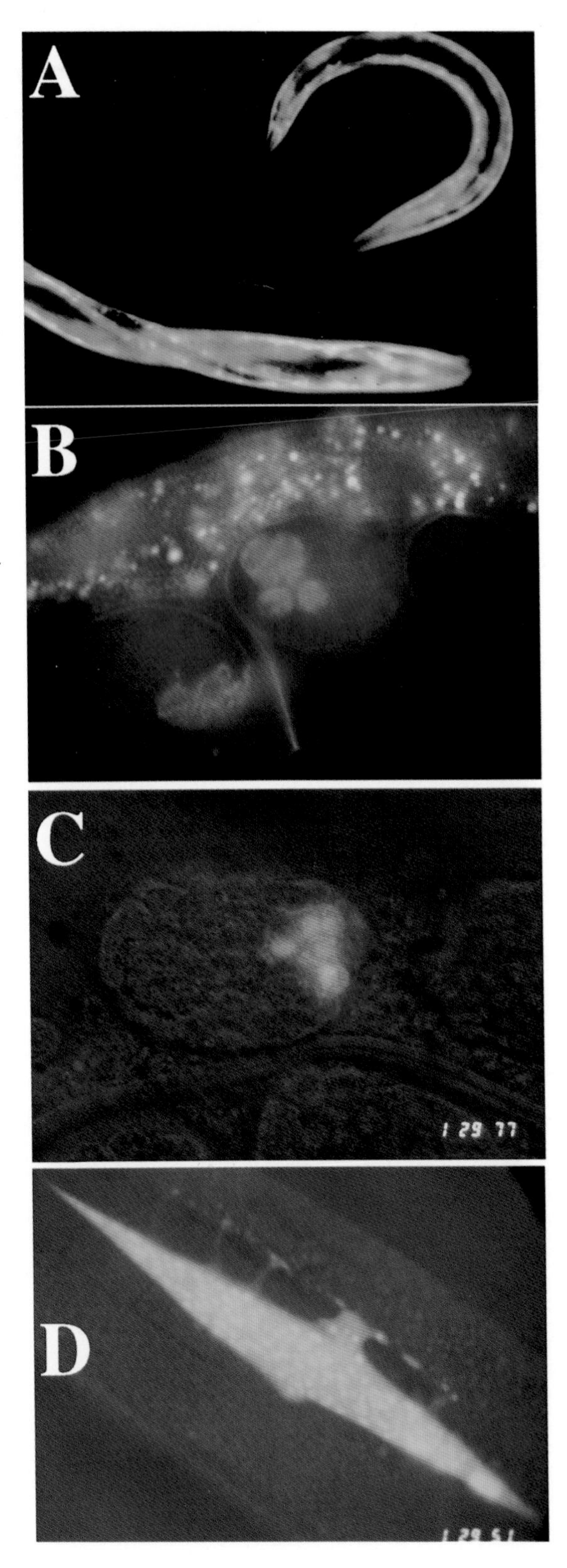

FIG. 8.2 Examples of GFP expression in C. *elegans*. All photos show trangenic lines carrying specific promoters driving S65C version of GFP. Panel *(a) myo-3* promoter (Okkema et al., 1993) is used to drive expression of partially nuclear-targeted GFP in body-wall muscles. Panel *(b)*: pes-10 promoter (Seydoux and Fire, 1994) is used to drive expression of mitochondrially targeted GFP in early somatic blastomes. Top (granular) signal is yellowish gut fluorescence of parent animal. Embryonic fluorescence (lower) is due to *gfp* construct. The *gfp* signal is present in punctate structures that are presumed to be mitochondria. More diffuse general staining in these cells may represent a combination of out-of-focus fluorescence, and *gfp*, which has not entered mitochondria. Panel *(c)*: *hlh-l* promoter (Krause et al., 1994) is used to drive *gfp* expression in myogenic precurser cells of the midstage C. *elegans* embryo. Panel *(d)*: High-magnification photo of mitochondrial-targeted *gfp* expressed in a single body-wall muscle cell. Faint projections from the cell are presumably the muscle arms that join the cell to the adjacent nerve chord.

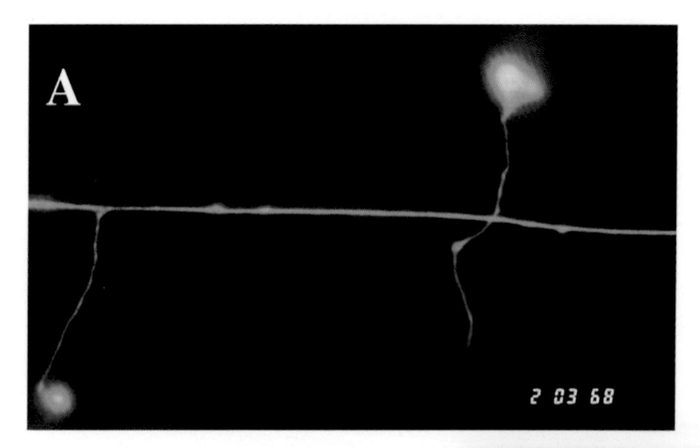

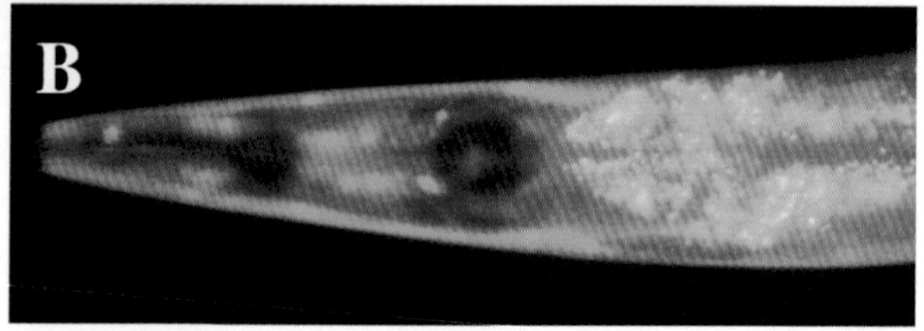

FIG. 8.3 Examples of two GFP applications in C. *elegans*. Panel *(a)*: Localization of unaltered GFP throughout the cytosol of touch neurons (see Chalfie et al., 1994). The cell bodies (slightly out of focus) are uniformly stained. Axon tracks for the two cells are clearly visible in the plane of focus. The GFP was expressed from the *mec-7* promoter using similar constructs to those of Chalfie et al. (1994), with modifications to increase activity (F64L S65T, and three internal introns) as described in the text. Panel *(b)*: A transgenic animal expressing two different *gfp* constructs, each driven from the body-muscle-specific *unc-54* promoter. This single-exposure simple photograph of an animal is illuminated using standard "DAPI" (Violet excitation) filter set. The two *gfp* constructs used are the *gfp* (S65C) expressed throughout the cytosol and the "Blue" *gfp* (Y66H Y145F) targeted to the nucleus by insertion into a construct with the SV40 nuclear localization signal appended to *lacZ* (Fire et al., 1990). At the left in the photo (anterior in the animal) are muscle cells with blue nucleus and green cytoplasm; strong blue fluorescence on the right side of the photo derives from autofluorescent granules in the gut.

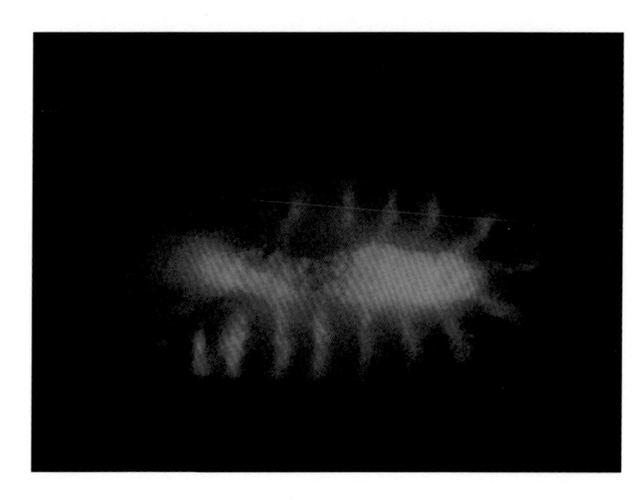

FIG. 9.2 Germband-extended embryo expressing mutant GFPS65T in the engrailed stripes. GFPS65T is distributed throughout the nuclei and the cytoplasm. Anterior is to the left. Imaging was on a Zeiss Axiophot epifluorescence microscope, using a 20X Neofluar objective and a standard Zeiss FITC band pass filter. [Courtesy of Felipe-Andres Ramirez-Weber and Thomas Kornberg.]

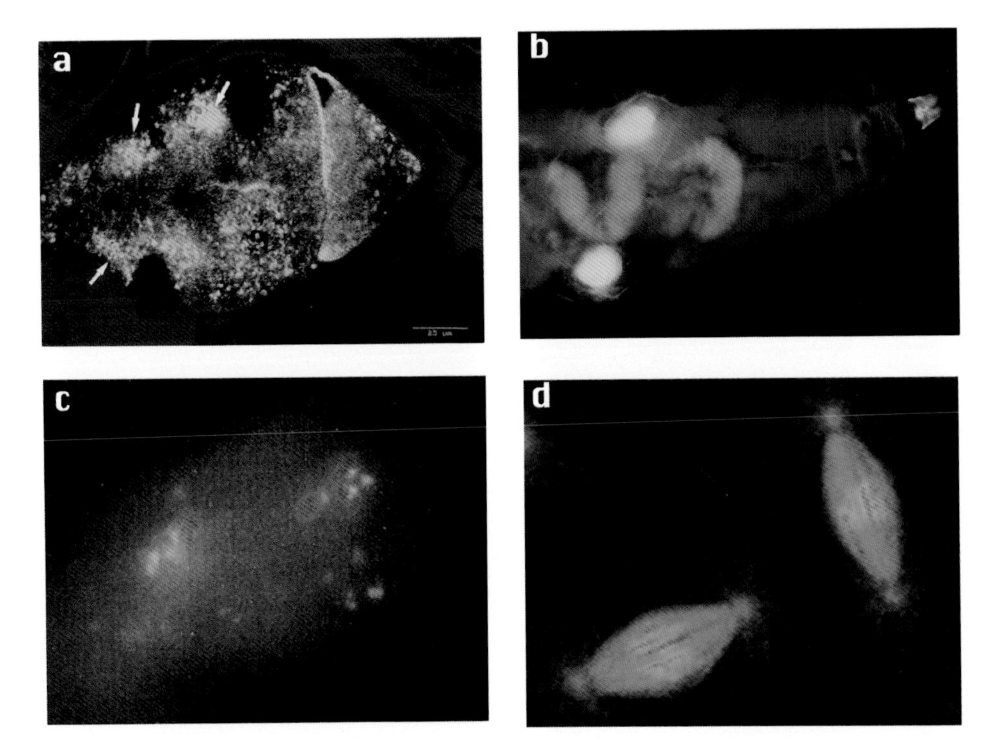

FIG. 9.5 GFP fusion proteins. *(a)* GFP-Exu particles in oogenesis. A fusion was made of GFP to the protein product of the *exuperantia* gene, required during oogenesis for correct localization of *bcd* mRNA to the developing oocyte. The fixed stage 9 egg chamber was stained with rhodamine-phalloidin (red) to detect the position of the actin-rich ring canals (arrows) relative to the GFP-Exu particles (green). The oocyte is on the right, connected via ring canals to the nurse cells that lie to the left. Imaging was with a BioRad MRC600 confocal unit attached to a Zeiss Axioskop microscope, using K1 and K2 filter blocks. (Reprinted with permission from *Nature*, Wang, S. and Hazelrigg, T. (1994). Implications for *bcd* mRNA localization from spatial destruction of *exu* protein in *Drosophila* oogenesis, 369:400-403. Copyright © 1994 Macmillan Magazine Limited.] *(b)* GFP-Exu in third instar male larva. The *exuperantia* gene is also expressed in the testis, where it is required for normal spermatogenesis (Hazelrigg et al., 1990). A live male larva was mounted in halocarbon oil on a slide and covered with a coverslip supported at its edges with additional coverslips. The signal from GFP-Exu in primary spermatocytes is visible through the body wall of the larva. Imaging was on a Zeiss Axiophot microscope, using a 40X plan-neofluar objective (N.A. 1.3) and a FITC 09 filter set. *(c)* MeiS332-GFP in male meiosis. A fusion gene was constructed with *meiS332*, a gene required for normal meiosis, and the distribution of the MeiS332-GFP fusion protein was observed in male meiosis. (Kerrebrock et al., 1995). Fixed testes were incubated with 7-AAD to detect the chromosomes. In this anaphase I cell, MeiS332-GFP (green) is localized to the centromere regions at the leading edges of the chromosomes (red). Imaging was with a Nikon Optiphot-2 microscope equipped with a Photometrics Image Point cooled CCD camera. [Courtesy of Terry Orr-Weaver. Reprinted with permission from Kerrebrock et al., 1995. Copyright © Cell Press.] *(d)* Ncd-GFPS65T in the metaphase spindles of a cycle 9 embryo. GFPS65T was fused at the C-terminus of the coding region of *ncd*, a member of the *Drosophila* kinesin gene family (Endow and Komma, 1996). Live embryos from mothers expressing Ncd-GFPS65T were injected with rhodamine-histone to visualize the chromosomes. The Ncd-GFPS65T fusion protein (green) is present in the spindles, and excluded from the chromosomes (blue), and can be seen in spindle fibers that extend across the chromosomes. Imaging was done on a BioRad MRC600 laser scanning confocal unit attached to a Zeiss Axioskop microscope, using a planapochromat 63X (N.A. 1.4) objective. [Courtesy of S. Endow. Reprinted with permission from Endow and Komma, 1996 and Company of Biologists, Ltd.]

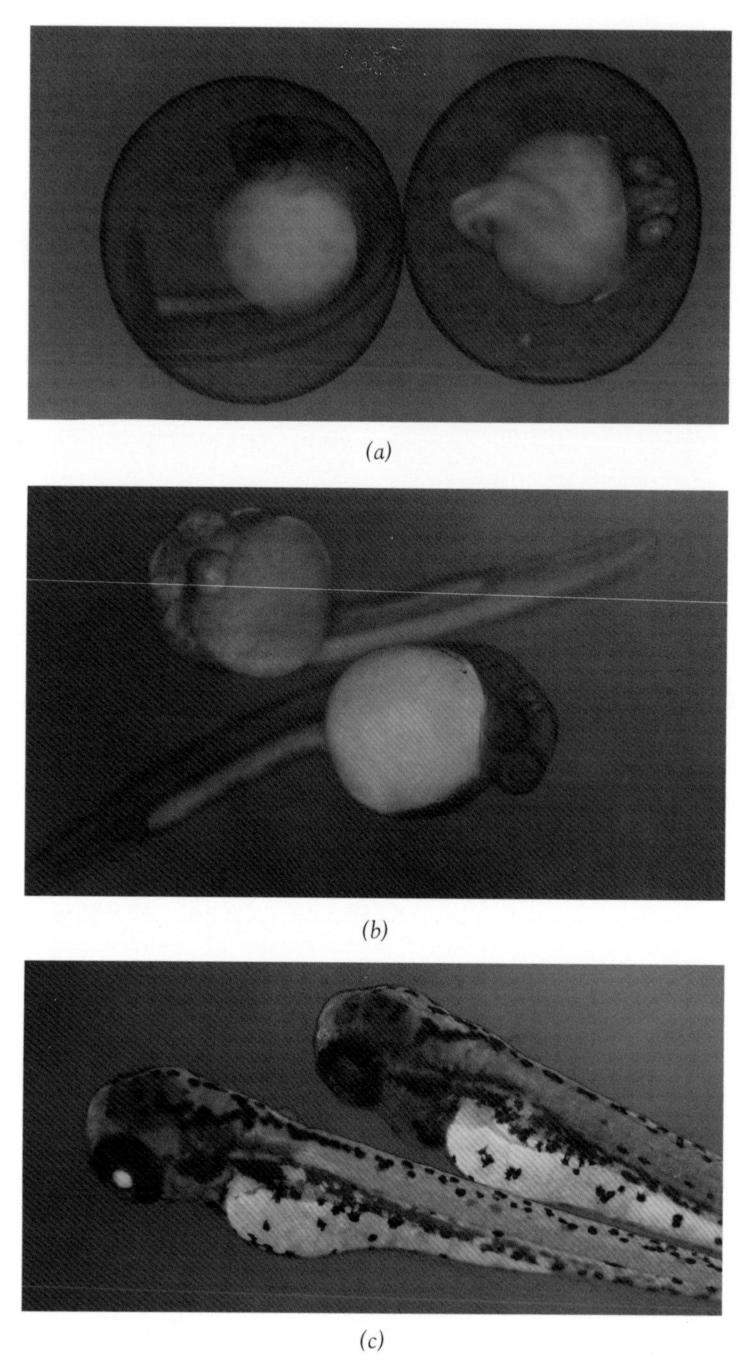

(a)

(b)

(c)

FIG. 11.1 Green fluorescent protein can be detected in stable zebrafish transgenics. Transgenic and nontransgenic siblings for a transgene containing the *Xenopus* ef1αa enhancer/promoter and the rabbit ß-globin first intron fused to the gfp gene. Sibling progeny of a heterozygous transgenic fish were observed with a Nikon microphot EPI-FL3, with a 370-420-nm excitation filter and a 455-nm long-pass emission filter. (*a*) Embryos at 24-h postfertization, still in their chorions. (*b*) Embryos at 32-h postfertilization. (*c*) Embryos at 72-h post-fertilization. Notice the strong fluorescence in the eye, but reduced fluorescence in the rest of the body. (Reprinted from Amsterdam et al., 1995)

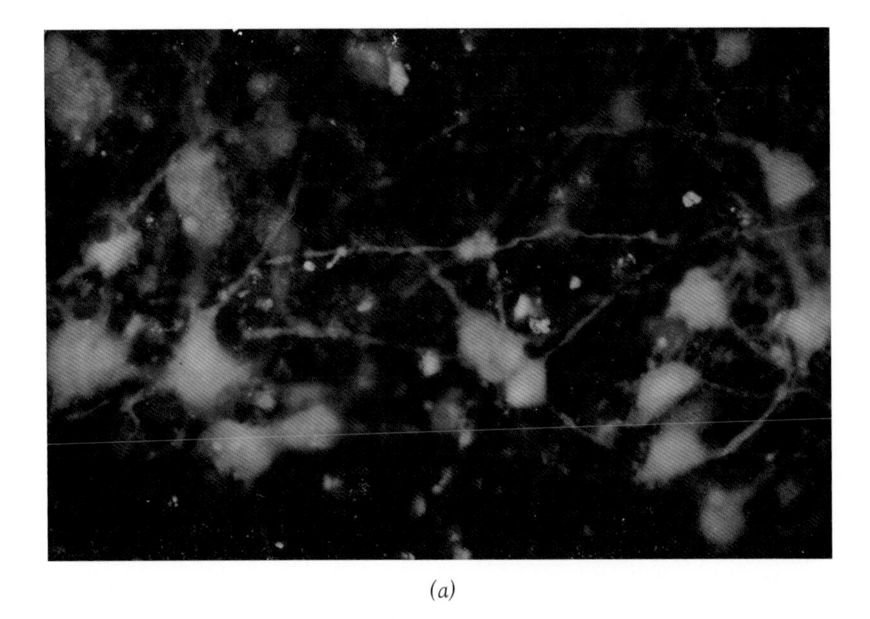

(a)

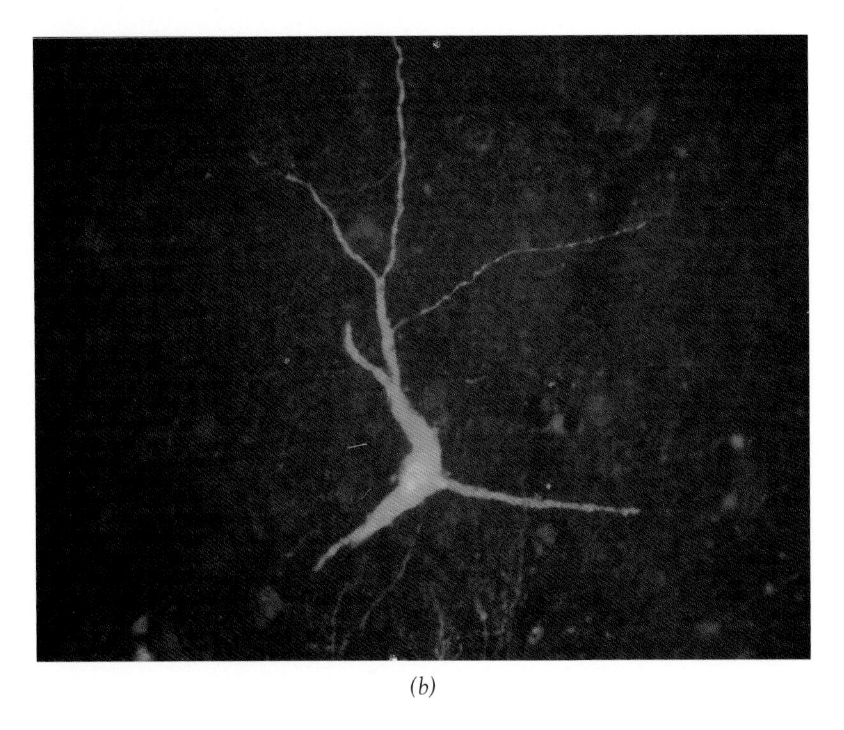

(b)

FIG 11.2 Green fluorescent protein can be detected specifically in neurons infected with a recombinant AAV(rAAV) in which a neuron-specific promoter is fused to the *gfp* gene. Adult rats received 5-mL injections of rAAV pTRNSE into the cervical enlargement of the spinal cord. pTRNSE contains a humanized *gfp* gene under the control of the neuron specific encolase (NSE) promoter. Animals were sacrificed and examined histologically using Zeiss FITC optics 2 weeks later. *(a)* Many infected neurons expressing GFP. (b) A single infected neuron expressing GFP. Note how the fluorescence fills the axonal and dentritic processes. [Photographs courtesy of A. Peel.]

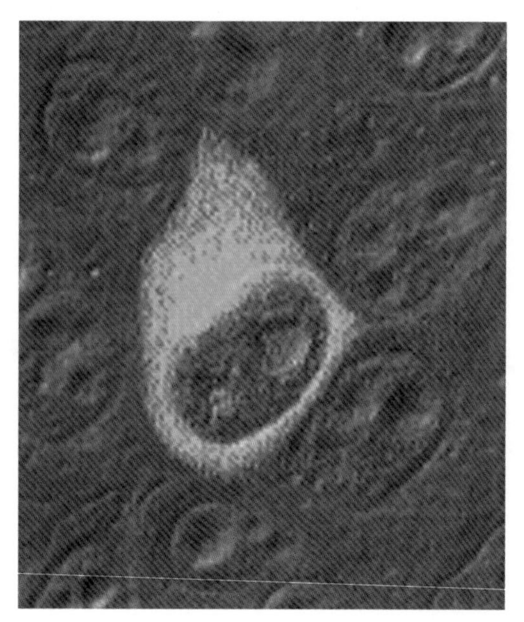

FIG 12.1 Co-expression of Rev protein in the nucleoli of living cells (blue) with Gag protein in the cytoplasm (green) using two different mutants sg25 (green) and sg50 (blue), developed by George Gaitanaris and Roland Stauber. Picture was taken by Roland Stauber and Eric Hudson. See Stauber et al. (1998).

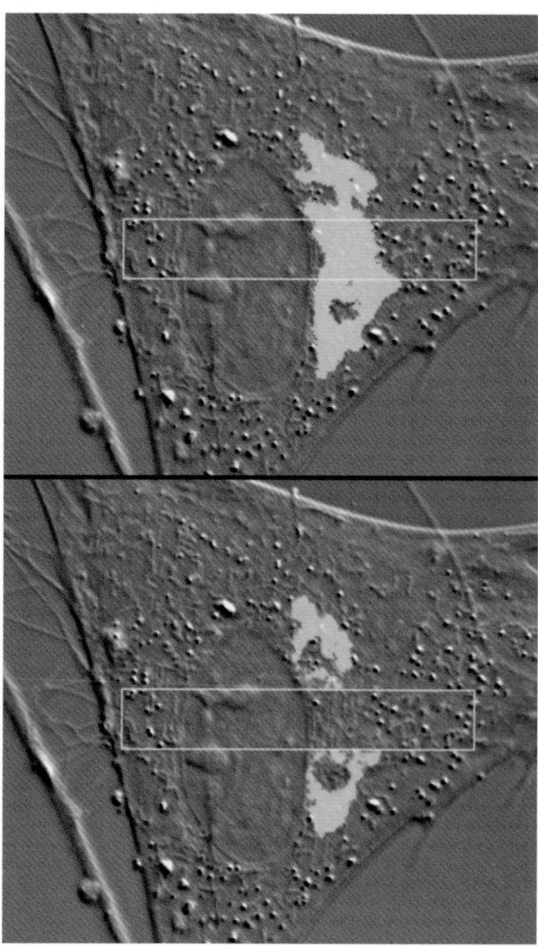

FIG 12.2 The HeLa cell expressing GFP Golgi chimera before (above) and immediately after (below) photobleaching of a small rectangular box. [Courtesy of Cole et al., 1996].

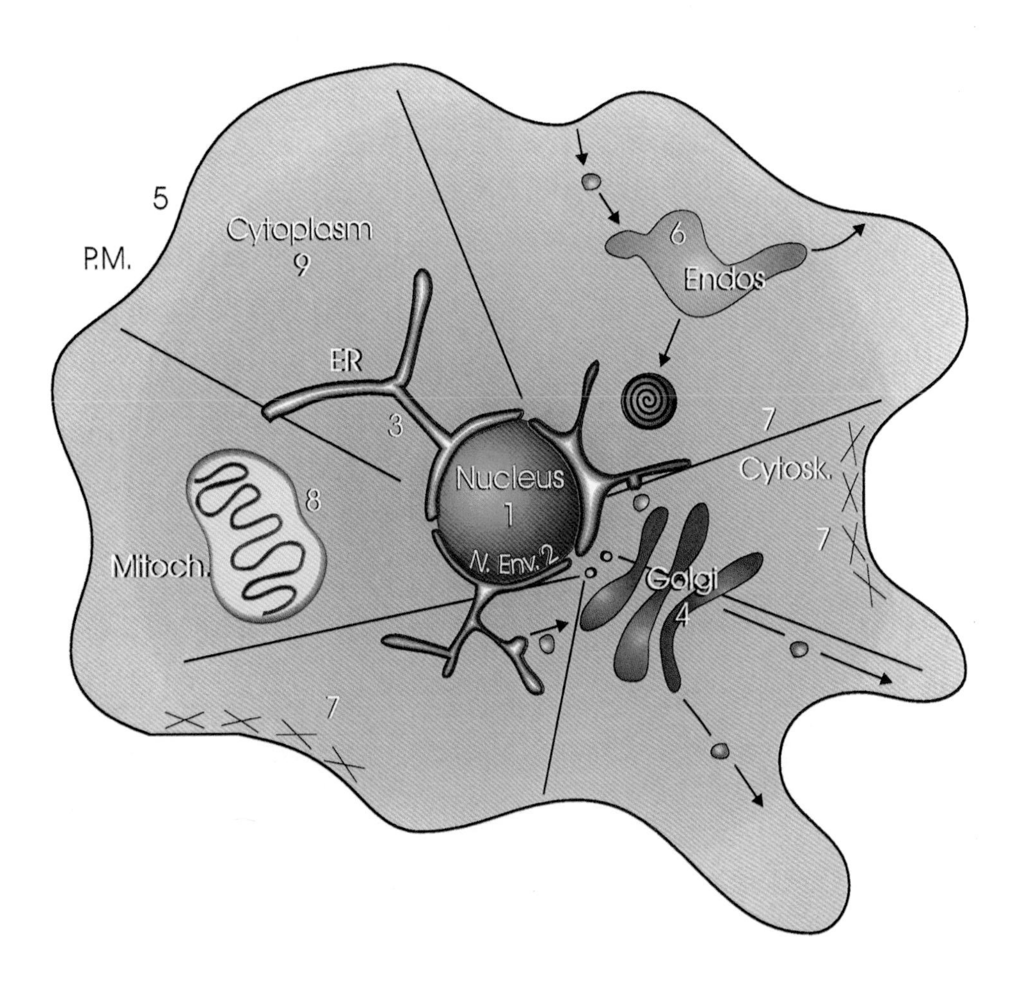

FIG 12.3 Examples of GFP chimeras and their subcellular localization.

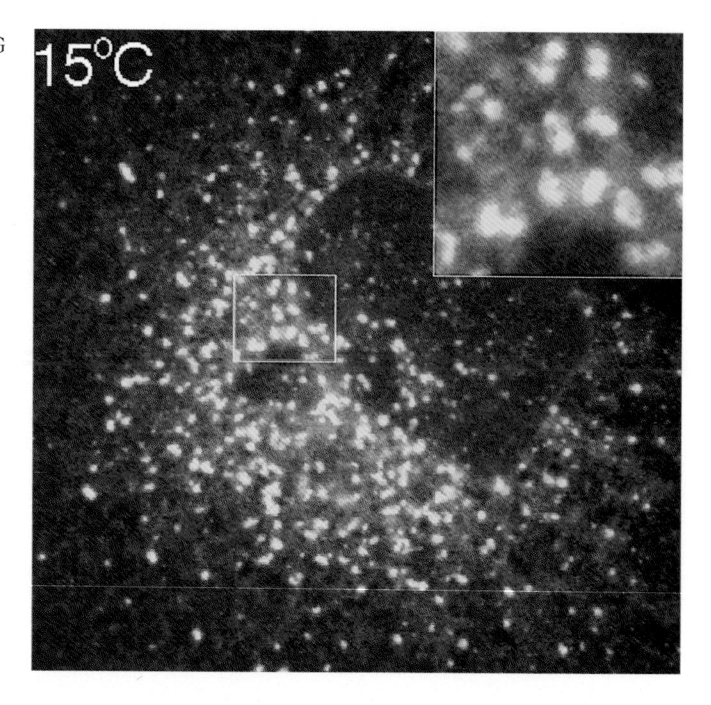

FIG 12.5 The GFP tagged VSVG protein expressed in Cos cells at 15°C and its colocalization with ßCOP protein. Green is VSVG-GFP, red represents ßCOP antibody staining, and yellow represents the overlap of the two. [Courtesy of Presley et al., 1996].

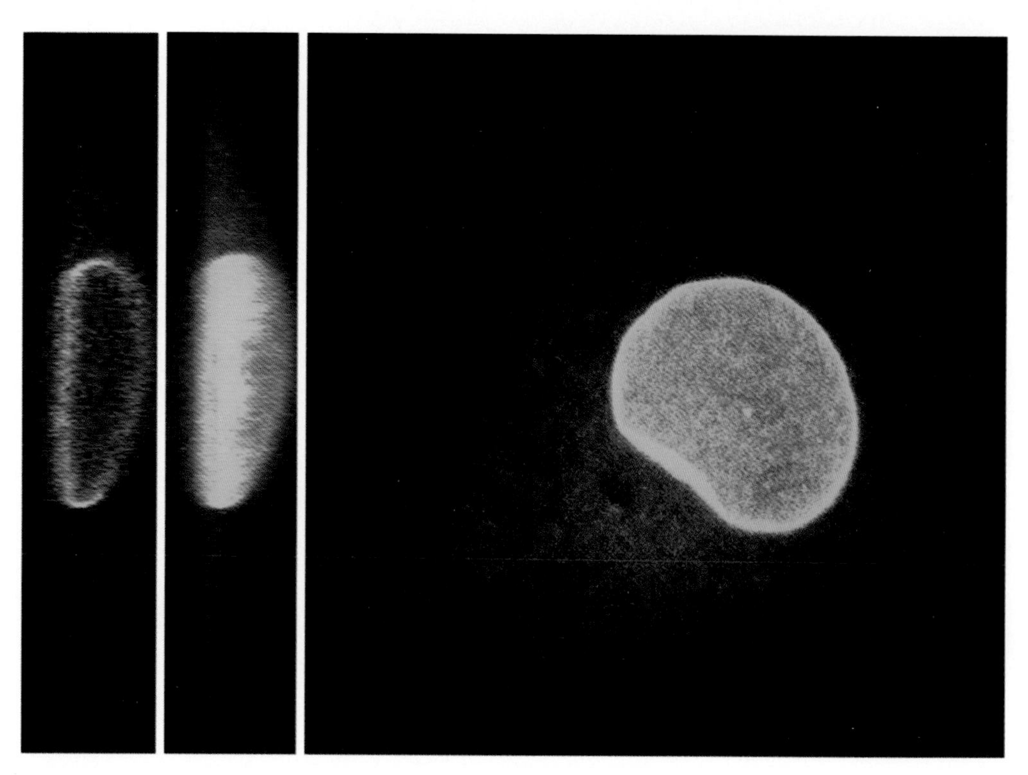

FIG 12.6. Localization of GFP tagged lamin B receptor to the inner nuclear envelope of Cos cells. [Courtesy of Ellenberg et al., 1996].

A stringent test for the proper function of a *gfp*-tagged gene is the ability to rescue a mutant in the corresponding chromosomal locus. This rescue has proven feasible for a wide variety of protein types including transcription factors, other nuclear components, cytosolic proteins, transmembrane factors, and nuclear components (unpublished data from our laboratory, Zhu and Hedgecock, personal communication). In setting up such a tagging experiment, several constructs are produced with *gfp* inserted at different sites; these are then tested for rescue of the corresponding mutant. If rescue is observed, the experimenter gains access to transformed animals in which the function of the molecule (as determined by phenotypic rescue) can readily be associated with the expression pattern as assessed by *gfp* observation.

8.3 APPLICATIONS FOR *gfp* IN *C. elegans*

8.3.A How Well Do *gfp* Fusions Mimic the Endogenous Gene Product?

Many of the experiments that one would like to carry out with *gfp* rely on the hope that *gfp*-tagged transgenes will behave similarly to the original chromosomal counterparts. In many cases, this has indeed been true. In other cases, transgenes do not faithfully represent the expression pattern of the exogenous gene (see Mello and Fire, 1995 for a further discussion of this topic). There are frequent examples in which a transgene can express in sets of cells not seen with the endogenous locus; in particular, posterior gut and pharynx are frequent sites for such ectopic expression (e.g., Krause et al., 1994). Conversely, in some cases, extensive flanking sequences are insufficient to give the complete physiological expression pattern. A related (and often problematic) property of *C. elegans* transgenes is mosaicism. Transgene expression in the correct cell type can be stochastic, appearing in only a fraction of animals or cells that carry and should express the transgene.

The determination of whether a given transgene correctly represents the endogenous locus should be carried out separately for each gene; nonetheless several general properties have been emerging.

1. Somatic expression of transgenes has been more straightforward to obtain than expression in the germline (e.g., Fire et al., 1990; Hope, 1991). Vectors for germline expression are currently a major technical goal in the field.
2. To maximize the chances that relevant cis-acting signals are included in the relevant expression constructs, it is advisable to include as much flanking sequence as possible. We often include 4–5-kb upstream of the gene as well as internal sequences and extensive downstream sequences. Internal control elements (e.g., Okkema et al., 1993) as well as control signals in 3′ regions (e.g., Ahringer and Kimble, 1991; Wightman et al., 1991) have been reported in many *C. elegans* genes. Thus, it is most desirable to use "tagged" constructs that contain the entire coding and

flanking regions of a gene, with *gfp* inserted internally with maintained reading frame both "reading in" and "reading out."

3. Since there is no guarantee (even with extensive flanking sequences present) that a given transgene will behave similarly to the endogenous locus, expression pattern should be assessed if possible by independent means. Frequently used methods for determining the expression patterns of endogenous genes are immunofluorescence detection of protein (Miller and Shakes, 1995) and in situ hybridization to RNA (Albertson et al., 1995; Seydoux and Fire, 1995).
4. In cases in which *gfp* has been inserted into an endogenous protein with retention of all (or a majority) of the protein coding sequences, it is possible to test the functional activity of the transgene by its ability to rescue a mutant in the endogenous gene. Although rescue of a mutation in the endogenous gene does not definitively prove that endogenous locus and transgene have identical expression patterns, it is a valuable consistency argument. In addition, this can provide a good argument that the intracellular localization of the original protein and the GFP tagged derivative might be similar.

8.3.B Following a Marked Cell Population with Time

The cell lineage of *C. elegans* was originally determined by painstaking experiments in which individual cells were followed in real-time using simple differential interference contrast (nomarski) microscopy (Sulston et al., 1983). The tools for observation in the earliest embryonic stages have been greatly improved using computer image acquisition systems (Thomas et al., 1996; Schnabel et al., 1997; Fire, 1994) that allow the whole history of an embryo to be recorded. These techniques still involve a large investment in analysis. In addition, the real-time tracking of cells is based only on their history and on the broad morphological features that can be distinguished using differential interference contrast microscopy.

Transgenic strains carrying *gfp* fusions active in specific cells will allow individual cell populations to be followed by virtue of their fluorescent properties; at the same time, this should allow the distribution and level of fluorescence to be followed as a function of the life history of the lineage. The technical means for carrying out this work (optimally) should involve a low light level fluorescence observation system connected to a computer image acquisition system that can collect fluorescence and nomarski images in multiple focal planes from a single sample at each of a large number of consecutive time points.

8.3.C Tracking Molecular Populations in the Living Embryo

In essence, the above discussion of tracking cell populations can be extended to the ability to track organelles and molecular populations, using *gfp*-tagged variants of proteins normally residing in these structures. The ability to follow chromosomes, membrane compartments, ribosomes, cytoskeletal elements are but a few examples of such applications. Given the available sequence of most

C. elegans genes and ease with which *gfp*-tagged versions of specific proteins can be introduced into transgenic lines, it seems likely that many such studies will be carried forward in the near future.

8.3.D Analyzing Mutant or Experimental Effects on Gene Expression or Protein Localization

The rich genetic history of *C. elegans* has provided us with many mutations affecting specific biological processes. The cloning of the corresponding genes and characterization of gene product distributions has been an initial goal; for many pathways this work is already well underway. Complementary to this work, it is of great interest to provide a detailed molecular characterization of defects or aberrations in each mutation. Green fluorescent protein based technology can be extremely useful in this respect, since individual biochemical and cellular processes can be tracked in the whole animal. Thus, for instance, the study of mutations affecting meiotic chromosome segregation (Hodgkin et al., 1979) could be greatly aided by the availability of strains in which each chromosome in meiosis was fluorescently labeled, allowing a real time analysis of effects on segregation. Of course, there will be a myriad of other examples.

8.3.E Isolation of New Mutations Using *gfp* Screens

Strains with populations of cells or molecular structures "tagged" with *gfp* provide an ideal starting point in screens for new mutations. Because *gfp* can be observed easily without killing or fixing animals, it is straightforward to carry out a mutagenesis on a population of animals carrying a *gfp*-tagged transgene and screen for alterations in the pattern of activity. Depending on the nature of the mutations to be isolated, the screen can be done in the generation following mutagenesis (looking for dominant mutations), in the second generation (looking for recessive zygotic effect mutations), or in the third generation (looking for recessive maternal effect mutations). Virtually any *gfp* strain could be the starting material for such screens. If the recovered mutations are expected to be sterile or lethal, then the screen will need to be performed in a clonal or semiclonal manner, so that siblings of interesting affected animals can be easily recovered.

Screening for alterations in the observed transgene activity pattern (e.g., Xie et al., 1995) can be extremely general, and are expected to yield several types of mutations. One class are mutations in the transgene sequences that lead to alterations in the pattern. Such mutations can be of use, but are often of less interest in attempts to understand the endogenous mechanisms. Mutations in chromosomal genes affecting the process being analyzed are generally of considerably greater interest. For this reason, it is advantageous to carry out transgene based screens so that the transgene is not present in the mutagenized animals (e.g., by subsequently crossing the transgene into the population) or alternatively to include genetic markers in the mutagenized strain that allow easy distinction between transgene and endogenous chromosomal mutations.

8.4 CONCLUSIONS

The full impact of *gfp* technology on our ability to address biological questions in C. *elegans* has only begun. In the next few years, our understanding of this organism and of the biological processes that can be studied in a transparent invertebrate system will undoubtedly move forward rapidly based on this new technology.

ACKNOWLEDGMENTS

We thank Marty Chalfie for introducing the use of *gfp* to C. *elegans*, H. Hutter, Xiaoping Zhu, Ed Hedgecock for communicating their unpublished data, Mary Montgomery for helpful suggestions on the use of *gfp* and on the manuscript, and A. Cubitt and R. Tsien for an early gift of antisera to *gfp*. Work in our laboratory is supported by the National Institutes of Health (Grant R01-GM37706 [AF], F32-HD07794 [WK]).

REFERENCES

Ahringer, J. and Kimble, J. (1991). Control of the sperm-oocyte switch in *Caenorhabditis elegans* hermaphrodites by the fem-3 3' untranslated region. *Nature (London)* 349: 346–348.

Albertson, D., Fishpool, R., and Birchall, P. (1995). Fluorescence in situ hybridization for the detection of DNA and RNA. *In Caenorhabditis elegans*: Modern Biological Analysis of an Organism. Methods in Cell Biology, Vol. 48, Epstein H. F. and Shakes D. C. Eds., Academic, San Diego, CA, pp. 340–364.

Babu, P. (1974). Biochemical genetics of *Caenorhabditis elegans*. *Mol. Gen. Genet.* 135:35–44.

Brenner, S. (1974). The genetics of *Caenorhabditis elegans*. *Genetics* 77:71–94.

Broverman, S., MacMorris, M. and Blumenthal, T. (1993). Alteration of *Caenorhabditis elegans* gene expression by targeted transformation. *Proc. Natl. Acad. Sci. USA* 90:4359–4363.

Chalfie, M., Tu, Y., Euskirchen, G., Ward, W., and Prasher, D. (1994). Green Fluorescent protein as a marker for gene expression. *Science* 263:802–805.

Clokey, G. V. and Jacobson, L. A. (1986). The autofluorescent "Lipofuscin granules" in the intestinal cells of *Caenorhabditis elegans* are secondary lysosomes. *Mech. Aging Develop.*, 35:79–94

Cormack, B., Valdivia., R. H., and Falkow, S. (1996). FACS optimized mutants of the green fluorescent protein (*gfp*). *Gene* 173:33–38.

Davis, B. O., Anderson, G. L., and Dusenbery, D. B. (1982). Total luminescence spectroscopy of fluorescent changes during aging in *Caenorhabditis elegans*. *J. Biochem.* 21:4081–4095

Epstein H. F. and Shakes D. C., Eds. (1995). *Caenorhabditis elegans*: Modern Biological Analysis of an Organism. Methods in Cell Biology, Vol 48, Academic, San Diego, CA.

Fire, A. (1994). A four dimensional digital image archiving system for cell lineage tracing and retrospective embryology. Computer applications in the biological sciences (CABIOS), 10:443–447.

Fire, A., Harrison, S., and Dixon, D. (1990). A modular set of *lacZ* fusion vectors for studying gene expression in *Caenorhabditis elegans*. *Gene* 93:189–198.

Heim, R. and Tsien, R. Y. (1996). Engineering green fluorescent protein for improved brightness, longer wavelengths and fluorescence resonance energy transfer. *Curr. Biol.* 6:178–182.

Heim, R., Cubitt, A., and Tsien, R. (1995). Improved Green Fluorescence. *Nature (London)* 373:663–664.

Heim, R., Prasher, D. and Tsien, R. (1994). Wavelength mutations and posttranslational autooxidation of green fluorescent protein. *Proc. Natl. Acad. Sci. USA* 91:12501–12504.

Hodgkin J., Horvitz, H.R. and Brenner, S. (1979). Nondisjunction mutants of the nematode *C. elegans*. *Genetics* 91: 67–94.

Hodgkin, J., Edgley, M., Riddle, D., and Albertson, D. (1988). Genetics. *In* The Nematode *Caenorhabditis elegans.*, Wood, W. B., Ed., Cold Spring Harbor Press, Cold Spring Harbor, NY, pp. 491–584.

Hodgkin, J., (1997). Protocol I Genetics. *In C. elegans II*. Riddle, D. L., Blumenthal, T., Meyer, B. J., and Priess, J. R., Eds., Cold Spring Harbor Press, Cold Springs Harbor, NY, pp. 881–1047.

Hope, I. A. (1991). 'Promoter trapping' in *Caenorhabditis elegans*. *Development*. 113:399–408.

Kimble, J. and Hirsh, D. (1979). The post-embryonic lineage of the hermaphrodite and male gonads in *Caenorhabditis elegans*. *Dev. Biol.* 70:396–417.

Krause, M., White Harrison, S., Xu, S., Chen, L., and Fire, A. (1994) Elements regulating cell and stage-specific expression of the *C. elegans* MyoD homolog hlh-1. *Dev. Biol.* 166: 133–148.

Mello, C. and Fire, A. (1995). DNA transformation. *In Caenorhabditis elegans:* Modern Biological Analysis of an Organism. Methods in Cell Biology, Vol. 48, Epstein, H. F. and Shakes, D. C., Eds. Academic, San Diego, CA, pp. 452–482.

Miller, D. M. and Shakes, D. C. (1995). Immunofluorescence microscopy. *In Caenorhabditis elegans*: Modern Biological Analysis of an Organism. Methods in Cell Biology. Vol. 48, Epstein H. F. and Shakes, D. C. Eds., Academic, San Diego, CA, pp. 365–395.

Moriyoshi, K., Richards, L. J., Akazawa, C., O'Leary, D. D. M., and Nakanishi, S. (1996). Labeling neuronal cells using adenoviral gene transfer of membrane-targeted GFP. *Neuron* 16, 255–260.

Okkema, P., White–Harrison, S., Plunger, V., Aryana, A. and Fire, A. (1993). Sequence requirements for myosin gene expression and regulation in *C. elegans*. *Genetics* 135:385–404.

Ormö, M., Cubitt, A. B., Kallio, K., Gross,. L. A., Tsien, R. Y., and Remington, S. J. (1996). Crystal structure of the Aequora victoria green fluorescent protein. *Science* 273:1392–1395.

Riddle, D. L., Blumenthal, T., Meyer, B. J., and Priess, J. R. (1997). *C. elegans II*. Cold Spring Harbor Laboratory, Cold Spring Harbor, New York.

Schnabel, R., Hutter, H., Moerman, D., and Schnabel, H. (1997). Assessing normal embryogenesis in Caenorhabditis elegans using a 4D microscope: Variability of development and regional specification. *Dev. Biol.* 184:234–265.

Sengupta, P., Chou, J. H. and Bargmann, C. I. (1996). odr-10 encodes a seven transmembrane domain olfactory receptor required for responses to the odorant diacetyl. *Cell* 84: 899–909

Seydoux, G. and Fire, A. (1994). Soma-germline asymmetry in the distributions of embryonic RNAs in *Caenorhabditis elegans*. *Development* 120:2823–2834.

Seydoux, G. and Fire, A. (1995). Whole mount in situ hybridization for the detection of RNA in *C. elegans* embryos. *In Caenorhabditis elegans:* Modern Biological Analysis of an

Organism. Methods in Cell Biology Vol 48, Epstein, H. F. and Shakes, D. C., Eds., Academic, San Diego, CA, pp. 323–339.

Stinchcomb, D. T., Shaw, J. E., Carr, S. H. and Hirsh, D. (1985). Extrachromosomal DNA transformation of *Caenorhabditis elegans*. *Mol. Cell Biol.* 5:3484–3496.

Sulston J. E., Schierenberg, E., White, J. G., and Thomson, J. N. (1983). The embryonic cell lineage of the nematode C. *elegans*. *Dev. Biol.* 100:64–119.

Sulston, J. E. and Horvitz, H. R. (1977). Postembryonic cell lineages of the nematode *Caenorhabditis elegans*. *Dev. Biol.* 82:41–55.

Thomas, C., Devries, P., Hardin, J., and White, J. (1996). Four-dimensional imaging—computer visualization of 3D movements in living specimens. *Science* 273:603–607.

Waterston, R. and Sulston, J. (1995). The genome of *Caenorhabditis elegans*. *Proc. Natl. Acad. Sci. USA* 92:10836–10840.

White, J. G., Southgate, E., Thomson, J. N. and Brenner, S. (1986). The structure of the nervous system of the nematode *Caenorhabditis elegans*. *Philos. Trans. R. Soc. London, Ser. B* 314:1–340.

Wightman, B., Burglin, T. R., Gatto, J., Arasu, P., and Ruvkun, G. (1991). Negative regulatory sequences in the lin-14 3′ untranslated region are necessary to generate a temporal switch during Caenorhabditis elegans development. *Genes Dev.* 5:1813–1824.

Wood, W. and the Community of C. *elegans* Researchers Eds. (1988). The Nematode *Caenorhabditis elegans*. Cold Spring Harbor Laboratory, Cold Spring Harbor, NY.

Xie G.F., Jia Y., and Aamodt, E. (1995). A C. *elegans* mutant screen based on antibody or histochemical staining. Genetic Analysis: *Biomol. Eng.* 12:95–100.

Yang, F., Moss, L. G., and Phillips, G. N. (1996). The molecular structure of Green Fluorescent Protein. *Nature Biotechnol.* 14:1246–1251.

Yang, T. T., Kain, S. R., Kitts, P., Kondepudi, A., Yang, M. M., and Youvan, D. C. (1996b). Dual color microscopic imagery of cells expressing the green fluorescent protein and a red-shifted variant. *Gene* 173:19–23.

9

The Uses of Green Fluorescent Protein in *Drosophila*

TULLE HAZELRIGG
Department of Biological Sciences, Columbia University, New York, NY

9.1 INTRODUCTION

Green fluorescent protein (GFP) is illuminating many fields of *Drosophila* research, providing answers to fundamental problems of cell biology, neurobiology, and developmental biology. GFP allows researchers to mark cells and follow their development, and to study protein trafficking and localization, in living cells. The well-developed genetics and molecular biology tools of *Drosophila*, and the P-element mediated transformation technology, make it possible to create transgenic animals expressing GFP in well-defined genetic backgrounds. In the case of GFP fusion proteins, often the GFP-tagged protein can be observed in flies missing the endogenous gene product, allowing one to determine if the GFP fusion protein is functional. Where GFP is used as a reporter gene, different cell types can be marked and followed throughout development in wild-type and mutant backgrounds.

9.2 STRATEGIES FOR EXPRESSING GFP IN *DROSOPHILA*

Transgenic *Drosophila* expressing GFP have been created by standard P-element mediated transformation (Spradling and Rubin, 1982; Ashburner, 1989). Various P-element vectors have been used, including expression vectors that provide promoters and polyadenylation sequences, and vectors that utilize the promoter

Green Fluorescent Protein: Properties, Applications, and Protocols, Edited by Martin Chalfie and Steven Kain
ISBN 0-471-17839-X

TABLE 9.1 GFP constructs expressed in *Drosophila*

Construct	Product	References
PUbnlsGFP	nlsGFP	Davis et al. (1995)
PUbGFP	GFP	Davis et al. (1995)

Construction

In both cases, wild-type *gfp* (plasmid pGFP10.1; Chalfie et al) was placed downstream of the polyubiquitin promoter, to express an mRNA with a polyubiquitin 5′ UTR, which contains an intron, and upstream of the polyadenylation signal of hsp27. Near its N-terminus, the protein product of PUbnlsGFP contains a 9 amino acid (aa) nuclear localization sequence (nls) adjacent to a 40 aa segment of β-galactosidase, followed by GFP. The PUbGFP is deleted for the nls and β-galactosidase protein sequences present in PUBnlsGFP. Both constructs were placed in the pCaSpeR4 transformation vector (Thummel and Pirrotta, 1992).

Construct	Product	References
pUAS-GFP	GFP	Yeh et al. (1995)

Construction

Wild-type *gfp* (from plasmid TU#65; Chalfie et al., 1994) was cloned into pUAST, a transformation vector designed to bring inserted coding sequences under regulation of the GAL4 transcription factor (Brand and Perrimon, 1993). In addition to being regulated by five adjacent UAS sequences, expression from this vector utilizes the *hsp70* TATA sequence and transcription start site, and at its 3′ end contains the SV40 small t intron and polyadenylation signals.

Construct	Product	References
pAct88F-GFP	GFP	Barthmaier and Fyrberg (1995)

Construction

PCR-amplified wild-type *gfp* (from plasmid TU#65, Chalfie et al., 1994), was fused downstream of a fragment containing 1420 NT of flanking DNA, the transcription start site, the 5′ UTR, and the AUG start codon of *Actin88F*. This fusion was subcloned into the P-element vector pCaSpeR-AUG-βgal (Thummel et al., 1988), replacing the Adh AUG-β-gal fragment. The pre-mRNA expressed from this construct contains introns in both the *Actin88F* 5′ UTR, and the SV40 small t 3′ UTR, and utilizes the SV40 small t polyadenylation signal.

Construct	Product	References
p[Cas,NGE]	GFP-EXU	Wang and Hazelrigg (1994)
p[Cas,CGE]	EXU-GFP	Wang and Hazelrigg (1994)

Construction

Restriction sites were introduced by standard *in vitro* mutagenesis just upstream of *exu's* initiating or stop codons, and PCR-amplified wild-type *gfp* (from cDNA clone pGFP10.1; Chalfie et al., 1994), was inserted at these sites. These in-frame fusions express proteins with GFP at either the N-or C-terminus of EXU. Both fusion genes were placed in the pCaSpeR4 transformation vector (Thummel and Pirrotta, 1992). Both constructs utilize *exu's* promoters and express pre-mRNAs containing the introns encoded by *exu*.

Construct	Product	References
pMei-S332-GFP	MEI-S332-GFP	Kerrebrock et al (1995)

Construction

PCR-amplified wild-type *gfp* (from plasmid TU#65, Chalfie et al, 1994) was inserted in-frame at a natural *Bam* HI site at the 5′ end of the coding region of a *mei-S332* genomic fragment cloned in the transformation vector CaSpeR4 (Thummel and Pirrotta, 1992). The resulting fusion protein contains GFP at the N-terminus. The RNA from this construct is expressed using *mei-S332's* own promoter and is predicted to contain the same 5′ and 3′UTRs present in *mei-S332* transcripts.

TABLE 9.1 (continued)

Construct	Product	References
pCaSpeR/*ncd-gfp*	Ncd-GFP	Endow and Komma (1996 and 1997)
pCaSpeR/*ncd-gfp**	Ncd-GFPS65T	Endow and Komma (1996 and 1997)

Construction

These constructs express fusions of Ncd to either wild-type GFP or mutant GFPS65T. The chimeric genes were constructed using genomic fragments containing *ncd's* promoter and 5′ UTR, PCR-amplified *ncd* coding sequence and 3′ UTR from a *ncd* cDNA, and PCR-amplified *gfp* coding sequence. Wild-type *gfp* was amplified from P[Cas, NGE] (Wang and Hazelrigg, 1994), and mutant *gfp** was amplified from *gfp*S65T (Heim et al., 1995). These fusion genes were placed in pCaSpeR3 (Thummel and Pirrotta, 1992). The proteins expressed from these constructs both contain GFP at the C-terminus of Ncd.

FI66GFP	GFP	Potter et al. (1996)

Construction

Wild-type *gfp* (from plasmid TU#64; Chalfie et al., 1994) was cloned into a modified pGMR (Hay et al., 1994), downstream of the promoter that contains multiple binding sites for the *glass* transcription factor, and upstream of the *hsp70* polyadenylation sequence. The modified pGMR present in the fly line analyzed in this paper contains an intron from the *ftz* gene in the 5′ UTR. Expression is driven in the photoreceptor cells of the developing and adult eye.

GFP-GMR	GFP	Plautz et al., (1996)

Construction

PCR-amplified wild-type *gfp* (from plasmid pGFP10.1; Chalfie et al., 1994) was cloned into pGMR (Hay et al., 1994), downstream of the promoter that contains multiple binding sites for the *glass* transcription factor, and upstream of the *hsp70* polyadenylation sequence. Expression is driven in the photoreceptor cells of the developing and adult eye.

UAS-GFP	GFP	Brand (1995)
UAS-Tau-GFP	Tau-GFP	Brand (1995)

Construction

Wild-type *gfp* (Prasher et al., 1992) was PCR-amplified and subcloned into pUAST (Brand and Perrimon, 1993) to create UAS-GFP. In the case of UAS-Tau-GFP, an in-frame fusion of bovine *Tau* (Butner and Kirschner, 1991; Callahan and Thomas, 1994) was made with PCR-amplified wild type *gfp* (from plasmid TU65; Chalfie et al., 1994).

hs-GFP-moe	GFP-Moesin tail	Edwards et al. (1997)

Construction

PCR-amplified wild-type *gfp* (plasmid pGFP10.1; Chalfie et al., 1994), and a PCR-amplified fragment of a *moesin* cDNA, were ligated into pCaSpeR-hs (Thummel and Pirrotta, 1992). Transcription is driven by the *hsp70* promoter. The mRNA encodes a fusion protein with GFP at the N-terminus linked to the C-terminal portion of Moesin, including a predicted extended α-helical region and an actin-binding domain.

TABLE 9.1 (continued)

Construct	Product	References
pUAST-βgal/GFP	β-galactosidase-GFP	Timmons et al. (1997)
pUAST-GFP/βgal	GFP-β-galactosidase	Timmons et al. (1997)

Construction

The *lacZ* coding region was PCR-amplified from pCaSpeR-AUG-βgal (Thummel et al., 1988), and inserted into the pUAST vector (Brand and Perrimon, 1993) along with PCR-amplified wild-type *gfp* (Chalfie et al., 1994), to create pUAST-βgal/GFP. In the case of the second construct, pUAST-GFP/βgal, a PCR-derived fragment (from pCaSpeR-AUG-βgal) containing the 5′ untranslated region and start codon from the Drosophila *Adh* gene was fused to PCR-amplified wild-type *gfp* and PCR-amplified *lacZ* coding sequence. This fusion gene was subsequently ligated into the pUAST vector. Thus pUAST-βgal/GFP expresses a fusion protein with β-galactosidase at the N-terminus, and pUAST-GFP/βgal expresses a protein with GFP at the N-terminus.

of an inserted gene (Thummel et al., 1988; Pirrotta, 1988; Brand and Perrimon, 1993; Thummel and Pirrotta, 1992). There have been no reported cases of toxicity from expressing GFP in flies.

Table 9.1 describes the various GFP constructs used successfully and published to date. In general, wild-type *gfp* (here referring to the original recombinant *gfp* used by Chalfie et al., 1994) or a mutant version of *gfp* (Heim et al., 1995) has been subcloned as either an original cDNA fragment or a PCR-amplified deoxyribonucleic acid (DNA) fragment with appended restriction sites. Creation of these restriction sites adjacent to a target gene's start or stop codons has allowed the creation of in-frame fusions of GFP at either the N- or C-terminus of the target gene's protein. Often additional linker amino acids are added due to the appended restriction sites; these additions have not interfered with the ability to detect a GFP signal.

9.3 FACTORS AFFECTING GFP SIGNAL DETECTION

Several parameters affect the ease of detection of GFP. These parameters include the type of GFP used (mutant versus wild-type forms), the concentration of GFP, the effects of background autofluorescence, and the time required for posttranslational modifications to form GFP's chromophore.

Recently, isolated mutant versions of GFP affect the excitation or emission properties of the protein, making it either a more brightly fluorescent protein, or altering the wavelengths of maximum excitation or emission (see Chapter 5). As in other organisms, GFPS65T (Heim et al., 1995) has a brighter signal than wild-type GFP in *Drosophila* embryos (Endow and Komma, 1996 and 1997; Timmons et al., 1997; F.-A., Ramirez-Weber and T. Kornberg, personal communication). No other modified versions of GFP have been used to date in flies, but they will no doubt be used to advantage in the future.

The concentration of GFP within a cell is a critical factor for its detection. Niswender et al (1995) estimated that intracellular concentrations must be at least 1 μM for wild-type GFP to be detected by epifluorescence, laser confocal, or two-photon microscopy. These considerations indicate that fairly strong promoters must drive the expression of GFP, although mutant GFPs that fluoresce brighter may be detected at lower concentrations. Other factors, such as mRNA stability and factors regulating translation, or the stability of fusion proteins, may also contribute to achieving critical levels of GFP. In cases where fusions are made with other proteins that are tightly localized within the cell, high local concentrations of the fusion protein may play an important role in the ability to detect GFP fluorescence. The examination of constructs with and without introns in *Caenorhabditis elegans* indicates that the inclusion of introns may lead to higher expression levels (see chapter 8). Whether the inclusion of an intron affects expression levels in *Drosophila* has not been systematically tested.

Most cellular autofluorescence occurs in the excitation and emission wavelengths of GFP, so background autofluorescence can obscure GFP signals (Aubin, 1979; Niswender et al., 1995). For instance, fluorescence from yolk in vitellogenic and mature oocytes, and young embryos, can make detection of GFP's signal difficult. This problem can be dealt with by selecting appropriate filters (see Appendix III), or by using new variants of GFP that excite or emit with different wavelengths than that producing the problematic autofluorescence. Also, targeting of GFP fusion proteins to the nucleus, when possible, may help, since in general there is little autofluorescence in nuclei (Aubin, 1979).

9.4 TIMING OF GFP's FLUORESCENCE

An important factor in assessing GFP's usefulness as a reporter gene is the time required for the post-translational modifications to create GFP's chromophore. While many situations do not require that this process occurs rapidly, there are some applications where one wants to monitor rapid changes in gene expression.

Davis et al. (1995) concluded that the posttranslational modifications needed to produce a fluorescent product require 3–5 h in *Drosophila* embryos. They made two different transgenes in which GFP was expressed from the polyubiquitin promoter. One construct, PUbnlsGFP, expressed a fusion protein beginning with a 9 aa segment of the Ubiquitin protein, containing a nuclear localization signal, followed by 40 aa of β-galactosidase fused to GFP (nlsGFP). The second construct, PUbGFP, was deleted for the nls-β-galactosidase sequence. In transformed animals the proteins behaved as expected: nlsGFP was concentrated in the nuclei (Fig. 9.1), while GFP, lacking a nuclear localization signal, was present in both nuclei and cytoplasm.

The acquisition of fluorescence of nlsGFP was determined in developing embryos. They compared embryos that received nlsGFP from mothers in which the transgene was expressed at rather low levels, but provided a fluorescent signal in very young embryos, to embryos in which nlsGFP was supplied only zygotically. When embryos with zygotically supplied protein first achieved similar levels of protein as the maternally-supplied embryos, nlsGFP was not fluorescent. A fluorescent signal was first detected in these zygotically supplied embryos

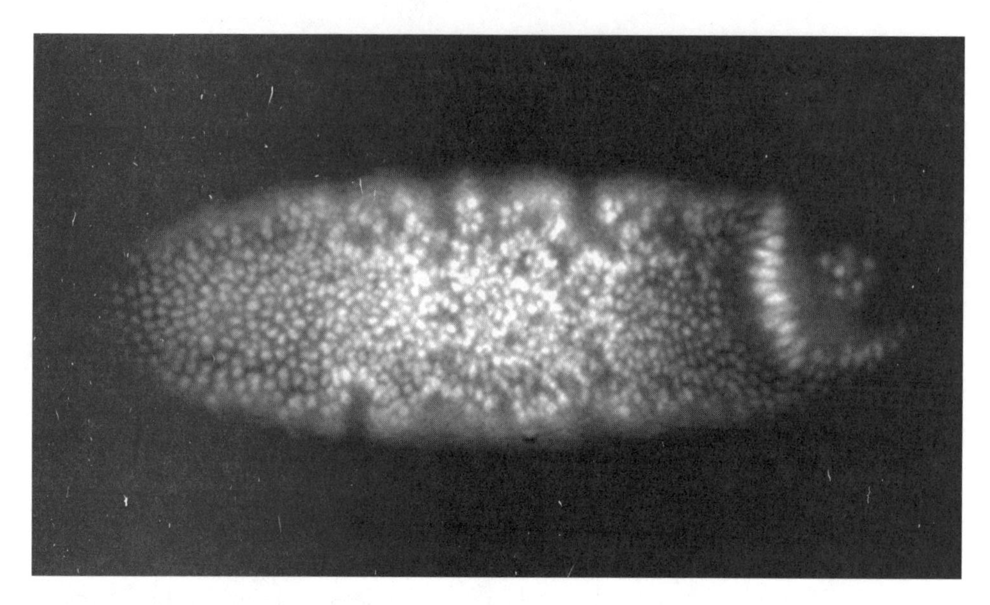

FIG. 9.1 nlsGFP in the nuclei of a live embryo at early gastrulation. The embryo was dechorionated and mounted in halocarbon oil on a gas-permeable membrane (Yellow Springs Instruments Co; see Davis et al., 1995). Anterior is to the left. Imaging was done on an Olympus IX70 inverted microscope, using a 20X Planapo (N.A. 0.75) objective and a FITC filter set. Images were collected with a Photometrics model PXL CCD camera, using the Delta Vision image analysis program (Applied Precision and Software). [Courtesy of Ilan Davis; Davis et al, 1995.]

several hours later. They calculated that in embryos it took about 3–5 h for nlsGFP to become fluorescent. Brand (1995) estimated a similar time lag of about 3 h for GFP signal detection.

It is clear that in some cases GFP can be detected sooner following the initiation of expression than that observed for nlsGFP by Davis et al. (1995). In T. Kornberg's laboratory, both wild-type GFP and mutant GFPS65T have been expressed in the embryo's *engrailed (en)* stripes (F. A.. Ramirez-Weber and T. Kornberg, personal communication). They found that while wild-type GFP could only be detected very faintly in the early *en* stripes, a readily detectable signal was observed with the mutant GFPS65T. Figure 9.2 shows a later germ-band-extended embryo expressing GFPS65T in *en* stripes.

We examined the onset of GFP fluorescence from N-terminal fusions of GFP to the Bicoid transcription factor from genes whose timing of transcription or translation could be controlled (Hazelrigg, Liu, Hong, and Wang, submitted]. Transcriptional control was obtained by driving GFP-Bcd expression from the *hsp70* promoter. In salivary glands, a nuclear fluorescent signal was detected within 40 min after a 10-min heat shock. Translational control was achieved by utilizing the *bcd* promoter and expressing a *gfp-bcd* mRNA with *bcd*'s own 5′ and 3′ untranslated regions (UTRs). Normally *bcd* mRNA is maternally supplied and localized to the anterior end of the oocyte (Berleth et al., 1988), and translational regulation of *bcd* mRNA ensures that it is not translated until early embryogenesis, after eggs are laid (Driever and Nusslein-Volhard, 1988). In female flies carrying the *gfp-bcd* transgene, no protein was detected in ovaries,

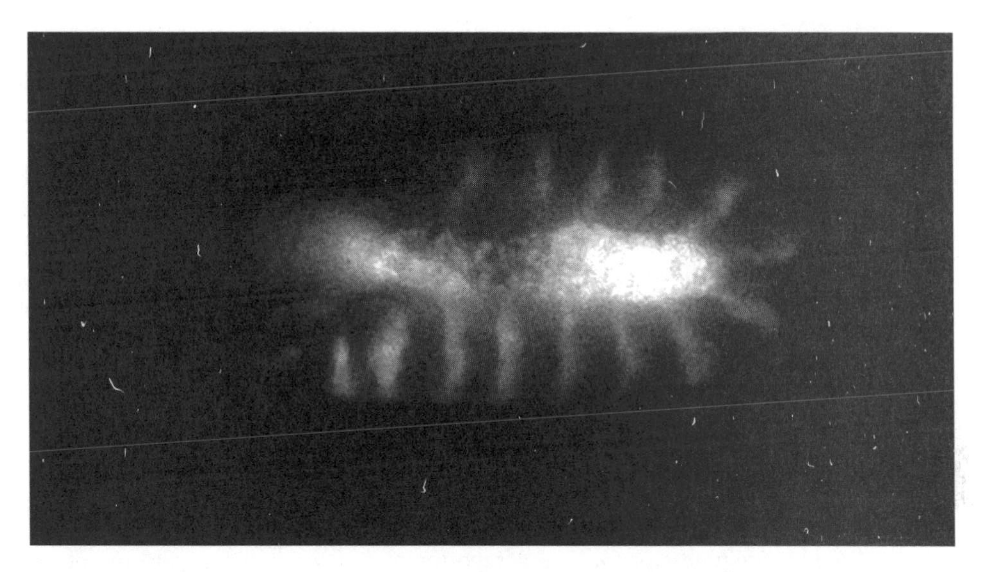

FIG. 9.2 Germband-extended embryo expressing mutant GFPS65T in the engrailed stripes. GFPS65T is distributed throughout the nuclei and the cytoplasm. Anterior is to the left. Imaging was on a Zeiss Axiophot epifluorescence microscope, using a 20X Neofluar objective and a standard Zeiss FITC band pass filter. [Courtesy of Felipe-Andres Ramirez-Weber and Thomas Kornberg.] Figure also appears in color section.

either by fluorescence microscopy or Western blotting, indicating the mRNA is regulated like the normal *bcd* mRNA. Nuclear fluorescence from GFP-Bcd protein was first detected in the anterior nuclei of early syncytial blastoderm embryos, 1–2 h after the initiation of translation.

9.5 MARKING CELLS WITH GFP

GFP has been expressed in many different *Drosophila* cell types. Its expression has been observed in germ cells and somatic cells in the gonads, in the nervous system, in salivary glands, in imaginal discs, in undifferentiated embryonic cells, and highly differentiated adult structures. When GFP is expressed without being fused to another protein, it entirely fills the cells in which it is expressed. This attribute has made GFP a useful reporter for gene expression, and a convenient marker to identify cells and observe their morphologies.

GFP has been expressed in several different cell types using the GAL4 enhancer trap system (Yeh et al., 1995). In this system (Brand and Perrimon, 1993), the expression of a transgenic reporter gene is regulated by binding sites (the UAS, or Upstream Activation Sequence) for the GAL4 transcriptional activator. The GAL4 activator is expressed in different cell types by transposing a P-element, containing GAL4 with a weak promoter, to different chromosomal sites, where flanking enhancer elements cause tissue-specific expression of GAL4.

After mobilizing a GAL4 P-element, Yeh et al. (1995) isolated 54 insertions that expressed GAL4 in specific temporal and spatial patterns in embryos, larvae, and adults (the expression patterns were determined by induction of transcription of a UAS-*lacZ* transgene, measured by X-Gal staining). Flies bearing these insertions were crossed to flies bearing an insertion of a P-element containing GFP regulated by flanking UAS elements, UAS-GFP.

GFP expression was thus targeted to a variety of tissues, including the embryonic peripheral nervous system, larval salivary glands, imaginal discs and neurons, and adult ovaries. GFP expression was observed by confocal microscopy in dissected, unfixed tissues, and also intact embryos and larvae. When UAS-GFP expression was compared to expression of a UAS-*lacZ* transgene regulated by the same Gal4 insertions, they found that GFP could be detected in the same cell types in which β-galactosidase was detected. Because GFP could be observed by confocal microscopy, they had better resolution of cell identities with GFP than with *lac-Z* staining using nonfluorescent substrates. In cases where live animals were imaged by confocal microscopy, larvae and embryos survived and did not appear to be adversely affected by the laser. GFP expressed in larval neurons filled the cell bodies and also axonal processes and nerve terminals.

The usefulness of GFP as a reporter to follow changing patterns of gene expression in living cells was shown by Yeh et al. (1995) in one case where GFP expression was targeted to a group of follicle cells in developing egg chambers. Dissected egg chambers were cultured over a period of 4 h, and periodically scanned by confocal microscopy for GFP expression. In isolated egg chambers that initially showed no GFP signal, the appearance of the signal could be monitored over time as the cells were cultured.

Fusions of GFP to β-galactosidase have been used to determine the efficiency of detecting each reporter in different types of cells (Timmons et al., 1997; Shiga et al., 1997). Timmons et al. (1997) analyzed expression of GAL4-inducible gene constructs expressing GFP fused to either the N- or C-terminus of β-galactosidase. They compared the ease of detection of GFP and β-galactosidase when expression of the *gfp-lacZ* constructs were regulated by a large set (120) of *Gal4* P-element insertions. They found that in general GFP and β-galactosidase detection patterns were quite similar, but there were some differences. Thus, in the nervous system, GFP was a better reporter than β-galactosidase, whereas in ovaries and testes β-galactosidase was a more reliable reporter. Also, imaginal discs were not equivalent in producing detectable GFP signals: GFP fluorescence was less frequently detected in eye imaginal discs than in the wing, haltere, or leg discs. β-galactosidase staining offered more sensitivity than GFP fluorescence in cases where the fusion protein was expressed at a low level, probably because of the enzymatic amplification afforded by β-galactosidase.

A GAL4 driven line expressing GFP in embryonic and larval motorneurons (Yeh et al., 1995) has been used in C.F. Wu's laboratory to identify these neurons in giant neuron cultures. These giant neurons are the products of cell-division-arrested neuroblasts from dissociated gastrulae (Wu et al., 1990; Saito and Wu, 1991). Figure 9.3 shows GFP expressed selectively in one neuron in culture. The expression of GFP did not affect the action potential recorded from these neurons (Zhao and Wu, personal communication).

GFP has been expressed from a flight-muscle-specific actin promoter, *Act88F* (Barthmaier and Fyrberg, 1995). Their purpose was to mark flight muscle precursor cells and follow muscle development in living animals. The GFP marked

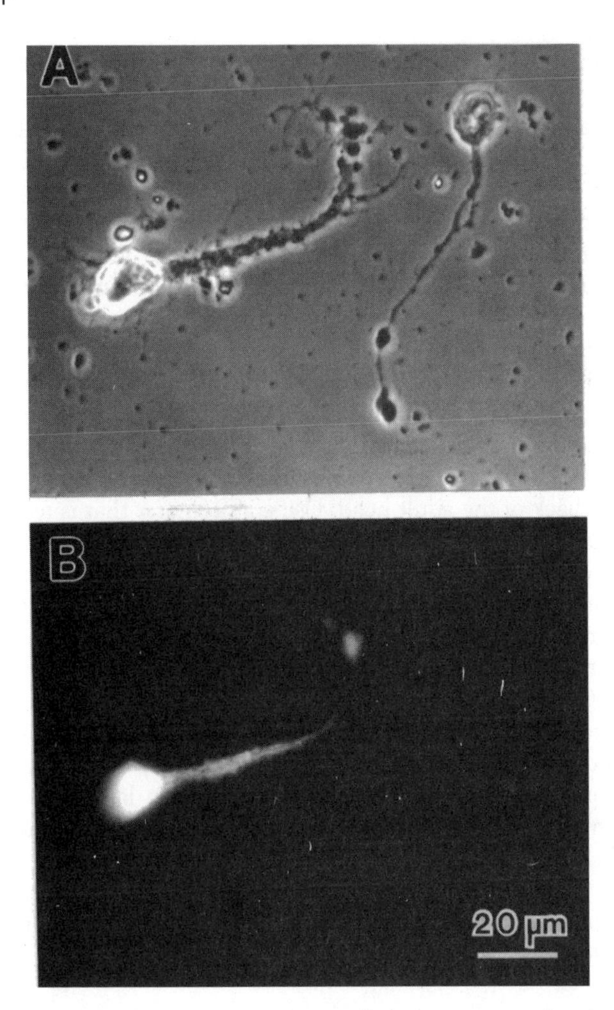

FIG. 9.3 GFP expressed in motorneurons in cell division-arrested embryonic neuroblasts in culture. Panel (*a*) is a phase contrast image of two giant neurons in culture, and panel (*b*) is an epifluorescent image of the same two neurons, showing that GFP is expressed selectively in one of these cells. Imaging was done on a Nikon Diaphot inverted microscope, using a Ph3 40X DL (N.A. 0.55) objective, and an FITC filter set. [Courtesy of Ming-Li Zhao and Chun-Fang Wu.]

muscle cells allowed them to identify these cells in mutant backgrounds and analyze the defects in muscle development produced by existing flightless mutations.

Flight muscles develop during pupation, beginning with myoblast fusion in early pupae. Flies transformed with *Act88F-gfp* expressed GFP faithfully in the indirect flight muscles and allowed the development of these muscles to be followed at all stages after myoblast fusion in early pupation. The GFP-marked muscle fibers could be seen through the cuticle, in live pupae and adults. Insights into the effects of Troponin-T and Troponin-I mutations on muscle development were gained by crossing the *Act88F*-GFP transgene into the mutant backgrounds, and following the degeneration of the flight muscles over time. This pilot study demonstrated that marking the flight muscles with GFP should facilitate analysis of a large extant collection of mutations affecting flight muscle development.

A GFP-β-galactosidase fusion has been used by Shiga et al., (1996) to study the developing tracheal system in live embryos. The fusion transgene contained a nuclear localization signal appended to *gfp* (to express GFPN), which was fused in frame to *lacZ* transcribed from a GAL4-inducible promoter. Expression of GFPN-β-galactosidase was driven by a second transgene that transcribed *Gal4* from the *breathless* promoter, expressed in the cells of the developing tracheal system (Glazer and Shilo, 1991). Using time lapse confocal microscopy to image serial sections, they observed the developing tracheal system over an 80 minute window during embryogenesis. They found that the embryos developed normally, allowing them to chart dynamic changes in the tracheal system over time.

GFP has also been expressed in the eyes and ocelli, the three simple eyes on the back of the adult head. Using a promoter that is responsive to the *glass* gene product, Plautz et al (1996) targeted expression of wild-type GFP to the photoreceptor cells. In live adult flies, GFP could be detected in each eye and the three ocelli. Axons extending from the ocelli were filled with GFP, and the fluorescent signal was detected through the adult cuticle. Fixed compound eyes showed GFP signal in the seven photoreceptor cells of the ommatidia. *glass*-driven expression of GFP was also used by Potter et. al. (1996) to direct expression in the photoreceptor neurons. Using confocal laser-scanning microscopy (CLSM) and two-photon laser scanning microscopy (TPLSM), observations were made on developing eye–brain complexes from third instar larvae, in newly dissected, unfixed samples (Fig. 9.4). They found that TPLSM was superior to CLSM in resolving the individual axons in bundles projecting from the eye disks to the optic lobe in the brain. In situations where imaging of GFP in intact tissues is desired, they concluded that TPLSM is better able to detect structures lying deep within the tissue.

9.6 TAGGING PROTEINS WITH GFP

GFP has been used to tag several *Drosophila* proteins, providing detailed information about the sites of localization of these proteins within cells. Using a GFP tag to determine a protein's cellular distribution has both advantages and disadvantages compared to using antibodies for this purpose. With a GFP-tagged protein, there is no background due to nonspecific binding of either primary or secondary antibodies. If protein null alleles exist for a gene, genetically engineered animals can be produced in which the GFP fusion gene is the only source of the gene product, so that in theory every protein molecule in the cell can be tagged with GFP. In these cases, the GFP tag should be a more efficient way of labeling the protein than antibody staining. The detection of a signal from a GFP fusion protein is sensitive to the intracellular concentration. As mentioned earlier, Niswender et al. (1995) estimated the minimum concentration for detection to be at least 1 μM for wild-type GFP; GFP variants may be detected at lower concentrations. Cases where a protein is normally present at very low levels within a cell may be more readily detected with antibody staining, where ampli-

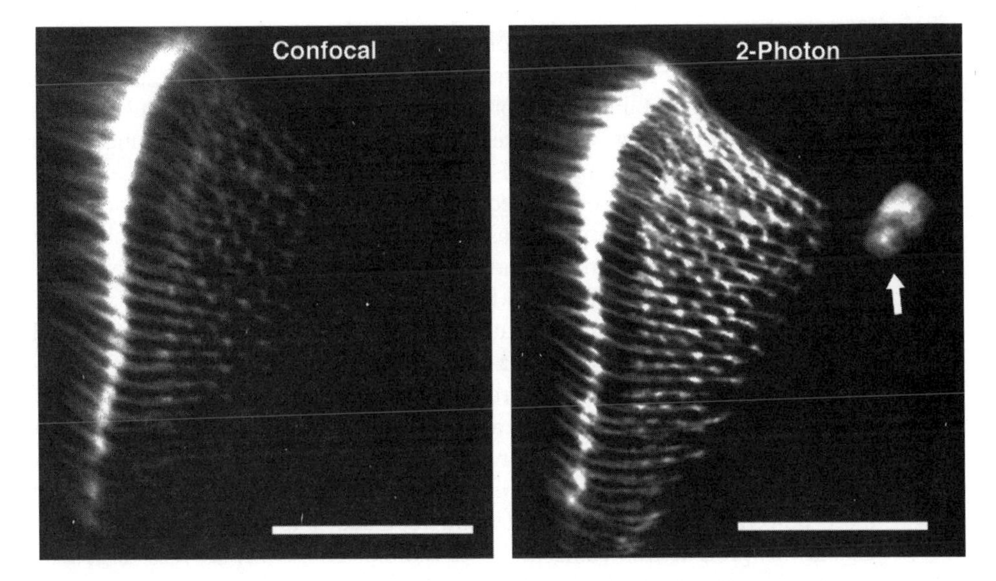

FIG. 9.4 Confocal versus two-photon imaging: *Drosophila* photoreceptors filled with GFP innervating the developing optic ganglia in a third instar larval brain. The left image was made using the BioRad MRC600 confocal microscope, with visible excitation at 488 nm. The same specimen was then imaged using a Molecular Dynamics Sarastro 2000 laser-scanning microscope converted for two-photon imaging with pulsed infrared (IR) excitation at 900 nm. The enhanced signal-to-noise (S/N) ratio of two-photon imaging translates into greater resolution of fine details. The enhanced penetration of the IR excitation allows imaging of deeper structures, such as the termination site of Bolwigs nerve (arrow). Both images are lookthrough projections of 20 sections scanned at 1-μm intervals. Scale bars, 50 μ. [Figure and figure legend courtesy of Steve Potter and Scott Fraser; Potter et al., 1996.]

fication of signal occurs due to the binding of multiple primary or secondary antibody molecules, increased further when enzymatic tags are coupled to the antibody.

The most important advantage GFP tagging has to antibody staining is that GFP tagged proteins can be detected in living cells. Structures that are labile to fixation may be seen with GFP fusion proteins in live cells. Detailed dynamic imaging of protein trafficking and localization in living cells can be obtained using laser confocal video microscopy.

GFP was first used to tag a protein to determine its subcellular sites in the case of the *Drosophila exuperantia (exu)* gene product (Wang and Hazelrigg, 1994). The *exu* gene is required for correct localization of bicoid (*bcd*) mRNA in developing oocytes (Berleth et al., 1988; St. Johnston et al., 1989). Antibody staining supported the view that *exu* must act in establishing localization of *bcd* mRNA, and not in later steps in which *bcd* mRNA is maintained at the anterior end of the mature oocyte (MacDonald et al., 1991; Marcey et al., 1991). However, antibody-staining had not yielded significant clues about the role of *exu* in the establishment steps. Fusion genes were engineered to express GFP fused either at the N- or C-terminus of Exu. In these constructs, the natural promoters of *exu* were used, so that expression was driven strongly in germ cells.

Both N- and C-terminal fusion genes rescued the function of an *exu* null allele. The ability to rescue function in a mutant background is an important criterion for using a GFP-tagged protein to learn about the subcellular localization of a protein; the genetic criterion imposes a strong demand for function, and function implies that the protein is probably localized correctly.

GFP-Exu fluorescence revealed the existence of subcellular particles, concentrated near and apparently passing through, the intercellular connections called ring canals that connect the nurse cells to the developing oocyte in egg chambers [Fig. 9.5(*a*)]. In mid to late stages of oogenesis, when *bcd* mRNA is being localized, these particles accumulate at the anterior of the oocyte, the site of *bcd* mRNA localization. Localization of the particles is dependent on intact microtubules. We hypothesized that the particles may transport either *bcd* mRNA, or a component of the anchor for *bcd* mRNA, along microtubules, targeting either of these components to the anterior oocyte cortex.

Time-lapse video analysis of GFP tagged proteins in living cells is one of the most powerful uses of GFP, providing detailed information about dynamic changes in protein trafficking and localization. The transport and targeting of the particles containing GFP-Exu has been observed in live cultured egg chambers by time-lapse video confocal microscopy (Theurkauf and Hazelrigg, submitted). This method allowed us to determine rates and directionality of particle movement, and to analyze more fully the role of the microtubule cytoskeleton in transport and anchoring of these particles in living cells.

Since the *gfp-exu* fusion genes are driven from the natural promoters of *exu*, these transgenes are also expressed during spermatogenesis. Strong expression occurs in primary spermatocytes and the fusion protein can be detected until late stages of spermiogenesis. The gonads of third instar male larvae already contain primary spermatocytes, so that GFP-Exu can be detected in males at this stage. Figure 9.5(*b*) shows the posterior end of a live male third instar larva; the gonads fluoresce brightly and are readily observed through the larval body wall.

GFP fusion proteins may in some cases be used as fluorescent tags to learn about the phenotypic effects of mutations in other genes on cellular properties. Mulligan et al. (1996) used the *gfp-exu* fusion gene to assess effects of mutations in the gene *stand still (stil)* on the distribution of maternal products during oogenesis. They found that *stil* mutations, which affect the structure of the ring canals and other actin-based cytoskeletal structures, also affected the number and size of GFP-Exu particles.

The GAL4 system has been used to drive expression of GFP fused to the microtubule-binding protein Tau (Brand, 1995). Confocal microscopy of living larvae revealed Tau-GFP in epidermal cells, outlining the microtubule cytoskeleton in these cells. In neurons, Tau-GFP was present throughout the cell bodies, extending into the axonal processes. The processes of individual neurons could be identified in live embryos, and traced back to their cell bodies. Figure 9.6 shows Tau-GFP expressed in the peripheral nervous system of a living embryo (Courtesy of A. Brand).

A useful GFP fusion protein for studying morphogenetic changes in cells was developed by Edwards et al. (1997). They fused GFP to the C-terminal region of Moesin, a protein that associates with the cortical actin cytoskeleton. This C-terminal region of Moesin (moe) contains its actin-binding domain, so that GFP-

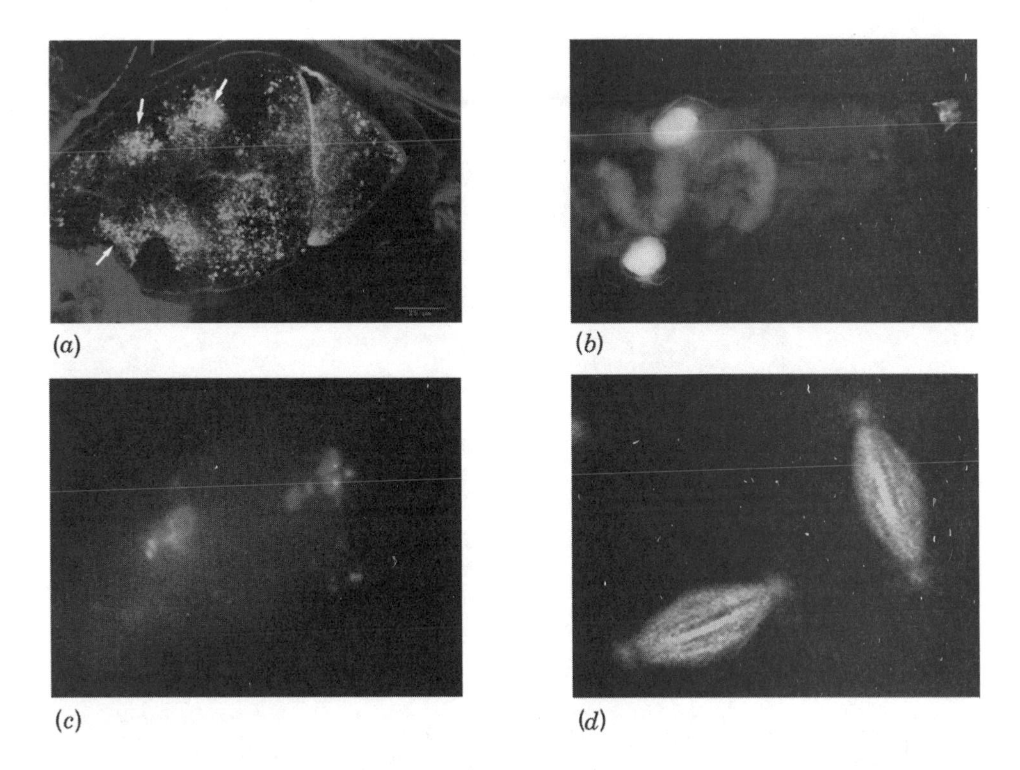

FIG. 9.5 GFP fusion proteins. (*a*) GFP-Exu particles in oogenesis. A fusion was made of GFP to the protein product of the *exuperantia* gene, required during oogenesis for correct localization of *bcd* mRNA to the developing oocyte. This fixed stage 9 egg chamber was stained with rhodamine-phalloidin to detect the position of the actin-rich ring canals (arrows) relative to the GFP-Exu particles. The oocyte is to the right, connected via ring canals to the nurse cells that lie on the left. Imaging was with a BioRad MRC600 confocal unit attached to a Zeiss Axioskop microscope, using K1 and K2 filter blocks. (Reprinted with permission from *Nature*, Wang, S. and Hazelrigg, T. (1994). Implications for *bcd* mRNA localization from spatial destruction of *exu* protein in *Drosophila* oogenesis, 369:400–403. Copyright © 1994 Macmillan Magazine Limited.] (*b*) GFP-Exu in third instar male larva. The *exuperantia* gene is also expressed in the testis, where it is required for normal spermatogenesis (Hazelrigg et al., 1990). A live male larva was mounted in halocarbon oil on a slide and covered with a coverslip supported at its edges with additional coverslips. The signal from GFP-Exu in primary spermatocytes is visible through the body wall of the larva. Imaging was on a Zeiss Axiophot microscope, using a 40X plan-neofluar objective (N.A. 1.3) and a FITC 09 filter set. (*c*) MeiS332-GFP in male meiosis. A fusion gene was constructed with *meiS332*, a gene required for normal meiosis, and the distribution of the MeiS332-GFP fusion protein was observed in male meiosis (Kerrebrock et al., 1995). Fixed testes were incubated with 7-AAD to detect the chromosomes. In this anaphase I cell, MeiS332-GFP (green) is localized to the centromere regions at the leading edges of the chromosomes (red). Imaging was with a Nikon Optiphot-2 microscope equipped with a Photometrics Image Point cooled CCD camera. [Courtesy of Terry Orr-Weaver. Reprinted with permission from Kerrebrock et al., 1995. Copyright © Cell Press.] . (*d*) Ncd-GFPS65T in the metaphase spindles of a cycle 9 embryo. GFPS65T was fused at the C-terminus of the coding region of *ncd*, a member of the *Drosophila* kinesin gene family (Endow and Komma, 1996). Live embryos from mothers expressing Ncd-GFPS65T were injected with rhodamine-histone to visualize the chromosomes. The Ncd-GFPS65T fusion protein is present in the spindles, and excluded from the chromosomes, and can be seen in spindle fibers that extend across the chomosomes. Imaging was done on a BioRad MRC600 laser scanning confocal unit attached to a Zeiss Axioskop microscope, using a planapochromat 63X (N.A. 1.4) objective. [Courtesy of S. Endow. Reprinted with permission from Endow and Komma, 1996 and Company of Biologists, Ltd.] Figure also appears in color section.

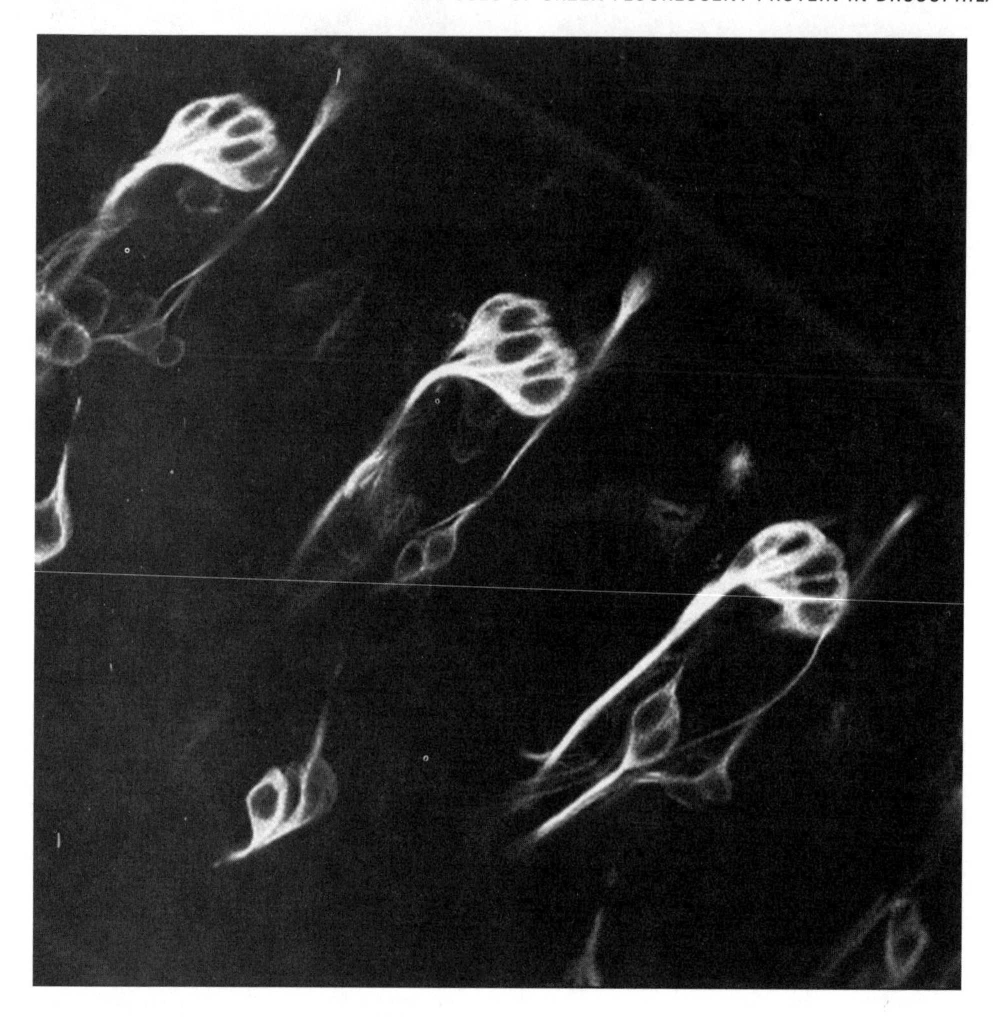

FIG. 9.6 Neurons in the *Drosophila* peripheral nervous system expressing a Tau-GFP protein fusion. Using the GAL4 system (Brand and Perrimon, 1993) to direct expression of GFP alone or GFP fused to the microtubule binding protein Tau, neuronal cell bodies and their axonal projections can be clearly and easily labeled in living animals (Brand, 1995). This figure shows three segments of a stage 17 embryo in which the cluster of chordotonal neurons and a subset of the more ventral neurons express Tau-GFP. The Tau-GFP fusion protein is excluded from the cell nucleus but is actively transported into axons. Individual axons can thus be traced to their respective cell bodies. The embryo was mounted in halocarbon oil and visualized with a BioRad MRC-1024 laser scanning confocal microscope, using a 60X oil immersion objective (N.A. 1.4). The image is a projection of three optical sections. [Figure and legend courtesy of Andrea Brand.]

moe is enriched in actin-rich regions of cells. Using the *hsp70* promoter to drive expression of GFP-moe in transgenic flies, they found that GFP-moe could be used for analyzing cell morphogenesis in a wide variety of somatic tissues in live embryos, larvae, pupae and adults. For instance, they took advantage of the fact that the polar follicle cells (PFCs), specialized pairs of follicles cells at the anterior and posterior ends of developing egg chambers, cease proliferation early in egg chamber development. This allowed them to essentially pulse label these cells with a heat shock, since the neighboring cells went on to divide many times and thus dilute GFP-moe, leaving the undivided PFCs with high levels of GFP-moe, and strongly fluorescent. They showed that the two anterior PFCs invariably become central cells in the group of follicle cells known as border cells, and observed the early development of a specialized PFC projection that is later inserted into the oocyte micropyle while this structure is being formed (Montell et al., 1992). Their observations on live embryos undergoing dorsal closure provided evidence for a critical role of an actin-rich purse string in the lateral epidermis (Young et al., 1993).

A GFP fusion protein has been used to show that the Mei-S332 protein associates with the centromere region of chromosomes (Kerrebrock et al., 1995) . The *mei-S332* gene plays an essential role in *Drosophila meiosis*, where it is required to maintain sister chromatid association through the first meiotic division until anaphase II, when sister chromatids normally separate (Davis, 1971; Goldstein, 1980; Kerrebrock et al., 1995). Mutations cause nondisjunction during meiosis in males and females, due to premature separation of sister chromatids at anaphase I. Sequence analysis of *mei-S332* revealed some interesting possible structural motifs, but did not in itself provide significant insights into function, as the predicted protein shared no significant similarities with proteins of known function.

Kerrebrock et al. (1995) constructed a chimeric gene expressing a fusion protein with GFP fused to the N-terminus of Mei-S332. Transformed copies of the fusion gene rescued a *mei-S332* mutant allele, demonstrating that the Mei-S332-GFP protein is functional. The subcellular localization of Mei-S332-GFP was analyzed in spermatogenesis. In prometaphase I, after chromosomes condense, Mei-S332-GFP is targeted to the centromere regions of each chromosome, which at this point are composed of two tightly paired sister chromatids. At anaphase I, as homologous chromosomes (= two paired sister chromatids) separate and move to opposite poles of the spindle, the GFP signal forms bright spots at the leading edges of the chromosomes where the centromeres lie [Fig. 9.5(c)]. This signal remains associated with the centromeres until anaphase II, and disappears at this time, when sister chromatids separate and move to opposite poles of the spindles. Consistent with the phenotypic effects seen in mutants, the temporal and spatial pattern revealed by the GFP tag indicates that Mei-S332 acts at the centromere regions of chromosomes to keep sister chromatids attached and to prevent separation of chromatids before this normally occurs. The precise disappearance of Mei-S332 from chromosomes at anaphase II is as striking as its precise centromeric appearance after chromosome condensation in prometaphase I, suggesting that the presence/absence of this protein at the centromeres is a key regulatory event in the orchestration of orderly meiotic chromosome behavior.

The *Drosophila ncd* gene encodes a member of the kinesin superfamily of proteins. Unlike more conventional kinesins, Ncd is a motor protein that moves along microtubules toward the minus ends (Walker et al., 1990). Mutants lacking normal *ncd* function suffer nondisjunction in female meiosis, and embryos from mutant mothers undergo non-disjunction in the early mitotic divisions. Immunohistochemical staining showed that Ncd protein is associated with spindles in female meiosis and early embryonic mitoses (Hatsumi and Endow, 1992). GFP tagged Ncd allowed Endow and Komma (1996 and 1997) to learn significant new details about the localization of Ncd in the mitotic spindles of young embryos, and the meiotic spindles of oocytes. Their fusion genes were constructed to express fusions with GFP at the C-terminus of Ncd: in one case with wild-type GFP and in a second case with mutant GFPS65T (Heim et al., 1995).

Both fusion genes expressing Ncd-GFP rescued function of the ca^{nd} null allele of *ncd*. The Ncd-GFP fusion proteins were localized in identical patterns; they estimated that the version with mutant GFPS65T was about sixfold to sevenfold brighter, by confocal imaging, than Ncd tagged with wild-type GFP. Striking images of spindle dynamics in mitotic cell cycles were obtained by time-lapse video analysis of Ncd-GFP in live embryos (Fig. 9.7). [A movie is available at S. Endow's web page (http://abacus.mc.duke.edu).] By injecting embryos containing Ncd-GFP with rhodamine-labeled histones or rhodamine-labeled tubulin, they were able to determine the distribution of Ncd relative to chromosomes and microtubules. In young embryos, Ncd-GFP is associated with spindles and centrosomes, as had previously been shown by antibody-staining for Ncd protein (Hatsumi and Endow, 1992; Endow et al., 1994). Metaphase spindles in live embryos contained Ncd-GFP associated with fibers that extended across the metaphase plate where the chromosomes lie [Fig. 9.5(*d*)]. Further analysis of fixed Ncd-GFP embryos showed that these fibers are labile and destroyed by fixation. Thus the ability to image Ncd-GFP in living embryos (as opposed to fixed embryos) was a critical factor for revealing the existence of these fibers and their association with Ncd.

Endow and Komma (1996) also demonstrated the utility of the GFP tag for analyzing phenotypic defects in mutants. They created a *ncd-gfp* fusion gene with a mutation that still allowed association of the mutant protein with spindles. Then, by time-lapse video confocal microscopy of live developing embryos containing the mutant Ncd-GFP, they obtained detailed dynamic images of the progressive defects in spindles, centrosome structure, and chromosome segregation brought about by this mutation (Endow and Komma, 1996).

9.7 FUTURE DIRECTIONS

What future light will GFP shed on *Drosophila* biology? In addition to the studies reviewed in this chapter, there are many more uses of GFP that can be envisioned. What follows are some thoughts about several specific applications.

The ability to mark individual cell types with GFP allows such cells to be identified in cell culture or intact animals. Chun-Fang Wu and coworkers showed that specific types of neurons can be identified in primary cell cultures

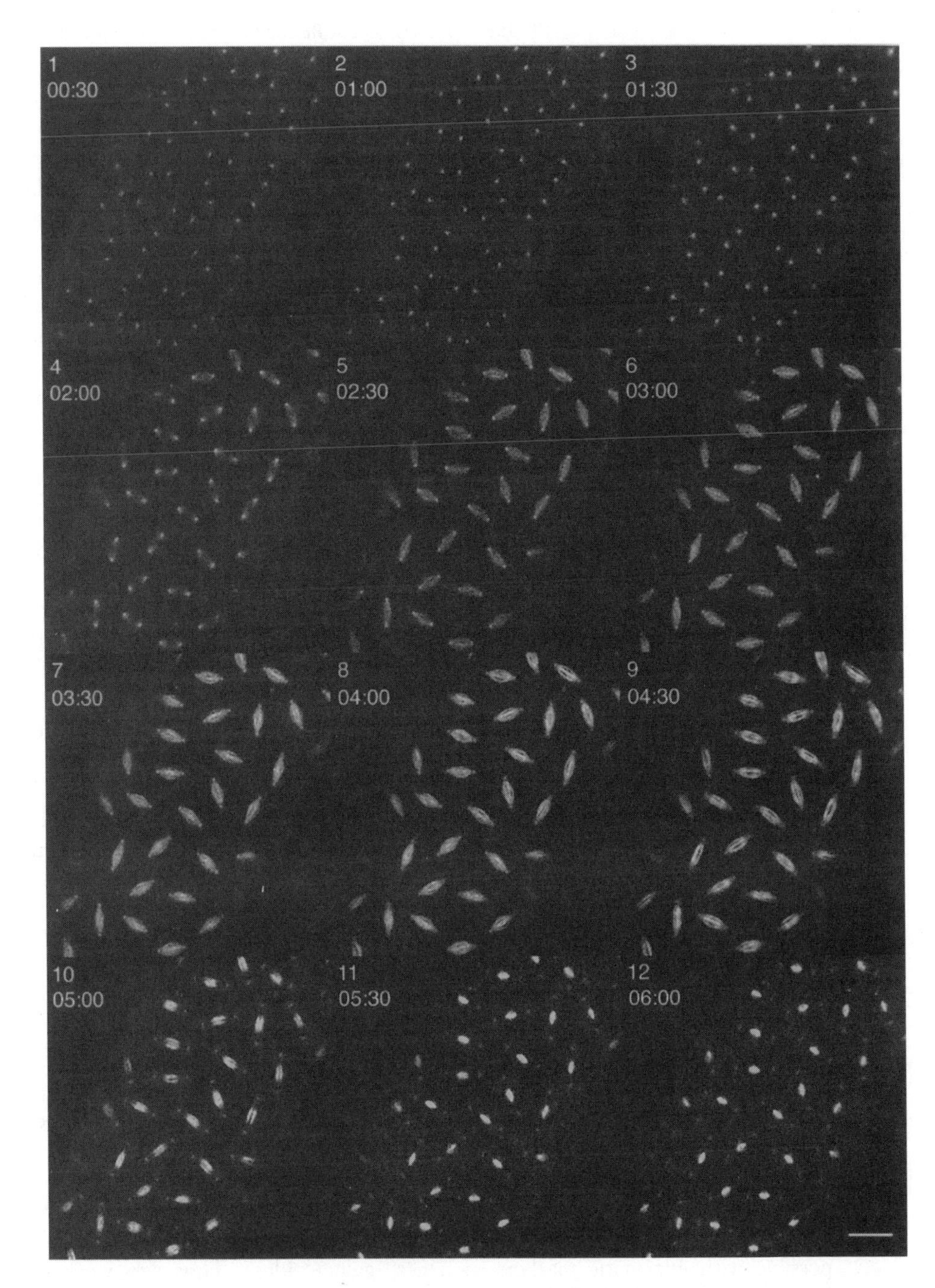

FIG. 9.7 Ncd-GFPS65T localized to the spindles throughout an embryonic mitotic cell cycle. These time-lapse images are of a living embryo in division cycle 9 (Endow and Komma, 1996.) The images shown were each collected 30 s apart; the time in minutes:seconds is indicated at the top of each frame. The mitotic stages are interphase (frames 1 and 2); nuclear envelope breakdown (frame 3); metaphase (frames 4–7); metaphase/early anaphase (frame 8); mid-anaphase (frame 9); late anaphase (frame 10); telophase (frames 11 and 12). Imaging was done on a BioRad MRC600 laser scanning confocal unit attached to a Zeiss Axioskop microscope, using a planapochromat 63X (N.A. 1.4) objective. For details about constructing time-lapse series, see Appendix IV.A.3. [Courtesy of Sharyn Endow; from Endow and Komma, 1996.]

and their electrophysiological characteristics studied. This technique could be extended to identify specific types of neurons in live embryos or larvae, and to record electrophysiologically from these cells in intact animals. In other cases, where tissue culture cells have been transformed with a GFP marker, it should be possible to identify and isolate these cells by fluorescence activated cell sorting (FACS).

In many instances reported in this chapter, GFP expressed alone, not as a fusion protein, has been observed to fill the entire cell in which it is expressed, and has proven to be a useful indicator of the morphology of a specific cell type. Morphological changes, such as cell shape changes during gastrulation and morphogenesis, and also cell migrations, may be studied in live animals with time-lapse confocal microscopy. The ability to detect GFP in nerve terminals (Brand, 1995; Yeh et al., 1995) suggests appealing applications in studying axon targeting and synapse specification. The TPLSCM analysis will no doubt prove to be advantageous over more traditional confocal microscopy, because of the finer resolution of structures lying deep within the embryo or tissue (Potter et al., 1996).

Genetic screens will be possible using GFP-marked cells to detect phenotypic alterations. For example, Barthmaier and Fyrberg (1995) proposed that the ability to mark the developing flight muscles with GFP should make possible screens for new mutations affecting muscle development. One advantage in this case is that such a screen is not dependent on behavioral changes. Similarly, the ability to mark specific types of neurons could provide a phenotypic screen for mutants (observation of morphological changes in the neurons) that otherwise may have been sought in difficult and time-consuming behavioral screens. GFP-labeling of cells should be of general use for genetic screens for *trans*-acting factors affecting the morphology and development of many different types of cells. Since GFP can be observed rapidly in unfixed tissues it offers advantages over the use of antibodies or enzyme reporter genes for these purposes. The rapidity and ease could greatly facilitate large-scale genetic screens. The recent development of epi-illuminators for use with dissecting microscopes should greatly facilitate such screens.

Green fluorescent protein should be a useful reporter in enhancer trap screens (Bellen et al., 1989; Bier et al., 1989). Yeh et al. (1995) and Timmons et al. (1997) demonstrated this usefulness with the GAL4 enhancer trap system (Brand and Perrimon, 1993). A major advantage, compared to enzymatic reporters, is that GFP allows dynamic changes in expression patterns to be followed in live tissues. Also, GFP makes possible the direct screening of live individuals carrying the new insertions, one generation earlier than is possible with enzymatic markers such as β-galactosidase. Thus, less work will be required to establish fly lines, and to prepare tissues to determine expression patterns, if the tissue type of interest can be imaged through the body wall of the animal.

Despite the extensive set of *Drosophila* genetic markers that exist, there are some situations where GFP will provide useful new markers. For instance, GFP will be a useful genetic marker to identify embryos of desired genetic constitutions. The construction of GFP marked balancer chromosomes that express GFP zygotically will allow one to identify homozygous embryos not carrying the balancer in question (e.g., non-GFP embryos). Another example where GFP could provide a useful genetic marker might be in the construction of transformation vectors for species other than *Drosophila melanogaster*, where visible markers are scarce.

The isolation of variants of GFP that are shifted in excitation and/or emission wavelengths (Heim et al., 1995; Delagrave et al., 1995; Heim and Tsien, 1996) has made possible the simultaneous detection of more than one GFP fusion protein within a single cell (Rizzuto et al., 1996). These variants should be useful for determining if proteins involved in similar functions are colocalized within a cell, and for determining the expression patterns of genes relative to each other. The development of fluorescence energy resonance transfer (FRET) holds the potential to be a powerful means of determining if proteins interact directly (Heim and Tsien, 1996).

The GFP fusion proteins reported in this chapter have all been localized like the fusion partners: to the nucleus, to chromosomes, to spindles, to subcellular particles, to the cytoskeleton. In most of these cases, the validity of the GFP tag to reflect subcellular localization was shown by requiring function of the fusion proteins, and was possible because of the existence of mutations in genes encoding the protein partners. However, GFP fusions can provide useful clues about the localization of a newly cloned gene's protein product, even in the absence of existing mutations in the gene in question. If a DNA fragment containing the entire gene exists, and enough of the sequence is known, then N- or C-terminal fusions can be constructed and transformed into flies. The entire time and resources required to make such constructs and transformant lines is less than that required to express the protein product in bacteria and produce reliable antibodies, and does not require the isolation of cDNAs. Indeed, GFP fusions with protein fragments may prove to be a useful way to identify domains of proteins required for localization to different subcellular compartments, or for specific functions where an appropriate assay exists that can take advantage of the GFP tag.

As more fusion genes expressing GFP tagged proteins are constructed, useful markers for subcellular structures and organelles will become available. For instance, Ncd-GFP binds to spindles, allowing time-lapse confocal imaging of the spindles in the early mitotic divisions of living embryos (Endow and Komma, 1996). GFP-moe highlights actin-rich regions of cells, allowing observations of dynamic changes in cell shape during tissue morphogenesis (Edwards et al., 1997). The construction of a GFP tagged histone, mitochondrial proteins, membrane proteins, additional microtubule-binding proteins such as Tau (Brand, 1995), tubulins, and so on, will yield a set of useful protein tags for studying cell biology. Thus, for example, instead of using a DNA stain to assess chromosome segregation defects caused by a mutation affecting mitosis, one could cross a GFP-Histone transgene into the mutant background to visualize the chromosomes. In the future, the ability to mark a variety of subcellular structures and organelles with GFP will make possible a *Drosophila* strain collection that will be a powerful cell biology tool kit, allowing analysis of living tissue without invasive measures, yielding dynamic data about processes over time.

REFERENCES

Ashburner, M. (1989). *Drosophila*, A Laboratory Handbook. Cold Spring Harbor Laboratory Press, Cold Springs Harbor, N.Y.

Aubin, J. E. (1979). Autofluorescence of viable cultured mammalian cells. *J. Histochem. Cytochem.* 27:36043.

Barthmaier, P. and Fyrberg, E. (1995). Monitoring Development and Pathology of *Drosophila* Indirect Flight Muscles Using Green Fluorescent Protein. *Dev. Biol.* 169:770–774.

Bellen, H. J., O'Kane, C. J., Wilson, C., Grossniklaus, U., Pearson, R. K., and Gehring, W. J. 1989. P-element-mediated enhancer detection: a versatile method to study development in *Drosophila*. *Genes Dev.* 3:1288–1300.

Berleth, T.,Burri., M., Thoma, G., Bopp, D., Richstein, S., Frigerio, M., Noll, M., and Nüsslein-Volhard, C. (1988). The role of localization of *bicoid* RNA in organizing the anterior pattern of the *Drosophila* embryo. *EMBO J.* 7:1749–1756.

Bier, E., Vaessin, H., Shepherd, S., Lee, K., McCall, K., Barbel, S., Ackerman, L., Caretto, R., Uemura, T., Grell, E., Jan, L. Y., and Jan Y. N. (1989). Searching for pattern and mutation in the *Drosophila* genome with a P-*lacZ* vector. *Genes Dev.* 3:1273–1287.

Brand, A. (1995). GFP in *Drosophila*. *TIG* 11:324–325.

Brand, A. H. and Perrimon, N. (1993). Targeted gene expression as a means of altering cell fates and generating dominant phenotypes. *Development* 118:401-415.

Chalfie, M., Tu, Y., Euskirchen, G. Ward, W. W. and Prasher, D. C. (1994). Green fluorescent protein as a marker for gene expression. *Science* 263:802–805.

Davis, B. K., (1971). Genetic analysis of a meiotic mutant resulting in precocious sister-centromere separation in *Drosophila melanogaster*. *Mol.. Gen. Genet.* 113:251–272.

Davis, I., Girdham, C. H., and O'Farrell, P. H. (1995). A nuclear GFP that marks nuclei in living *Drosophila embryos*; maternal supply overcomes a delay in the appearance of zygotic fluorescence. *Dev. Biol.* 170:726–729.

Delagrave, S., Hawtin, R. E., Silva, C. M., Yang, M. M., and Youvan, D. C. (1995). Red-shifted excitation mutants of the green fluorescent protein. *Bio.Tech.* 13:151–154.

Driever, W., and Nusslein-Volhard, C. (1988). A gradient of *bicoid* protein in *Drosophila* embryos. *Cell* 54:83–93.

Edwards, K. A., Demsky, M., Montague, R. A., Weymouth, N., and Kiehart, D. P. (1997). GFP-Moesin illuminates actin cytoskeleton dynamics in living tissue and demonstrates cell shape changes during morphogenesis in *Drosophila*. *Dev. Biol.* 191:103–117.

Endow, S. A., Chandra, R., Komma, D. J., Yamamoto, A. H., and Salmon, E. D., (1994). Mutants of the *Drosophila ncd* microtubule motor protein cause centrosomal and spindle pole defects in mitosis. *J. Cell Sci.* 107:859–867.

Endow, S. and Komma, D. (1996). Centrosome and spindle function of the *Drosophila* Ncd microtubule motor visualized in live embryos using Ncd-GFP fusion proteins. *J. Cell Sci.* 109:2429–2442.

Endow, S. and Komma, D. (1997). Spindle dynamics during meiosis in *Drosphila* oocytes. *J. Cell Biol.* 137:1321–1336.

Glazer, L., and Shilo, B-Z. (1991). The *Drosophila* FGF-R homolog is expressed in the embryonic tracheal system and appears to be required for directed tracheal cell extension. *Genes Dev.* 5:697–705.

Goldstein, L. S. B. (1980). Mechanisms of chromosome orientation revealed by two meiotic mutants in *Drosophila melanogaster*. *Chromosoma* 78:79–111.

Hatsumi, M. and Endow, S. (1992). The *Drosophila* ncd microtubule motor protein is spindle-associated in meiotic and mitotic cells. *J. Cell Sci.* 103:1013–1020.

Hay, B. A., Wolff, T., and Rubin, G. M. (1994). Expression of baculo virus P35 prevents cell death in *Drosophila. Development* 230:2121–2129.

Hazelrigg, T., Watkins, W. S., Marcey, D., Tu, C., Karow, M., and Lin, X. (1990). The *exuperantia* gene is required for *Drosophila* spermatogenesis as well as anteroposterior polarity of the developing oocyte, and encodes overlapping sex-specific transcripts. *Genetics* 126:607–617.

Heim, R., Cubitt, A. B., and Tsien, R. Y. (1995). Improved green fluorescence. *Nature (London)* 373:663–664.

Heim, R. and Tsien, R. (1996). Engineered green fluorescent protein for improved brightness, longer wavelengths and fluorescence resonance energy transfer. *Curr. Biol.* 6:178–182.

Kerrebrock, A. W., Moore, D. P., Wu, J. S., and Orr-Weaver, T. L. (1995). Mei-S332, a *Drosophila* protein required for sister-chromatid cohesion, can localize to meiotic centromere regions. *Cell* 83:247–256.

MacDonald, P. M., Luk, S. K.-S, and Kilpatrick, M. (1991). Protein encoded by the *exuperantia* gene is concentrated at sites of *bicoid* mRNA accumulation in *Drosophila* nurse cells but not in oocytes or embryos. *Genes Dev.* 5:2455–2466.

Marcey, D., Watkins, W.S., and Hazelrigg, T. (1991). The temporal and spatial distribution pattern of maternal *exuperantia* protein: evidence for a role in establishment but not maintenance of *bicoid* mRNA localization. *EMBO J.* 10:4259–4266.

Montell, D. J., Rorth, P., and Spradling, A. C. (1992). *Slow border cells*, a locus required for a developmentally regulated cell migration during oogenesis, encodes *Drosophila* C/EBP. *Cell* 71:51–62.

Mulligan, P. K., Campos, A. R., and Jacobs, J. R., (1996). Mutations in the gene *stand still* disrupt germ cell differentiation in *Drosophila* ovaries. *Devel. Genet.* 18:316–326.

Niswender, K. D., Blackman, S. M., Rohde, L., Magnuson, M. A., and Piston, D. W. (1995). Quantitative imaging of green fluorescent protein in cultured cells: comparison of microscopic techniques, use in fusion proteins and detection limits. *J. Microsc.* 180: 109–116.

Pirrotta, V. (1988). Vectors for P-mediated transformation in *Drosophila*. In Vectors, A Survey of Molecular Cloning Vectors and Their Uses. Rodrigues, R. L. and Denhardt, D. T., Eds., Butterworths, New York, pp. 437–456..

Plautz, J. D., Day, R. N., Dailey, G. M., Welsh, S. B., Hall, J. C., Halpain, S., and Kay, S. A. (1996). Green fluorescent protein and its derivatives as versatile markers for gene expression in living *Drosophila melanogaster*, plant and mammalian cells. *Gene* 173:83–87.

Potter, S. M., Wang, C.-W., Garrity, P. A., and Fraser, S. E. (1996). Intravital imaging of green fluorescent protein using two-photon laser-scanning microscopy. *Gene* 173:25–31.

Prasher, D. C., Eckenrode, V. K., Ward, W. W., Prendergast, M. J., and Cormier, M. J. (1992). Primary structure of the *Aequorea victoria* green fluorescent protein. *Gene* 111:229–233.

Rizzuto, R., Brini, M., De Giorgi, F., Rossi, R., Heim, R., Tsien, R., and Pozzan, T. (1996). Double labeling of subcellular structures with organelle-targeted GFP variants *in vivo*. *Curr Biol.* 6:182–188.

Saito, M. and Wu, C.-F. (1991). Expression of ion channels and mutational effects in giant *Drosophila* neurons differentiated from cell division-arrested embryonic neuroblasts. *J. Neurosci.* 11(7):2135–2150.

Shiga, Y., Tanaka-Matakatsu, M., and Hayashi, S. (1996). A nuclear GFP/β-galactosidase fusion protein as a marker in living *Drosophila*. *Develop. Growth Differ.* 38:99–106.

Spradling, A. C., and Rubin, G. M. (1982). Transposition of Cloned P-elements into *Drosophila* Germ Line Chromosomes. *Science* 218:341–347.

St. Johnston, D., Driever, W., Berleth, T., Richstein, S., and Nusslein-Volhard, C. (1989). Multiple steps in the localization of *bicoid* RNA to the anterior pole of the *Drosophila* oocyte. *Development (Suppl.)* 107:13–19.

Thummel, C. S., Boulet, A. M., and Lipshitz, H.D. (1988). Vectors for *Drosophila* P-element-mediated Transformation and Tissue Culture Transfection. *Gene* 74:445–456.

Thummel, C. S. and Pirrotta, V. (1992). New PCaSpeR P element vectors. *Drosophila Information Service* 71:150.

Timmons, L., Becker, J., Barthmaier, P., Fyrberg, C., Shearn, A., and Fyrberg, E. (1997). Green Fluorescent Protein/β-galactosidase double reporters for visualizing *Drosophila* gene expression patterns. *Dev. Genetics* 20:338–347.

Walker, R. A., Salmon, E. D., and Endow, S. (1990). The *Drosophila claret* segregation protein is a minus-end directed motor protein. *Nature (London)* 347:780–782.

Wang, S. and Hazelrigg, T. (1994). Implications for *bcd* mRNA localization from spatial distribution of *exu* protein in *Drosophila* oogenesis. *Nature (London)* 369:400–403.

Wu, C.-F., Sakai, K., Saito, M., and Hotta, Y., (1990). Giant *Drosophila* neurons differentiated from cytokinesis-arrested embryonic neuroblasts. *J. Neurobiol.* 21(3):499–507.

Yeh, E., Gustafson, K., Boulianne, G. L. (1995). Green fluorescent protein as a vital marker and reporter of gene expression in *Drosophila*. *Proc. Natl. Acad. Sci.* 92:7036–7040.

Young, P. E., Richman, A. M., Ketchum, A. S., and Kiehart, D. P. (1993). Morphogenesis in *Drosophila* requires nonmuscle myosin heavy chain function. *Genes Dev.* 7:29–41.

10

The Uses of Green Fluorescent Protein in Plants

JIM HASELOFF AND KIRBY R. SIEMERING
Division of Cell Biology, MRC Laboratory of Molecular Biology, Hills Road, Cambridge, UK

10.1 INTRODUCTION

Marker genes have proved extremely useful for reporting gene expression in transformed plants. The β-glucuronidase *gusA* or (*GUS*) gene (Jefferson et al., 1987) has been used extensively. Transformed cells or patterns of gene expression within plants can be identified histochemically, but this is generally a destructive test and is not suitable for assaying primary transformats, or for following the time course of gene expression in living plants, or as a means of rapidly screening segregating populations of seedlings. The green fluorescent protein (GFP) from *Aequorea victoria* shares none of these problems, as its intrinsic fluorescence can be seen in living cells. In addition, there has been intense interest in its use as a marker for transgenic plants.

Unmodified *gfp* has been successfully expressed at high levels in tobacco plants using the cytoplasmic ribonucleic acid (RNA) viruses potato virus X (Baulcombe et al., 1995) and tobacco mosaic virus (Heinlein et al., 1995). In these experiments, the gene was directly expressed as a viral mRNA in infected cells, and spectacularly high levels of GFP fluorescence were seen.

In contrast to the efficient RNA virus-mediated expression of GFP, variable results have been obtained with transformed cells and plants. Although green fluorescence has been seen in *gfp* transformed protoplasts of citrus (Niedz et al., 1995) and maize (Hu et al., 1995; Sheen et al., 1995), we and others have seen

Green Fluorescent Protein: Properties, Applications, and Protocols, Edited by Martin Chalfie and Steven Kain
ISBN 0-471-17839-X

no fluorescence in *Arabidopsis* and other transformed plant species (Haseloff and Amos., 1995; Haseloff et al., 1997). Hu and Cheng (1995) reported that no signal was seen in *gfp* transformed *Arabidopsis thaliana* protoplasts. Reichel et al. (1996) also failed to detect fluorescence in *gfp* transformed *Arabidopsis*, tobacco, or barley protoplasts. Sheen et al. (1995) also saw no expression of a CAB2-driven *gfp* gene in transgenic *Arabidopsis* plants and Pang et al (1996) saw little or no expression in *gfp* transformed wheat, corn, tobacco, and *Arabidopsis* plants. There appears to be a need for substantial improvement of the wild-type *gfp* gene for use in plants. In this chapter, we describe some of the pitfalls affecting *gfp* expression and detection in plants, and describe modified forms of the gene and new techniques which have helped to overcome these problems.

10.2 CRYPTIC SPLICING OF *gfp* mRNA IN *ARABIDOPSIS*

Useful expression of the *gfp* cDNA (cyclic deoxyribonucleic acid) in plants requires that (a) the GFP apoprotein be produced in suitable amounts within plant cells, and (b) the nonfluorescent apoprotein undergoes efficient posttranslational modification to produce the mature GFP. The high levels of GFP fluorescence seen in plants infected with suitable RNA virus vectors (Baulcombe et al., 1995; Heinlein et al., 1995) demonstrate that the protein can undergo efficient posttranslational modification in plants. However, the expression of integrated copies of the gene has proved problematic. We have used *Agrobacterium* mediated transformation to produce transgenic *Arabidopsis* plantlets containing a cauliflower mosaic virus 35S promoter-driven *gfp* cDNA (Haseloff et al., 1997). However, at no stage during the transformation procedure did we detect GFP related fluorescence, using an ultraviolet (UV) lamp illumination and epifluorescence microscopy. Therefore we used polymerase chain reaction (PCR) based methods to verify the correct insertion of the gene, and to check mRNA transcription and processing in these transformed plantlets. Samples of DNA and mRNA were separately extracted, and *gfp* sequences were amplified via PCR from the separate extracts and analysed. While the expected full-length *gfp* product was obtained after amplification of the integrated gene, RT-PCR of *gfp* mRNA sequences gave rise to a truncated product.

The shortened RT-PCR product was cloned and sequenced, and a deletion of 84 nucleotides was found between residues 380–463 of the GFP coding sequence (Fig. 10.1). The missing sequence bears close similarity to known plant introns and it is likely that expression of *gfp* in *Arabidopsis* is curtailed by aberrant mRNA splicing, with an 84 nucleotide sequence being recognized as a cryptic intron. This explanation would also account for the efficient expression of *gfp* from RNA virus vectors that replicate in the cytoplasm, and thus evade splicing. The nucleotide sequences bordering the deletion are shown in Figure 10.1, and demonstrate similarity to known plant introns. Matches were found for sequences that are conserved at the 5′ and 3′ splice sites of plant introns (reviewed in Luehrsen et al., 1994), and for conserved branch point nucleotides in plant introns (Liu and Filipowicz., 1996; Simpson et al., 1996). The excised *gfp* sequence contains a high AU content (68%) that has also been shown to be important for recognition of plant introns (Hanley and Schuler, 1988; Wiebauer et al., 1988; Goodall and Filipowicz, 1989; Goodall and Filipowicz, 1991). It is likely that this 84 nucleotide region of the jellyfish *gfp* cDNA sequence is efficiently recognized as an intron when transcribed in *Arabidopsis*, resulting in an in frame deletion and the production of a defective protein product, which is predicted to be 28 aa shorter. Subsequently, an artificial neural network program has been used to correctly predict the presence of the *Arabidopsis* cryptic intron in the *gfp* coding sequence (Hebsgaard et al., 1996), and similar cryptic splicing has been seen now in other plant species (Schuler, personal communication). It should be noted that the borders of the cryptic intron do not coincide with any of the natural spliced junctions found after processing of the *gfp* mRNA in *Aequorea victoria* (Prasher et al., 1992). No full-length *gfp* mRNA is detectable by RT-PCR, and so misprocessing must be close to complete in transformed *Arabidopsis* plantlets. It has been claimed that *gfp* fluorescence has been detected in bombarded *Arabidopsis* tissues (Sheen et al., 1995). However, in these experiments, leaf tissue was treated with methanol prior to microscopic examination, where methanol causes rapid and irreversible bleaching of GFP (Ward et al., 1980). Local wounding due to particle bombardment can cause punctate patterns of bright autofluorescence, and this type of experiment needs to be interpreted with care. The same authors saw no expression of a CAB2-driven *gfp* gene in transgenic *Arabidopsis* plants (Sheen et al., 1995). It is likely that elimination of cryptic splicing is essential for proper expression of the *gfp* gene in *Arabidopsis*.

10.3 REMOVAL OF THE CRYPTIC INTRON

It has proved necessary to destroy this cryptic intron to ensure proper expression in plants. We have altered the codon usage for GFP, deliberately mutating recognition sequences at the putative 5′ splice site and branch point and decreasing the AU content of the intron. All of the sequence modifications affected only codon usage, and this modified gene, *mgfp4*, encodes a protein product that is identical to that of the jellyfish (Fig. 10.2). When the *mgfp4* sequence was inserted behind the 35S promoter and introduced into *Arabidopsis* using the root transformation technique (Valvekens et al., 1988), bright green fluorescent plant cells were detected within 2–3 days of cocultivation. As cell proliferation continued, the brightest clumps of callus and developing shoot tissue were so intensely fluorescent that they were clearly visible by eye, using a 100-W long wavelength hand-held UV lamp (UV Products, B100AP). We have also adapted an inverted fluorescence microscope (Leitz DM-IL) to allow more sensitive, higher magnification observation of cells in sterile culture during transformation and regeneration. the microscope was fitted with a filter set (Leitz-D excitation BP355–425, dichroic 455, emission LP460) suitable for the main 395-nm excitation and 509-nm emission peaks of GFP, and we used a 7-mm threaded extension tube with a 4X objective (EF 4/0.12) to give a greater working distance above the microscope stage. This allows the convenient direct observation of transformed tissues and plantlets within sealed inverted Petri dishes.

The ease with which fluorescent proteins can be monitored in living tissues allows new approaches for improving transformation and regeneration of intractable or slow-growing plant species. During our own regeneration experiments, we

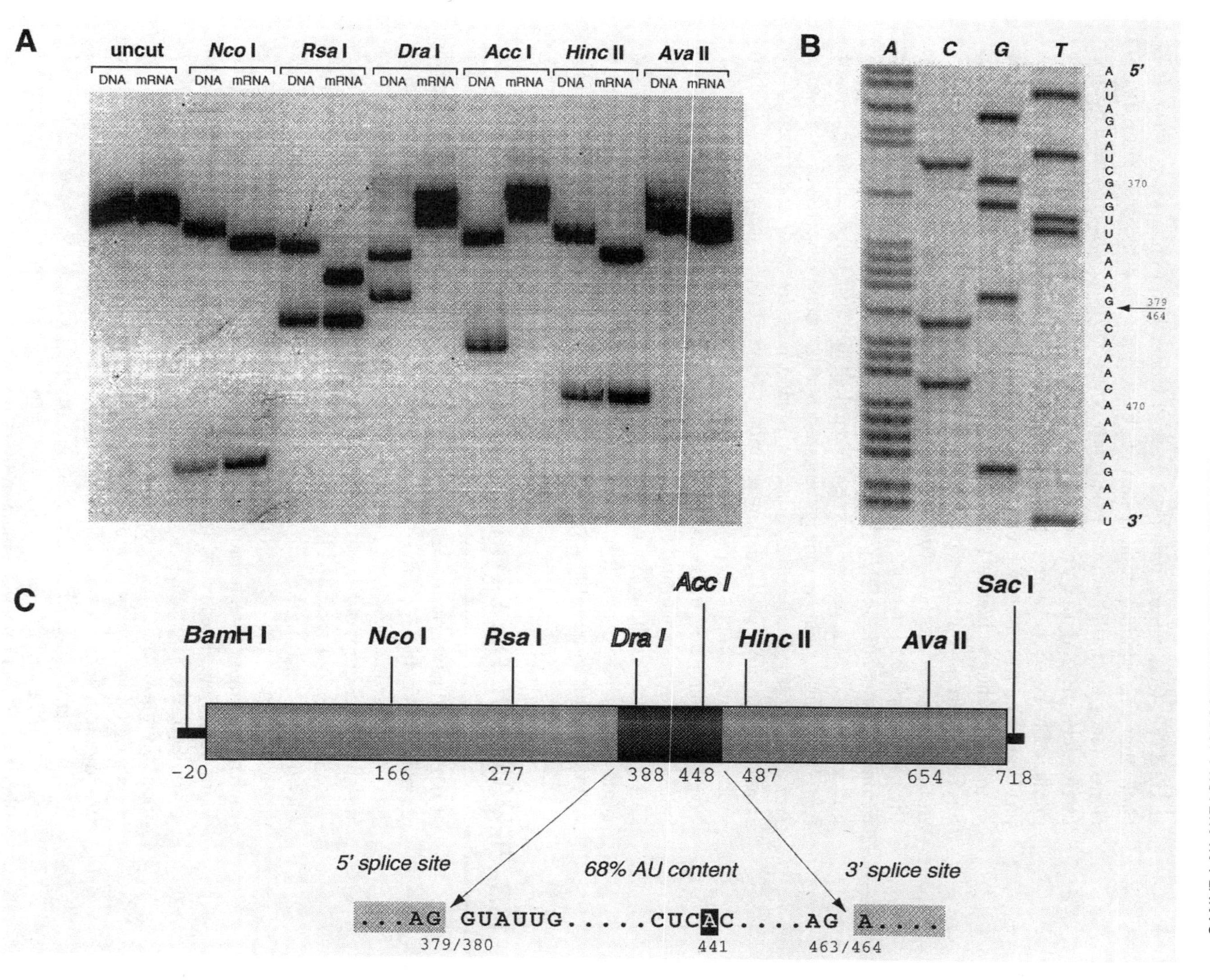
A
uncut
Nco I
Rsa I
Dra I
Acc I
Hinc II
Ava II
DNA mRNA
B
A C G T
5'
AAUAGAAUCGAGUUAAAAGACAAACAAAAGAAU
3'
370
379
464
470
C
BamH I
Nco I
Rsa I
Dra I
Acc I
Hinc II
Ava II
Sac I
-20
166
277
388
448
487
654
718
5' splice site
68% AU content
3' splice site
...AG GUAUUG.....CUCAC....AG A....
379/380
441
463/464

observed a wide range of GFP fluorescence intensities in 35S-*mgfp4* transformed plantlets, which we expect arose from position-dependent modulation of gene expression in different transformants. It proved difficult to regenerate fertile plants from the brightest transformants, with cells remaining as a highly fluorescent callus or mass of shoots after several months of culture. It is possible that high levels of GFP expression were mildly toxic or interfered with differentiation. This toxicity is of special concern with a fluorescent molecule such as GFP, which would be expected to generate free radicals upon excitation, and which undergoes oxidative modification and could possess catalytic properties. The conditions that we have used for plant regeneration should provide a stringent test for any deleterious effect due to GFP. The 35S promoter was used to drive expression of the protein at high levels throughout the plant, including meristematic cells, and regeneration took place under continual illumination, allowing the possibility for GFP mediated phototoxicity. Despite poor regeneration of the brightest transformants, we managed to obtain over 50 separate transgenic *Arabidopsis* lines, most of which contained levels of GFP that were easily detectable by microscopy.

10.4 EXPRESSION OF GFP IN OTHER PLANTS

Some expression of wild-type *gfp* has been seen in plant protoplasts of tobacco (Reichel et al., 1996), *Citrus sinensis* (Niedz et al., 1995) and maize (Hu and Chang, 1995; Sheen et al., 1995; Pang et al., 1996), and so aberrant splicing of *gfp* mRNA may not be as efficient in other plant species, as in transgenic *Arabidopsis*. However, the *mgfp4* gene has proved useful for expression studies in other plants, which share features involved in intron recognition (Luehrsen et al., 1994). Experiments with tobacco and barley protoplasts (Reichel et al., 1996)

FIG. 10.1 (opposite) Cryptic splicing of *gfp* transcripts in transgenic *Arabidopsis thaliana*. (*a*) Restriction endonuclease digestion of PCR fragments derived from *gfp* DNA and mRNA sequences. Sequences corresponding to the integrated *gfp* gene and to mRNA transcripts were isolated and separately amplified using PCR techniques, and incubated with various restriction endonucleases. The radiolabeled fragments were fractionated by electrophoresis in a 5% polyacrylamide gel, and are shown labeled with the source of the amplified sequences (DNA or mRNA) and the name of the restriction endonuclease used for digestion, or not (uncut). The mRNA derived sequences appeared to lack sites for *Dra* I and *Acc* I, and to contain a corresponding deleted region of 80–90 nucleotides. Restriction endonuclease fragments that are smaller than those expected of the gene sequence have been indicated with a white asterisk. (*b*) Sequence analysis of cloned *gfp* mRNAs. Autoradiograph and sequence of the amplified *gfp* mRNA sequence. Nucleotides 380–463 are absent from the transcribed sequence, and the site of this 84 nucleotide deletion is arrowed. (*c*) Schematic diagram of the *gfp* gene sequence, which shows the positions of restriction endonuclease sites used for the analysis of PCR amplified mRNA transcripts, and the location of the cryptic intron, shown with dark shading. Sequences that are similar to those normally found at splice sites and branch points of plant introns. are shown below. Splice sites are arrowed, and the putative lariant branch point is shown in reverse type (Haseloff et al., 1997).

```
                                                                                          M1  S   K   G   E
                                                                                          atg agt aaa gga gaa      gfp
 BamHI   RBS         TICS                                                                 atg agt aaa gga gaa      mgfp4
gg atc caa gga gat ata aca atg aag act aat ctt ttt ctc ttt ctc atc ttt tca ctt ctc cta tca tta tcc tcg gcc gaa ttc agt aaa gga gaa   m-gfp5

                           M   K   T   N   L   F   L   F   L   I   F   S   L   L   L   S   L   S   S   A   E   F
                                  Arabidopsis thaliana basic chitinase signal sequence

E   L   F   T   G   V   V   P   I   L   V   E   L   D   G20 D   V   N   G   H   K   F   S   V   S   G   E   G   E   G   D   A   T
                                                                                                                          gfp
                                                                                                                          mgfp4
                                                                                                                          m-gfp5

  40 K   L   T   L   K   F   I   C   T   T   G   K   L   P   V   P   W   P   T   L60 V   T   T   F   S   Y   G   V   Q   C   F
                                                                                                                          gfp
                                                                                                                          mgfp4
                                                                                                                          m-gfp5

                         80 H   D   F   F   K   S   A   M   P   E   G   Y   V   Q   E   R   T   I   F   F100K   D   D   G
                                                                                                                          gfp
                                                                                                                          mgfp4
                           g cgg cac gac ttc ttc aag agc gcc atg cct gag gga tac gtg cag gag agg acc atc ttc ttc aag gac gac ggg   m-gfp5

                                                  120 N   R   I   E   L   K   G   I   D   F   K   E   D   G   N   I   L
                                                                              gt att gat ttt aaa gaa gat gga aac att ctt   gfp
                                  g gga gac acc ctc gtc aac agg atc gag ctt aag gga atc gat ttc aag gag gac gga aac atc ctc   mgfp4
                                  g gga gac acc ctc gtc aac agg atc gag ctt aag gga atc gat ttc aag gag gac gga aac atc ctc   m-gfp5

              140 L   E   Y   N   Y   N   S   H   N   V   Y   I   M   A   D   K   Q   K   N   G160I   K   V   N   F   K   I   R   H   N
                                                                        ac aaa caa aag aat gga atc aaa gtt aac ttc aaa att aga cac aac   gfp
c cac aag ttg gaa tac aac tac aac tcc cac aac gta tac atc atg gcc gac aag caa aag aac ggc atc aaa gcc aac ttc aag acc cgc cac aac   mgfp4
c cac aag ttg gaa tac aac tac aac tcc cac aac gta tac atc atg gcc gac aag caa aag aac ggc atc aaa gcc aac ttc aag acc cgc cac aac   m-gfp5

                                                                                                      A               T

                    S   V   Q   L   A   D180H   Y   Q   Q   N   T   P   I   G   D   G   P   V   L   L   P   D   N   H   Y200L   S   T
                                                                                                                          gfp
                                                                                                                          mgfp4
c gaa gac ggc ggc gtg caa ctc gct gat cat tat caa caa aat act cca att ggc gat ggc cct gtc ctt tta cca gac aac cat tac ctg tcc aca   m-gfp5
                    G

                                                  220 L   E   F   V   T   A   A   G   I   T   H   G   M   D   E   L
                                                                                                                          gfp
                                                                                                                          mgfp4
                                                                                                                          mgfp5-
```

demonstrated that *mgfp4* derived sequences are expressed at much higher levels than the wild-type gene in these species. The *mgfp4* is also expressed efficiently in soybean cells (Plautz et al., 1996). There have been reports of improved GFP expression in mammalian cells after alteration of gene codon usage (Haas et al., 1996; Zolotukhin et al., 1996). Increased levels of expression have been attributed to improved rates of translation due to optimized codon usage. However, this "humanization" of *gfp* also leads to alteration of the cryptic intron sequence, and expression of sGFP (Haas et al., 1996) been shown to result in 20-fold increased fluorescence in maize protoplasts (Chiu et al., 1996). The increased levels of expression may be due, at least in part, to an effect on RNA processing. Other workers have also found it necessary to deliberately alter the codon usage of *gfp* for efficient expression in plants. Transient expression of the synthetic *pgfp* gene gave rise to about 20-fold more fluorescence than wild-type *gfp* in maize and tobacco protoplasts (Pang et al., 1996).

It is possible that altered mRNA sequences affect posttranscriptional processing in animal cells as well. However, introns found in animals, including *A. victoria* (Prasher et al., 1992), share a conserved polypyrimidine tract adjacent to the 3′ splice site, reviewed in (Green, 1992), and introns in yeast cells possess a requirement for additional conserved sequences (UACUA<u>A</u>C) located at the branch point (Langford et al., 1984). The lack of these additional features may help to minimize recognition of the cryptic intron and aberrant processing of *gfp* mRNA in fungal and animal cells.

10.5 LOCALIZATION OF GFP IN PLANT CELLS

In transgenic *Arabidopsis* cells, GFP is found throughout the cytoplasm, but appears to accumulate within the nucleoplasm (Haseloff and Amos 1995; Chiu et al., 1996; Grebenok et al., 1997; Haseloff et al., 1997; Kohler et al., 1997). It is excluded from vacuoles, organelles, and other bodies in the cytoplasm, and is excluded from the nucleolus (Fig. 10.3). A similar subcellular distribution of GFP was seen in all *Arabidopsis* cell types examined in our experiments, and red autofluorescent chloroplasts provide an effective counter-fluor for

FIG. 10.2 (opposite) Sequence comparison of *gfp* and the modified *mgfp5-ER* gene. The sequence of *gfp* is as described for the *gfp10* cDNA (Prasher et al., 1992), except that codon 80 contains a change from CAG to CGG resulting in replacement of a glutamine with arginine, as noted by Chalfie *et al.* (1994). Both the *gfp* and *mgfp5-ER* gene cassettes are flanked by restriction endonuclease sites for *Bam*HI and *SacI*, a ribosome binding site (RBS) for bacterial expression and the sequence AACA upstream of the start codon for improved plant translation. The cryptic plant intron present in *gfp* (Haseloff et al., 1997) is shown underlined with the 5′ and 3′ splice sites arrowed. Nucleotide sequence alterations present in *mgfp5-ER* are shown outlined in gray. Most alterations are silent and all amino acid substitutions are shown in reverse type below the nucleotide sequence. The *mgfp5-ER* gene cassette contains additional sequences shown in bold face type, which comprise a 5′ terminal signal peptide and 3′ HDEL sequence. An *Eco*RI site was used to link the signal peptide and coding sequences.

GFP in the upper parts of the plant. Cytoplasmic streaming and the movement of organelles could be observed in these living cells. In addition to cell ultrastructure, the architecture of the intact tissue was also clearly discernible, and the arrangement of different cell types could be seen in longitudinal optical sections of root tips and cotyledons. For example, cells within the epidermis of the cotyledon contain few mature chloroplasts and could be distinguished from layers of neighboring mesophyll cells, and flies of developing cells around the primary root meristem are clearly evident (Fig. 10.3).

While the *mgfp4* gene was proving useful as a marker in transgenic *Arabidopsis*, it was also clear from the initial studies that it could bear improvement. While we were able to generate 35S-*mgfp4* transformed cells and calli that were intensely fluorescent, and easily detectable by eye under long wavelength UV illumination, it proved difficult to regenerate fertile plants from the brightest transformants. It is possible that very high levels of GFP expression are mildly toxic or interfere with regeneration, perhaps due to the fluorescent or catalytic properties of the protein. In jellyfish photocytes, where high levels of GFP are well tolerated, the protein is found sequestered in cytoplasmic granules (Davenport and Nichol, 1955). In contrast, the mature protein is found through-

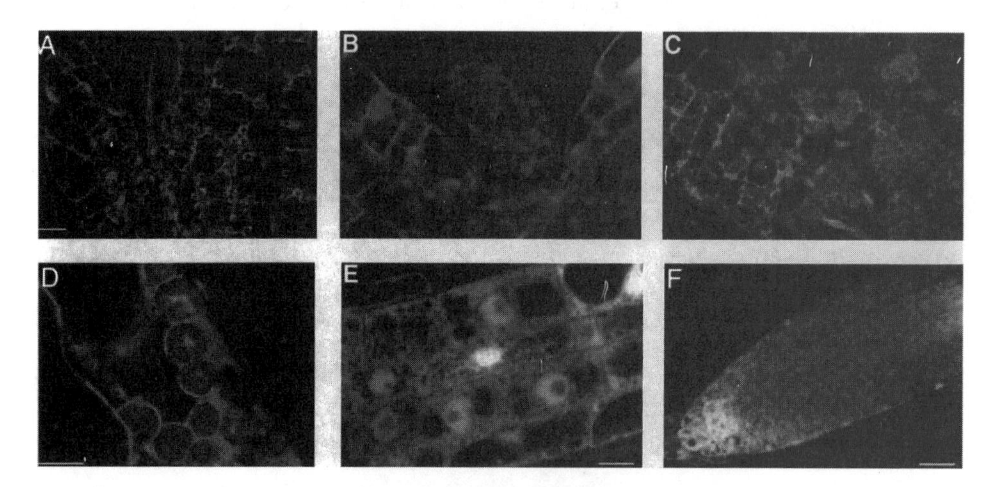

FIG. 10.3 Confocal images of 35S-*mgfp4* transformed *Arabidopsis* plants. The 35S-*mgfp4* transformed seedlings were grown in sterile agar culture and mounted intact in water for confocal microscopy. Images were collected using a BioRad MRC-600 instrument equipped with Nikon Optiphot microscope and Nikon planapo 60X water immersion lens. The GFP and chlorophyll were excited using the 488- and 568-nm lines, respectively, of a 25-nm krypton–argon ion laser. The green and red emissions were collected in separate channels and combined using Adobe Photoshop. (*a*) The shoot apical meristem is shown. Individual vacuolate cells that each contain a layer of green fluorescent cytoplasm containing red autofluorescent chloroplasts, can be distinguished. (*b*) A emerging leaf primordia, positioned at the shoot apex between two cotyledons. (*c*) Mesophyll cells within a cotyledon show large numbers of mature chloroplasts. (*d*) An optical section of a single hypocotyl cortex cell, showing mature chloroplasts. (*e*) Cells from within the root meristem. The GFP accumulates within nuclei, but is excluded from nucleoli, and is found throughout the cytoplasm where various endomembrane compartments are shown in negative relief. (*f*). Median longitudinal optical section of a root tip.

out the cytoplasm and accumulates within the nucleoplasm of transformed *Arabidopsis* cells. If GFP is a source of fluorescence-related free radicals, for example, it might be advisable to target the protein to a more localized compartment within the plant cell.

10.6 SUBCELLULAR TARGETING OF GFP

We have fused several targeting peptides to GFP, and directed the protein to different subcellular compartments. The targeted forms of the *mgfp4* gene were initially tested by expression in *Saccharomyces cerevisiase*. The modified genes were introduced into yeast cells on a multicopy vector and expressed fluorescent protein was visualized using confocal microscopy (Fig. 10.4). Unmodified protein is normally found throughout the cytoplasm and nucleoplasm of yeast cells. The addition of a peptide containing the SV40 T-antigen NLS (amino acids APKKKRKVEDPR) to the N- or C-terminus of the protein does little to alter its distribution (not shown). However, if the NLS-GFP protein is fused to a larger protein, such as that encoding β-galactosidase, the fusion protein is exclusively found in the nucleoplasm [Fig. 10.4(*d*)]. We have also fused to GFP a mitochondrial targeting sequence from the yeast cytochrome *c* oxidase IV protein (amino acids MLSLRQSIRFFKPATRTLCSSR). This sequence confers mitochondrial localization to GFP in both yeast [Fig. 10.4(*b*)] and *Arabidopsis* cells. Kohler et al. (1997) fused a similar localization sequence to *mgfp4* and demonstrated the utility of the encoded fluorescent protein as a precise marker for mitochondria in *Arabidopsis*. In addition, we fused the yeast carboxypeptidase Y (amino acids MKAFTSLLCGLGLSTTLAKA) and *Arabidopsis* basic chitinase (amino acids MKTNLFLFLIFSLLLSLSSA) signal sequences to GFP, and have successfully targeted the protein to the secretory pathway [Figs. 10.4*c* and 10.5].

It would be highly advantageous to produce relatively high levels of fluorescence for routine screening of GFP expression in transgenic plants and we tested the targeted forms of GFP in *Arabidopsis*. The modified genes were placed behind the 35S promoter, introduced into *Arabidopsis* by *Agrobacterium* mediated root transformation (Valvekens et al., 1988), and we tested for localization of the protein and fluorescence intensity in regenerated plants. The one variant that showed a substantial improvement over unmodified GFP was one that was targeted to the endoplasmic reticulum (ER) (Haseloff et al., 1997). This targeted form of GFP contains an N-terminal signal peptide derived from an *Arabidopsis* vacuolar basic chitinase and the C-terminal amino acid sequence HDEL (Fig. 10.2), to ensure entry into the secretory pathway and retention of the protein within the lumen of the ER. By using this modified gene (*mgfp4-ER*), it has been possible to regenerate intensely fluorescent and fertile plantlets consistently. Fluorescence within these plants could be readily observed by eye using a long wavelength UV lamp. The *mgfp4-ER* expressing plants were examined by confocal microscopy, and fluorescent protein was found mainly within the endomembrane system. The protein is excluded from the nucleus, shows a perinuclear distribution, and is found associated with the ER that forms a characteristic reticulate network in highly vacuolate cells. In highly cytoplasmic meristematic

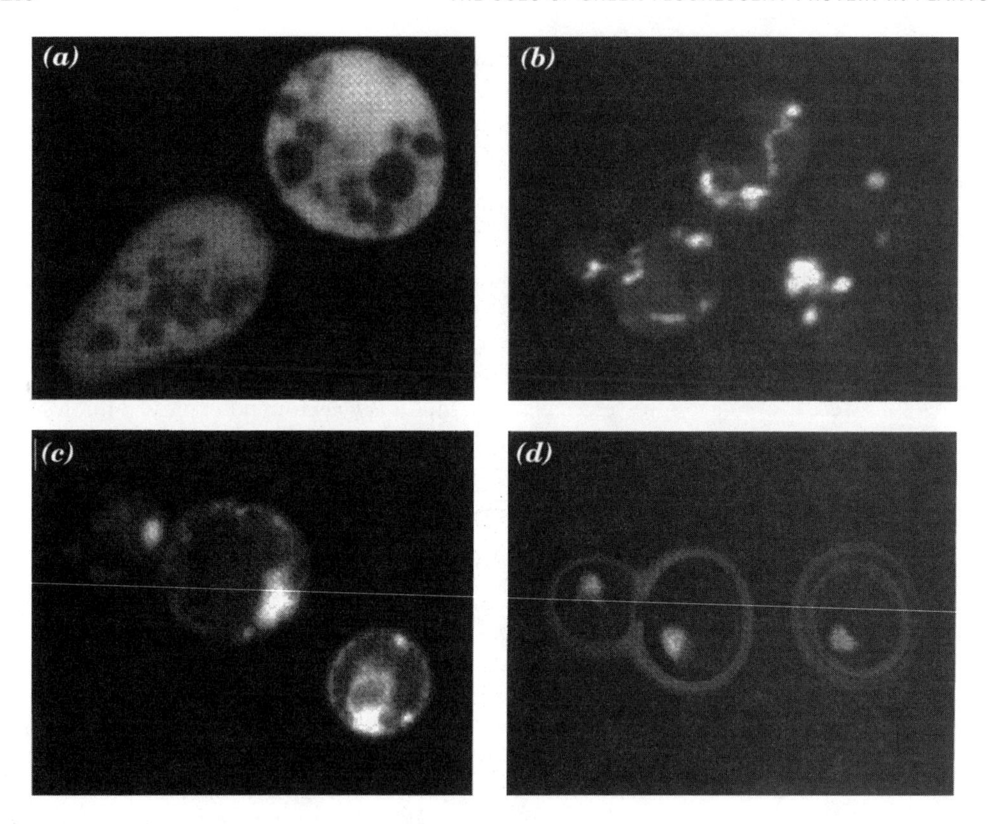

FIG. 10.4 Localization of GFP in yeast. Various peptide targeting sequences were fused to GFP in order to direct the protein to different subcellular compartments. The modified proteins were expressed in yeast cells and visualized by confocal microscopy. (*a*) Unmodified GFP is found throughout the cytoplasm and nucleoplasm. (*b*) Fusion of N-terminal sequences from yeast cytochrome oxidase B subunit IV results in mitochondrial localization of GFP. (*c*) N-terminal fusion of the signal sequence from yeast carboxypeptidase Y and C-terminal fusion of the amino acids HDEL results in retention of GFP within the endoplasmic reticulum. (*d*) N-terminal fusion of a nuclear localization sequence (NLS) from the SV40 T-antigen ensurers nuclear import of GFP, however, only larger forms of GFP are efficiently retained within the nucleus. In this case, NLS-GFP has been fused in *Escherichia coli* β-galactosidase. This results in exclusive localization of the fusion protein within nucleis. The outlines of the yeast cells, obtained using phase contrast optics, are superimposed in this image.

cells, the nuclei and orientation of cell divisions can be clearly distinguished. Localization of the modified protein to cytoplasmic organelles was also evident, to what appear to be large leucoplasts or proplastids. For example, an optical section of a hypocotyl epidermal cell is shown in Figure 10.5 and this includes a thin portion of cytosol that is pressed between the cell wall and vacuole. Such hypocotyl cells in *mgfp4-ER* transformed seedlings appear to contain a spectrum of developing plastids that range from the brightly green fluorescent to those that take on a yellow, orange, or red appearance in dual channel confocal micrographs (Fig. 10.6). We presume that this is due to increasing chlorophyll synthesis, and that the green fluorescent plastids may be the maturing precursors of

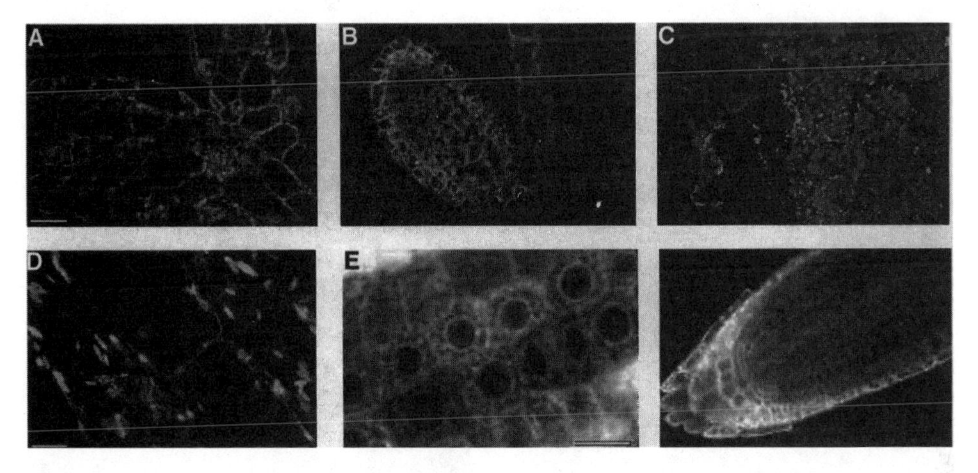

FIG. 10.5 Images of 35S-*mgfp4-ER* transformed seedlings. The ER-localized form of GFP was visualized in transgenic *Arabidopsis* seedlings using the procedures described for FIGure 10.2. (*a*) An optical section of the apical meristem showing the junction of the hypocotyl and cotyledons, and (*b*) an emerging first leaf. (*c*) Cells within the mesophyll of a cotyledon, packed with mature red autofluorescent chloroplasts. (d) A view of epidermal cells within the hypocotyl, showing the reticulate distribution of GFP within the endomembrane system and the appearance of green fluorescence within maturing plastids. Mature chloroplasts are brightly red autofluorescent in these cells. (*e*) Cells within the root meristem clearly display the characteristic perinuclear distribution expected for the ER-localized GFP. This perinuclear distribution is also seen in the shoot (panel A). (*f*) Median longitudinal optical section of a root tip.

chloroplasts in these cells. These green fluorescent plastids are also found within the chloroplast-free epidermal cells of leaves and cotyledons, but are not found within the underlying mesophyll cells that are packed with mature chloroplasts. It seems likely that these organelles are proplastids and are capable of developing into chloroplasts, but we cannot exclude the possibility that they are some specialized form of leucoplast.

The accumulation of *mgfp4-ER* protein within leucoplasts or developing proplastids, in addition to its entry into the secretory pathway and retention in the endoplasmic reticulum, may indicate misrecognition of the N-terminal signal peptide. Proplastid accumulation of GFP is not seen in the 35S-*mgfp4* transformed plants. If the *mgfp4-ER* encoded signal peptide is inefficiently recognized prior to docking and cotranslational transport of the protein into the lumen of the ER, a proportion of GFP bearing fused terminal sequences may be produced in the cytoplasm. If so, it is possible that the neglected signal peptide may act as a transit sequence for plastid entry. Alternatively, there may be some direct exchange between developing plastids and the endomembrane system. We see no free cytoplasmic fluorescence, and the protein is sorted very efficiently to the ER or to plastids.

It is unclear whether the beneficial effects of targeting GFP to the ER are due to increased levels or safer accumulation of mature GFP within cells. For example, if accumulation of fluorescent protein leads to the generation of free radicals in illuminated cells, it is conceivable that removing GFP from the nucleus could

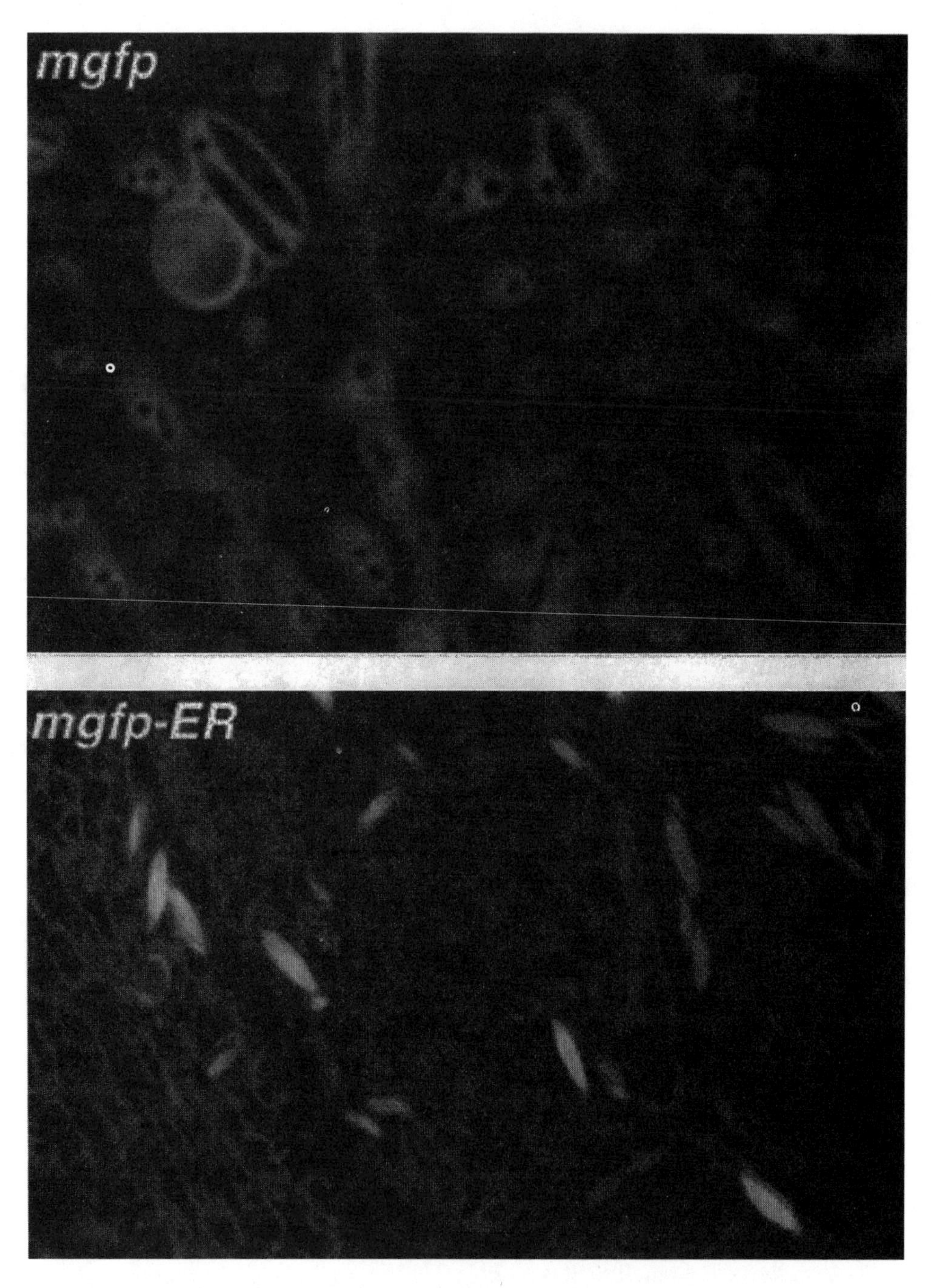

FIG. 10.6 Subcellular distribution of mgfp4-ER. The distribution of GFP and GFP-ER is shown in cells of the hypocotyl epidermis of transformed *Arabidopsis*. The cytosol forms a thin layer at the periphery of these highly vacuolate cells. The cytoplasmic form (*mgfp*) is excluded from endomembrane components and plastids within the cytosol, and forms a negative stain for these components. A single chlorplast, with its red fluorescent chlorophyll contents, can be seen in this image. Several nonfluorescent cigar-shaped bodies that appear to be some kind of plastids are also evident. In contrast, *mgfp-ER* is found within the endoplasmic reticulum and unexpectedly within the plastid-like organelles. The distribution of these labeled plastids is mainly limited to epidermal cells of the shoot, and varying degrees of chlorophyll fluorescence can be detected within the organelles, indicating that they may be developing pro-plastids.

protect cells from DNA damage due to such short-range highly reactive species. However, it is also possible that the fusion of peptide targeting sequences may improve the properties of the protein itself, or that the localization of GFP to the lumen of the ER may improve its maturation and accumulation. The maturation of the GFP aproprotein is sensitive to temperature, and the apoprotein readily misfolds under certain conditions (Siemering et al., 1996). The lumen of the ER is known to contain components, such as chaperones and peptidyl prolyl isomerases that aid protein folding (Fischer, 1994), and secretion and retention of GFP within the ER may allow improved formation and accumulation of the mature fluorescent protein. These improvements have allowed us to routinely generate transgenic *Arabidopsis* plants that contain high levels of GFP fluorescence (Fig. 10.7).

10.7 IMPROVED MATURATION OF GFP

The green fluorescent protein is normally produced within photocytes of the jellyfish *A. victoria*, and must undergo a series of posttranslational maturation steps to produce the fluorescent form of the protein. Expression of GFP in a number of heterologous systems has been described as poor or variable. For

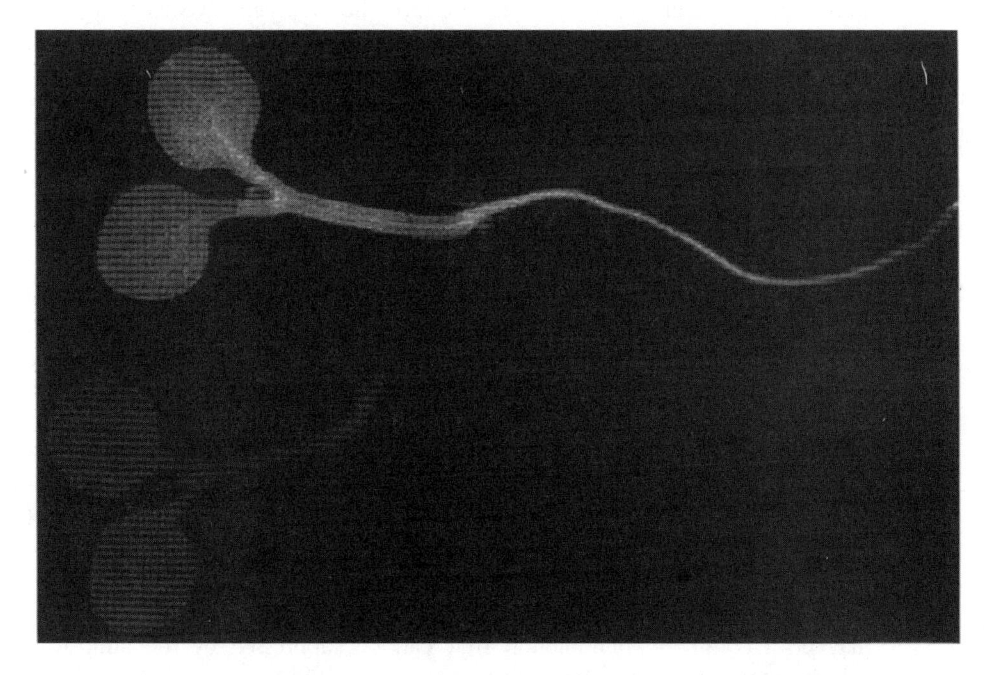

FIG. 10.7 Transgenic 35S-*mgfp4-ER Arabidopsis* seedlings. Both 5-day-old wild-type (left) and 35S-*mgfp4-ER* transgenic (right) seedlings were mounted in water on a Leitz DM-IL inverted fluorescence microscope and illuminated with long-wavelength UV light (Leitz-D filter set, excitation BP355-425, dichroic 455, emission LP460). Seedlings were visualized using a 4X objective (EF 4/0.12) and a Sony DXC-930P videocamera with F100-MPU framestore. A montage of the entire seedlings was assembled from collected videoimages using Adobe Photoshop.

example, strong promoters and decreased incubation temperatures have been required for efficient expression of *gfp* in mammalian cells (Kaether and Gerdes, 1995; Ogawa et al., 1995; Pines, 1995). Other researchers found that development of fluorescence is favored by a lower incubation temperature during expression of GFP in bacteria (Webb et al., 1995) and yeast (Lim et al., 1995). These observations suggested that expression of GFP in heterologous cells may be far from optimal. We have clearly demonstrated that maturation of the wild-type GFP is temperature sensitive, due to a defect in the folding of the GFP apoprotein. We have produced mutant forms of GFP that have improved folding and spectral properties (Siemering et al., 1996; Haseloff et al., 1997). These new GFPs are cured of the cryptic intron and are expressed brightly in plant cells.

The *mgfp4* gene was subjected to random mutagenesis, expressed in *E. coli* at 37°C and colonies were screened for increased fluorescence. Brighter mutants were isolated, and mapped by recombination with the wild-type *mgfp4* gene. Sequencing of the brightest mutant (GFPA) revealed two amino acid differences, V163A and S175G. The mutant GFP produced up to 35-fold increased fluorescence in bacterial cells, while the difference in protein levels was not nearly enough to account for this. The result suggested that a large proportion of wild-type GFP that is expressed in cells at 37°C is nonfluorescent. Experiments with a GFP–nucleoplasmin fusion protein have indicated that maturation of GFP to the fluorescent form may be sensitive to temperature during expression in the yeast. *S. cerevisiae* (Lim et al., 1995). To test whether the same could be true of expression in *E. coli* and whether the substitutions present in GFPA enhance maturation by suppressing any such sensitivity, we examined expression of GFP and GFPA over a range of different temperatures. Strains expressing GFP or GFPA were grown overnight at temperatures ranging between 25° and 42°C. For each culture, the fluorescence values were measured and normalized against the amount of recombinant protein present in the cells to give a measure of the proportion of intracellular GFP that is fluorescent at different temperatures. The proportion of GFP that is fluorescent steadily decreases with increasing incubation temperatures (Fig. 10.8), indicating that either mature GFP or the maturation pathway leading to its formation is temperature sensitive. Mature GFP is a highly stable protein whose fluorescence *in vitro* in unaffected by temperatures up to 65°C (Bokman and Ward, 1981), and we confirmed that the fluorescence of the mature protein is unaltered in bacterial cells at 42°C. Therefore, higher incubation temperatures must interfere with the posttranslational maturation of GFP, rather than causing inactivation of the mature protein. We confirmed that expression of GFP is also temperature sensitive in yeast and demonstrated that this is suppressed by the substitutions present in GFPA. These results indicate that the thermosensitivity of GFP maturation may be a common phenomenon that can be suppressed by the amino acid substitutions present in GFPA (Siemering et al., 1996).

The posttranslational maturation of GFP presumably involves initial folding of the apoprotein into an active conformation, to allow the cyclization and oxidation reactions that form the chromophore (Cody et al., 1993; Heim et al., 1994; Cubitt et al., 1995). The mature protein must then be correctly folded to maintain its fluorescent properties, to protect the chromophore from solvent

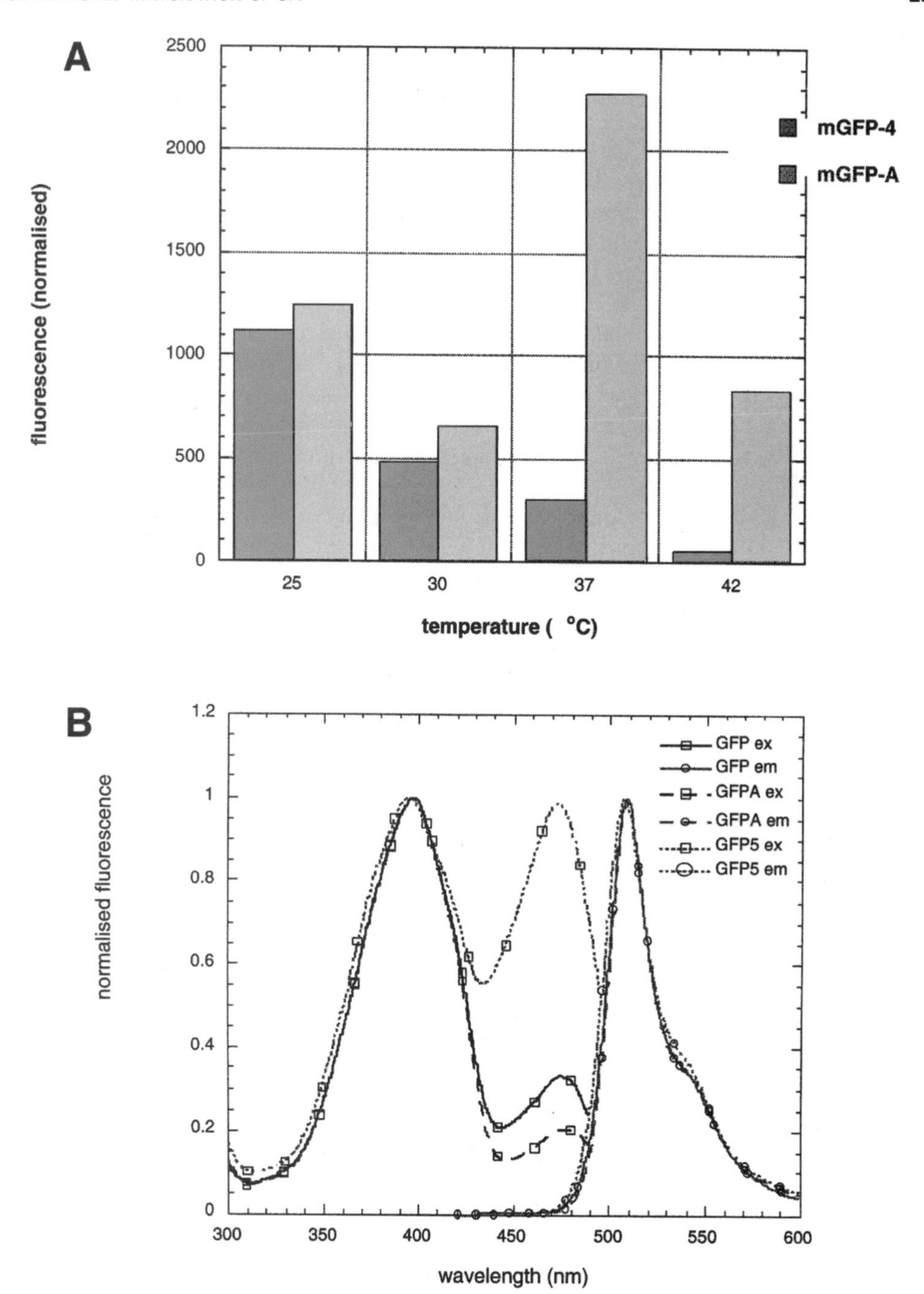

FIG. 10.8 Improved mutants of GFP. (*a*) Improved thermotolerance of GFP. Bacterial cells expressing GFP and GFPA (V163A, S175G) were grown at different temperatures. The GFP fluorescence values were measured and normalized with respect to the amount of intracellular recombinant protein for cultures grown at 25, 30, 37, and 42°C (Siermering et al., 1996). (*b*) Excitation and emission spectra of GFP, GFPA (V163A, S175G), and GFP5 (V163A, 1167T, S175G). Protein concentrations were 23.5 μg mL^{-1} in PBS (PH 7.4). All spectra have been normalised to a maximum value of 1.0 (Siemering et al., 1996).

effects (Ward et al., 1980). In principle, any of these processes could be sensitive to temperature and thus be responsible for the observed thermosensitivity of GFP during maturation.

Since the oxidation reaction involved in GFP chromophore formation appears to require molecular oxygen (Heim et al., 1994), oxidation rates can be measured after growth under anaerobic conditions by measuring the development of fluorescence after admission of air. We measured the rates of oxidation of GFP and GFPA expressed in anaerobically grown yeast at both 25°C and 37°C. The time constant measured for the oxidation of GFP at 37°C (5.9 ± 0.1 min) was found to be approximately three-fold faster than that measured at 25°C (16.2 ± 0.3 min), indicating that the posttranslational oxidation of the GFP chromophore is not the step responsible for the temperature sensitivity of maturation. In confirmation of this conclusion, the time constants derived for GFPA were somewhat slower than those measured for GFP (Siermering et al., 1996).

In contrast, we found that the folding of GFP is clearly temperature sensitive and the substitutions present in GFPA enhance proper folding at increased temperatures. We examined the solubilities of the two proteins during expression in *E. coli* at 25° and 37°C. Fluorescence was found almost exclusively in the soluble fraction. At 25°C, both GFP and GFPA were found predominantly in the soluble fraction, indicating that proper folding of both proteins is relatively efficient at this temperature. At 37°C, however, the majority of GFP was found as nonfluorescent protein in the insoluble fraction, whereas most of GFPA was still present in the soluble fraction. To obtain information on which species in the maturation pathway of GFP misfolds at higher temperatures, we examined the absorption spectrum of denatured protein isolated from inclusion bodies. If GFP undergoes cyclization of the chromphore prior to aggregation, protein from inclusion bodies should show an absorption in the near UV/blue region that is characteristic of the GFP chromophore in either the mature or chemically reduced state (Ward et al., 1980; Inouye and Tsuji, 1994). On the other hand, if unmodified GFP (apo-GFP) is the aggregating species, no such absorption should be observable in this region. GFP was purified from the inclusion bodies of bacterial cells grown at 37°C and, as a positive control, from the soluble fraction of cells grown at 25°C. Protein derived from cells grown at 25°C showed a characteristic absorption peak similar to that of acid-denatured GFP (Ward et al., 1982). By contrast, protein purified from inclusion bodies of cells grown at 37°C showed no such absorption, indicating that the aggregating species has not formed a chromophore. Taken together, the results indicate that the temperature sensitivity of GFP maturation is due primarily to the failure of the unmodified apoprotein to fold into its catalytically active conformation at higher temperatures. Furthermore, the amino acid substitutions present in GFPA suppress this defect by enhancing proper folding at elevated temperatures.

10.8 MODIFICATION OF FLUORESCENCE SPECTRA

The fluorescence excitation spectrum of GFP and GFPA exhibits peaks at wavelengths of 395 and 475 nm, with the 395-nm peak predominating. This property is useful for simple detection of the protein using a long-wavelength UV source. Ultraviolet illumination is not efficiently detected by the human eye and a suitable long-wavelength UV lamp can be used to excite GFP for simple observation of transformed plant material without obscuring the green emission. However, efficient blue light excitation (~470 nm) is essential for use with imaging devices such as confocal microscopes or cell sorters, which are equipped with argon laser sources.

Recently, it has been demonstrated that the relative amplitudes of the excitation peaks of GFP can be altered by means of mutagenesis (Heim et al., 1994; Delagrave et al., 1995; Ehrig et al., 1995; Heim et al., 1996). These mutations appear to affect the microenvironment of the chromophore so as to influence the equilibrium between two spectroscopic states of the chromophore (Heim et al., 1994; Ehrig et al., 1995). One of these mutations, I167T, has been shown to increase the amplitude of the 475-nm excitation peak relative to that of the 400-nm peak (Heim et al., 1994). We recombined the I167T substitution with the substitutions present in GFPA to increase the amplitude of the 475-nm peak relative to the 395-nm excitation peak, to produce a variant (GFP5), which has two excitation peaks (maxima at 395 and 473 nm) of almost exactly equal amplitude and an emission spectrum (λ_{max} = 507 nm) largely unchanged from that of wild type. The GFP5 variant retains a thermotolerant phenotype, and bacterial cells grown at 37°C fluoresce 39- and 111-fold more intensely than cells expressing GFP, when excited at 395 and 473 nm, respectively.

The broad excitation spectrum of GFP5 allows both efficient UV and blue light excitation of the protein. For example, the expression of *gfp5* gene fusions can be rapidly scored after transformation of microbial colonies or plant tissues by simple inspection with a UV lamp. The same material is well suited for laser scanning confocal microscopy. In addition, plants are highly autofluorescent and the use of a dual wavelength excitation mutant like GFP5 also enables faint signals to be easily distinguished from autofluorescence during microscopy, by alternating the excitation sources. For example chloroplasts are intensely fluorescent but are less efficiently excited by UV light. We routinely use long-wavelength UV excitation for visual and microscopic screening of transformed tissues. Autofluorescence can also be an advantage. For example, UV light excites a faint blue fluorescence in *Arabidopsis* cell walls, and this "counterstain" allows roots growing in agar culture to be easily located and scored for GFP fluorescence. Recently, screened we have several thousand *Arabidopsis* transformants for root specific "enhancer-trap" expression patterns, and this feature was very useful (J. Haseloff and S. Hodge, unpublished results). In contrast, widely used GFP variants that contain the S65T mutation (Heim et al., 1995; Cormack et al., 1996) provide optimized properties for blue light excitation, but are not useful for detection by long-wavelength UV light.

It is possible to manipulate the fluorescence spectra of GFPA by introducing additional substitutions into the protein without deleteriously affecting its improved folding characteristics. The Y66H substitution dramatically blue shifts both the excitation and emission spectra of GFP to give a "blue fluorescent protein" (Heim et al., 1994). The GFPA containing the Y66H substitution, was found to have identical fluorescence spectra to those of the corresponding GFP(Y66H) protein (excitation maximum = 384 nm, emission maximum = 448 nm), and gave rise to 29-fold more fluorescence when expressed at 37°C and three-fold more fluorescence when expressed at 25°C. In addition, a number of workers obtained GFP variants that show brighter fluorescence in heterologous cell types, and it is likely that the improved properties of these proteins is due largely to improved folding. For example, the V163A mutation present in GFPA has also been generated independently by at four different groups (Crameri et al., 1996; David and Vrestra, 1996; Heim et al., 1996; Kohler et al., 1997) and this residue may play a pivotal role in folding of the protein. Cormack et al. (1996) introduced random amino acid substitutions throughout the 20 residues flanking the chromophore of GFP. They used fluorescence activated cell sorting to select variants that fluoresced 20- to 35-fold more intensely than wild type, and noted that the mutant proteins had improved solubility during expression in bacteria. The mutant proteins presumably have improved folding properties. One of these variants (GFPmut1, (Cormack et al., 1996)) contains two amino acid differences, F64L and S65T, located within the central α helix of the protein, adjacent to the chromophore. The V163A and S175G mutations that we have isolated are positioned on the outer surface of the protein (Ormö et al., 1996; Yang et al., 1996) and recombination of these two sets of mutations appears to result in markedly improved fluorescence in bacterial, plant and animal cells (Zoenicka-Goetz et al., 1996; Zoenicka-Goetz et al., 1997; J. Haseloff and K. R. Siemening unpublished results). It is possible that the mutations affect separate steps of the folding or maturation process, and that their benefit is additive.

10.9 MODIFIED GFP GENES FOR PLANT EXPRESSION

Expression of the wild-type *gfp* gene has given poor results in a number of plant systems, and we have found it necessary to alter the gene for our experiments with transgenic *Arabidopsis* plants. As outlined above, we have (a) altered codon usage to remove a cryptic plant intron, (b) added peptide sequences to allow targeting of the protein to the lumen of the endoplasmic reticulum, and mutated the protein to (c) improve folding of the apo-GFP during posttranslational maturation (V163A, S175G), and (d) provide equalised UV and blue light excitation (I167T). These alterations have all been incorporated into a single highly active form of the gene (*mgfp5-ER*), which we now routinely use for monitoring gene expression and marking cells in live transgenic plants (Siemering et al., 1996; Haseloff et al., 1997).

Removal of the cryptic intron appears to be essential for *gfp* expression in *Arabidopsis*, and other workers have observed improved expression in plants using *gfp* genes containing "humanized" or synthetic codon usage (Chiu et al., 1996; Pang et al., 1996) Altered codon usage of the *gfp* gene appears to be a crucial requirement for efficient expression in plant cells. Improvements in the folding, spectral properties, and subcellular localization of the protein provide secondary improvements that allow the accumulation of high levels of fluorescent protein in plant cells.

10.10 IMAGING GFP IN PLANT CELLS

GFP can be visualized directly in living plant tissue, unlike commonly used markers such as β-glucuronidase, which requires a prolonged and lethal histochemical staining procedure (Jefferson et al., 1987). The GFP is therefore finding application in three broad areas (1) for the dynamic visualization of labeled protein within the cells, and at a larger scale, (2) for the selective labeling and monitoring of whole plant cells within growing plant tissue, and (3) for the identification of individual transgenic plants expressing GFP. For example, different peptide domains can be fused to GFP to allow the decoration of particular structures within cells and/or to observe the subcellular distribution of the fusion protein. In addition, use of an active GFP marker gene allows transgenic cells to be scored by simple observation during a plant transformation experiment, throughout regeneration to the adult plant and its progeny. The use of tissue specific promoters to drive expression of GFP also allows the selective labeling of particular cell types within intact transformed plants. In these cases, it is beneficial to express GFP at high levels within the marked cells to aid detection, and to minimize any deleterious effects of GFP expression. We have found the optimized *mgfp5-ER* gene very useful for this type of experiment. The dynamic properties of labeled cells or subcellular features can be resolved at high resolution in whole plant tissues using fluorescence microscopy techniques, however, the use of intact tissue imposes some additional constraints on the imaging process.

Intact plant tissue proves a difficult subject for fluorescence microscopy as it consists of deep layers of highly refractile cell walls and aqueous cytosol and contains various autofluorescent and light scattering components. There are two approaches to the difficulties imposed by these conditions: to fix and to clear the tissue with a high refractive index mounting medium, or to directly image living tissue using suitably corrected microscope optics. In our experience, it has proved difficult to effectively clear *Arabidopsis* wholemounts without causing artifacts or losing GFP fluorescence, and there are considerable advantages to working with living tissues. Thus we have mainly pursued the second approach. The natural autofluorescence and depth of intact plant tissue means that out of focus blur often obscures high-magnification views obtained with a conventional epifluorescence microscope. However, the technique of laser scanning confocal microscopy can be used to optically section GFP expressing plant tissues. Confocal imaging allows precise visualization of fluorescent signals within a narrow plane of focus, with exclusion of out-of-focus blur, and the technique permits the reconstruction of three dimensional (3D) structures from serial optical sections. *Arabidopsis* seedlings can simply be mounted in water for microscopy, and examined using a long-working distance water immersion objective to minimize the effects of spherical aberration when focusing deep into an aqueous sample (Haseloff et al., 1995). Young seedlings (3–7 days old) can be grown on agar culture media, and then placed in a drop of water (100–200 μL) on a glass slide. A glass coverslip is lowered gently to flatten and cover the seedling. Even with the use of a specialized water immersion objective such as the Nikon 60X planapochromat, N.A. 1.2 (working distance 220 μm), image quality degrades

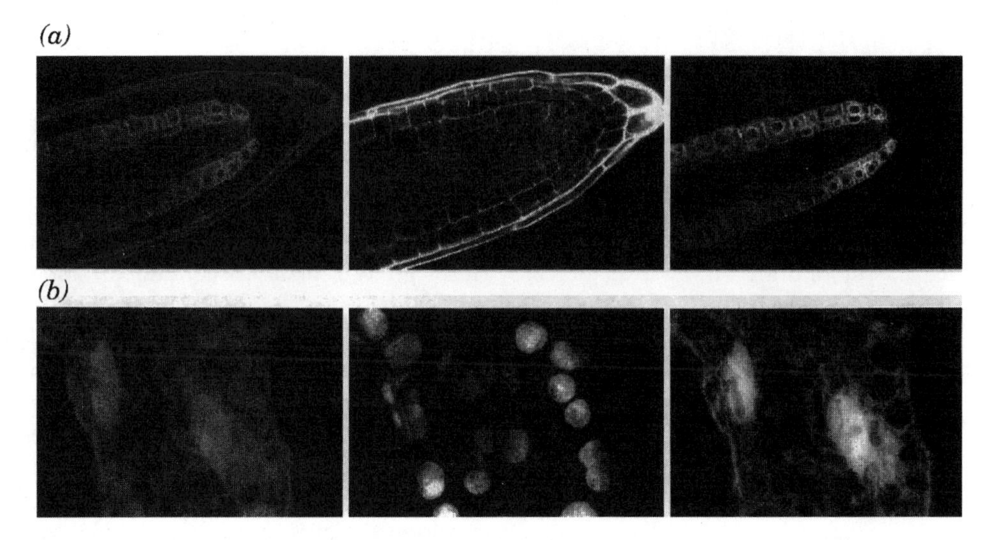

FIG. 10.9 Dual channel imaging of GFP/chlorophyll/propidium iodide. A BioRad MRC-600 confocal microscope equipped with an argon–krypton mixed-gas laser and K1/K2 filter blocks was used for green and ref fluorescence imaging of GFP expressing Arabidopsis seedlings. (*a*) separate images of a 35S-*mgfp4* transformed cotyledon mesophyll cell showing the green fluorescence channel with the GFP signal distributed throughout the nucleoplasm and cytoplasm, the red fluorescent signal of the chloroplasts, and the combined dual channel image. There is little spillover between the two channels. (*b*) Separate green, red, and combined fluorescence images are also shown for a propidium iodide stained root tip of a GFP expressing enhancer trap line J0571 (Haseloff and Hodge, unpublished results). The *mgfp5-ER* gene is expressed strongly in the root cortex and endodermis of this line. Propidium iodide provides a distinct counterfluor that outlines cells in the living root tip.

rapidly for optical sections deeper than 50–80 μM within the tissue. Ideally, the tissue of interest should be positioned immediately below the cover slip and depression slides should be avoided unless this is ensured. Despite these limitations, the small size of *Arabidopsis* seedlings allows very useful imaging and, for example, median longitudinal optical sections can be easily obtained from intact root tips (e.g., Fig. 10.9).

The blue 488-nm wavelength line of the commonly used argon ion or krypton–argon lasers is ideal for exciting GFP, and this can be used in combination with a fluorescein/rhodamine or Texas Red® filter set for dual channel imaging of GFP and chlorophyll for photosynthetic tissues. For nonphotosynthetic tissues, a fluorescent counterstain that can be distinguished from GFP is often very useful. For example, *Arabidopsis* seedlings can be placed in a solution of 10 μg mL^{-1} propidium iodide for 5–20 min, before being directly mounted in water for confocal microscopic examination of roots. Propidium iodide is red fluorescent and highly charged and does not enter living cells. It stains the

Texas Red is a registered trademark of Molecular Probes, Inc., Eugene, OR.

walls of living cells within the root tip and fills dead cells (van der Berg et al., 1995) (Fig. 10.10). In a similar way the red fluorescent dye nile red can be used to stain neutral lipids, and rhodamine 6B can be used to stain the casparian strip and lignified cells within living roots (J. Haseloff, unpublished results).

10.11 VISUALIZING SUBCELLULAR DYNAMICS

The expression of GFP within an organism produces an intrinsic fluorescence that colors normal cellular processes, and high-resolution optical techniques can be used noninvasively to monitor the dynamic activities of these living cells. For time-lapse studies, it is very important that GFP fluorescence be bright, to minimize levels of illumination that can cause phototoxicity and photobleaching during observation. The modified *mgfp5-ER* gene that is described above has proved very useful for generating highly fluorescent transgenic *Arabidopsis* plants that are suitable for intensive time-lapse studies. During confocal microscopy experiments, we have routinely observed high rates of cytoplasmic streaming within living specimens, and we have used short-term time-lapse observations to gain a better understanding of the relative movements of cellular components. *Arabidopsis* seedlings that expressed the *mgfp5-ER* gene at high levels were simply mounted in water for confocal microscopy, which allowed observation for up to 2 h. Hypocotyl epidermal cells form ideal specimens for viewing the various components of the cytoplasm. The cells are large, highly vacuolate and surface borne. An extremely thin layer of cytoplasm is squeezed between the wall and the vacuole of these cells. The thinness of the layer greatly limits the movement of cytoplasmic components to within a single plane of focus of the microscope, and objects can be rapidly tracked across a portion of the cell without the need for refocusing. A seedling can easily be mounted so that the hypocotyl is pressing closely against the microscope coverslip, and the layer of cytoplasm beneath the outer wall of an epidermal cell will be only a few microns from the surface, allowing high optical resolution.

To follow rapid movements in cell it is necessary to use a correspondingly fast sampling rate. We have collected time-lapse confocal images at up to two frames per second, which requires almost continual laser scanning with a BioRad MRC-600 microscope. Living specimens have been examined for up to an hour without appreciable phototoxic or bleaching effects, but this is only possible with bright samples, which allow attenuation of the exciting laser light. A short segment of a time-lapse experiment is shown in Figure 10.11. A section of hypocotyl epidermis was monitored at a rate of 0.5 frames per second for about 20 min and representative confocal images are shown for a 4.5-min period. Cellular components are clearly recognizable in the optical sections, and their identity is indicated in a schematic diagram [Fig. 10.11(*a*)]. The cells contain green fluorescent proplastids and highly reticulate endomembranes. The nuclei are outlined due to the peripheral distribution of the ER, and the reticulate surfaces of a nucleus can be seen in the cell that is central to the field of view. A cross-section of a nucleus can also be seen in the adjacent lower cell. Chloroplasts are red autofluorescent, and characteristically small and spheroid in these hypocotyl epidermal cells.

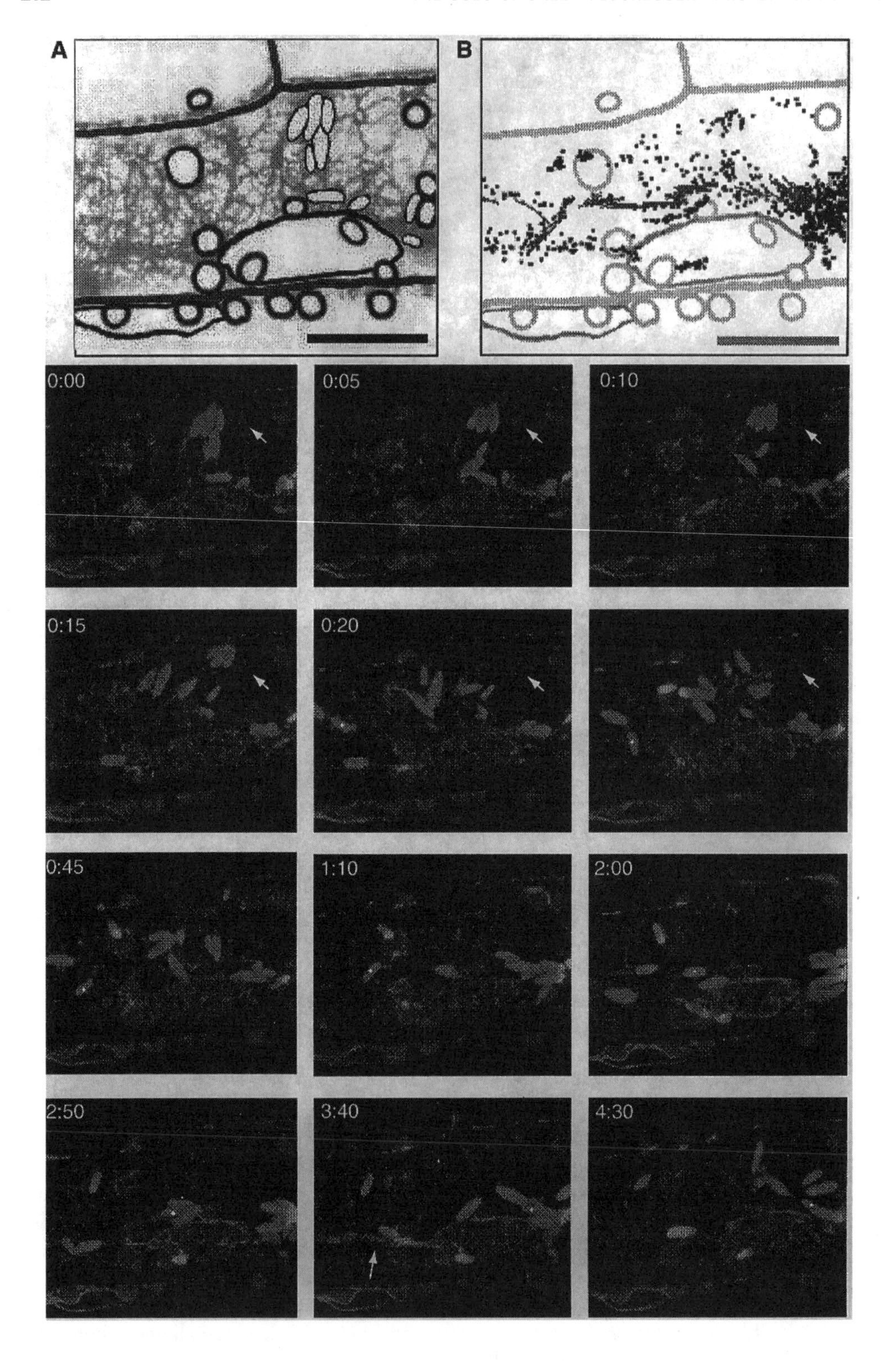
A
B
0:00
0:05
0:10
0:15
0:20
0:45
1:10
2:00
2:50
3:40
4:30

When a time course of images is played at videorate, proplastids and what appears to be vesicular material move vigorously and erratically through the cells. The distribution of all proplastids was plotted frame by frame though this experiment and the path of one example is shown in Figure 10.11(*b*). These plastids more with uneven velocities, up to 20 μ/s, along irregular paths that may correspond to underlying cytoskeletal elements such as actin. In contrast, the endoplasmic reticulum, which is presumably associated with cortical microtubules, undergoes relatively slower rearrangement. A relatively stable feature of the ER is indicated with an arrow in Figure 10.11, panels 0:00–0:25, while nearby proplastids undergo substantial movement. Chloroplasts and nuclei moved only slowly during the 20-min time course of the experiment. These cells contain an ER retained form of GFP, and we expect the protein to be cycled in vesicles between the lumen of the ER and the cis golgi. A rapid and irregular movement of small vesicle-like particles is seen throughout cells during the time course. Although these small movements are difficult to see in still images, we also see the transient formation of extended filamentous structures (Fig. 10.11, panel 3:40), which are comprised of a larger amount of this fluorescent material, and are associated with rapid movement of both vesicular-like material and proplastids. The location of the cis golgi in these micrographs is unclear, although small regions of punctate fluorescence can be seen associated with endomembranes.

FIG. 10.10 (opposite) Time-lapse confocal microscopy of subcellular processes. A transgenic *Arabidopsis* seedling expressing the *mgfp5-ER* gene was mounted in water and a small segment of the hypocotyl epidermis was observed using a BioRad MRC-600 laser scanning confocal microscope. The laser light was attenuated by 99% using a neutral density filter, and the confocal aperture was stopped down. Two channel, single scan images were collected at the rate of 1/2 s for 20 min, and transferred to an Apple Macintosh computer. The large data file was then converted to full-color numbered PICT files using the program PicMerge, and finally converted to a Quicktime movie for analysis and videorate playback. A section corresponding to 4.5 min of the original observation was chosen and representative frames are presented here. Each frame is marked with the time (minute:second) that had elapsed from the first chosen frame. Two schematic diagrams are shown. A key for the identities of cellular structures and organelles is shown in diagram A. Nuclei (N), chloroplasts (C), endoplasmic reticulum (ER), proplastids or leucoplasts (P), and the position of the cell wall (CW) are shown (scale bar = 10 μ). In the second diagram (B), the positions of proplastids throughout the 4.5-min period of the experiment is shown. The plastids were located in each frame, and their cumulative positions within the cell were plotted frame by frame, indicated by black dots. The position of one plastid is also plotted, with a series of red lines representing the successive orientations of the long axis of the plastid. This particular organelle is indicated by a white cross on the timed confocal images, from its appearance in the field of view at 0:20. The images 0:00–0:25 each contain an arrow that indicates a ring-like feature within the ER that provides a morphological landmark. Image 3:40 contains an arrow that indicates the formation of a transient filamentous structure that appears associated with rapid vesicular and plastid traffic.

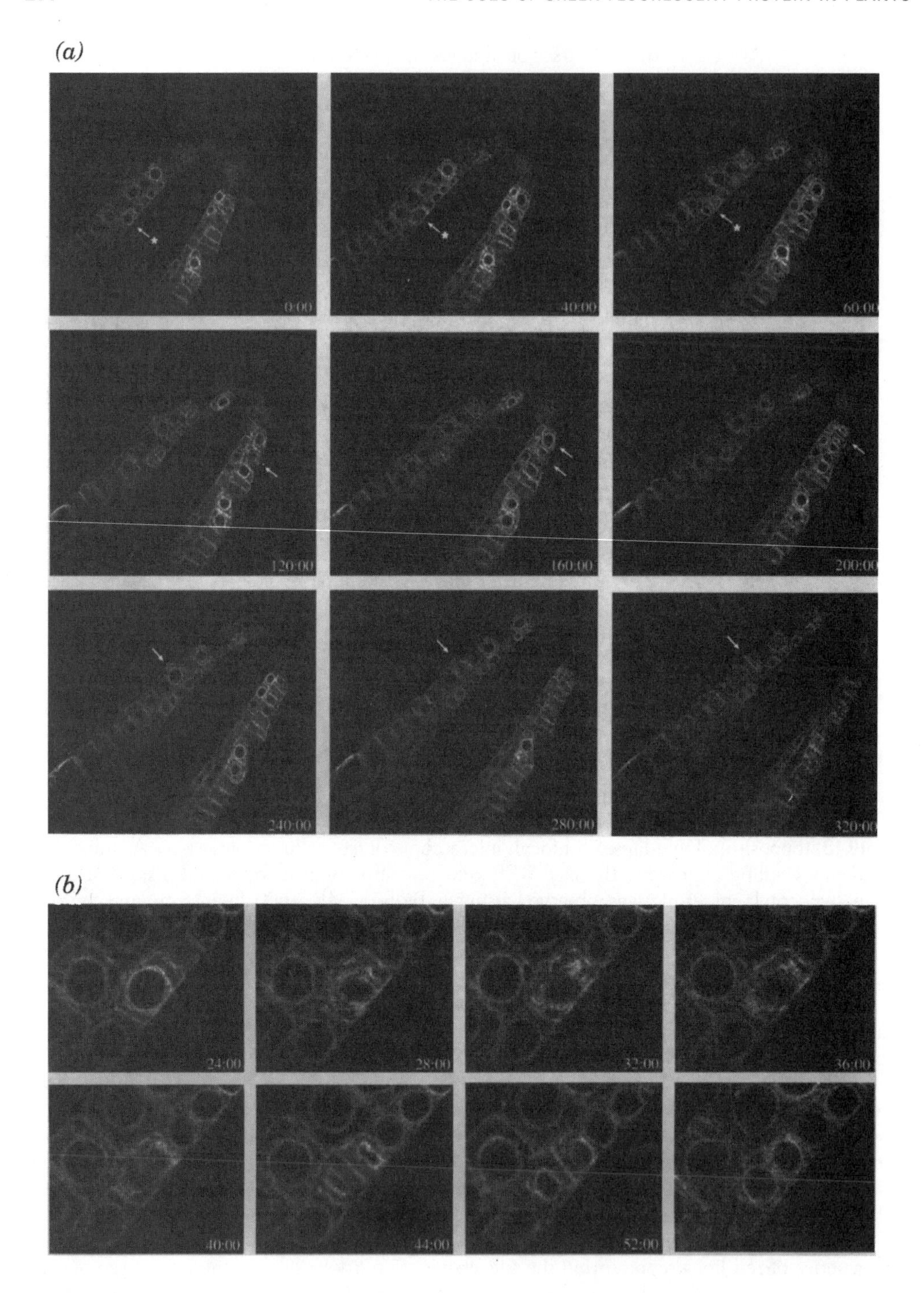
(a)
0:00
40:00
60:00
120:00
160:00
200:00
240:00
280:00
320:00
(b)
24:00
28:00
32:00
36:00
40:00
44:00
52:00

10.12 MARKING CELLS WITHIN THE PLANT

The expression of GFP can be limited to particular cell types within a plant, to provide a means for visualizing the behavior of individual cells within the living organism. We have developed a scheme for targeted gene expression in plants, which is based on a method widely used in *Drosophila* (Brand and Persimon, 1993). We have randomly inserted a gene for a foreign transcription activator (a derivative of the yeast *GAL4* gene) into the *Arabidopsis* genome using *Agrobacterium tumefaciens* mediated transformation. We have thus generated a library of *Arabidopsis* lines, which each express the *GAL4* derivative in a particular pattern, dependent on adjacent genomic DNA sequences. The inserted DNA has also been engineered to contain a GAL4-responsive *mgfp5-ER* gene, and so interesting patterns of *GAL4* gene expression are immediately and directly visible, with each *GAL4* expressing cell marked by green fluorescence within the endoplasmic reticulum. Importantly, *GAL4* expression within these lines allows precise targeted ectopic gene expression. A chosen target gene can be cloned under the control of a *GAL4* responsive promoter, separately transformed into *Arabidopsis*, and maintained silently in the absence of *GAL4*. Genetic crossing of this single line with any of the library of *GAL4* containing lines allows a specific activation of the target gene in particular tissue and cell types, and the phenotypic consequences of misexpression, including those lethal to the organism, can be conveniently studied (J. Haseloff, S. Hodge, and H. M. Goodman, unpublished results).

We have used *in vivo* detection of GFP to develop a new and efficient enhancer-trap screening procedure. As our particular interest is in the cells of the *Arabidopsis* root tip, we have modified the plant transformation protocol to include an auxin induction of roots from regenerating shootlets. More than 7500 transformants were then generated, planted in grid patterns in sterile culture dishes and directly screened by fluorescence microscopy for GAL4 driven GFP expression within roots. Several hundred lines with interesting patterns of root expression were chosen, documented, transferred to soil and grown to seed, to both amplify and self-hybridize the lines. Consequently, we have a collection of 250 *Arabidopsis* lines with distinct and stable patterns of GAL4 and GFP expression in the root. These GAL4-GFP lines provide a valuable set of markers, where particular cell types are tagged and can be visualised with unprecedented ease and clarity in living plants.

FIG. 10.11 (opposite) Time-lapse confocal microscopy of root development. Seeds of the *Arabidopsis* enhancer trap line J0571 (J. HAseloff and S. Hodge, unpublished results) were germinated and grown in agar medium on a coverglass. After 10 days of growth, an emerging lateral root was visualized by confocal time-lapse microscopy. The root tip was imaged through the coverglass of the tissue culture vessel. A median longitudinal optical section was collected every 2 min over a 6-h period. (*a*) Representative frames from a 320-min period are shown, labeled with the time of collection (minute: second). Cell divisions and growth of the labeled cortex and endodermis cell layers are evident. Individual cells in the process of mitotic division are arrowed. One endodetermis cell is marked with an asterisk, and its behavior is shown in more detail. (*b*) Frames collected at shorter intervals are shown for the marked cell. The times of image collection are indicated (minute: second). The ER-localization of the GFP marker allows clear visualization of nuclear division and phragmoplast formation in these cells.

The GAL4-GFP enhancer-trap screen was designed to yield markers for the *Arabidopsis* root meristem, which is our choice of a model system. The simple and well characterized architecture of the root (Dolan et al., 1993) enables simple analysis of GAL4 mediated perturbations of cell fate within the meristem. The *Arabidopsis* root meristem consists of a plate of quiescent cells surrounded by initials that divide to produce distal root cap cells, and also lay down continuous cell files proximally. Behind the tip, the newly formed cells of the root undergo differentiation and expansion to build a conserved arrangement of cell types within the mature root. We have generated GAL4-GFP lines that precisely mark particular cell types within the meristem, and one example is shown in Figure 10.11. *Arabidopsis* line J0571 exhibits GAL4 directed expression of GFP within the cortex and endodermis of the root, including the initials shared by these two cell files. Five-day-old seedlings can be briefly counterstained with 10 $\mu g\ mL^{-1}$ propidium iodide and mounted in water for confocal microscopy. Optical sectioning allows very simple and precise imaging of the GFP labeled cells within the root meristem [Fig. 10.11(*a*)]. The behavior of these cells within the developing root meristem can be observed using time-lapse techniques. The GFP expressing seedlings can be planted in sterile agar media and grown in coverslip-based vessels. The roots grow down through the media and then along the surface of the coverslip. The roots are then ideally positioned for microscopic imaging through the base of the vessel. A series of images are shown in Figure 10.11(*b*) that illustrate 2 h during the growth of a root tip of *Arabidopsis* line J0571. Confocal optical sections were collected at 2-min intervals. The cortical and endodermal cell files and their initials are clearly seen due to the expression of GAL4 driven *mgfp5-ER* in this line. The localization of GFP to the endoplasmic reticulum, and its consequent perinuclear distribution, ensures that the cell nuclei are clearly evident in these meristematic cells. In addition, the processes of cell division can be seen within the living plant. The breakdown of the nuclear membrane, segregation of chromosomes and formation of the daughter nuclei and cell wall plate are reflected in changes of the distribution of the ER localized GFP. Also, the cell nuclei appears to possess a larger volume prior to cell division, consistent with an extra, newly replicated DNA complement. This characteristic may be useful for scoring DNA replication within living cells.

Such GFP expressing lines allow the simple, noninvasive observation of events within living plants at an unprecedented level of detail. The GFP can now be used as a cellular marker to illuminate the defective behavior of mutant plants, or the perturbations induced by reverse genetic techniques.

10.13 CONCLUSIONS

In order to overcome problems with the expression of GFP in plant cells, and with the safe accumulation and detection of GFP in whole *Arabidopsis* plants, we have engineered improvements to the *gfp* gene. The modified gene contains (a) altered codon usage to remove a cryptic plant intron; (b) added peptide sequences to allow targeting of the protein to the lumen of the endoplasmic reticulum, and mutations that (c) improve folding of the apoprotein during posttranslational maturation (V163A, S175G); and (d) provide equalised UV and blue light excitation (I167T). This highly modified variant (*mgfp5-ER*) is proving useful as a safe and bright marker in transgenic plants. We expect that the *mgfp5-ER* gene and its derivatives will also be useful in work with transgenic fungi and animals, where at least some similar problems may be encountered.

A major use for GFP will be as a replacement for the β-glucuronidase gene, which is widely used as a reporter for promoter and gene fusions in transformed plants. The *GUS* gene product can be localized or quantified using histochemical techniques, but these are generally destructive tests (Jefferson et al., 1987). In contrast, GFP can be directly seen in living tissues. For example, high levels of fluorescence intensity are obtained in GFP transformed bacterial and yeast colonies, allowing simple screening for GFP expression with the use of a hand-held UV lamp. Such an assay for gene expression in living plants will be a very useful tool for plant transformation and breeding experiments. Many transformation techniques give rise to regenerating tissues that are variable or chimeric, and require testing of the progeny of the primary transformants. Potentially, GFP expressing tissues could be monitored using *in vivo* fluorescence, avoiding any need for destructive testing, and the appropriate transformants could be rescued and directly grown to seed. Similarly, *in vivo* fluorescence will be an easily scored marker for field testing in plant breeding, allowing transgenes linked to the GFP gene to be easily followed, and provide a potential alternative to antibiotic resistance markers.

Unlike enzyme markers, GFP can be visualized at high resolution in living cells using confocal microscopy. The images are not prone to fixation or staining artifacts, and can be of exceptional clarity. Moreover, the activities of living cells, such as cytoplasmic streaming, are clearly evident during microscopy. Ordinarily, movement within a sample is a nuisance, placing constraints on the use of sometimes lengthy techniques for noise reduction during confocal microscopy, such as frame averaging. However, we have shown that it is also possible to monitor dynamic events by time-lapse confocal microscopy, and this combination of a vital fluorescent reporter with high-resolution optical techniques shows much promise for use in cell biological and physiological experiments.

Genetic systems such as that of *Arabidopsis* provide a large resource of potentially informative mutants, and there has been much recent improvement in techniques for determining the molecular basis of a particular phenotype. The use of fluorescent proteins will provide further tools for examining the biology of mutant cells. The ability to simply and precisely monitor both particular cells and subcellular structures that have been highlighted with a fluorescent signal will improve both the screening for particular abnormal phenotypes and the characterization of dynamic process.

REFERENCES

Baulcombe, D. C., Chapman, S., and Cruz, S. S. (1995). Jellyfish green fluorescent protein as a reporter for virus infections. *Plant J* 7:1045–1053.

Bokman, S. H. and Ward, W. W. (1981). Renaturation of *Aequorea* green fluorescent protein. *Biochem. Biophys. Res. Commun.* 101:1372–1380.

Brand, A. H. and Perrimon, N. (1993). Targeted gene expression as a means of altering cell fates and generating dominant phenotypes. *Development* 118:401–415.

Chalfie, M., Tu, Y., Euskirchen, G., Ward, W. W., and Prasher, D. C. (1994). Green fluorescent protein as a marker for gene expression. *Science* 263:802–805.

Chiu, W. L., Niwa, Y., Zeng, W., Hirano, T., Kobayashi, H., and Sheen, J. (1996). Engineered GFP as a vital reporter in plants. *Curr. Biol.* 6:325–330.

Cody, C. W., Prasher, D. C., Westler, W. M., Prendergast, F. G., and Ward, W. W. (1993). Chemical-structure of the hexapeptide chromophore of the aequorea green-fluorescent protein. *Biochemistry* 32:1212–1218.

Cormack, B. P., Valdivia, R. H., and Falkow, S. (1996). FACS-optimized mutants of the green fluorescent protein (GFP). *Gene* 173:33–38.

Crameri, A., Whitehorn, E. A., Tate, E., and Stemmer, W. P. C. (1996). Improved green fluorescent protein by molecular evolution using dna shuffling. *Nature Biotechnol.* 14: 315–319.

Cubitt, A. B., Heim, R., Adams, S. R., Boyd, A. E., Gross, L. A., and Tsien, R. Y. (1995). Understanding, improving and using green fluorescent proteins. *Trends Biochem. Sci.* 20:229–233.

Davenport, D. and Nichol, J. A. C. (1955). Luminescence in Hydromedusae. *Proc. R. Soc., Ser. B.* 144:399–411.

David, S. and Viestra, R. (1996). Soluble derivatives of green fluorescent protein (GFP) for use in *Arabidopsis thaliana. Weeds World* 3:43–48.

Delagrave, S., Hawtin, R. E., Silva, C. M., Yang, M. M., and Youvan, D. C. (1995). Red-shifted excitation mutants of the green fluorescent protein. *Bio-Technology* 13: 151–154.

Dolan, L., Janmaat, K., Willemsen, V., Linstead, P., Poethig, S., Roberts, K., and Scheres, B. (1993). Cellular organisation of the *Arabidopsis thaliana* root. *Development* 119:71–84.

Ehrig, T., Okane, D. J., and Prendergast, F. G. (1995). Green fluorescent protein mutants with altered fluorescence excitation spectra. *FEBS Lett.* 367:163–166.

Fischer, G. (1994). Peptidyl-prolyl isomerases and their effectors. *Angew. Chem. Int. Ed. Engl.* 33:1415–1436.

Goodall, G. J. and Filipowicz, W. (1989). The AU-rich sequences present in the introns of plant nuclear pre-mRNAs are required for splicing. *Cell* 58:473–483.

Goodall, G. J. and Filipowicz, W. (1991). Different effects of intron nucleotide composition and secondary structure on pre-mRNA splicing in monocot and dicot plants. *EMBO J.* 10:2635–2644.

Grebenok, R. J., Pierson, E., Lambert, G. M., Gong, F. C., Afonso, C. L., Haldeman-Cahil, R., Carrington, J. C., and Galbraith, D. W. (1997). Green fluorescent protein fusions for efficient characterization of nuclear targeting. *Plant J.* 11:573–586.

Green, M. R. (1991). Biochemical mechanisms of constitutive and regulated pre-mRNA splicing. *Ann. Rev. Cell Biol.* 7:559–599.

Haas, J., Park, E. C., and Seed, B. (1996). Codon usage limitation in the expression of hiv-1 envelope glycoprotein. *Curr. Biol.* 6:315–324.

Hanley, B. A. and Schuler, M. A. (1988). Plant intron sequences: evidence for distinct groups of introns. *Nucleic Acids Res.* 16:7159–7176.

Haseloff, J. and Amos, B. (1995). GFP in plants. *Trends Gen.* 11:328–329.

Haseloff, J., Siemering, K. R., Prasher, D. C., and Hodge, S. (1997). Removal of a cryptic intron and subcellular localisation of green fluorescent protein are required to mark transgenic *Arabidopsis* plants brightly. *Proc. Natl. Acad. Sci. USA* 94:2122–2127.

Hebsgaard, S. M., Korning, P. G., Tolstrup, N., Engelbrecht, J., Rouze, P., and Brunak, S. (1996). Splice site prediction in *Arabidopsis thaliana* pre-mRNA by combining local and global sequence information. *Nucleic Acids Res.* 24:3439–3452.

Heim, R., Prasher, D. C., and Tsien, R. Y. (1994). Wavelength mutations and posttranslational autoxidation of green fluorescent protein. *Proc. Natl. Acad. Sci. USA* 91: 12501–12504.

Heinlein, M., Epel, B. L., Padgett, H. S., and Beachy, R. N. (1995). Interaction of tobamovirus movement proteins with the plant cytoskeleton. *Science* 270:1983–1985.

Hu, W. and Cheng, C. L. (1995). Expression of *Aequorea* green fluorescent protein in plant cells. *FEBS Lett.* 369:331–334.

Inouye, S. and F. I. Tsuji (1994). Evidence for redox forms of the *Aequorea* green fluorescent protein. *FEBS Lett.* 351:211–214.

Jefferson, R. A., Kavangh, T. A., and Bean, M. W. (1987). GUS fusions: β-glucuronidase as a sensitive and versatile gene fusion marker in higher plants. *EMBO J.* 6:3901–3907.

Kaether, C. and Gerdes, H. H. (1995). Visualization of protein-transport along the secretory pathway using green fluorescent protein. *FEBS Lett.* 369:267–271.

Kohler, R. H., Zipfel, W. R., Webb, W. W., and Hanson, M. R. (1997). The green fluorescent protein as a marker to visualize plant mitochondria *in vivo*. *Plant J.* 11:613–621.

Langford, C. J., Klinz, F.-J., Donath, C., and Gallwitz, D. (1984). Point mutations identify the conserved, intron-contained TACTAAC box as an essential splicing signal in yeast. *Cell* 36:645–653.

Lim, C. R., Kimata, Y., Oka, M., Nomaguchi, K., and Kohno, K. (1995). Thermosensitivity of green fluorescent protein fluorescence utilized to reveal novel nuclear-like compartments in a mutant nucleoporin nsp1. *J. Biochem.* 118:13–17.

Liu, H. X. and Filipowicz, W. (1996). Mapping of branchpoint nucleotides in mutant pre-mRNAs expressed in plant cells. *Plant J.* 9:381–389.

Luehrsen, K. R., Taha, S., and Walbot, V. (1994). Nuclear pre-mRNA processing in higher plants. *Prog. Nucleic Acid Res. Mol. Biol.* 47:149–193.

Niedz, R. P., Sussman, M. R., and Satterlee, J. S. (1995). Green fluorescent protein—an in vivo reporter of plant gene-expression. *Plant Cell Rep.* 14:403–406.

Ogawa, H., Inouye, S., Tsuji, F. I., Yasuda, K., and Umesono, K. (1995). Localization, trafficking, and temperature-dependence of the *Aequorea* green fluorescent protein in cultured vertebrate cells. *Proc. Nat. Acad. Sci USA* 92:11899–11903.

Ormö, M., Cubitt, A. B., Kallio, K., Gross, L. A., Tsien, R. Y., and Remington, S. J. (1996). Crystal structure of the *Aequorea victoria* green fluorescent protein. *Science* 273: 1392–1395.

Pang, S. Z., DeBoer, D. L., Wan, Y., Ye, G., Layton, J. G., Neher, M. K., Armstrong, C. L., Fry, J. E., Hinchee, M. A. W., and Fromm, M. E. (1996). An improved green fluorescent protein gene as a vital marker in plants. *Plant Physiol.* 112:893–900.

Pines, J. (1995). GFP in mammalian cells. *FEBS Lett.* 11:326–327.

Plautz, J. D., Day, R. N., Dailey, G. M., Welsh, S. B., Hall, J. C., Halpain, S., and Kay, S. A. (1996). Green fluorescent protein and its derivatives as versatile markers for gene expression in living *Drosophila melanogaster*, plant and mammalian cells. *Gene* 173:83–87.

Prasher, D. C., Eckenrode, V. K., Ward, W. W., Prendergast, F. G., and Cormier, M. J. (1992). Primary structure of the *Aequorea Victoria* green-fluorescent protein. *Gene* 111: 229–233.

Reichel, C., Mathur, J., Eckes, P., Langenkemper, K., Reiss, B., Koncz, C., Schell, J., and Maas, C. (1996). Enhanced green fluorescence by the expression of an *Aequorea victoria* green fluorescent protein mutant in mono- and dicotyledonous plant cells. *Proc. Natl. Acad. Sci. USA* 93:5888–5893.

Sheen, J., Hwang, S. B., Niwa, Y., Kobayashi, H., and Galbraith, D. W. (1995). Green-fluorescent protein as a new vital marker in plant-cells. *Plant J.* 8:777–784.

Simering, K. R., Golbik, R., Sever, R., and HAseloff, J. (1996). Mutations that suppress the thermosensitivity of green fluorescent protein. *Curr. Biol.* 6:1653–1663.

Simpson, C. G., Clark, G., Davidson, D., Smith, P., and Brown, J. W. S. (1996). Mutation of putative branchpoint consensus sequences in plant introns reduces splicing efficiency. *Plant J.* 9:359–380.

Valvekens, D., Van Montagu, M., and Van Lijsebettens, M. (1988). *Agrobacterium tumefaciens*-mediated transformation of *Arabidopsis thaliana* root explants by using kanamycin selection. *Proc. Natl. Acad. Sci. USA* 85:5536–5540.

van der Berg, C., Willemsen, V., Hage, W., Wiesbeek, P., and Scheres, B. (1995). Cell fate in the *Arabidopsis* root meristem determined by directional signalling. *Nature (London)* 378:62–65.

Ward, W. W., Cody, C. W., Hart, R. C., and Cormier, M. J. (1980). Spectrophotomeric identity of the energy transfer chromophores in Renilla and Aequorea green fluorescent proteins. *Photochem. Photobiol.* 31:611–615.

Webb, C. D., Decatur, A., Teleman, A., and Losick, R. (1995). Use of green fluorescent protein for visualization of cell-specific gene expression and subcellular protein localization during sporulation in *Bacillus subtilis*. *J. Bacteriol.* 177:5906–5911.

Wiebauer, K., Herrero, J.-J., and Filipowicz, W. (1988). Nuclear pre-mRNA processing in plants: distinct modes of 3′-splice-site selection in plants and animals. *Mol. Cell. Biol.* 8:2042–2051.

Yang, F., Moss, L. G., and Phillips, G. N. J. (1996). The molecular structure of green fluorescent protein. *Nature Biotech.* 14:1246–1251.

Zoenicka-Goeta, M., Pines, J., Ryan, K., Siemering, K. R., Haseloff, J., Evans, M. J., and Gurdon, J. B. (1996). An indelible lineage marker for *Xenopus* using a mutated green fluorescent protein. *Development* 122:3719–3724.

Zoenicka-Goetz, M., Pines, J., Siermering, K. R., Haseloff, J., and Evans, M. J. (1997). Following cell fate in the living mouse embryo. *Development* 124: 1133–1137.

Zolotukhin, S., Potter, M., Hauswirth, W., Guy, J., and Muzyczka, N. (1996). A "humanized" green fluorescent protein cDNA adapted for high level expression in mammalian cells. *J. Virol.* 70:4646–4654.

11

The Uses of Green Fluorescent Protein in Transgenic Vertebrates

ADAM AMSTERDAM AND NANCY HOPKINS

Center for Cancer Research and Department of Biology, Massachusetts Institute of Technology, Boston, MA

11.1 INTRODUCTION

The ability to express exogenous DNA in vertebrate animals has been invaluable to a wide range of biological studies. Our understanding of processes as diverse as gene expression, cell lineage relationships, and gene function have all benefited from the use of "transgenic" animals. Furthermore, transgenesis is a powerful tool for techniques such as mutagenesis, genetic alteration of animals for commercial uses, and human gene therapy. Many transgenic studies utilize reporter genes, such as the bacterial *lacZ* and chloramphenicol acetyltransferase *cat* genes, that can be used both to show where the gene was expressed and to quantify the level of expression. However, the analysis of these reporter genes usually involves killing the animal, precluding many types of experiments. The discovery that the gene encoding green fluorescent protein (GFP) could be used as a reporter for gene expression in living animals (Chalfie et al., 1994) has potentially made many of these experiments possible. This chapter will briefly summarize different types of transgenesis in a few vertebrate organisms, the different biological questions best addressed by the use of these different types of transgenic vertebrates, and the ways in which the *gfp* gene could be particularly useful in such studies. This background discussion will be followed by a review of the successful uses of

Green Fluorescent Protein: Properties, Applications, and Protocols, Edited by Martin Chalfie and Steven Kain

ISBN 0-471-17839-X

GFP in transgenic vertebrates to date and what issues remain to be resolved in order for GFP to be best utilized in these powerful systems.

11.2 TRANSGENESIS IN VERTEBRATES: METHODS AND USES

The word "transgenic" is used to refer to an organism whose cells contain exogenous DNA. There are two predominant distinctions: whether the DNA is integrated into the host's chromosome or not (transient vs. stable transgenics) and whether or not all of the cells of the organisms contain the DNA (mosaic vs. non-mosaic). Often the term "transgenic" is only used for the case where DNA has integrated into the host genome and is present in every cell in the animal, a situation best confirmed by passage through the germ line. However, it is equally appropriate to think of animals in which only some cells contain the foreign DNA, which may or may not be stably maintained, as transgenic. For many types of experiments, such transgenics are the only feasible option, and for others they are actually more appropriate than transgenics in which every cell harbors the integrated transgene.

Transgenics can be defined methodologically into two major classes: those organisms into which the DNA was introduced (G_0, the founder generation) and those (F_1, F_2, etc., subsequent generations) who have inherited stably integrated copies of it through the germ line. The former can be divided into three phenomenological classes: transient and mosaic, stable and mosaic, and stable and non-mosaic. The last is often treated as functionally equivalent to a germ line transgenic, since both involve integration of the transgene in all of their cells (see Table 11.1).

11.2.A Transient Transgenics

Transient transgenics, where the foreign DNA is not necessarily integrated into the host's chromosome, can be achieved either by the delivery of plasmid DNA or by the use of episomal viruses, such as adenovirus, herpes simplex virus, or vaccinia virus. Such methods generally only deliver DNA to some of the cells of the animal, and this DNA is not necessarily stable over time. These are most

TABLE 11.1 Methods for making different types of transgenics

Type of Transgenic	Methods
G_0 transient	DNA microinjection; episomal virus; particle-mediated gene transfer
G_0 stable mosaic	Integrating virus; cell transplantation
G_0 stable non-mosaic	Microinjection into egg pronucleus (mouse) sperm nuclear transplantation (frog)
Germline stable non-mosaic	Inheritance of integrated transgene through germline of G_0 transgenic

useful either for quickly testing the tissue specificity of cis-acting elements or for testing the effects of ectopic expression of a given gene where it is either unnecessary or impossible to express it in the entire organism.

One example of transient transgenesis is the injection of plasmid DNA into one-cell stage frog or fish embryos. Such DNA is highly replicated during blastula stages and is unevenly distributed among the rapidly dividing cells of the developing embryo. In some proportion of the cells the DNA can integrate, and most, but not all, nonintegrated DNA appears to be destroyed during gastrulation (Rusconi and Schaffner, 1981; Stuart et al., 1988). Thus, distribution of the exogenous DNA is widespread but uneven, highly variable, and unstable over time. Nevertheless, these methods can be used to demonstrate temporal and tissue- and region-specific restriction of a cis-acting element (Krieg and Melton, 1985; Mohun et al., 1986; Brakenhoff et al., 1991; Westerfield et al., 1992) by using fusion genes of the cis-acting DNA to be tested with a reporter gene. GFP could prove to be a very useful reporter in such experiments as the expression pattern of the transgene could be observed in live embryos. Instead of histochemical procedures that require many processing steps, one would only need to observe the embryos by fluorescence microscopy. More importantly, the same animals could be observed at several time points, which could help overcome the problem of variability in the mosaicism of the transgene from embryo to embryo.

Transient mosaic expression by plasmid microinjection can also be used to investigate the biological effect of widespread ectopic expression of a given gene product (Giebelhaus et al., 1988; Christian and Moon, 1993; Hammerschmitt et al., 1996). There are other methods for delivering plasmid DNA later in the development of an animal that are more appropriate for examining the effect of a specific gene product in a more select population of cells later in development. Wolff et al. (1990) demonstrated that plasmid DNA could be injected directly into the muscle tissue of adult mice and that 10–30% of the cells in the injected area would take up and express this DNA. Furthermore, extrachromosomal DNA was stable in these cells and continued to express for at least 2 months. Another method for delivering DNA into cells in living tissue is particle-mediated gene transfer, or biolistics, in which micron-sized gold particles are coated with DNA. These particles are then used to bombard target tissue at a very high speed, resulting in the uptake of the DNA in many of the cells (Pecorino and Lo, 1992). Recombinant episomal viruses, such as vaccinia virus (Moss, 1991), herpes simplex virus (HSV, Geller and Breakefield, 1988), and adenovirus (Stratford-Perricaudet and Perricaudet, 1991) , also have been used as gene delivery vectors. With these methods, as with microinjected DNA, expression is variable and mosaic, and the correct interpretation of the results of ectopic gene expression requires knowing which cells are expressing the gene. While a coexpression marker such as *lacZ*, or in situ hybridization for the gene of interest itself, can indicate the expression pattern of the gene at the time of analysis, sometimes it can be useful to see the expression before the time point of the final assay. Thus a reporter gene whose expression could be assessed in real time in the same animal could be very useful for monitoring the extent and persistence of gene expression.

11.2.B Stable Mosaic Transgenics

While transient transgenics are generally both mosaic and unstable, it is possible to make stable mosaics—animals in which only some cells are transgenic, but the DNA is integrated in their chromosomes and thus stable in these cells and their descendants. This form of transgenesis is very useful for establishing lineage relationships of cells or investigating the effects of a transgene or genetic alteration in a subset of cells. It is also a potential vehicle for gene therapy, in which a gene is introduced into some of the cells of an animal (human animals being the ultimate goal) to correct a genetic defect. The primary means of achieving stable mosaic transgenesis are infection with integrating viruses (such as retroviruses or adeno-associated virus) and cell transplantation.

Retroviruses infect cells and insert a copy of their genome into the chromosome. Most retroviruses (though not certain lentiviruses) require that the infected cell go through mitosis before integration can occur. Their natural life cycle then allows them to replicate and spread from cell to cell, though they can be engineered to be replication-defective. Replication-defective recombinant retroviruses that express a visible marker can be used to infect a small number of cells and thus mark clones of their descendants (Price et al., 1987). Similarly, one can investigate the consequences of the ectopic expression of a gene of interest in a spatially limited area of a developing organism by expressing it from either a replication defective virus (Ishibashi et al., 1995) or a replication-competent virus, which can increase the target area (Morgan et al., 1992). Another virus that can be used for gene transfer is adeno-associated virus (AAV, Muzyczka, 1994). This virus is nonpathogenic and its genome can persist extrachromosomally but can also integrate into the host cell genome at some frequency. Unlike retroviruses, AAV does not appear to require cell division for integration. Recombinant AAV has been used to stably express genes in a number of primary tissues, such as hematopoietic cells and neurons and glia in the brain (Zhou et al., 1994; Kaplitt et al., 1994). Both retroviruses and AAV have been proposed as vehicles for gene therapy, thus the production of GFP might be a good reporter for effective gene transfer and maintenance of expression.

Alternatively, one can make a chimera by transplanting transgenic cells from a transgenic to a nontransgenic organism, thus creating clones of transgenic cells. Such transplantations are easily done in amphibian and fish embryos at many stages of development. In mice, it is also possible to transplant cells between embryos at multiple stages of development, though one can also make chimeras by the transplantation of embryonic stem (ES) cells, which can be genetically altered in culture. Thus, not only can one make clones of cells that ectopically express a gene of interest, but one can make clones of cells with directed mutations via homologous recombination in the ES cells. In order to mark the transplanted cells, they need to possess some characteristic distinct from the host cells; expression of GFP could be such a mark.

11.2.C Stable Non-Mosaic Transgenics

Finally, one can make stable, non-mosaic transgenics in which the transgene is stably integrated into the same chromosomal location(s) in every cell of the organism. This method can be a better way to study the role of cis-acting sequences on gene expression and the effects of ectopic gene expression because the lack of mosaicism makes interpretation easier. Furthermore, it is the best way to analyze the roles of specific genes through mutagenesis, either by the expression of dominant negative alleles (Herskowitz 1987; Stacey et al., 1988; Kroll and Amaya, 1996), targeted recessive mutations by homologous recombination in murine ES cells (Schwartzberg et al., 1989; Thompson et al., 1989; Zijlstra et al., 1989), or insertional mutagenesis (Meisler, 1992; Gaiano et al., 1996).

In some situations, stable non-mosaic transgenesis can be achieved in G_0 animals, while in others, only mosaic G_0 animals can be made. In the latter case, transmission of an integrated transgene through the germ line is required to make a stable non-mosaic transgenic; however, even in the former case it is preferable to get such "germ line transgenics" to be certain that the integrated DNA is non-mosaic. The use of G_0 animals is preferred in these cases only if expression of the transgene might be lethal to the animal, requiring analysis of the consequence of the transgene's expression before the animals are sexually mature.

In mice, stable non-mosaic transgenics can be made by injection of DNA into the pronucleus of a fertilized egg; in some proportion of injected eggs the DNA will be incorporated into the genome. Integration usually occurs at the one cell stage, thus most of these animals are non-mosaic; however, as stated above, only when transgenes have been inherited through the germ line can one be sure that the animal is non-mosaic. One can also make transgenic mice by infecting preimplantation embryos with a retrovirus. While the infected animals will be mosaic, if the integrated provirus is inherited through the germ line, the offspring will inherit the viral sequences in every cell and thus be non-mosaic. Additionally, one can use chimeras made by the transplantation of genetically altered ES cells; some of these cells will contribute to the germline and thus some proportion of the progeny of these chimeras will contain the genetic alteration.

Stable non-mosaic transgenic fish are made primarily by two methods—DNA microinjection and retroviral infection (Stuart et al., 1988; Lin et al., 1994). In both cases, the founder fish are mosaic, and thus the transgene must be inherited through the germ line. DNA microinjection has been used to make transgenics in many fish species, such as zebrafish, medaka, carp, rainbow trout, and Atlantic salmon (reviewed in Maclean and Rahman, 1994), and often are capable of expressing genes of interest (reviewed in Iyengar et al., 1996). At this time, retrovirally induced transgenesis has only been reported in zebrafish, and transgene expression has been observed only in mosaic G_0 animals, not after passage through the germ line (N. Gaiano, M. Allende, K. Kawakami and N. Hopkins, unpublished observations).

Transgenic frogs have also been made by the inheritance through the germ line of microinjected DNA (Etkin and Pearman, 1987), but the generation time of *Xenopus* is quite long (8 months) and transgene expression was not observed. Recently, Kroll and Amaya (1996) reported a novel way to generate transgenic frogs such that the G_0 animals were predominantly non-mosaic. Sperm nuclei were isolated, incubated with plasmid DNA and restriction enzyme, and injected into unfertilized eggs. While some of the resulting animals appear mosaic, 30–40% appear to be uniformly transgenic and are capable of expressing the transgene.

11.3 ISSUES AFFECTING THE SUCCESSFUL USE OF GFP IN TRANSGENIC VERTEBRATES

The ability to use GFP for any of the uses described above will require that the particular application produces enough GFP fluorescence to be detectable. Thus the real issues for the successful application of GFP in transgenic vertebrates are absolute expression levels, which mutant versions of GFP are most easily detectable, and which detection methods to use. Additionally, the animal itself is an issue. Zebrafish embryos are particularly well suited to GFP expression due to their transparency and development outside their mother. *Xenopus* embryos also develop externally and, while not as transparent as zebrafish, can reveal GFP expression nicely as well (Kroll and Amaya, 1996). However, the full potential of GFP as a vital marker may not be fully realized in mice and other mammals. Because these embryos develop inside their mothers, development of a single animal cannot be watched over time. Instead, individual embryos are removed from the uterus at discrete time points and analyzed only at that developmental stage. Since this prevents further development of the embryo, this imposes the same limits upon the analysis of gene expression as the use of a nonvital marker. However, there may be specific applications of GFP that could be useful in experiments involving transgenic mice, such as those involving explants that are then manipulated in culture.

11.3.A Expression levels

Most reports of GFP expression in vertebrates have used very strong regulatory elements to drive expression, such as the ef1α enhancer/promoter in fish (Amsterdam et al., 1995), cytomegalovirus (CMV) in mammals and frogs (Ikawa et al., 1995a,b, Kroll and Amaya, 1996). The use of tissue specific promoters has mostly been used in transient assays (see exceptions below), where variability and mosaicism of expression has made it difficult to assess whether GFP was always detected everywhere that it was expressed. In one case, where GFP expression was subsequently assessed immunohistochemically, it appeared that a majority of the expressing cells were not visible via fluorescence of the native protein (Peel et al., 1997). By using transplanted cells from zebrafish embryos injected with bacterially expressed recombinant GFP, we have estimated that detection of fluorescence requires nearly 10 times more GFP when only expressed in single cells than when ubiquitously expressed (Amsterdam et al., 1996). Consistent with this observation, cells transplanted from stable transgenic zebrafish embryos expressing detectable GFP ubiquitously (see Section 11.4.A) into nontransgenic embryos could not be detected by epifluorescence microscopy (Amsterdam and Hopkins, unpublished observations). Thus, while a systematic and truly quantitative analysis of gene expression levels required for detection of fluorescence has not been done, clearly levels of expression are a concern. It is noteworthy, however, that toxic effects from high levels of expression of GFP have not been observed in any of these animals; thus expressing too much GFP does not appear to be a problem.

11.3.B GFP variants

The variants of GFP with altered spectral properties and brighter fluorescence (see Chapter 5) will be advantageous, but to date there are few reports of their use in vertebrate animals. The few reports we have are encouraging. The Ser65 to Thr mutant (Heim et al., 1995) was used in transgenic frogs by Kroll and Amaya (1996), while Zernicka-Goetz et al. (1996) observed very strong and long lasting fluorescence when using another variant, GFP.RN3, for RNA injections into frog embryos. This variant incorporates three amino acid substitutions to alter the absorption characteristics: Phe64Leu, Ser65Thr, and Ile167Thr. This variant also incorporates two amino acid substitutions thought to improve the folding of the protein at higher temperatures and a number of silent codon changes. Another variant selected on the basis of FACS optimization when expressed in bacteria (GFP mut2, Cormack et al., 1996) was used for tissue specific expression in both injected zebrafish embryos and stable transgenic zebrafish (Long et al., 1997; Meng et al., 1997). This particular variant has three amino acid changes in the chromophore region: Ser65Ala, Val68Leu, and Ser72Ala. In addition to variants with amino acid substitutions, altering the codon usage has appeared to improve expression in mammalian systems. A "humanized" GFP (Zolotukhin et al., 1996), was used by Peel et al. (1997) for expression in the rat spinal cord. This *gfp* gene has the Ser65Thr substitution, and was further modified by altering 88 codons to codons used more frequently in mammalian cells (see Chapter 5). Both this variant and one in which 169 codons were altered (Haas et al., 1996) have also been used in retroviral vectors to infect mammalian tissue culture cells, and each additional "humanization" resulted in brighter fluorescence (Levy et al., 1996; Muldoon et al., 1997).

The GFP variants may be useful not only because they are brighter, but also because many of them excite maximally in the blue spectrum rather than in the ultraviolet (UV). This alteration allows for the use of fluorescein filter sets for observation, which provides less autofluorescence in embryos from some species and is less likely to cause UV-induced damage to the organism during observation. Additionally, Zernicka-Goetz et al. (1996) injected RNA for another variant that emits in the blue spectrum and observed blue fluorescence. This finding suggests that two spectrally distinguishable reporters could be expressed in the same animal.

11.3.C Visualization Methods

Most reports on the use of GFP in vertebrates have used direct observation with fluorescence microscopes. A systematic comparison of different filter combinations for viewing in any given organism has not been reported. When using wild-type GFP, filters that use the UV excitation peak and take advantage of the full emission spectrum (e.g., 370–420-nm excitation and 455-nm long pass emission) have been used in fish and mice (Amsterdam et. al., 1995; Ikawa et al., 1995a,b), as have filters that excite primarily in the blue and have a higher wavelength emission barrier (e.g. Chroma or Omega's "GFP" filters, Wu et al., 1995; Chiocchetti et al., 1997). Fluorescein filters have been used as well (Lo et al.,

1994), though these are only able to excite wild-type GFPs minor absorbance peak in the visible spectrum and cut out much of the emission. Alternatively, for the red-shifted variants, such as the Ser65 to Thr mutant (Heim et al., 1995) or mut2 (Cormack et al., 1996), fluorescein filters are optimal. Our experience with wild-type GFP in zebrafish embryos is relevant here (Amsterdam and Hopkins, unpublished observations). The filter sets that excite at lower wavelengths (especially in the UV) and use long-pass emmission filters at lower wavelengths (e.g., 455 nm) appear to give much stronger signals, but also allow for much brighter autofluorescence from the yolk. However, this autofluorescence appears more yellow than green, so it can generally be distinguished from the signal of interest. The filters that excite at higher wavelengths and have emmision filters that cut out some of GFPs emmision peak provide somewhat less illumination of the GFP, though yolk autofluorescence is much less. However, the autofluorescence in this case is much more similar in color to the actual GFP signal, especially when band-pass emmision filters were used.

As noted above, many of the GFP variants have different excitation and emission peaks than the wild-type protein, thus they are more efficiently detected with different filter sets. For example, the red-shifted variants, such as the Ser65Thr mutant (Heim et al., 1995) or mut2 (Cormack et al., 1996), are optimally observed using fluorescein filters.

Most embryos can be observed using either inverted fluorescent microscopes, such as a Nikon Diaphot, or a noninverted fluorescence microscope designed for large working distances (the distance between the lens and the sample) such as a Nikon microphot-SA. The use of such microscopes with a 4X or 5X objective provides a large enough field of view to see entire early embryos; however, later stage embryos will not entirely fit in the field of view.

In some instances, fluorescence was enhanced with the use of a cooled CCD camera. Digital manipulation of the image can suppress the background and highlight the true signal. Digital image recording also added the ability to quantitate expression levels. However, it is unclear whether or not this actually allowed detection of GFP that could not be seen at all without it. Potter et al. (1996) used two-photon laser-scanning microscopy to detect fluorescent dyes in cultured neurons. They suggest that GFP could be similarly detected and that this method is far more sensitive than epifluorescence or confocal microscopy.

11.4 SUCCESSFUL APPLICATIONS OF GFP IN TRANSGENIC VERTEBRATES

Stable transgenics expressing GFP both ubiquitously and tissue-specifically have been made successfully in zebrafish, mice, and frogs. Tissue-specific expression has been seen in transiently transgenic zebrafish, transiently and stably transgenic frogs, and in virally infected rat neurons. The GFP has been used to assess alternative gene delivery systems, including biolistics, several viruses, and direct DNA injection into muscle tissue (see Table 11.2)

TABLE 11.2 Examples of GFP expression in transgenic vertebrates

Type of Transgenic and Method of Transgene Delivery	Organism		
	Fish	Frog	Mammal
G_0 Transient, mosaic: DNA microinjection	Wild-type GFP Muscle-specific expression with mouse myosin light chain enhancer/ promoter (Moss et al., 1996) GFPmut2 Independent regulation of expression in blood progenitors, enveloping layer, and CNS neurons with deletion analysis of GATA-2enhancer/ promoter (Meng et al., 1997)		Wild-type GFP Muscle cell expression when injected into adult muscle driven by CMV enhancer/ promoter or creatine kinase promoter (Bartlett et al., 1996)
G_0 Transient, mosaic: Particle-mediated gene transfer			Humanized GFPSer65Thr Adenoassociated virus vector infection in adult rate spinal cord neuron-specific expression with PDGF-β-chain promoter and neuron-specific enolase promoter (Peel et al., 1997)
G_0 Transient, mosaic: Viral vector		Wild-type GFP Tectal neurons infected in live animals with recombinant vaccinia virus; expression driven by vaccinia early/late promoter (Wu et al., 1995)	

TABLE 11.2 Examples of GFP expression in transgenic vertebrates (continued)

Type of Transgenic and Method of Transgene Delivery	Organism		
	Fish	Frog	Mammal
G_0 Stable, mosaic: Viral vector			Wild-type GFP Neurons and glia in rat, ferret and tree shrew brain slices with CMV enhancer/promoter (Lo et al., 1994)
G_0 Stable, non-mosaic: Sperm nuclear transplantation		Wild-type GFP Ubiquitous expression with CMV enhancer/promoter and muscle-specific expression with cardiac actin enhancer/promoter (Kroll and Amaya, 1996)	
Germline Stable, non-mosaic: DNA microinjection	Wild-type GFP Ubiquitous expression with Xenopus ef1α enhancer/promoter (Amsterdam et al., 1995) GFPmut2 Expression in erythroid progenitor cells driven by GATA-1 enhancer/promoter (Long et al., 1997)		Wild-type GFP Ubiquitous expression from four-cell stage on with CMV enhancer/ β-actin promoter in transgenic mice (Ikawa et al., 1995 a,b) Green Lantern GFP Taste-cell restricted expression driven by gustducin promoter (Margolskee, personal communication) Wild-type GFP Expression in the adult mouse liver driven by the hemopexin promoter and expression throughout embryos driven by the β-1 integrin distal promoter (Chiocchetti et al., 1997)

11.4.A Stable Non-Mosaic Transgenics

Stable transgenic lines expressing detectable GFP have been made in zebrafish, frogs, and mice. The Xenopus ef1α enhancer/promoter designed for early gene expression in frogs (Johnson and Krieg, 1994) was used in conjunction with a rabbit β-globin intron to express wild-type GFP in transgenic zebrafish generated by DNA microinjection (Amsterdam et al., 1995). In all five transgenic lines containing this construct (and in two out of four in which the ß-globin intron was omitted) transgenic embryos could be unambiguously identified by 24-h postfertilization, even while still in their chorions [Fig. 11.1(*a*)]. Importantly, a single integrated copy of this transgene was sufficient to produce detectable fluorescence. Fluorescence in most lines appeared ubiquitous between 20 and 36 h after fertilization [Fig. 11.1(*a,b*)]. After this, fluorescence slowly decreased in most tissues, except the lens of the eye [Fig. 11.1(c)]. The absence of detectable fluorescence until 20-h post-fertilization is neither due to a lack of transcription before this point in time or a general inability to detect GFP in early embryos. In injected (G_0) embryos, fluorescence can be detected as early as 4-h postfertilization, only an hour after the beginning of zygotic transcription. This demonstrates both that the enhancer/promoter is transcriptionally active at this time, and that GFP is readily detectable. Furthermore, in situ hybridization on embryos from the transgenic lines demonstrated the presence of GFP RNA as early as 10-h postfertilization (Y. Grinblat, A. Amsterdam, H. Sive, and N. Hopkins, unpublished observations). Thus, it appears that there is a significant delay between the onset of transcription and the appearance of detectable fluorescence. Presumably it takes this much time to accumulate sufficient GFP for detection in the transgenic embryos; fluorescence is probably seen earlier in the injected embryos because, due to the high number of plasmid molecules per cell, there is much more GFP produced. This delay may be due to a requirement for both the accumulation of GFP protein and its correct posttranslational modification for chromophore formation. Thus the delay might be shorter if a variant such as the Ser65Thr were used, as the kinetics of chromophore formation seem to be faster in vitro.

Transgenic zebrafish have also been made which express GFP in a tissue-specific manner. Long et al. (1997) used 5.4 kb of sequence upstream of the *GATA-1* gene to drive the expression of the GFPmut2 variant (Cormack et al., 1996). Fluorescence was specifically seen in erythroid progenitor cells in the embryo, first in the intermediate cell mass, then in both the heart endocardium and circulating blood cells. In transgenic adults, fluorescence could still be observed in circulating blood cells, as well as in the head kidney, which is thought to be the site of adult hematopoiesis. Thus fluorescence is observed in a pattern recapitulating the expression of the endogenous *GATA-1* gene, the first demonstration in germ-line transgenic zebrafish of any reporter gene expressed in the pattern of an endogenous tissue-specific gene.

Transgenic *Xenopus* expressing GFP were generated by the sperm nuclear transplantation method (Kroll and Amaya, 1996). Though these transgenes were analyzed in founder frogs instead of those in which the transgene had been inherited through the germ-line, expression did appear to be non-mosaic in the majority of the embryos that express the transgene at all. Of the embryos

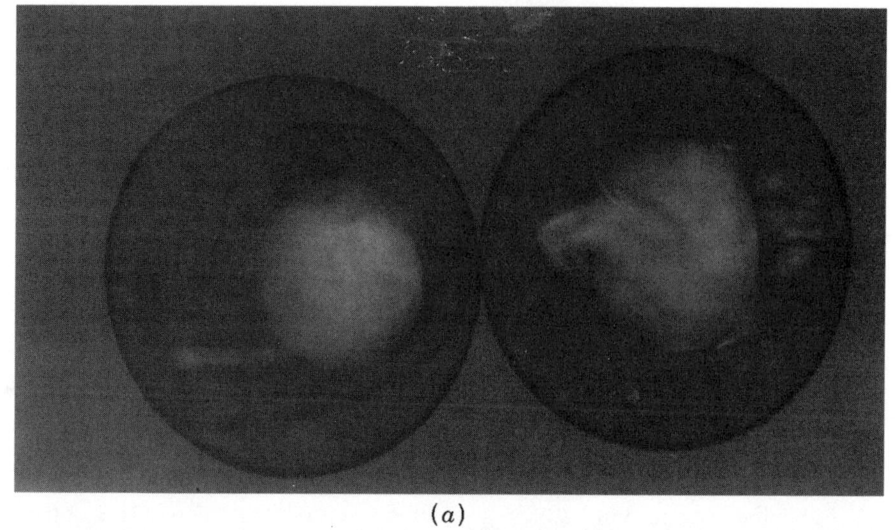

(a)

(b)

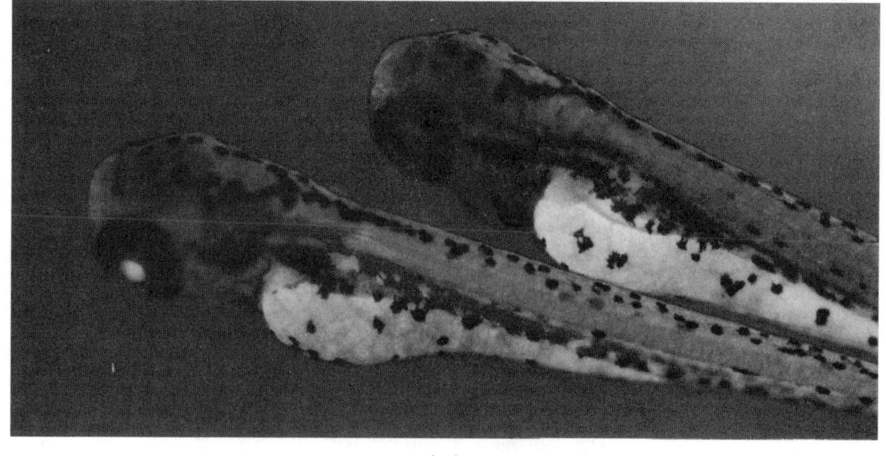

(c)

produced by this method, 35% expressed the transgene non-mosaically, 20% expressed mosaically, and 45% did not express at all, presumably because the transplanted sperm nuclei had not incorporated plasmid DNA. Southern analysis of genomic DNA from these frogs indicated that embryos expressing the transgene non-mosaically contained integrated plasmid while those that failed to express it did not.

Both ubiquitous and tissue-specific expression of GFP could be observed in transgenic frogs made by this method. A transgene in which a nuclear localized red-shifted Ser65Thr GFP was driven by the CMV enhancer/promoter produced ubiquitous expression; fluorescence could be seen in all nuclei of the developing embryo, as confirmed by colocalization of this fluorescence with DAPI staining. In another plasmid used to make transgenics, the enhancer/promoter of the cardiac actin gene (Mohun et al., 1986) was used to drive the expression of wild-type GFP. Tissue-specific non-mosaic fluorescence was observed in somites and cardiac muscle of developing embryos. This expression was followed over several months, during which time tadpoles continued to express GFP non-mosaically in myotomes.

Transgenic mice were generated by pronuclear injection and germ-line transmission of a plasmid containing the CMV enhancer and β-actin promoter (Ikawa et al., 1995a) driving wild-type GFP. Transgenic offspring were easily identified postnatally by the observation of fluorescence in their fingers or tails in all three transgenic lines produced. Both microscopic examination of sections and spectrophotometric analysis of protein extracts from tissues from newborn transgenic mice revealed that expression was nearly, but not entirely ubiquitous. For example, while fluorescence was observed in all tissues examined, in the kidney most fluorescence appeared to be localized to the glomeruli. Muscle tissue appeared to have the most fluorescence; heart and pancreas had nearly as much, while lung and kidney had far less.

Preimplantation embryos from these lines could also be identified with 100% confidence based on observation of fluorescence of morula stage embryos, and fluorescence could actually be detected as early as the four cell stage (Ikawa et al., 1995b). The brief exposure to UV light required to observe fluorescence did not appear to affect the development of the embryos, though it is known that longer exposures can. Ikawa et al. (1995b) postulate that the use of red-shifted variants of GFP could allow detection while exposing the embryos to light in the visible spectrum, thus making the procedure safer.

FIG. 11.1 (opposite) Green fluorescent protein can be detected in stable zebrafish transgenics. Transgenic and nontransgenic siblings for a transgene containing the *Xenopus* ef1αa enhancer/promoter and the rabbit β-globin first intron fused to the *gfp* gene. Sibling progeny of a heterozygous transgenic fish were observed with a Nikon microphot EPI-FL3, with a 370–420-nm excitation filter and a 455-nm long-pass emission filter. (*a*) Embryos at 24-h postfertilization, still in their chorions. (*b*) Embryos at 32-h postfertilization. (*c*) Embryos at 72-h post-fertilization. Notice the strong fluorescence in the eye, but reduced fluorescence in the rest of the body (Reprinted from Amsterdam et al., 1995). Figure also appears in color section.

Tissue-specific expression of GFP has also been achieved in transgenic mice. When the gustducin promoter was used to drive the *gfp* gene, expression of GFP was restricted to the taste cells of the mice (Margolskee, personal communication). This study utilized the Green Lantern GFP (available from Gibco/BRL Life Technologies), which incorporates the Ser65Thr mutation as well as many silent changes to "humanize" the codon usage. When detected with antibodies, GFP protein could be seen very clearly in the taste cells of fixed tissue from young mice. Fluorescence from the GFP itself was much harder to see; only faint fluorescence could be seen in fixed tissues that were not antibody-stained, and it was not clear if as many cells could be seen to express GFP by its own fluorescence as opposed to by antibody detection.

Both tissue-specific expression and ubiquitous expression of GFP in transgenic mice were demonstrated by Chiocchetti et al. (1997), using wild type GFP fused to the second intron and final exon of the human β-globin gene. First, 700 bp of the hemopexin promoter, which is strongly active in the adult liver and weakly active in some cells in the brain, was used to drive either *gfp* or *lacZ*. Both reporters were expressed in the correct tissue-specific manner. However, the fluorescent signal in the GFP transgenics, detected in paraformaldehyde-fixed sections, was much stronger than the Xgal staining in the *lacZ* transgenics, and a far greater number of cells were observed to be fluorescent in the former than X-gal staining in the latter. This was true even when comparing lines in which an equal concentration of GFP protein and β-galactosidase protein were found in liver extracts. Transgenic mice were also made in which either wild type *gfp* or *lacZ* were driven by 1.5 kb of the β-1 integrin distal promoter, which is ubiquitously active in embryogenesis. As with the hemopexin-driven transgenes, in any given tissue fluorescence in the GFP lines was much stronger than X-gal staining in *lacZ* lines. Thus it appears that wild type GFP can give greater sensitivity as a reporter than β-galactosidase in fixed sections from both embryonic and adult mouse tissues.

11.4.B Transient Transgenics: DNA Microinjection

Green fluorescent protein has also been used in transient expression in zebrafish to demonstrate tissue-specificity of cis-acting sequences. Moss et al. (1996) demonstrated that a plasmid containing the wild-type GFP gene driven by the rat myosin light chain enhancer/promoter was able to produce fluorescence exclusively in muscle cells. The number of fluorescent somitic muscle fibers varied widely from embryo to embryo, but fluorescence was never seen in any other tissue. This expression could be observed as early as 24 h after fertilization and fluorescent muscle cells could still be observed in 6-week-old fish.

Another study utilized GFP as a reporter to evaluate the regulatory sequences upstream of the *GATA-2* gene (Meng et al., 1997). Using 7.3 kb of upstream sequence to drive GFPmut2, fluorescence was observed mosaically but exclusively in ventral mesoderm and ectoderm during gastrulation, blood progenitors in the intermediate cell mass (ICM), enveloping layer (EVL) cells, and neurons in the central nervous system. This faithfully reproduces the expression pattern of the endogenous *GATA-2* gene (Detrich et al., 1995), with the possible exception of the EVL expression, which was not detected by whole mount in situ hybridization. Deletions and mutations were then made in this cis-acting sequence to identify sequences that affected expression in each of these tissues individually, demonstrating the potential of GFP as a tool for the detailed analysis of developmentally controlled transcriptional regulatory elements.

11.4.C As a Reporter for Alternative DNA Delivery Techniques

Green fluorescent protein has also been a popular reporter for other DNA delivery techniques, including experimental studies of gene therapy approaches. For example, two studies have used GFP to demonstrate the efficacy of gene transfer into neurons for the purpose of examining the effects of ectopic expression of other genes in these cells. Lo et al. (1994) demonstrated that particle-mediated gene transfer (bombardment of tissue with DNA coated gold particles) into brain slices from developing rats, ferrets, and tree shrews under conditions in which these tissues can remain healthy and continue to differentiate could be used to deliver and express a *gfp* transgene. By using the CMV enhancer/promoter to drive the wild-type *gfp* gene, fluorescence could be detected in neurons and glia in the brains slices, and expression was sufficiently high to see dendritic and axonal processes. Lo et al. (1994) suggest that this will allow the identification of live transfected neurons in a physiological setting, opening up the possibility of investigating the effects of expression of cotransfected genes with methods such as time-lapse videomicroscopy and patch-clamp recording.

Toward a similar end, Wu et al. (1995) used GFP encoding recombinant vaccinia virus to infect the tectum in developing frogs. The viral genome contained the vaccinia early/late promoter driving the wild-type *gfp* gene. By injecting low titers of virus (so that few cells were infected) they were able to detect fluorescence in individual neurons for 3–4 days beginning about 1 day after infection, allowing individual infected neurons to be followed in time.

Several studies have utilized GFP as a reporter for long-term gene expression in potential gene therapy vectors. Some of these have involved the infection of tissue culture cells with retroviruses expressing GFP variants (Cheng et al., 1996; Levy et al., 1996; Muldoon et al., 1997). Each of these studies utilized red shifted variants, either the Ser65Thr or RSGFP4 (Phe64Met, Ser65Gly, Gln69Leu; Delagrave et al., 1995), expressed from a murine retrovirus and was able to show that infected cells were fluorescent. Furthermore, the use of variants in which more codons were changed to human codon bias (primarily by the use of either C or G in the third position) improved the amount of fluorescence and the number of cells that were fluorescent. While none of these studies infected cells in living animals, the successful expression and detection of GFP variants in tissue culture cells with retroviral vectors holds out promise for this goal.

Bartlett et al. (1996) combine the successful uptake, retention, and expression of plasmid DNA in injected skeletal muscle observed by Wolff et al. (1990) with the fact that AAV vectors appear to integrate into the host cell chromosome at some frequency. Using plasmids in which the AAV integrative inverted repeats (the sequences necessary to achieve integration) flank a wild-type *gfp* gene, they were able to investigate the extent and duration of gene expression following injection into the skeletal muscle of mice. The GFP expression was seen in nearly 100% of the muscle fibers in the region of the injection by 2 weeks after injection, and expression was even higher after 6 weeks. This study was also able to demonstrate that such a gene delivery system produces more gene product when the muscle-specific creatine kinase promoter is used to drive expression, rather than the usually strong CMV enhancer/promoter.

In another study using actual AAV vectors to infect target cells, Peel et al. (1997) were able to demonstrate long-term and cell-type-specific gene expression in the spinal cord of adult rats. Recombinant AAV encoding the "humanized" GFP (the Ser65 to Thr GFP variant with the human codon bias) driven by one of two neuron-specific promoters produced fluorescence in neurons, but not glial cells, for at least 15 weeks. Many neurons could be infected and express detectable fluorescence [Fig. 11.2(*a*)]. Often, enough GFP was produced in these cases to fill the axonal and dendritic processes [Fig. 11.2(*b*)]. Conversely, the use of an astrocyte-specific promoter driving GFP in a recombinant virus restricted observed fluorescence to this cell type. This finding was useful for demonstrating the efficacy of AAV for gene transfer and tissue-specific and long-term expression in neurons. However, when GFP was stained for immunohistochemically, it was clear that many more cells contained GFP protein than were observed to be fluorescent. Thus most of the cells did not contain enough GFP to detect by fluorescence in spinal cord explants. It is noteworthy that in both of these studies, GFP was not used as a vital marker; the animals had to be killed for analysis as fluorescence could not be observed through the skin and fur.

11.5 PROSPECTS FOR THE USE OF GFP IN TRANSGENIC VERTEBRATES

The successful applications of GFP in transgenic vertebrates so far provide a good indication of what experiments are possible now and which will require more technical advances, such as more sensitive detection methods, better expression vectors, and additional GFP variants. Transient, mosaic expression assays appear to be particularly effective, probably because the high number of copies of the plasmid per cell allows for a high level of GFP expression. However, expression in stable transgenics has also been observed when strong promoters have been used, even when only a single copy of the transgene was present.

11.5.A Identifying cis-Acting Sequences

The examples above suggest that GFP, particularly the brighter variants, will prove to be a very good reporter for defining cis-acting sequences in both frogs and fish. The sperm nuclear transplant experiments indicated that GFP expression in transgenic frogs was as extensive and specific as that of other reporters. Thus it could prove to be the simplest reporter for delineating sequences required for temporally and tissue-specific gene expression, as fluorescence can be observed repeatedly over time. In fish, one should be able to identify cis-acting sequences by the microinjection of GFP-containing plasmids into one-cell stage embryos. The results with the GATA-2 upstream sequences suggest that GFP could be very useful in identifying the distinct sequences responsible for different aspects of a gene's regulation, even though expression is mosaic.

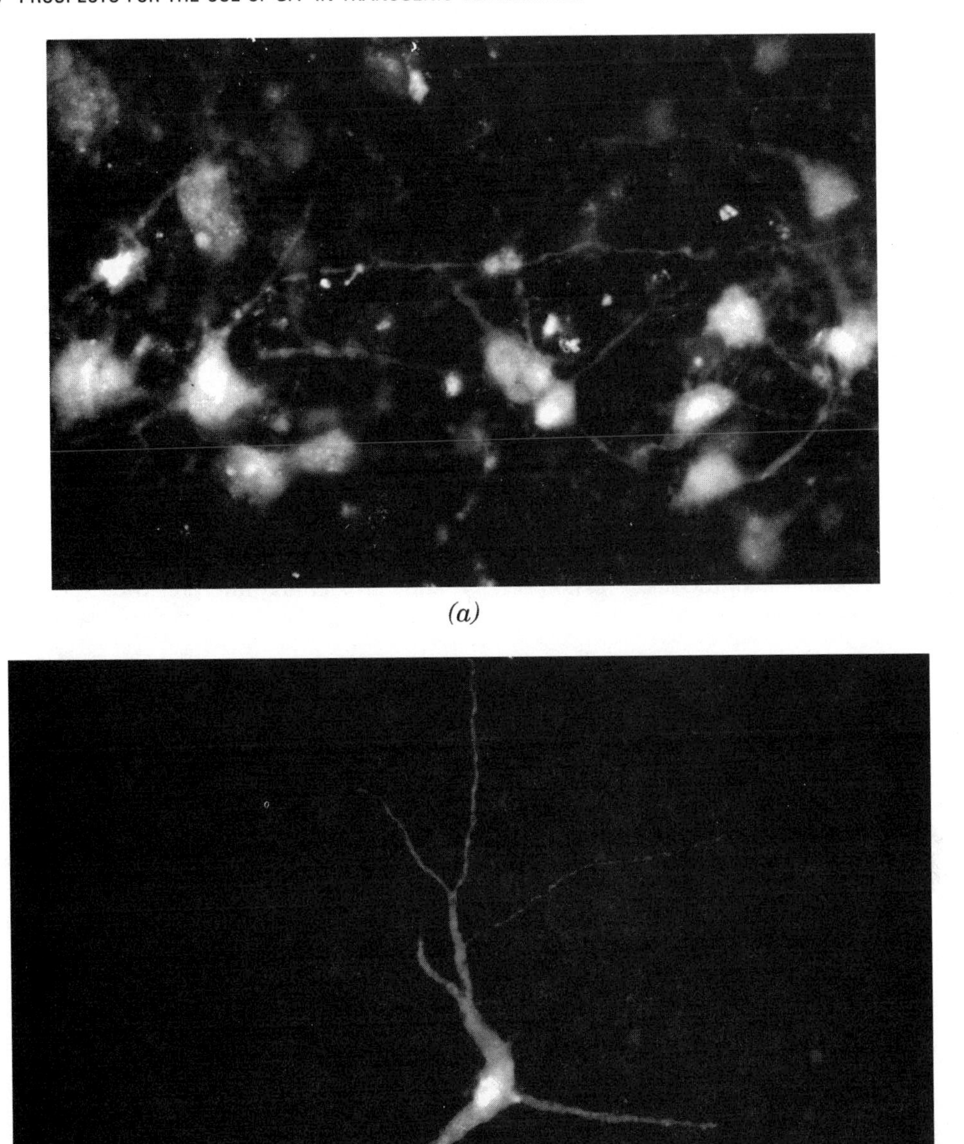

(a)

(b)

FIG. 11.2 Green fluorescent protein can be detected specifically in neurons infected with a recombinant AAV(rAAV) in which a neuron-specific promoter is fused to the *gfp* gene. Adult rats received 5-mL injections of rAAV pTRNSE into the cervical enlargement of the spinal cord. pTRNSE contains a humanized *gfp* gene under the control of the neuron specific enolase (NSE) promoter. Animals were sacrificed and examined histologically using Zeiss FITC optics 2 weeks later. (*a*) Many infected neurons expressing GFP. (*b*) A single infected neuron expressing GFP. Note how the fluorescence fills the axonal and dendritic processes. [Photographs courtesy of A. Peel.] Figure also appears in color section.

11.5.B As a Coexpression Marker

Green fluorescent protein could also be very useful as a coexpression marker in mosaic transgenics. Coexpression of GFP with another gene of interest would identify the cells in which that gene was being expressed. For widespread expression in embryos, one could inject one-cell stage embryos with a plasmid encoding GFP and the gene of interest, both driven by an ubiquitously active promoter. Alternatively, one could use tissue specific regulatory sequences to drive both the gene of interest and GFP so that expression will be restricted to one cell type but still be widely distributed. In either case, while expression will be mosaic, the actual cells expressing the gene of interest should be identifiable by their fluorescence.

To investigate the effects of a gene's overexpression in more discrete areas, delivery systems like the vaccinia virus, AAV, or muscle injection could be used. These would be especially appropriate if one wanted to conduct some kind of assay in living tissue (e.g., electrophysiology) that could be removed from the animal. One could know which cells in the explant express the gene of interest because of the coexpression of GFP and conduct and interpret the experiments accordingly.

11.5.C As a Marker for Transgenic Animals or Cells

The ability to use GFP to unambiguously identify stable transgenics will also be of some use. In the case of the transgenic frogs, if one wants to use the sperm nuclear transplant method to determine the effects of ectopic gene expression, incorporation of GFP into the transgene will make it possible to distinguish transgenics from nontransgenics before analysis. While about one-half of the embryos generated by this method can be expected to be transgenic, eliminating those that are not (or are mosaic) will simplify the analysis of the effects of the transgene. In the case of establishing transgenic lines of fish or mice, the inclusion of GFP in the transgene to facilitate the identification of transgenic offspring would save some amount of time and animal-raising space. However, it is important to note that in both cases, such identifiable transgenics have only been made by DNA microinjection. It is not clear yet whether transgenics made by retroviral infection will produce detectable GFP.

The existence of ubiquitously expressing transgenic lines could prove useful for transplantation experiments, including both lineage analysis and wild-type/mutant chimeras. As stated above, the zebrafish lines described by Amsterdam et al. (1995) do not express strongly enough for this purpose. However, it is possible that the use of a brighter variant of GFP could make transgenics that produce enough fluorescence to do this. It is currently unclear if the ef1α enhancer/promoter will continue to express ubiquitously in such lines (in the lines with the wild-type GFP, fluorescence did seem to fade somewhat), but other regulatory regions could also be tried, such as the carp β-actin enhancer/promoter in conjunction with boundary elements (Caldovic and Hackett, 1995).

The feasibility of other applications of GFP is still difficult to assess. For example, as stated above, it is currently unclear whether the expression of detectable GFP from retroviruses in transgenic animals will be possible, which may have more to do with working out aspects of expression from retroviruses in general than with GFP per se. Similarly, it is unclear whether stable transgenics in which the promoters of specific genes are used to drive GFP will produce enough fluorescence for detection. It still needs to be determined whether the level of expression of most endogenous genes (from their endogenous locus) will provide sufficient expression of GFP. The use of the bright GFP variants and possibly better fluorescence imaging technology can only make these possibilities more likely.

REFERENCES

Amsterdam, A., Lin, S., and Hopkins, N. (1995). The *Aequorea victoria* green fluorescent protein can be used as a reporter in live zebrafish embryos. *Dev Biol.* 171:123–129.

Amsterdam, A., Lin, S., Moss, L. G., and Hopkins, N. (1996). Requirements for green fluorescent protein detection in transgenic zebrafish embryos. *Gene* 173:99–103.

Bartlett, R. J., Secore, S. L., Singer, J. T., Bodo, M., Sharma, K., and Ricordi, C. (1996). Long-term expression of a fluorescent reporter gene via direct injection of plasmid vector into mouse skeletal muscle: comparison of human creatine kinase and CMV promoter expression levels in vivo. *Cell Traspl.* 5:411–419.

Brakenhoff, R. H., Ruuls, R. C., Jacobs, E. H. M., Schoenmakers, J. G. G., and Lubsen, N. H. (1991). Transgenic *Xenopus laevis* tadpoles: a transient *in vivo* model system for the manipulation of lens function and lens development. *Nucleic Acids Res.* 19:1279–1284

Caldovic, L. and Hackett, P. B. Jr. (1995). Development of position-independent expression vectors and their transfer into transgenic fish. *Mol. Mar. Biol. Biotechnol* 4:51–61.

Chalfie, M., Tu, Y., Euskirchen, G., Ward, W. W., and Prasher, D. (1994). Green fluorescent protein as a marker for gene expression. *Science* 263:802–805.

Cheng, L., Fu, J., Tsukamoto, A., and Hawley, R. G. (1996). Use of green fluorescent protein variants to monitor gene transfer and expression in mammalian cells. *Nature Biotechnol.* 14:606–609.

Chiocchetti, A., Tolosano, E., Hirsch, E., Silengo, L., and Altruda, F. (1997) Green fluorescent protein as a reporter of gene expression in mice. *Biochim. Biophys. Acta* 1352:193–202.

Christian, J. L. and Moon, R. T. (1993). Interactions between *Xwnt-8* and Spemann organizer signaling pathways generate dorsoventral pattern in the embryonic mesoderm of *Xenopus. Genes Dev.* 7:13–28.

Cormack, B. P., Valdivia, R. H., and Falkow, S. (1996). FACS-optimized mutants of the green fluorescent protein (GFP). *Gene* 173:33–38.

Delagrave, S., Hawtin, R. E., Silve, C. M., Yang, M., and Youvan, D. C. (1995). Red-shifted excitation mutants of the green fluorescent protein. *Bio Technol.* 13:151–154.

Etkin, L. D. and Pearman, B. (1987). Distribution, expression and germ line transmission of exogenous DNA sequences following microinjection into *Xenopus laevis* eggs. *Development* 99:15–23.

Gaiano, N., Amsterdam, A., Kawakami, K., Allende, M., Becker, T., and Hopkins N. (1996). Insertional mutagenesis and rapid cloning of essential genes in zebrafish. *Nature (London)* 383:829–832.

Geller, A. I. and Breakefield, X. O. (1988). A defective HSV-1 vector expresses *Escherichia coli* β-galactosidase in cultured peripheral neurons. *Science* 241:1667–1669.

Giebelhaus, D. H., Eib, D. W., and Moon, R. T. (1988). Antisense RNA inhibits expression of membrane skeleton protein 4.1 during embryonic development of *Xenopus*. *Cell* 53:601–615.

Haas, J., Park, E.-C. , and Seed, B. (1996). Codon usage limitation in the expression of HIV-1 envelope glycoprotein. *Curr. Biol.* 6:315–324.

Hammerschmidt, M., Serbedzija, G. N., and McMahon, A. P. (1996). Genetic analysis of dorsoventral pattern formation in the zebrafish: requirement of a BMP-like ventralizing activity and its dorsal repressor. *Genes Dev.* 10:2452–2461.

Heim, R., Cubitt, A. B., and Tsien, R. Y. (1995). Improved green fluorescence. *Nature (London)* 373:663–664.

Herskowitz, I. (1987). Functional inactivation of genes by dominant negative mutations. *Nature (London)* 329:219–222.

Ikawa, M., Kominami, K., Yoshimura, Y., Tanaka, K., Nishimune, Y., and Okabe, M. (1995a). Green fluorescent protein as a marker in transgenic mice. *Dev. Growth Differ.* 37:455–459.

Ikawa, M., Kominami, K., Yoshimura, Y., Tanaka, K., Nishimune, Y., and Okabe, M. (1995b). A rapid and non-invasive selection of transgenic embryos before implantation using green fluorescent protein (GFP). *FEBS Lett.* 375:125–128.

Ishibashi, M., Moriyoshi, K., Sasai, Y., Shiota, K., Nakanishi, S., and Kageyama, R. (1995). Persistent expression of helix–loop–helix factor HES-1 prevents mammalian neural differentiation in the central nervous system. *EMBO J.* 13:1799–1805.

Iyengar, A., Muller, F., and Maclean, N. (1996). Regulation and expression of transgenes in fish — a review. *Transgenic Res.* 5:147–166.

Johnson, A. D. and Krieg, P. A. (1994). pXex, a vector for efficient expression of cloned sequences in *Xenopus* embryos. *Gene* 147:223–226.

Kaplitt, M. G., Leone, P., Samulski, R. J., Xiao, X., Pfaff, D. W., O'Malley, K. L., and During, M. J. (1994). Long-term gene expression and phenotypic correction using adeno-associated virus vectors in the mammalian brain. *Nature Genet.* 8:148-154.

Krieg, P. A. and Melton, D. A. (1985). Developmental regulation of a gastrula-specific gene injected into fertilized *Xenopus* eggs. *EMBO J.* 4:3463–3471.

Kroll, K. L. and Amaya, E. (1996). Transgenic *Xenopus* embryos from sperm nuclear transplantations reveal FGF signaling requirements during gastrulation. *Development* 122: 3173–3183.

Levy, J. P., Muldoon, R. R., Zolotukhin, S., and Link, C. J. (1996). Retroviral transfer and expression fo a humanized, red-shifted green fluorescent protein gene into human tumor cells. *Nature Biotechnol.* 14:610–614.

Lin, S., Gaiano, N., Culp, P., Burns, J. C., Freidmann, T., Yee, J.-K., and Hopkins, N. (1994). Integration and germ-line transmission of a pseudotyped retroviral vector in zebrafish. *Science*: 666–669

Lo, D. C., McAllister, K., and Katz, L. C. (1994). Neuronal transfection in brain slices using particle-mediated gene transfer. *Neuron* 13:1263–1268.

Long, Q., Meng, A., Wang, H., Jessen, J. R., Farrell, M. J., and Lin, S. (1997) GATA-1 expression pattern can be recapitulated in living transgenic zebrafish using GFP reporter gene. *Development* 124:4105–4111.

Maclean, N. and Rahman, A. (1994). Transgenic fish. *In* Animals with Novel Genes, Maclean, N., Ed., Cambridge University Press, Cambridge, England, pp. 63–105.

Meisler, M. (1992). Insertional mutation of 'classical' and novel genes in transgenic mice. *TIGS* 8:341–344.

Meng, A., Tang, H., Ong, B. A., Farrell, M. J., and Lin, S. (1997) Promoter analysis in living zebrafish embryos identifies a cis-acting motif required for neuronal expression of GATA-2. *Proc. Natl. Acad. Sci. USA* 94:6267–6272.

Mohun, T. J., Garrett, N., and Gurdon, J. B. (1986). Upstream sequences required for tissue-specific activation of the cardiac actin gene in *Xenopus laevis* embryos. *EMBO J.* 5:3185–3193.

Morgan, B. A., Izpisua-Belmonte, J.-C., Duboule, D., and Tabin, C. J. (1992). Targeted misexpression of *Hox-4.6* in the avian limb bud causes apparent homeotic transformations. *Nature (London) 358:236–239*

Moss, B. (1991). Vaccinia virus: a tool for research and vaccine development. *Science* 252: 1662–1667.

Moss, J. B., Price, A. L., Raz, E., Driever, W., and Rosenthal, N. (1996). Green fluorescent protein marks skeletal muscle in murine cell lines and zebrafish. *Gene* 173:89–98.

Muldoon, R. R., Levy, J. P., Kain, S. R., Kitts, P. A., and Link, C. J. (1997). Tracking and quantitation of retroviral-mediated transfer using a completely humanized, red-shifted green fluorescent protein gene. *Biotechniques* 22:162–167.

Muzyczka, N. (1994). Adeno-associated virus (AAV) vectors: will they work? *J. Clin. Invest.* 94:1351.

Pecorino, L. T. and Lo, D. C. (1992). Having a blast with gene transfer. *Curr. Biol.* 2:30–32.

Peel, A. L., Zolotukhin, S., Schrimsher, G. W., Muzyckza, N., and Reier, P. J. (1997). Efficient transduction of green fluorescent protein in spinal cord neurons using adeno-associated virus vectors containing cell-type specific promoters. *Gene Therapy* 4:16–24

Potter, S. M., Wang, C.-M., Garrity, P. A. and Fraser, S. E. (1996). Intravital imaging of green fluorescent protein using two-photon laser-scanning microscopy. *Gene* 173:25–31.

Price, J., Turner, D., and Cepko, C. (1987). Lineage analysis in the vertebrate nervous system by retrovirus-mediated gene transfer. *Proc. Natl. Acad. Sci. USA* 84:156–160.

Rusconi, S. and Schaffner, W. (1981). Transformation of frog embryos with a rabbit β-globin gene. *Proc. Natl. Acad. Sci. USA* 78:5051–5055.

Schwartzberg, P., Goff, S. P., and Robertson, E. J. (1989). Germ-line transmission ofa c-abl mutation produced by targeted gene disruption in ES cells. *Science* 246:799–802.

Stacey, A., Bateman, J., Choi, T., Mascara, T., Cole, W., and Jaenisch, R. (1988). Perinatal lethal osteogenesis imperfecta in transgenic mice bearing an engineered mutant pro-α(I) collagen gene. *Nature (London)* 332:131–136

Stratford-Perricaudet, L. D. and Perricaudet, M. (1991). Adenovirus. *In* Human Gene Transfer, Cohen-Haguenauer, O., Boiron, M., and Libbey, J., Eds. Eurotext Ltd., London, pp. 51–61.

Stuart, G. W., McMurray, J., and Westerfield, M. (1988). Replication, integration and stable germ-line transmission of foreign sequences into early zebrafish embryos. *Development* 103:403–412.

Thompson, S., Clarke, A. R., Pow, A. M., Hooper, M. L., and Melton, D. W. (1989). Germline transmission and expression of a corrected HPRT gene produced by gene targeting in embryonic stem cells. *Cell* 56:313–321.

Westerfield, M., Wegner, J., Jegalian, B. G., DeRobertis, E. M., and Puschel, A. W. (1992). Specific activation of mammalian *Hox* promoters in mosaic transgenic zebrafish. *Genes Dev.* 6:591–598.

Wolff, J. A., Malone, R. W., Williams, P., Chong, W., Acsadi, G., Jani, A., and Felgner, P. L. (1990). Direct gene transfer in mouse muscle in vivo. *Science* 247:1465–1468.

Wu, G.-Y., Zou, D.-J., Koothan, T., and Cline, H. (1995). Infection of frog neurons with vaccinia virus permits in vivo expression of foreign proteins. *Neuron* 14:681–684.

Zernicka-Goetz, M., Pines, J., Ryan, K., Siemering, K. R., Haseloff, J., Evans, M. J., and Gurdon, J. B. (1996). An indelible lineage marker for *Xenopus* using a mutated green fluorescent protein. *Development* 122:3719–3724.

Zhou, S. Z., Cooper, S., Kang, L. Y., Ruggieri, L., Heimfeld, S., Srivastava, A., and Broxmeyer, H. E. (1994). Adeno-associated virus 2-mediated high efficiency gene transfer into immature and mature subsets of hematopoietic progenitor cells in human umbilical cord blood. *J. Exp. Med.* 179:1867–1875.

Zijlstra, M., Li, E., Saijadi, F., Subramani, S., and Jaenisch, R. (1989). Germline transmission of a disrupted β2-microglobulin gene produced by homologous recombination in embryonic stem cells. *Nature (London)* 342:435–438.

Zolotukhin, S., Hauswirth, W. W., Guy, J., and Muzyczka, N. (1996). A 'humanized' green fluorescent protein cDNA adapted for high level expression in mammalian cells. *J. Virol.* 70:4646–4654

12

The Uses of Green Fluorescent Protein in Mammalian Cells

JENNIFER LIPPINCOTT-SCHWARTZ
Department of Cell Biology and Metabolism, NICHD, NIH, Bethesda, MD

12.1 INTRODUCTION

The green fluorescent protein (GFP) from the jellyfish *Aequorea victoria* and its variants have been expressed as fusion products with other proteins in many cell types, and are unique fluorescent reporters for monitoring dynamic processes in cells and organisms (Chalfie et al., 1994). In mammalian cells, GFP chimeras have been used in a wide variety of applications including time-lapse imaging, double-labeling and photobleach experiments, to study the subcellular distribution, function, and expression of proteins (Table 12.1). They have also been used in gene transfer and expression studies, as well as in monitoring viral infection and pathogenesis. Future work with GFP chimeras promises to revolutionize our thinking about the dynamics of cytoplasmic, cytoskeletal and organellar proteins, and their intracellular associations.

12.2 GFP AND GFP VARIANTS

Green fluorescent protein and its variants have been expressed as fusion products with other proteins in many mammalian cell types including BHK, CHO, COS, GH3, HeLa, NIH 3T3, cc12, NRK, PTK1, and mouse cells (Olson et al., 1995; Cole et al., 1996; Rizzuto et al., 1996; Ogawa et al., 1995; Kaether and Gerdes, 1995). Several properties of GFP make it an excellent reporter molecule. Proper folding of GFP into a functional fluorophore, for example, does not require

Green Fluorescent Protein: Properties, Applications, and Protocols, Edited by Martin Chalfie and Steven Kain
ISBN 0-471-17839-X

TABLE 12.1 Applications of GFP in mammalian cells

Application	References
1. Gene targeting/expression	Phillips et al., 1995 Dhandayuthapani et al., 1995; Zolotukhin et al., 1996; Muldoon et al., 1997; Cheng et al., 1996; Zhang et al., 1996; Subramanian and Srienc, 1996; Levy et al., 1996; Bartlett et al., 1995; Moriyoshi et al., 1996; Peel et al., 1997.
2. Viral infection	Dorsky et al., 1996; Dhandayuthapani et al., 1995
3. Protein localization/dynamics	Endow et al, 1995; Ludin et al., 1996; Olson et al., 1995; Maniack et al., 1995; Moores et al., 1995; Waddle et al., 1996; Roberts et al., 1996; Rizzuto et al., 1995; Bauer et al., 1996; Cole et al., 1996; Htun et al., 1996; Ogawa et al., 1995; Stauber et al.,1995; Hanakam et al., 1996; Hampton et al., 1996; Lee et al., 1996; Terasaki et al., 1996; Pines, 1995; Marshal et al., 1995; Brown et al., 1995; Ellenberg et al., 1997.
4. Protein quantitation	Terasaki et al., 1996; Niswender et al., 1995; Potter et al., 1996; Patterson et al., 1997.
5. Protein trafficking	Presley et al., 1996; Cole et al., 1995; Kaether and Gerdes, 1995; Girotti and Banting, 1996; Conrad et al., 1996.
6. Double labeling with GFP color variants	DeGiorgi et al., 1996; Rizzuto et al., 1996; Cubitt et al., 1995; Yang et al., 1996
7. FRET	Heim and Tsien, 1996; Mitra et al., 1996;
8. Photobleaching (FRAP and FLIP)	Cole et al., 1996; Terasaki et al., 1996; Ellenberg et al., 1997.
9. Cell sorting/selection/lineage Analysis	Mosser et al., 1997; Ikawa et al., 1995; Analysis Niswender et al., 1995; Raz et al., 1996; Steele-Mortimer et al., 1995

specific cofactors or additional proteins from jellyfish (Chalfie et al., 1994; Heim et al., 1994; Inouye and Tsuji, 1994). Moreover, GFP fluorescence is very stable in the presence of denaturants and proteases, as well as over a range of pH and temperatures (Ward, 1981; Ward et al., 1982).

Wild-type GFP (wtGFP) absorbs ultraviolet (UV) and blue light with a maximum peak of absorbance at 395 nm and a minor peak at 470 nm, and emits green light at 509 nm with a shoulder at 540 nm (Ward et al., 1980). The detection limit of wtGFP is approximately 1 μM above cellular autofluorescence (Niswender et al., 1995). Random and site-directed mutagenesis have produced useful GFP mutants with single and multiple amino acid changes that are brighter and have excitation and emission spectra different from wtGFP (Heim et al., 1995; Cormack et al., 1996; Crameri et al., 1996; Delgrave et al., 1995; Ehrig et al., 1995; Zhang et al., 1996). A commonly used GFP variant that allows detection of lower fluorescence levels is S65T-GFP. This variant contains the amino acid substitution of Ser65 to Thr (Heim et al., 1995) and has a red-shifted excitation spectra whose fluorescent signal is several-fold greater than wtGFP. A similiar mutant to S65T is GFPmut1 (Cormack et al., 1996). Other red-shifted GFP variants have been introduced containing multiple amino acid substitutions, including the double mutant EGFP of Phe64 to Leu and Ser 65 to Thr (F64L/S65T) (Yang et al., 1996b) and the triple mutant αGFP (F100S/M154T/V164A) (Crameri et al., 1996). These GFP variants fluoresce many fold more intensely than wtGFP when excited at 488 nm. Ormö et al. (1996) reported an additional red-shifted mutant (T203/S65G/V68L/S72A), which is capable of limited spectral separation from other GFPs. Since the major excitation peak of red-shifted variants includes the excitation wavelength of commonly used filter sets and the emission wavelength of the argon ion laser used in fluorescence activated cell sorter (FACS) machines and the confocal scanning laser microscope (i.e., 488 nm), these variants are more efficient as well as brighter than wtGFP. All of these GFPs are more resistant to photobleaching than fluorescein when excited at 488 nm.

The GFP variants with blue-emission spectra include the double mutant P4-3 (Y66H/Y145F) (Heim and Tsien, 1996) and the blue-emission triple mutants BFP5 (F64M/Y66H/V68I) (Mitra et al., 1996) and EBFP (F64L/S65T/Y66H/Y145F) (Yang et al., 1998). When used together with red-shifted variants, these mutants allow multiple labeling of different protein species, which is useful in protein colocalization experiments (Yang et al., 1996b; Rizzuto et al., 1996). They also offer the potential for assessing differential gene expression by flow cytometry (Ropp et al., 1996) and for measuring protein–protein interactions through fluorescence resonance energy transfer analysis (FRET) (Heim and Tsien, 1996; Mitra et al., 1996).

12.3 OPTIMIZING GFP EXPRESSION IN MAMMALIAN CELLS

There has been great interest in optimizing the expression and brightness of GFP in mammalian cells. Optimal brightness has been obtained by the development of GFP variants with different folding, spectral, and photobleaching properties, while high expression levels of GFP fusion proteins have been obtained by optimal codon usage (Zolotukhin et al., 1996; Chiu et al., 1996; Yang et al., 1996a). The latter has been essential for studies using GFP in many types of mammalian cells.

As with any fluorescent probe, the extinction coefficient and quantum yield of GFP underlies its brightness and photostability (Patterson et al., 1997). Unlike other fluorescent probes, GFP is synthesized by the cell and must fold properly in order to be fluorescent. Its folding and structural properties thus must be considered in optimizing its expression and brightness. Conditions that interfere with or facilitate the folding of GFP have been investigated by many labs revealing optimal chromophore formation to be dependent on temperature and pH for many GFP variants (Patterson et al., 1997; Cubitt et al., 1995; Chattoraj et al., 1996).

Different GFP variants offer advantages and disadvantages for imaging in living cells. Of the available GFP mutants, S65T and EGFP appear most useful for optimal brightness and photostability. Their intrinsic brightness at 488 nm allows detection of approximately 200 nM above average cellular autofluorescence (or ~10,000 GFP molecules in the cytoplasm) (Patterson et al., 1997). For GFP molecules that are spatially localized within the cell (e.g., in a small membrane area), however, as few as 10–20 molecules is sufficient for generating a detectable signal (Niswender et al., 1995).

A major drawback of wt and αGFP for quantitative studies using confocal microsocopy is that they exhibit excitation-induced photoconversion of their chromophore (Chalfie et al., 1994; Chattoraj et al., 1996; Cubitt 1995; Patterson et al., 1997). During the photoconversion process, the absorption at 397 nm of these variants decreases while the 475-nm absorption increases. This change makes data analysis of quantitative signals very difficult. Thus, even though EGFP and S65T photobleach about two times faster than wtGFP at 488 nm (but five times more slowly than fluorescein; Patterson et al., 1997), they are more useful in quantitative imaging studies, because they do not photoconvert.

In studies using blue-emission GFP variants, which excite with near-UV light, EBFP is more difficult to use than wtGFP or αGFP variants. This difficulty is because EBFP at 380 nm has decreased photostability and low quantum yield relative to the wtGFP or αGFP variants at 397 nm (Patterson et al., 1997). Nevertheless, as a second label used in conjunction with a green GFP, EBFP is superior since its emission spectra (peak 440 nm) is easily separated from other GFP variants (504 nm for wtGFP and 506 nm for αGFP). Moreover, overlap between EBFP emission and absorption of S65T and EGFP is likely to be useful in FRET experiments (Heim and Tsien 1996; Mitra et al., 1996).

Several studies have revealed a requirement for incubations at temperatures less than 37°C to produce optimal fluorescence (Kaether and Gerdes, 1995; Ogawa et al., 1995). Investigation of temperature effects on GFP folding (Patterson et al., 1997) found that many GFP variants are brighter when expressed at 28°C than at 37°C. This difference is due to permanent chromophore deformation at higher temperatures rather than simple misfolding of the protein (Patterson et al., 1997). Two variants, αGFP and EGFP, do not have this temperature dependence (Patterson et al., 1997) and therefore may be preferable to wtGFP and S65T for imaging in living cells and in organisms where chromophore formation at physiological temperatures is necessary.

In addition to temperature, optimal chromophore formation of GFP also depends on pH. Fluorescence of wtGFP is stable from pH 6–8, but fluorescence of S65T, EGFP, and EBFP is significantly reduced at pH values below 7 (Patterson et al., 1997). Caution, therefore, should be made using these GFP variants for cellular imaging of structures with acidic pH levels, including endosomes and lysosomes that have pH values as low as pH 4.6 (Mellman et al., 1986). At such low pH levels, GFP signals may be very dim.

12.4 APPLICATIONS OF GFP IN MAMMALIAN CELLS

Green fluorescent protein technology has been used in a wide variety of applications for monitoring reporter proteins expressed in living mammalian cells. Addition of GFP to proteins is usually benign with no apparent disruption of function, despite its relatively large size. Since no exogenously added substrate or cofactors are necessary for detecting GFP fluorescence, cells are exposed to minimal invasive treatment. Applications using GFP range from development of gene therapy vectors to studies of protein localization/dynamics (Table 12.1).

12.4.A Gene Targeting Using Viral Vector Systems

Viral vectors expressing GFP are very powerful tools for gene transfer and expression studies (Muldoon et al., 1997). The GFP has been used as a reporter gene for retroviral-mediated stable gene transfer and expression (Cheng et al., 1996). Viable GFP expressing cells were selected by FACS based on their fluorescence signal with the brightest cells sorted and expanded for cloning. Selection of different populations of stably expressing cell lines (including high and low titer viral producer clones) using this approach was simple and efficient. GFP expression vectors of this type delivered into different mammalian cells have been shown to give bright cellular fluorescence within 16–24 h of transfection with the relative sensitivity of GFP greater than that of β-galactosidase catalyzed conversion of the X-gal substrate (Zhang et al., 1996). Given optimal temperature and pH, nonfluorescent pools of GFP generally are found only during the initial periods of gene expression (Subramanian and Srienc, 1996), presumably due to folding requirements of the GFP molecule.

Retroviral transfer of GFP also promises to be an invaluable tool for studying the fate of normal and tumor cells in gene therapy protocols. Levy et al. (1996) showed retroviral vectors containing a red-shifted GFP gene could efficiently be transferred into human tumor cells and murine fibroblasts. Stable bright green fluorescence could be detected by fluorescence microscopy or FACS analysis after several days.

The GFP viral vector systems have been used to monitor production and release of therapeutic molecules from cells and tissues. Bartlett et al. (1995) developed an adenovirus vector delivery system using GFP inserted downstream from the human muscle creatine kinase promoter and found efficient GFP expression in skeletal muscle injected with the vector. In contrast to traditional reporter methods including β-galactosidase, firefly luciferase, or chloramphenicol amino transferase (CAT) assays (which require cell lysis), GFP expression could be monitored consecutively over several days.

In addition to the enomorous potential of viral reporter genes carrying GFP in clinical studies for tracking the expression of gene products, such molecules also are extremely valuable for basic research. Moriyoshi et al. (1996), for example, used an adenovirus vector to transfer GFP into postmitotic neuronal cells *in vivo* to study cell migration and development of neuronal connections. Adeno-associated virus vectors expressing GFP were also used to target GFP to spinal neurons (Peel et al., 1997), allowing the fate of neurons to be followed and their response to various transducers analyzed.

12.4.B Viral Infection and Pathogenesis

Viral vectors containing GFP can be used to monitor viral infection and pathogenesis with no need for processing of cells to detect infected cells. Using this approach, Dorsky et al. (1996) identified human immunodeficiency virus (HIV)-1 infected cells in tissue using GFP tagged HIV-1. GFP under the control of HIV-1 LTR promoter was readily detected in virally infected cells either by fluorescence microscopy or by fluorescence-activated cell sorting. With the same goal in mind, Dhandayuthapani et al. (1995) used a mycobacterial shuttle-plasmid vector carrying GFP cDNA to assess mycobacterial interactions with macrophages.

12.4.C Flow Cytometry

Screening and selection of cells by flow cytometry can be greatly facilitated using GFP expression, since it provides an easy method for fluorescent labeling of viable cells. This method eliminates the task of characterizing cell lines through standard biochemical methods involving protein analysis. As an example, Mosser et al. (1997) generated a dicistronic mRNA encoding both a gene of interest and the gene for GFP. Clone selection involved the simple monitoring for GFP fluorescence using the fluorescence-activated cell sorter. Quantitative detection from two different genes within single mammalian cells has also been achieved with GFP and its red-shifted variant using multiparameter flow cytometry (Anderson et al., 1996).

12.4.D Protein Localization

Green fluorescent protein chimeras provide a major advance over previous methods for studying the intracellular localization and dynamics of proteins. Most other techniques currently available require fixation and permeabilization methods to gain access within the cell to the protein of interest. Numerous problems in specimen preparations could arise as a result of fixation including the danger of extracting or damaging antigen, the possibility that antigen might redistribute to an artifactual location, and the possibility that labeling efficiencies within different cell structures will differ. With GFP chimeras these problems are avoided because the protein of interest is viewed in a living, unperturbed cell. The GFP reporter, itself, usually does not interfere with the normal functioning of the tagged protein and can be added to either the C- or N-terminus of target proteins. Since the GFP fluorophore is relatively photostable, little photodamage occurs during imaging. These properties make GFP an ideal reporter protein.

12.4.E Protein Quantitation

In addition to offering a simple way to localize proteins within living cells, use of GFP fusion proteins allows accurate quantitatation of proteins. The GFP attached to protein A, for instance, has been used for western blotting and provides a sensitive and specific assay system for rapid and easy quantitative screening of proteins (Aoki et al., 1996). The concentration of GFP chimera expressed within cells can be quantitated by comparing the intensity of its fluorescence with fluorescence of a known standard, for example, fluorescein, using a sensitive camera system (e.g. a cooled CCD camera) (Terasaki et al., 1996; Niswender et al., 1995; Potter et al., 1996). Much of this work with GFP in mammalian cells uses the S65T variant (Heim et al., 1995; see Chapter 5). Since the extinction coefficient and quantum yield for S65T GFP are 39,200 $M^{-1}cm^{-1}$ and 0.68 (Heim et al., 1995), while for fluorescein they are 75,000 $M^{-1}cm^{-1}$ and 0.71 (Tsien and Waggoner, 1995), a known population of fluorescein-labeled molecules will be two times as fluorescent as the same population of GFP chimeras. Terasaki et al. (1996) used this approach to measure the quantity of GFP-KDEL protein in sea urchin oocytes, and to determine the rate of synthesis of GFP-KDEL proteins. By microinjecting a known amount of fluorescent dextran as a reference standard they estimated that 2.2 million GFP-KDEL molecules per minute were synthesized in a single oocyte. A similar process could be used with mammalian cells.

12.4.F Fluorescence Resonance Energy Transfer

The GFP chimeras also have been used to examine the proximity of different protein domains by FRET in solution (Heim and Tsien, 1996; Mitra et al., 1996). Fluorescence resonance energy transfer measures the transfer of photon energy from one fluorophore to another molecule when both are located within a few nanometers of each other (Stryer, 1978; Tsien et al., 1993; Uster and Pagano, 1986). If the energy of the excited fluorophore coincides with the energy needed to excite the absorber, then energy is transferred. This transfer results in loss of fluorescence intensity of donor and fluorescence emission from the acceptor. The working scale of FRET is less than or equal to 100 Å, in contrast to conventional light microscopy, which is a few tenths of a micron (Stryer, 1978). The availability of several different mutants of GFP opens the possiblity of using FRET to probe inter- and intramolecular distances in proteins *in vivo*. Heim and Tsien (1996) attached the GFP mutants, Y66H/Y145F and S65C, to the same protein by a 25 residue cleavable spacer and used the first as donor and the second as acceptor in FRET experiments. Proteolytic cleavage of the spacer resulted in the two protein domains diffusing apart, causing loss of green emission by the acceptor S65C domain and enhancement of blue emission from the donor domain. The FRET applied to GFP chimeras promises to greatly enhance our understanding of the spatial distribution of proteins and will enable monitoring of protein-protein associations in living cells.

12.4.G Time-Lapse Imaging

In addition to being an important tool for localizing proteins and studying their proximal interactions *in vivo*, GFP chimeras can be used to examine protein dynamics, including the lifetime, sorting, and intracellular pathways of proteins. Previous work examining the dynamics of proteins within cells and their response to cellular perturbations (including drug treatments, temperature shifts, and microinjection of antibodies) relied on static images of large populations of cells. Piecing together a specified cellular response with such a "snapshot" approach to imaging cells is often difficult. GFP chimeras, by contrast, allow one to follow the dynamics of proteins in real time in a single cell and in so doing have provided new perspectives on protein interactions and transport pathways (see Section 12.5).

12.4.H Double Labeling/Ratio Imaging

Use of GFP mutants that fluoresce or are excited at different wavelengths offers the possibility of double labeling to compare the distribution and dynamics of two different populations of proteins simultaneously within cells (Heim et al., 1994; Cubitt et al., 1995; Rizzuto et al., 1996; Yang et al., 1996b; De Giorgi et al., 1996; Stauber et al., 1997) (Fig. 12.1). For this method, cells are doubly transfected with proteins attached to different GFP variants that have different excitation or emission spectra and are imaged with alternative filter sets (Rizzuto et

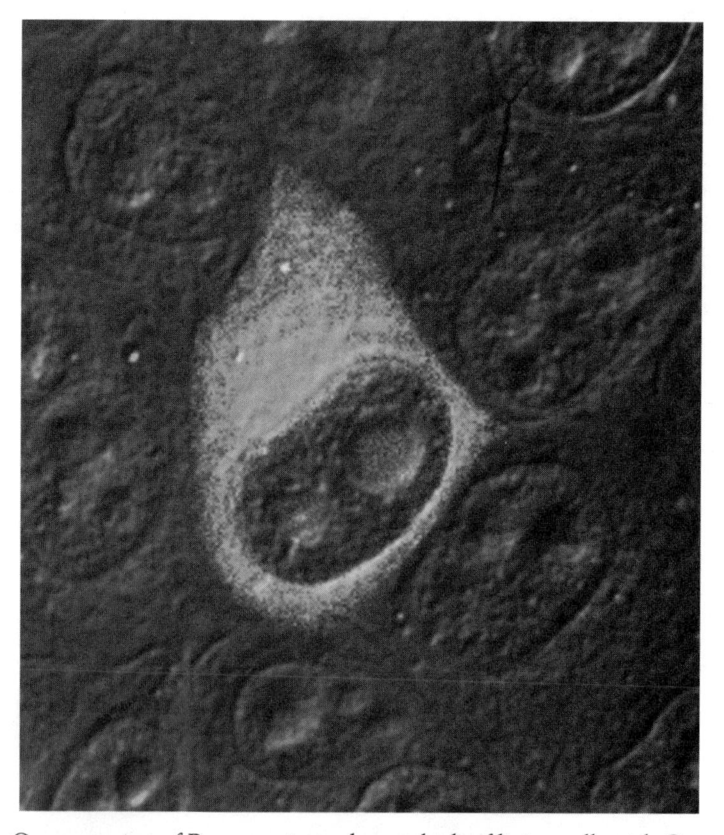

FIG 12.1 Co-expression of Rev protein in the nucleoli of living cells with Gag protein in the cytoplasm using two different mutants sg25 and sg50, developed by George Gaitanaris and Roland Stauber. Picture was taken by Roland Stauber and Eric Hudson. See Stauber et al. (1998). Figure also appears in color section.

al., 1996). Pairs of images can be quantitated using digital image processing techniques to see if the ratio of intensity of the two populations changes with time. Such ratio imaging approaches have already been standardized and used with rhodamine and fluorescein tags in the endosomal system (Mayor et al., 1993), and promise to be an important application of GFP variants.

12.4.I Photobleaching Recovery

The mobility and environment of GFP tagged proteins within cells can also be studied using fluorescence recovery after photobleaching (FRAP) (Cole et al., 1996). Until now, experimental access to intracellular membrane proteins has been blocked by the plasma membrane, making it extremely difficult to study the diffusional mobilities of these proteins. Since GFP chimeras are expressed endogenously, permeabilization or microinjection techniques to label intracellular sites for photobleaching experiments are not needed. In FRAP, fluorescent proteins in a small area (~2 μm) are irreversibly bleached by an intense laser flash and recovery is measured using an attentuated laser beam (Edidin, 1994; Fig 12.2). Mobility parameters are then derived from the kinetics of fluorescent recovery. The immobile fraction of fluorescent proteins, for example, equals the difference in complete versus incomplete recovery. Several types of interactions have been shown by FRAP to constrain or immobilize the lateral diffusion of proteins (Edidin, 1992). GFP chimeras are excellent probes for FRAP studies, since no evidence of photodamage or photoinduced cross-linking of the GFP chimeras has been seen (Cole et al., 1996). An additional photobleaching technique, fluorescence loss in photobleaching, (FLIP), has been used with GFP chimeras to examine the extent to which regions outside a photobleached region can contribute to recovery of fluorescence at a bleached site (Cole et al., 1996). This approach allows the extent of continuity within a given compartment to be studied and provides a means for defining compartmental boundaries within cells.

The above list emphasizes the broad applications of GFP technology and why GFP chimeras have become powerful tools for uncovering numerous dynamic cellular events. The following sections highlight examples where GFP chimeras have provided interesting new insights into *in vivo* protein function and dynamics (see Fig. 12.3).

12.5 PROTEIN FUNCTION AND DYNAMICS REVEALED FROM GFP FUSION PROTEINS

12.5.A Cytoskeletal Interactions and Dynamics

Proper organization and assembly of cytoskeletal elements is required for many cellular processes including membrane transport, mitosis, and the establishment of cell shape and polarity. To function in such diverse processes, cytoskeletal elements undergo dramatic changes and are extensively regulated. Normal microtubule organization, for example, is believed to be dynamically unstable,

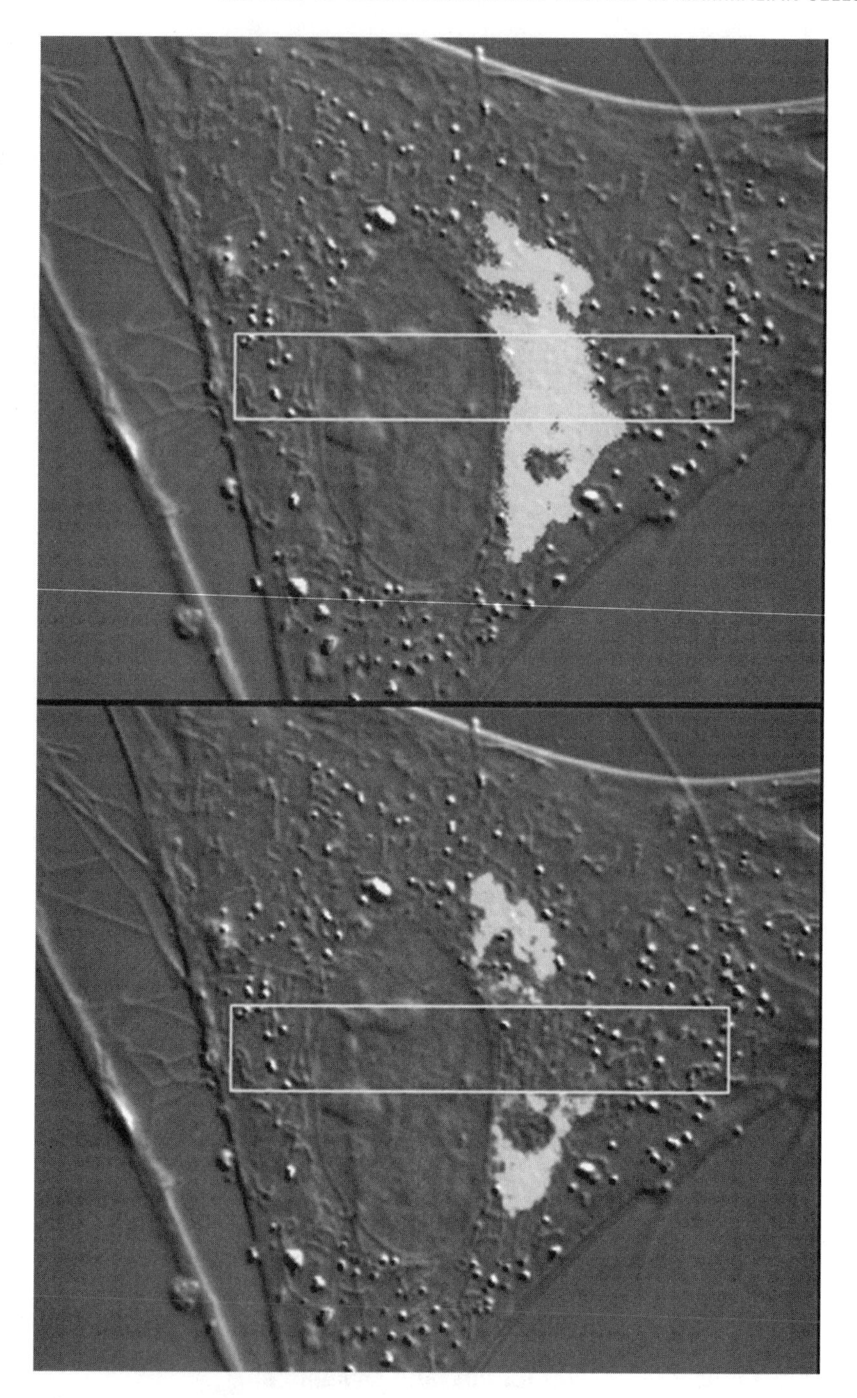

FIG 12.2 The HeLa cell expressing GFP Golgi chimera before (above) and immediately after (below) photobleaching of a small rectangular box. [Courtesy of Cole et al., 1996]. Figure also appears in color section.

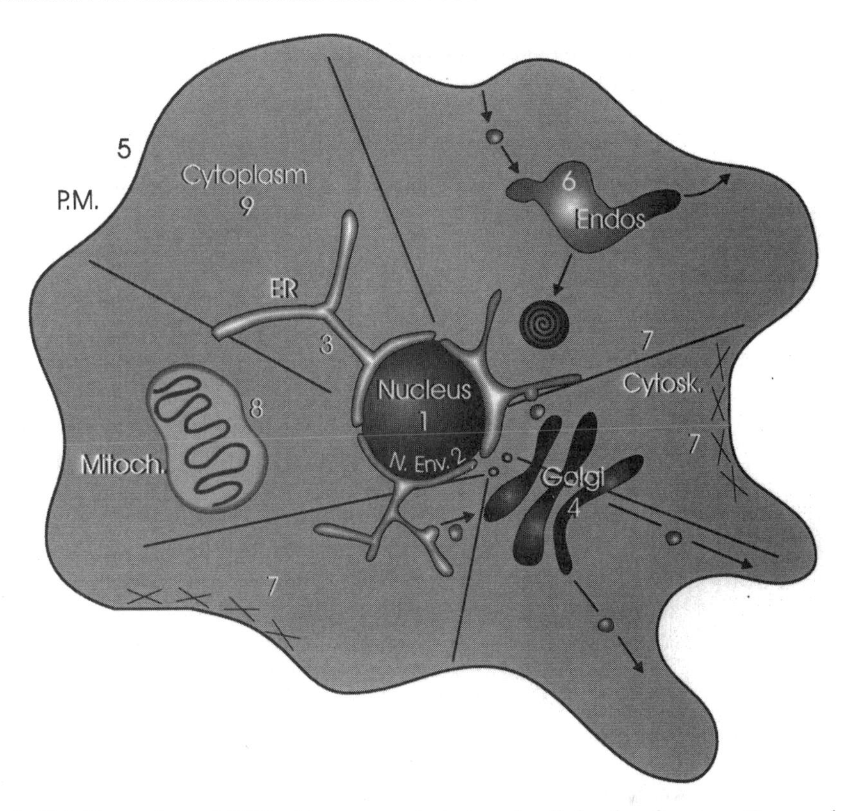

FIG 12.3 Examples of GFP chimeras and their subcellular localization. Figure also appears in color section.

with rapid changes between phases of depolymerization and regrowth. Given the ease of localizing and following GFP chimeras in real time, answers to a host of questions regarding the dynamics and interactions of the cytoskeletal system are now being addressed using GFP as a reporter.

Several laboratories have used GFP chimeras to study microtubule dynamics. Ludin et al. (1996) and Kaech et al. (1996) generated GFP chimeras of Map2C and Tau 34 to study microtubules. Time-lapse recordings of living cells revealed overexpression of Map2 and Tau did not affect rates of microtubule growth and shrinkage. The MAP induced microtubule bundles, however, were found to influence changes in cell shape and thus are likely to contribute to the morphological plasticity of many cell types. Olson et al. (1995) used the behavior of Map4-GFP chimeras to analyze the effects of MAP4 and its different subdomains on microtubule dynamics in living cells. Map4 is an ubiquitous microtubule-associated protein thought to facilitate the polymerization and stability of microtubules in interphase and mitotic cells (Olmsted et al., 1991). Different subdomains of Map4 that were tagged with GFP and expressed in cells had distinct effects on microtubule organization and dynamics. The entire basic domain of Map4, for example, reorganized microtubules into bundles and stabilized these arrays against depolymerization with nocodazole. Interestingly, sequences outside the basic domain were found to affect the microtubule binding characteristics

TABLE 12.2 Examples of GFP Chimeras and their Subcellular Localization in Mammalian Cells

Compartment	Chimera	Reference
1. Nucleus	Glucocorticoid receptor	Htun et al., 1996
		Ogawa et al., 1995
		Carey et al., 1996
	Rev protein	Stauber et al., 1995
	VHL protein	Lee et al., 1996
	Hisactophilin	Hanakam et al., 1996
	Calreticulin	Roderick et al., 1997
	CENP-B	Shelby et al., 1996
2. Nuclear envelope	Lamin B receptor	Ellenberg et al., 1997
3. Endoplasmic reticulum	KDEL	Terasaki et al., 1996
	11β-HSD	Bauer et al., 1996
	HMG-CoA reductase	Hampton et al., 1996
	α-1protease inhibitor	Brown et al., 1995
	tso45 VSVG protein	Presley et al., 1996
4. Golgi complex	KDEL receptor	Cole et al., 1995
	Mannosidase II	Cole et al., 1996
	Galactosyltransferase	Cole et al., 1996
	TGN 38	Girotti and Banting, 1996
	hCgB	Kaether and Gerdes, 1995
		Wacker et al., 1997
5. Plasma membrane	GLUT4	Tavare (unpublished results)
	GPI-proteins	Conrad et al., 1996
	GAP 43	Moriyoshi et al., 1996
6. Endosomes	rab5	Roberts et al., 1996
7. Lysosome	MHC Class II	Wubbolts et al., 1996
8. Cytoskeleton	Map4	Olson et al., 1995
	Map2c	Ludin et al., 1996
	Tau34	Kaech et al., 1996
9. Mitochondria	Mitochondrial import signal	Rizutto et al., 1995
	Raf-1	Wang et al., 1996
	Ornithine transcarbamylase	Yano et al., 1997
10. Cytoplasm	mRNA	Philips et al., 1995
	cyclins A and B	Pines, 1995
	Salmonella	Steele-Mortimer et al., 1995
	mycobacteria	Dhandayuthapani et al., 1995
	HIV-1 protein, VP22	Elliot and O'Hare, 1997
11. Peroxisomes	Peroxisome targeting signal 1	Weimer et al., 1997
	Peroxisome targeting signal 1 and 2	Kalish et al., 1996

of other domains. Since microtubules labeled with rhodamine-tubulin were simultaneously visualized with the GFP chimeras after microinjection of rhodamine-tubulin into cells, the relative affinity of different GFP-Map4 variants for microtubules could also be analyzed.

12.5.B Nuclear Targeting and Function

The mechanisms involved in translocation, targeting, and retention of nuclear proteins are poorly understood. Recent studies using GFP tagged nuclear targeted proteins promise to be invaluable tools for exploring the organization, function, and architecture of the interphase nucleus and have already provided important clues into the mechanisms involved in these processes. An excellent example is the application of GFP to the study of steroid receptors, which are hormone-dependent activators of gene expression that normally reside in the cytoplasm, but upon hormone binding translocate into the nucleus where they activate genes. Since nuclear translocation and targeting of steroid receptors to regulatory sites in chromatin is not well understood, the ability to follow ligand dependent changes in receptor affinity in single cells over extended periods of time with GFP receptor chimeras is extremely worthwhile.

Several groups have attached GFP to glucocorticoid receptor (GR) in order to study the process of nuclear targeting and translocation (Htun et al., 1996; Ogawa et al., 1995; Carey et al., 1996). GR-GFP was found dispersed throughout the cytoplasm in its unbound state and translocated to the nucleus upon hormone binding where it was transcriptionally active. Time-lapse videomicroscopy of this redistribution process by Hager and colleagues revealed an unusual intranuclear architecture that included GR-GFP in discrete foci that were excluded from nucleoli (Htun et al., 1996) (Fig 12.4). Three-dimensional (3D) reconstruction of the nuclear GR binding sites by these authors revealed a reproducable nuclear organization of target sites. Interestingly, activation with hormone ago-

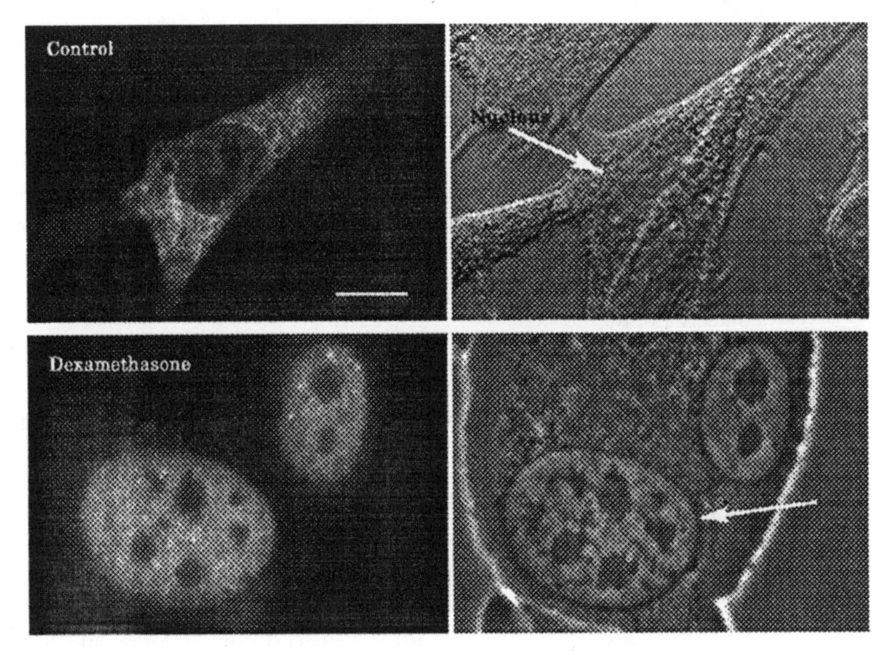

FIG 12.4. Distribution of GFP tagged glucocorticoid receptors before and after hormone binding. [Reprinted with permission from Htun et al., 1996].

nist gave subnuclear focal localization, while induction with a hormone antagonist gave complete translocation but no subnuclear targeting. Htun et al. (1996) also used a tandem array of integrated reporter genes to see targeting of GR-GFP to an actual gene locus in living cells. These results provide the first evidence of nuclear compartmentalization of steroid receptors. More recently, Roderick et al. (1997) provided evidence for the interaction of GRs with calreticulin, a calcium binding protein, by showing enhanced nuclear targeting of GFP-calreticulin in dexamethasone-treated cells.

Carey et al. (1996) used GR-GFPs to analyze nuclear import and tested the effects of various components of the nuclear transport machinery. Specifically, they examined the effect of several Ran mutants on nuclear translocation of GR-GFP. Ran is a GTPase that is thought to mediate nuclear translocation. Cytosolic GR-GFP expressed in BHK cells rapidly translocated into the nucleus (half-time 5 min) when dexamethasone was added. Coexpression of Ran mutants that were constitutively inactive, however, dramatically reduced the accumulation of GR-GFP in nuclei, while an effector domain mutant did not. These results are the first to relate data obtained in in vitro assays to nuclear transport of endogenous cargo in living cells.

Stauber et al. (1995) analyzed nucleus to cytoplasm trafficking of Rev and transdominant Rev proteins using GFP chimeras (Fig.12.1). Rev protein is expressed by HIV and regulates expression of *gag/pol* and *env* genes. Transdominant mutants of Rev protein (TDRev) inhibit Rev function. GFP was fused to Rev and TDRev to study the trafficking and interactions of these proteins in living human cells. Both GFP tagged proteins retained nucleolar localization and function. Upon actinomycin D treatment, however, Rev-GFP was transported to the cytoplasm within 1.5 h, while TDRev was retained within the nucleus. Significantly, TDRev inhibited transport of Rev-GFP from the nucleus to the cytoplasm when RevGFP and TDRev were coexpressed. This result suggested that Rev and TDRev formed heteromultimers in the nucleolus that prevented the export of Rev from the nucleus to the cytoplasm.

Lee et al. (1996) used GFP chimeras to analyze the cellular localization of the von Hippel–Lindau (VHL) tumor suppressor gene product. Germline mutation of the VHL gene is associated with the inherited VHL cancer syndrome. Interestingly, the subcellular localization of VHL gene product is dependent on the density at which cells are grown, with pVHL accumulating in the nuclei of sparsely populated cells and redistributing into the cytosol as cells grow to confluence. Lee et al. (1996) found that GFP tagged pVHL shuttles back and forth from the cytosol to the nucleus in sparse but not in confluent conditions. Addition of the transcription inhibitor actinomycin D caused a rapid increase of GFP-pVHL in the nucleus of sparsely grown cells. These results suggest that pVHL is retained in the cytosol in confluent cells. Under sparse conditions, pVHL is not retained and can migrate between the nucleus and cytosol in a transcription-dependent manner.

Protein shuttling from the plasma membrane to the nucleus using GFP tagged hisactophilins was studied by Hanakam et al. (1996). Histactophilins are myristolated proteins that are rich in histidine residues and bind actin and membrane proteins. Lowering the pH of the cell caused nuclear translocation of GFP hisactophilin in less than 1 min. Since DNA replication is extremely pH sensitive (Moolenar, 1986), Hanakan et al. (1996) proposed that hisactophilin or related proteins maintain pH homeostasis by buffering the intranuclear space. The rapid nuclear translocation of these proteins make them valuable tools for studying nuclear import in living cells.

The motile and mechanical properties of centromeres, which hold daughter chromatids together in mitotic chromosomes, has been studied by Shelby et al. (1996) in mammalian cells. CENP-B-GFP fusion protein that targets to centromeres was used in time-lapse imaging studies to follow centromeres during mitosis. Centromeres were found to elongate during mitosis in a microtubule-dependent manner before dispersing through the nucleus during chromatin decondensation in telophase.

12.5.C Membrane Trafficking and GFP Chimeras

Significant advances have been made over the last 10 years in elucidating biochemical mechanisms of membrane transport. The morphological counterparts, including the nature of transport intermediates and how they translocate through the cytoplasm, however, have remained elusive, since morphological characteristics of secretory traffic have been based largely on static electron microscopic and immunofluorescence images taken from fixed specimens.

Several recent studies following the trafficking of GFP tagged soluble and membrane components through the secretory pathway offer new perspectives on the properties of intracellular membrane traffic. As one example, the nature of the transport intermediates that carry proteins from the ER to the Golgi complex has been controversial, some studies suggest that a stable intermediate compartment serves as a source for vesicles carrying cargo to the Golgi complex; while other studies propose that the intermediate compartment itself are the transport vehicles (analogous to endosomes). To resolve this issue Presley et al. (1997) used GFP tagged ts045 VSVG protein to study transport intermediates from the ER to the Golgi. The ts045 VSVG protein is a viral membrane glycoprotein that misfolds and accumulates in the ER at 40°C but trafficks through the secretory pathway at 32°C (Bergmann, 1989). When the temperature was lowered to 15°C for 3 h, VSVG-GFP redistributed into large pre-Golgi intermediates that were localized at multiple peripheral sites and contained the peripheral coat protein known as βCOP (Pepperkok et al., 1993) (Fig 12.5). These structures have previously been identified as the intermediate compartment between the ER and Golgi complex. Upon warming cells from 15 to 32°C, Presley et al. (1997) found that pre-Golgi intermediates containing VSVG-GFP translocated along microtubule tracks toward the Golgi complex at speeds up to 1 $\mu m/s^{-1}$. Long tubule processes extended from these structures before movement began. Disruption of microtubules blocked their movement and resulted in a large accumulation of VSVG-GFP in stationary, peripheral pre-Golgi structures. These results favor a model of ER to Golgi transport involving intermediates that translocate along microtubules minus end-directed before fusing with the Golgi complex.

Kaether and Gerdes (1995) tagged the secretory protein chromogranin B with GFP to study its movement from the Golgi to the plasma membrane. When cells were arrested at 20°C for several hours, a large pool of GFP chromogranin B

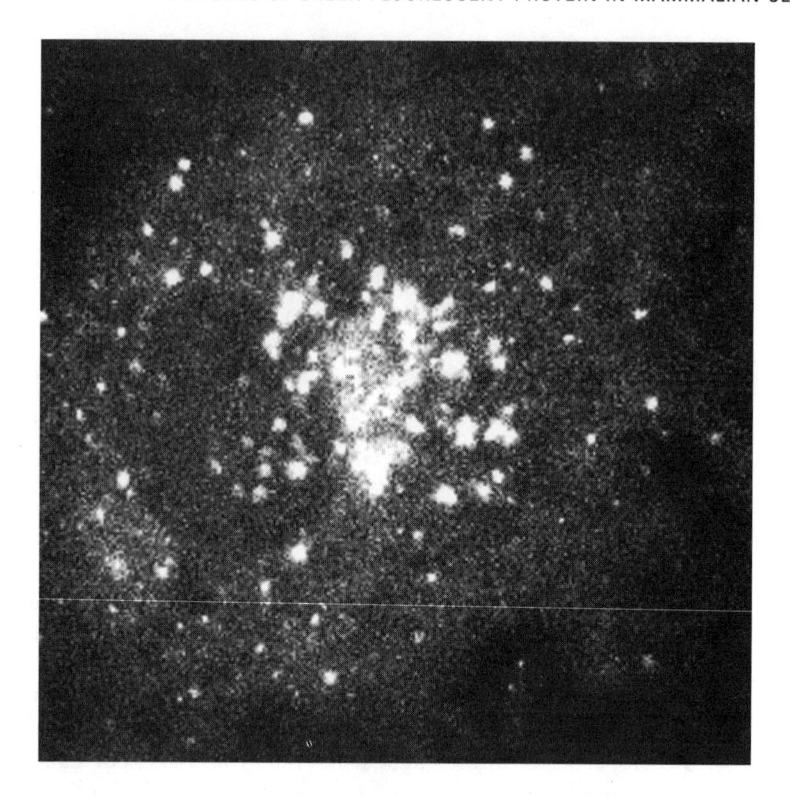

FIG 12.5. The GFP tagged VSVG protein expressed in Cos cells at 15°C and its colocalization with βCOP protein. Green is VSVG-GFP, red represents βCOP antibody staining, and yellow represents the overlap of the two. [Courtesy of Presley et al., 1996]. Figure also appears in color section.

accumulated in the trans Golgi network (TGN). This pool moved rapidly to the cell surface when the cells were warmed to 37°C. Time-lapse imaging of this redistribution process allowed the visualization of transport intermediates (Wacker et al., 1997). These intermediates moved bidirectionally along microtubules enroute to the plasma membrane at maximal velocities of 1 $\mu m\ s^{-1}$. The transport intermediates were devoid of TGN 38 and transferrin as well as markers of the endosomal/lysosomal pathway. GFP-chromogranin B within the structures was released into the extracellular millieu as a fluorescent protein within 20–30 min after transport from the TGN.

Studies using GFP chimeras have also begun to provide insight into mechanisms of vesicle fusion and budding. Rab proteins are small GTP binding proteins involved in vesicle targeting and fusion (Zerial and Stenmark, 1993). Using GFP-tagged rab proteins, Stahl and co-workers studied vesicle fusion in living cells (Roberts et al., 1996). A GTPase defective rab5 mutant, rab5:Q79L, was tagged with GFP and expressed in CHO and BHK cells. Giant cytoplasmic vesicles, characteristic of early endosomes, were formed that had immunoreactivity for GFP, rab5 and transferrin receptor. By time-lapse videomicroscopy, these enlarged endosomal vesicles were found to be produced by fusion of smaller vesicles through two distinct mechanisms that differed in temporal and spatial characteristics. "Explosive" fusion was rapid and was characterized by fusion pores expanding rapidly before the docked vesicles coalesced. "Bridge" fusion, by contrast, was relatively slow and required several seconds for completion of vesicle coalescence. Transfer of membrane always occurred through membrane bridges, where rab5-GFP was enriched, and proceeded from the smaller to the larger vesicle. These results are unexpected and provide important clues for the role of rab5 in endosomal fusion.

12.5.D Organelle Dynamics

GFP chimeras have been a powerful tool for visualizing subcellular organelles *in vivo*. The first application of this type analyzed mitochondrial dynamics and targeting. Rizzuto et al. (1995) attached GFP to a mitochondrial import signal and found extensive labeling of mitochondria. With this reporter they studied changes in mitochondrial morphology under a variety of physiological and pathological conditions, including drugs that collapse the mitochondrial membrane potential (Venerando et al., 1996). They also looked at mitochondrial intermixing and fusion during cellular fusion and heterokaryon formation, and performed double labeling of mitochondria and nucleus (Rizzuto et al., 1996). More recently, Yano et al. (1997) used a GFP ornithine trans fusion protein to analyze mitochondral protein import and its requirements in living cells. To probe mitochondrial targeting mechanisms and function, Wang et al. (1996) used a GFP-Raf-1 fusion protein, which phosphorylates substrates involved in the regulation of apoptosis. They demonstrated that Bcl-2 targets GFP-Raf-1 to mitochondrial membranes. Within mitochondria, kinase activation of Raf-1 could phosphorylate BAD or other protein substrates involved in apoptosis.

GFP tagged chimeras have been used to study the morphology and dynamics of the ER. The ER is comprised of an extensive array of interconnecting membrane tubules and cisternae, which extend throughout the cell. Its diverse functions include: synthesis, folding, assembly, and degradation of proteins; biosynthesis and metabolism of lipids; and, detoxification, compartmentalization of the nucleus, regulation of ion gradients, and membrane transport (Lippincott-Schwartz, 1994). Using photobleaching recovery techniques, Cole et al. (1996) showed that GFP chimeras in the ER of HeLa cells diffuse rapidly throughout the entire compartment, including the nuclear envelope.

GFP chimeras have begun to provide insights into the role of the ER in detoxification, lipid metabolism, and protein degradation. Bauer et al. (1996) localized GFP tagged 11 β-hydroxysteroid dehydrogenase (11β-HSD) exclusively on the cytoplasmic surface of the ER. Since 11β-HSD2, which confers aldosterone specificity to mineralocorticoid target cells by protecting the mineral corticoid receptor from occupancy by endogenous glucocorticoids, does not contain any known ER retrieval signal, use of different GFP chimera constructs of this protein should help reveal the responsible motifs. Additional studies by Brown et al. (1995) and Hampton et al. (1996) looked at constitutive and regulated protein degradation in the ER, respectively, using GFP tagged α1-proteinase inhibitor or hydroxymethlyglutaryl-CoA reductase. These studies set the stage for more detailed work focusing on the molecular basis of ER degradation.

The nuclear envelope is continuous with the ER and functions to separate genetic material in the nucleus from protein synthesis machinery in the cytoplasm. The outer nuclear envelope contains ribosomes on its surface and connects with the rough ER, while the inner nuclear envelope is associated with chromatin and the nuclear lamina. Regulation of RNA and protein transport through the pore complexes of the nuclear envelope is essential for gene expression and DNA replication. The GFP tagged proteins that target to the nuclear envelope and pore complex promise to provide tremendous insights into the properties of these structures, as well as their assembly and disassembly during the cell cycle. Recently, Ellenberg et al. (1997) tagged GFP to lamin B receptor (Fig. 12.6), a protein of the inner nuclear membrane (Schuler et al., 1994), and used photobleaching recovery techniques to examine its diffusional mobility in interphase and mitotic cells. Their results showed lamin B receptor is immobilized in the inner nuclear envelope during interphase, but highly mobile within ER membranes during mitosis.

GFP chimeras have also been made with membrane proteins residing within the Golgi complex, an organelle that plays an important role in the processing and sorting of newly synthesized proteins that are exported out of the ER system. How proteins comprising the Golgi complex are retained in this organelle despite a continuous flow of protein and lipid through the secretory pathway is not understood. To address this question, Cole et al. (1996) used FRAP to measure the diffusional mobilities of several mammalian GFP tagged Golgi membrane proteins (including galactosyltransferase, mannosidase II and the KDEL receptor, see Fig. 12.2). Their results revealed rapid and extensive diffusion of all GFP-Golgi chimeras within Golgi membranes, indicating that Golgi retention of these proteins does not depend on protein immobilization. Rapid and widespread loss

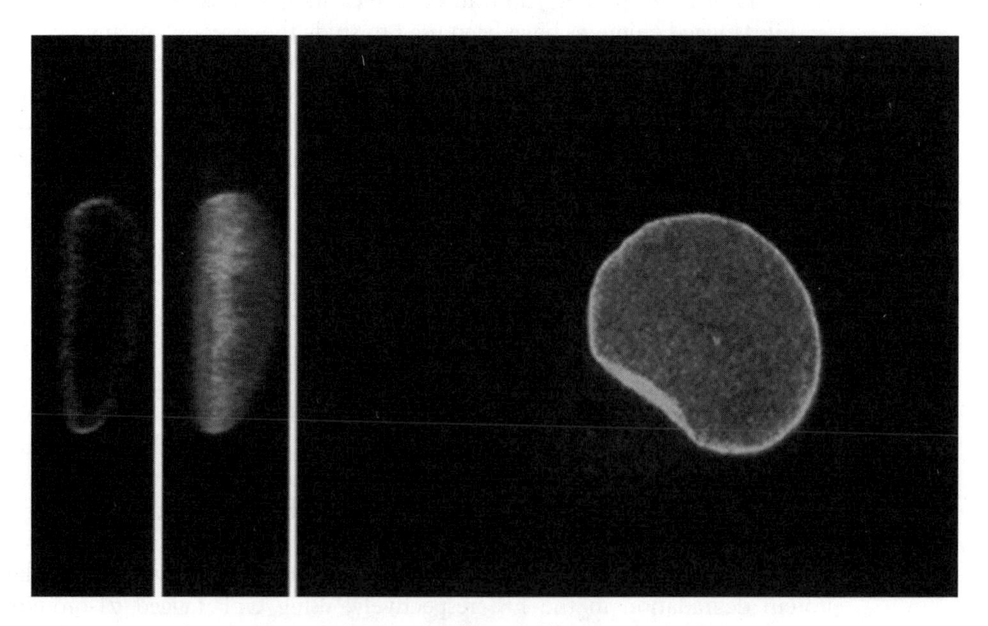

FIG 12.6. Localization of GFP tagged lamin B receptor to the inner nuclear envelope of Cos cells. [Courtesy of Ellenberg et al., 1996]. Figure also appears in color section.

of fluorescence from the Golgi upon repeated bleaching of a small zone within the complex (i.e., FLIP), furthermore, implied significant lateral diffusion rather than vesicular transport of the GFP tagged proteins between Golgi stacks. These results raise the question of how Golgi membranes maintain their identity amidst rapid diffusion of resident components and what role such dynamic properties play in Golgi membrane structure and function.

GFP-Golgi chimeras have also been used to study the dynamics of Golgi membranes *in vivo* and during perturbation by the drug brefeldin A (BFA) (Sciaky et al., 1997). In untreated cells expressing GFP-Golgi chimeras, thin membrane tubules containing GFP chimera rapidly extended out and disconnected from stable perinuclear Golgi elements. Such tubules are likely to mediate retrograde traffic from the Golgi to the ER. Upon BFA treatment, which blocks an essential activation step in the membrane interaction of ADP ribosylation factor or ARF (Klausner et al., 1992), these tubule processes formed more frequently and failed to detach from Golgi structures. A dynamic tubule network extending through the cytoplasm was soon generated and persisted for 5–10 min before rapidly (15–30 s) emptying its contents into the ER. These results would have been difficult to obtain with conventional immunofluorescent methods in fixed cells and suggest that Golgi membranes are actively absorbed into the ER during BFA treatment rather than simply mixing diffusively with ER membranes.

The trans Golgi network (TGN) is a major sorting station at the distal end of the Golgi complex and contains both resident and secretory components. Girotti and Banting (1996) visualized the dynamics of this compartment using GFP tagged to a TGN resident protein, TGN 38. They found that stably expressed N- or C- terminally tagged TGN38 localized to the TGN and recycled through endosomal membranes. This result demonstrated that TGN38 maintains its steady-state residency in the Golgi complex while circulating within the endosomal system.

The plasma membrane is an extremely dynamic structure that is continually remodeled by endocytosis and exocytosis. GFP chimeras have been useful in studying the dynamic distribution and trafficking of plasma membrane-targeted proteins in single living cells. Using time-lapse imaging techniques, for instance, Wubbolts et al. (1996) visualized the translocation of GFP-MHC Class II from lysosomes to the plasma membrane including the fusion of transport vesicles with the plasma membrane. GFP tagged GLUT4, the insulin regulatable glucose transporter, was studied by Dobson et al. (1996) who showed it resided in a perinuclear compartment of untreated CHO cells, but translocated to the plasma membrane upon addition of insulin. Reinternalization of GLUT4 upon insulin washout was shown to require an internalization signal in the N-terminus of GLUT4. The dynamics of different subdomains of the plasma membrane, including coated pits and caveolae, has been studied by Conrad et al. (1996), who compared the distribution of a GFP tagged GPI anchored protein (thought to be enriched in caveolae) with GFP tagged to the transmembrane anchor of LDL receptor, which contains a coated pit targeting signal. GFP chimeras also promise to faciliate studies of ion channel expression and localization on the plasma membrane, as reported for GFP tagged *N*-methyl-D-aspartate receptor (Marshall et al., 1995).

GFP fusion proteins to peroxisomal proteins have been used as a versatile reporter system for the peroxisomal compartment, which plays a central role in fatty acid metabolism and detoxification. Kalish et al. (1996) used GFP peroxisomal fusion proteins to analzye peroxisomal targeting signals. More recently, Wiener et al. (1997) analyzed peroxisomal dynamics using time-lapse confocal microscopy. They found a pool of peroxisomes exhibited directional movement that was microtubule dependent.

12.6 SUMMARY AND PERSPECTIVES

GFP and its variants are now recognized as powerful tools for uncovering dynamic processes in mammalian cells. They serve as excellent fluorescent markers for studies of protein localization, function and dynamics, as well as for studies of gene expression and cell differentiation. The applications of GFP are numerous including time-lapse imaging, FRET, FRAP, FLIP, and double-labeling studies. Such applications have the potential to greatly expand our knowledge of *in vivo* protein localization and dynamics, including protein–protein interactions, protein diffusional mobility, compartmental boundaries and protein sorting. Newer applications of GFP will depend on further optimization of GFP expression (e.g., by optimal codon usage) (Haas et al., 1996; Chiu et al., 1996; Zolotukhin et al., 1996), continued development of GFP variants with altered spectral properties (Heim and Tsien, 1996; Crameri et al., 1996; Delgrave et al., 1995), and investigation of the properties of the GFP chromophore itself (Patterson et al., 1997; Palm et al., 1997). The advancement of microscopic imaging systems is also important for newer applications of GFP in mammalian systems. Advances include more sensitive and quicker camera systems, superior software packages for analyzing digital information, and the use of novel microscope techniques, for example, two photon laser-scanning microscopy (TPLSM) for fluorescence imaging with 3D resolution (Potter et al., 1996; Williams et al., 1994; Sandison et al., 1995). By localizing light to a focal point, TPLSM provides an optical probe appropriate to a wide variety of future applications with GFP, including uncaging of bioeffector molecules and three-dimensional imaging of cellular kinetics and dynamics in thick tissue sections.

REFERENCES

Anderson, M. T., Tjioe, I. M., Lorincz, M. C., Parks, D. R., Herzenberg, L. A., Nolan, G. P., and Herzenberg, L. A. (1996). Simultaneous fluorescence-activated cell sorter analysis of two distinct transcriptional elements within a single cell using engineered green fluorescent proteins. *Proc. Natl. Acad. Sci. USA* 93 (16):8508–8511.

Aoki, T., Takahashi, Y., Koch, K. S., Leffert, H. L. and Watabe, H. (1996). Construction of a fusion protein between protein A and green fluorescent protein and its application to western blotting. *FEBS Lett.* 384 (2):193–197.

Bartlett, R. J., Singer, J. T., Shean, M. K., Weindler, F. W. and Secore, S. L. (1995). Expression of a fluorescent reporter gene in skeletal muscle cells from a virus-based vector: implications for diabetes treatment. *Transplantation Proc.* 27:3344.

Bauer, K. D., Naray-Fejes-Toth, A., and Fejes-Toth, G. (1996). Subcellular localization of the type 2, 11 beta-hydroxysteroid dehydrogenase: A green fluorescent protein study. *J. Biol. Chem.* 271 (26):15436–15442.

Bergmann, J. E. (1989). Using temperature sensitive mutants of VSV to study membrane protein biogenesis. *Methods Cell Biol.* 32:85–110.

Brown, J. S., Pfenninger, O., Dahl, R., Wang, J., Betz, W. J., and Howell, K. E. (1995). Green fluorescent protein-alpha-1 proteinase inhibitor, a model to study ER retention and degradation. *Mol. Cell Biol.* 6:437.

Carey, K. L., Richards, S. A. Lounsbury, K. M., and Macara, I. G. (1996). Evidence using a green fluorescent protein-glucocorticoid receptor chimera that the Ran/TC4 GTPase mediates an essential function independent of nuclear protein import. *J. Cell. Biol.* 133(5):985–996.

Chattoraj, M., King, B. A., Bublitz, G. U., and Boxer, S. G. (1996). Ultra-fast excited state dynamics in green fluorescent protein: multiple states and proton transfer. *Proc. Natl. Acad. Sci. USA* 93:8362–8367.

Chalfie, M., Tu, Y., Euskirchen, G., Ward, W. W., and Prasher, D. C. (1994). Green fluorescent protein as a marker for gene expression. *Science* 263:802–805.

Cheng, L., Fu, J., Tsukamoto, A., and Hawley, R. G. (1996). Use of green fluorescent protein variants to monitor gene transfer and expression in mammalian cells. *Nature Biotechnol.* 14:606–609.

Chiu, W., Niwa, Y., Zeng, W., Hirano, T., Kobayashi, H., and Sheen, J. (1996). Engineered GFP as a vital reporter in plants. *Curr. Biol.* 6:325–330.

Cole, N. B., Smith, C. L., Sciaky, N., Terasaki, M., Edidin, M., and Lippincott-Schwartz, J. (1996). Diffusional mobility of Golgi proteins in membranes of living cells. *Science* 273:797–801.

Conrad, P. A., Smart, E. J., Dougherty, D., Anderson, R. G. W. Bloom, G. S., and Lacey, S. W. (1996). GPI anchors target green fluorescent proteins on the plasma membrane to caveolae. *Mol. Biol. Cell* 7:275a.

Cormack, B. P., Valdivia, R. and Falkow, S. (1996) FACS-optimized mutants of the green fluorescent protein (GFP). *Gene* 173:33–38.

Crameri, A., Whitehorn, E. A., Tate, E., and Stemmer, P. C. (1996). Improved GFP by molecular evolution using DNA shuffling. *Nature Biotech.* 14:315–319.

Cubitt, A. B., Heim, R., Adams, S. R., Boyd, A. E., Gross, L. A., and Tsien, R. Y. (1995). Understanding, improving and using green fluorescent proteins. *TIBS* 20:448–455.

De Giorgi, F., Brini, M., Bastianutto, C., Marsault, R., Montero, M., Pizzo, P., Rossi, R., and Rizzuto, R. (1996). Dual-color microscopic imagery of cells expressing the green fluorescent protein and a red-shifted variant. *Gene* 173:19–23.

De Hostos, E. L., Rehfuess, C., Bradtke, B., Waddell, D. R., Albrecht, R., Murphy, J., and Gerisch, G. (1993). *Dictyostellium* mutants lacking the cytoskeletal protein coronin are defective in cytokinesis and cell motility. *J. Cell Biol.* 120:163–173.

Delgrave, S., Hawtin, R. E., Silva, C. M., Yang, M. M., and Youvan, D. C. (1995). Red shifted excitation mutants of GFP. *Bio Technol.* 13:151–154.

Dhandayuthapani, S., Via, L. E., Thomas, C. A., Horowitz, P. M., Deretic, D. and Deretic, V. (1995). Green fluorescent protein as a marker for gene expression and cell biology of *mycobacterial* interactions with macrophages. *Mol. Microbiol.* 17:901–912.

Dobson, S. P., Smart, E. J., Dougherty, D., Anderson, R. G. W., Bloom, G. S., and Lacey, S. W. (1996) GPI anchors target green fluorescent proteins on the plasma membrane to caveolae. *Mol. Biol. Cell* 7:275.

Dorsky, D. I., Wells, M., and Harrington, R. D. (1996). Detection of HIV-1 infection with a green fluorescent protein reporter system. *J. Acquir. Immune Defic. Syndr. Hum. Retrovirol.* 13 (4):308–313.

Edidin, M. (1992). Patches, posts and fences: proteins and plasma membrane domains. *Trends Cell Biol.* 2:376.

Edidin, M. (1994). Fluorescence photobleaching and recovery, FPR, in the analysis of membrane structure and dynamics. *In* Mobility and proximity in biological membranes, Damjanovich, S., Edidin, M., Szollosi, J., and Tron, L., Eds., CRC, Boca Raton, FL, pp. 109–135.

Ehrig, T., O'Kane, D.J., and Prendergast, F. G. (1995). Green fluorescent protein mutants with altered fluorescent spectra. *FEBS Lett.* 367:163–166.

Fra, A. M., and Sitia, R. (1993). The endoplasmic reticulum as a site of protein degradation. *Subcellular Biochem.* 21:143–168.

Gerdes, H-H., and Kaether, C. (1996). Green fluorescent protein: applications in cell biology. *FEBS Lett.* 389:44–47.

Girotti, M. and Banting, G. (1996). TGN38-GFP hybrid proteins expressed in stably transfected eukaryotic cells provide a tool for the real-time, in vivo study of membrane traffic pathways and suggest a possible role for rat TGN 38. *J. Cell Sci.* 109:2915–2926.

Hampton, R. Y. Koning, A., Wright, R., and Rine, J. (1996). In vivo examination of membrane protein localization and degradation with green fluroescent protein. *Proc. Natl. Acad. Sci. USA* 93 (2):828–833.

Hanakam, F., Albrecht, R., Eckerskorn, C., Matzner, M., and Gerisch, G. (1996). Myristoylated and non-myristoylated forms of the pH snesor protein hisactophilin II: intracellular shuttling to plasma membrane and nucleus monitored in real time by a fusion with green fluorescent protein. *EMBO J.* 15:2935–2943.

Heim, R., Prasher, D. C., and Tsien, R. Y. (1994). Wavelength mutations and posttranslational autoxidation of green fluorescent protein. *Proc. Natl. Acad. Sci. USA* 91:12501–12504.

Heim, R. and Tsien, R. Y. (1996). Engineering green fluorescent protein for improved brightness, longer wavelengths and fluorescence resonance energy transfer. *Curr. Biol.* 6:178–182.

Heim, R., Cubitt, A. B., and Tsien, R. Y. (1995). Improved green fluorescence. *Nature (London)* 373:663–664

Htun, H., Barsony, J. Renyi, I., Gould, D. L., and Hager, G. L. (1996). Visualization of glucocorticoid receptor translocation and intranuclear organization in living cells with a green fluorescent protein chimera. *Proc. Natl. Acad. Sci. USA* 93 (10):4845, 4850.

Ikawa, M., Kominami, K., Yoshimura, Y., Tanaka, K., Nishimune, Y., and Okabe, M. (1995). A rapid and non-invasive selection of transgenic embryos before implantation using green fluorescent protein (GFP). *FEBS Lett.* 375:125–128.

Inouye, S., and Tsuji, F. I. (1994). *Aequorea* green fluorescent protein: expression of the gene and fluorescent characteristics of the recombinant protein. *FEBS Lett.* 341:277–280.

Kaech, S., Ludin, B., and Matus, A. (1996). Cytoskeletal plasticity in cells expressing neuronal microtubule-associated proteins. *Neuron* 17 (6):1189–1199.

Kaether, C. and Gerdes, H.-H. (1995). Visualization of protein transport along the secretory pathway using green fluorescent protein. *FEBS Lett.* 369:267–271.

Kalish, J. E., Keller, G. A., Morell, J. C., Mihalik, S. J., Smith, B., Cregg, J. M., and Gould, S. J. (1996). Characterization of a novel component of the peroxisomal protein import apparatus using fluorescent peroxisomal proteins. *EMBO J.* 15 (13):3275–3285.

Klausner, R. D., Donaldson, J. G., and Lippincott-Schwartz, J. (1992). Brefeldin A: insights into the control of membrane traffic and organelle structure. *J. Cell Biol.* 116:1071–1080.

Lee, S., Neumann, M., Pause, A., Zhou, S., Chen, D., Humphrey, J. S., Gnarra, J. R., Pavlakis, G., Linehan, W. M., and Klausner, R. D. (1996). Analysis of the cellular localization of the von Hippel-Lindau (VHL) tumor suppressor gene product coupled to the green fluorescent protein. *Mol. Cell Biol.* 7:174.

Levy, J. P. Muldoon, R. R., Zolotukhin, S., and Link, C. J. (1996). Retroviral transfer and expression of a humanized, red-shifted green fluorescent protein gene into human tumor cells. *Nature Biotechnol.* 14:610–614.

Lippincott-Schwartz, J. (1994). The Endoplasmic reticulum-Golgi membrane system. *In* The liver: biology and pathobiology, 3rd ed., Arias, I. M., Boyer, J., Fausto, N., Jacoby, W. B., Schacter, D., and Shafritz, D. A. Eds., Raven PL, New York.

Ludin, B., Doll, T., Meili, R., Kaech, S., and Matus, A., (1996). Application of novel vectors for GFP-tagging of proteins to study microtubule-associated proteins. *Gene* 173:107–111.

Lybarger, L., Dempsey, D., Franek, K. J., and Chervenak, R. (1996). Rapid generation and flow cytometric analysis of stable GFP-expressing cells. *Cytometry* 25 (3):211–220.

Maniak, M., Rauchenberger, Albrecht, R., Murphy, J., and Gerixch, G. (1995). Coronin involved in phagocytosis: dynamics of particle-induced relocalization visualized by a green fluorescent protein tag. *Cell* 83:915–924.

Marshall, J. R., Molloy, R., Moss, G. W. J., Howe, J. R., and Hughes, T. E. (1995). The jellyfish green fluorescent protein : a new tool for studying ion channel expression and function. *Neuron*. 14:211–215.

Mayor, S., Presley, J. F., and Maxfield, F. R. (1993). Sorting of membrane components from endosomes and subsequent recycling to the cell surface occurs by a bulk flow process. *J. Cell Biol.* 121:1257–1269.

Mellman, I., Fuchs, R., and Helenius, A. (1986). Acidification of the endocytic and exocytic pathways. *Annu. Rev. Biochem.* 55:663–700.

Mitra, R. D., Silva, C. M., and Youvan, D.C. (1996). Fluorescence Resonance Energy Transfer between blue-emitting and red shifted excitation derivatives of the greeen fluorescent protein. *Gene* 173:13–17.

Moolenar, W. H. (1986). Effects of growth factors on intracellular pH regulation. *Annu. Rev. Physiol.* 48:363–376.

Moores, S. L., Sabry, J. H., and Spudich, J. A. (1995). Myosin dynamics in live *Dictyostelium* cells. *Proc. Natl. Acad. Sci. USA* 93:443–446.

Moriyoshi, K., Richards, L. J., Akazawa, C., O'Leary, D. D., and Nakanishi, S. (1996). Labeling neural cells using adenoviral gene transfer of membrane-targeted GFP. *Neuron* 16 (2):255–260.

Mosser, D. D., Caron, A. W., Bourget, L., Jolicoeur, P., and Massie, B. (1997). Use of a dicistronic expression cassete encoding the green fluorescent protein for screening and selection of cells expressing inducible gene products. *Biotechniques* 22 (1):150–4.

Muldoon, R. R., Levy, J. P., Kain, S. R., Kitts, P. A., and Link, C. J. Jr. (1997). Tracking and quantitation of retroviral-mediated transfer using a completely humanized, red-shifted green fluorescent protein gene. *Biotechniques* 22 (1):162–7.

Niswender, K. D., Blackman, S. M., Rohde, L., Magnuson, M. A., and Piston, D. W. (1995). Quantitative imaging of green fluorescent protein in cultured cells: comparison of microscope techniques, use in fusion proteins and detection limits. *J. Microsc.* 180:109–116.

Ogawa, H., Inouye, S., Tsuji, F. I., Yasuda, K., and Umesono, K. (1995). Localization, trafficking and temperature-dependence of the Aequorea green fluorescent protein in cultured vertebrate cells. *Proc. Natl. Acad. Sci. USA* 93:4845–4850.

Olson, K. R., McIntosh, J. R., and Olmsted, J. B. (1995). Analysis of MAP4 function in lviing cells using green fluorescent protein (GFP) chimeras. *J. Cell Biol.* 130:639–650.

Olmsted, J. B. (1991). Non-motor microtubule associated proteins. *Curr. Opinion Cell Biol.* 3:52–58.

Ormö, M., Cubitt, A. B., Kallio, K., Gross, L. A, Tsien, R. Y., and Remington, S. J. (1996). Crystal structure of the *Aequeorea victoria* green fluorescent protein. *Science* 273:1392–1395.

Palm, G. J., Zdanov, A., Gaitanaris, G. A., Stauber, R., Pavlakis, G. N., and Wlodawe, A. (1997). The structural basis for spectral variations in green fluorescent protein. *Nature Struc. Biol.* 4:361–365.

Patterson, G. H., Knobel, S. M., Sharif, W. D., Kain, S. R., and Piston, D. W. (1997). Use of green fluorescent protein (GFP) and its mutants in quantitative fluorescence microscopy. *Biophys. J.* 73:2782–2790.

Peel, A. L., Zolotukhin, S., Schrimsher, G. W., Muzyczka, N., and Reier, P. J. (1997). Efficient transduction of green fluorescent protein in spinal cord nurons using adeno-associated virus vectors containing cell type-specific promoters. *Gene Ther.* 4 (1):16–24.

Pepperkok, R. et al. (1993). βCOP is essential for biosynthetic membrane transport from the endoplasmic reticulum to the Golgi complex in vivo. Cell 74, 71–82.

Phillips, J. L. Meinkoth, J., and Eberwine, J. H. (1995). A model system for characterizing mRNA localization and translation in neurites using green fluorescent protein. *Mol. Cell Biol.* 6:310a.

Pines, J. (1995). GFP in mammalian cells. *Trends Genet.* 11:326–327.

Potter, S. M., Wang, C-M., Garrity, P. A., and Fraser, S. E. (1996). Intravital imaging of GFP using two photon laser scanning microsopy. *Gene* 173:25–31.

Presley, J. , Zaal, K, Schrorer, T., Cole, N. B., and Lippincott-Schwartz, J. (1997). ER to Golgi transport visualized in living cells. *Nature (Lond.)* 389:81–85.

Raz, E., Driever, W., and Rosenthal, N. (1996). Green fluorescent protein marks skeletal-muscle in murine cell lines and zebra fish. *Gene* 173:89–98.

Rizzuto. R., Brini, M., Pizzo, P., Murgia, M., and Pozzan, T. (1995) Chimeric green fluorescent protein as a tool for visualizing subcellular organelles in living cells. *Curr. Biol.* 5: 635–642.

Rizzuto, R., Brini, M., DeGiorgi, F., Rossi, R., Heim, R., Tsien, R. Y., and Pozzan, T. (1996). Double labeling of subcellular structures with organelle-targeted GFP mutants in vivo. *Curr. Biol.* 6 (2):183–188.

Roberts, R., Chua, M., Barieri, M. A., Li, G-P. Heuser, J., and Stahl, P. D. (1996). A novel form of vesicle fusion, "bridge" fusion, observed in the prescence of Rab5:Q79L, a GTPase defective activator of endosome fusion. *Mol. Biol. Cell.* 7:592a.

Roderick, H. L., Campbell, A. K., and Llewellyn, D. H. (1997). Nuclear localization of calreticulin *in vivo* is enhanced by its interaction with glucocorticoid receptors. *FEBS Lett.* 405 (2):181–185.

Ropp, J. D., Donahue, C. J., Wolfgang-Kimball, D., Hooley, J. J., Chin, J. Y., Cuthbertson, R. A., and Bauer, K. D. (1995). Aequorea green fluorescent protein analysis by flow cytometry. *Cytometry* 21 (4):309–317.

Ropp, J. D., Donahue, C. J., Wolfgang-Kimball, D., Hooley, J. J., Chin, J. Y., Cuthbertson, R. A., and Bauer, K. D. (1996). Aequorea green fluorescent protein: simultaneous analysis of wild-type and blue-fluorescing mutant by flow cytometry. *Cytometry* 24 (3):284-248.

Sandison, D. R., Piston, D. W., Williams, R. M., and Webb, W. W. (1995). Resolution, background rejection, and signal-to-noise in widefield and confocal microscopy. *Applied Optics* 34:3576–3588.

Schuler, E., Lin, F., and Worman, H. J., (1994). Characterization of the human gene encoding LBR, an integral protein of the nuclear envelope inner membrane. *J. Biol. Chem.* 269:11312–11317.

Sciaky, N., Presley, J., Smith, C., Zaal, K., Cole, N., Terasaki, M., Siggia, E., and Lippincott-Schwartz, J. (1997). Golgi tubule traffic and the effects of brefeldin A visualized in living cells. *J. Cell Biol.* 139:1137–1155.

Shelby, R. D., Hahn, K. M., Sullivan, K. F. (1996). Dynamic elastic behavior of alpha-satellite DNA domains visualized in situ in living human cells. *J. Cell, Biol.* 135 (3):545–57.

Siemering, K. R., Golbik, R., Sever, R., and Haseloff, J. (1996). Mutations that suppress the thermosensitivity of green fluorscent protein. *Curr. Biol.* 6 (12):1653–1665.

Stauber, R. H., Carney, P., Horie, K., Hudson, E. A., Gaitanaris, G. A., Pavlakis, G. N. (1998). Development and applications of enhanced green fluorescent protein mutants. *Bio Techniques* 24:462–471.

Stauber, R., Gaitanaris, G. A., and Pavlakis, G. N. (1995). Analysis of trafficking of Rev and transdominant Rev proteins in living cells using green fluorescent protein fusions: transdominant Rev blocks the export of Rev from the nucleus to the cytoplasm. *Virology* 213 (2):439–449.

Stearns, T. (1995). The green revolution. *Curr. Biol.* 5:262–264.

Steele-Mortimer, O., Mills, S. D., Stein, M. A., and Finley, B.B. (1995). Development of a system to study the effects of *Salmonella Typhimurium* infection on host cells *in vitro* using the green fluorescent protein. Mol. Biol. Cell 6:215a.

Stryer, L. (1978). Fluorescence energy transfer as a spectroscopic ruler. *Annu. Rev. Biochem.* 47:819–846.

Subramanian, S. and Srienc, F. (1996). Quantitative analysis of transient gene expression in mammalian cells using the green fluorescent protein. *J. Biotechnol.* 49 (1–3):137–51.

Terasaki, M., Jaffe, L. A., Hunnicutt, G. R., and Hammer, J. A. (1996). Structural change of the endoplasmic reticulum during fertilization: evidence for loss of membrane continuity using green fluorescent protein. *Dev. Biol.* 179:320–328.

Tsien, R. Y. and Waggoner, A. (1995). Fluorophores for confocal microsocopy. *In* Handbook of biological confocal microscopy. Pawley, J. B., Ed., 2nd ed., Plenum, New York, pp. 267–279.

Tsien, R. Y., Bacski, B. J., and Adams S. R. (1993). FRET for studying intracellular signaling. *Trends Cell Biol.* 3:242–245.

Uster, P. S. and Pagano, R. E. (1986). Resonsance energy transfer microsocpy: observations of membrane-bound fluorescent probes in model membranes and in living cells. *J. Cell Biol.* 103:1221–1234.

Venerando, R., Miotto, G., Pizzo, P., Rizzuto, R., and Siliprandi, N. (1996) Mitonchondrial alterations induced by aspirin in rat hepatocytes expressing mitochondrially targeted green fluorescent protein. *FEBS Lett.* 382 (3):256–260.

Wacker, I., Kaether, C., Kromer, A., Migala, A., Almers, W., and Gerdes, H-H. (1997). Microtubule-dependent transport of secretory vesicles visualized in real time with a GFP-tagged secretory protein. *J. Cell Sci.* 110:1453–1463.

Waddle, J. A., Karpova, T. S., Waterston, R. H., and Cooper, J. A. (1996). Movement of cortical actin patches in yeast. *J. Cell Biol.* 132 (5):861–870.

Wang, H. G., Rapp, U., and Reed, J. C. (1996). Bcl-2 targets the protein kinase Raf-1 to mitochondria. *Cell* 87:629–638.

Ward, W. W., (1981). Properties of the coelenterate green-fluorescent protein. In Bioluminescence and Chemiluminescence. DeLuca, M. A., and McElroy, W. D., Eds., Academci, New York, pp. 235–242.

Ward, W. W., Prentice, H. J., Roth, A. F., Cody, C. W. and Reeves, S. C. (1982). Specral perturbations of the Aequorea green-fluorescent protein. *Photochem. Photobiol.* 35:803–808.

Ward, W. W., Cody, C. W., Hart, R. C., and Cormier, M. J. (1980). Spectrophotometric identity of the energy transfer chromophores in *Renilla* and *Aequorea* green fluorescent proteins. *Photochem. Photobio.* 31:611–615.

Wiemer, E. A., Wenzel, T., Deerinck, T. J., Ellisman, M. H., and Subramani, S. (1997). Visualization of the peroxisomal compartment in living mammalian cells: dynamic behavior and association with microtubules. *J. Cell Biol.* 136 (1):71–80.

Williams, R. M., Piston, D. W., Webb, W. W. (1994). Two-photon molecular excitation provides intrinsic 3-dimensional resolution for laser-based microscopy and microphotochemistry. *FASEB J.* 8:804–813.

Wubbolts, R., Fernandez-Borja, M., Oomen, L., Verwoerd, D., Janssen, H., Calafat, J., Tulp, A., Dusseljee, S., and Neefjes J. (1996). Direct vesicular transport of MHC class II molecules from lysosomal structures to the cell surface. *J. Cell. Biol.* 135:611–622.

Yang, T. T., Cheng, L., and Kain, S.R. (1996a). Optimized codon usage and chromophore mutations provide enhanced sensitivity with the green fluorescent protein. *Nucleic Acids Res.* 24 (22):4592–4593.

Yang, T. T., Kain, S. R., Kitts, P., Kondepudi, A., Kondepudi, A., Yang, M. M., and Youvan, D. C. (1996b). Dual color microscopic imaging of cells expressing the green fluorescent protein and a red shifted variant. *Gene* 173:19–23.

Yang. T. T., Sinai, P., Green, G., Kitts, P. A., Chen, Y.-T., Lybarger, L., Chervenak, R., Patterson, G. H., Piston, D. W., and Kain, S. R. (1998). Improved fluorescence and dual color detection with enhanced blue green variants of the green fluorescent protein. *J. Biol. Chem.*, in press.

Yang, F., Moss, L. G., and Phillips G. N., Jr, (1996c). The molecular structure of green fluorescent protein. *Nature Biotechnol.* 14:1246–1251.

Yano, M., Kanazawa, M., Terada, K., Namchai, C., Yamaizumi, M., Hanson, B., Hoogenraad, N., and Mori, M. (1997). Visualization of mitochondrial protein import in cultured mammalian cells with green fluorescent protein and effects of overexpression of the human import receptor Tom20. *J. Biol. Chem.* 272 (13):8459–8465.

Zerial, M. and Stenmark, H. (1993). RabGTPases in vesicular transport. *Curr Opinion Cell. Biol.* 5:613–620.

Zernicka-Goetz, M., Pines, J., McLean Hunter, S., Dixon, J. P., Siemering, K. R., Haseloff, J., and Evans, M. J. (1997). Following cell fate in the living mouse embryo. *Development* 124 (6):1133–1137.

Zhang, G., Gurtu, V., and Kain, S. R. (1996). An enhanced green fluorescent protein allows sensitive detection of gene transfer in mammalian cells. *Biochem. Biophys. Res. Commun.* 227 (3):707–711.

Zolotukhin, S., Potter, M., Hauswirth, W. W., Guy, J., and Muzyczka, N. (1996). A humanized green fluorescent protein cDNA adapted for high-level expression in mammalian cells. *J. Virol.* 70:4646–4654.

PART FOUR

Methods and Protocols

Methods and Protocols

SHARYN A. ENDOW
Department of Microbiology and Immunology, Duke University Medical Center
Durham, NC

DAVID W. PISTON
Department of Molecular Physiology and Biophysics, Vanderbilt University,
Nashville, TN

Although GFP has now been successfully expressed and visualized in a wide number of divergent organisms, the rules governing expression in some organisms have only recently become apparent. Efficient expression is essential to detection, but detection of GFP can also be problematic if care is not taken to use proper conditions for visualization. Here we present methods and protocols to express, visualize, and document GFP for those desiring to use GFP for different experimental purposes. Specific problems that researchers have encountered in the expression or visualization of GFP are described, together with approaches that have been used to overcome these problems. General methods for molecular cloning and gene expression are not covered – we refer readers requiring further information regarding these methods to the excellent protocols books that are available (Ausubel et al., 1996; Sambrook et al., 1989). The guidelines presented here should be generally useful for expressing GFP in any organism, with special attention to the biological and physical properties of GFP that have made its detection elusive in instances in the past.

Green Fluorescent Protein: Properties, Applications, and Protocols, Edited by Martin Chalfie and Steven Kain
ISBN 0-471-17839-X

PROTOCOL I EXPRESSION OF GREEN FLUORESCENT PROTEIN

Various strategies have been developed to optimize the expression of GFP in the cells or organisms into which *gfp* has been introduced. Protocol I discusses considerations for designing constructs, together with mutants of *gfp* with altered or enhanced fluorescence properties. Vectors and constructs designed for specific purposes are collected in Table A.I.1 and available mutants of *gfp* are compiled in Table A.I.2. Protocol I.E provides a protocol for high-yield purification of recombinant *gfp* from bacteria.

I.A Constructs

I.A.1 Nonfusion Constructs

GFP and its variants have been expressed alone or as fusion proteins in many prokaryotic and eukaryotic cells (Table A.I.1). Nonfusion constructs were first made to demonstrate expression of the protein in bacteria and *Caenorhabditis elegans* (Chalfie et al., 1994). The initial *gfp* plasmids included the original bacterial plasmid, pGFP10.1, the bacterial expression plasmid, TU#58, and the *C. elegans* expression vectors, TU#60-64.

The bacterial plasmid, pGFP10.1, was constructed by inserting a polymerase chain reaction (PCR) fragment amplified from $\lambda gfp10$ (Prasher et al., 1992) into pBS(+) (Stratagene). The primers used for PCR flanked the EcoRI sites in the *gfp* gene, and *gfp* was cloned into pBS(+) as an EcoRI fragment. A missense mutation, Q80R, is present in pGFP10.1 and in all of the *gfp* plasmids derived from it. This change presumably arose during the PCR amplification, but it has no detectable effect on the spectral properties of the protein compared to wild type (Chalfie et al., 1994). The plasmid TU#65 is a pBluescript II KS(+) (Stratagene) derivative containing *gfp* amplified from pGFP10.1.

The bacterial expression plasmid, TU#58, was constructed by ligating a PCR-amplified NheI-EcoRI fragment from pGFP10.1 to pET3a (Studier et al., 1990) with the conversion of the initial Met of GFP to an Ala and the addition of an initiator Met to the Ala. Expression of GFP can be induced in the host strain, *BL21(DE3)pLysS*, by addition of 0.8 mM IPTG to the culture medium.

The originally reported construct for expression of GFP in *C. elegans*, TU#64, consisted of *gfp* under the regulation of the promoter for the *C. elegans mec-7* gene, which encodes a β-tubulin abundant in the touch receptor neurons but less abundant in other neurons. The *gfp* gene in TU#64 was obtained by PCR amplication of pGFP10.1. The TU#60-63 expression vectors were made by replacing *lacZ* with a *gfp* fragment from TU#65 in four plasmids made for *C. elegans* expression studies (Fire et al., 1990). The plasmid TU#60 is now available commercially as pGFP (CLONTECH). The high copy number plasmid pGFP expresses GFP as a β-galactosidase fusion protein from the *Escherichia coli lac* promoter.

Constructs to express GFP as a nonfusion protein have subsequently been made to obtain high-level expression of GFP for protein purification, or to utilize GFP as a reporter for specific promoters in different organisms. GFP can serve as an excellent reporter in many applications, including fluorescence-activated cell sorting (FACS) (Cormack et al., 1996; Galbraith et al., 1995; Ropp et al., 1995) and photobleaching recovery experiments (Cole et al., 1996). GFP has also been used as a reporter to follow cell lineage, or as a marker for transfection.

I.A.2 Fusion Constructs

Fusion of GFP to a protein under study was first demonstrated to be feasible in studies with the *Drosophila* protein Exuperantia (Wang and Hazelrigg, 1994) and has rapidly become an accepted method for cellular localization of proteins as well as for monitoring dynamic processes in the cell. Fusions of *gfp* both to the N- and C-terminus of *exu* were made and shown to rescue the maternal-effect lethal phenotype of an *exu* null mutant, serving as a precedent for other constructs. Most of the fusion proteins reported to date have been made with GFP fused in-frame to the N- or C-terminus of the protein of interest, rather than inserted at an internal site within the protein. The *gfp* gene can, however, be inserted into internal restriction sites in the gene coding region and the resulting constructs screened for activity, e.g., mutant rescue. Several examples of functional internally tagged constructs have been recovered (Xu, Kelly, Harfe and Fire, personal communication).

The first generation of fusion constructs that were made for expression of GFP, including those first used to demonstrate that GFP could be expressed as a fusion protein and retain its fluorescent properties (Wang and Hazelrigg, 1994), utilized the initial plasmids as templates to amplify *gfp* for fusion to the target gene. Gene coding regions can be joined to *gfp* by taking advantage of existing restriction enzyme sites in the target sequences, but more commonly requires PCR amplification of both the gene of interest and *gfp*. For PCR reactions, the use of error-correcting DNA polymerases, for example, Pfu (Stratagene) or Vent (New England BioLabs), is recommended to reduce or eliminate PCR-induced missense mutations in the amplified product. Mutations can also be introduced into constructs by base changes in the oligonucleotide primers used for PCR. These changes are frequently deletions of 1–2 bases at the ends of the primers (Fire, personal communication), which can cause frame-shifts in gene fusions. It is therefore critical to sequence PCR-amplified fragments in constructs, particularly the regions corresponding to the PCR primers.

In addition to directly ligating together PCR fragments corresponding to the gene of interest and *gfp*, gene fusions can be made by cloning the target gene and *gfp* in-frame into neighboring restriction sites of a plasmid. The fragment containing the gene fusion can then be excised and subcloned into an expression vector. Expression constructs are then analyzed by restriction enzyme mapping and sequencing to determine insert orientation and fidelity of PCR amplification.

I.A.3 Introns and GFP Expression

The presence of introns in the 5′ or 3′ untranslated region (UTR) of *gfp* or *gfp* gene fusions may be important for optimal expression of GFP (see Chapters 8 and 9 on *C. elegans* and *Drosophila* for further discussion regarding the effect of introns in the UTRs on GFP expression).

The presence of a cryptic splice site in *gfp* that is recognized by the *Arabidopsis* RNA splicing machinery results in excision of an 84-nucleotide "intron" between nucleotides 400 and 483, causing reduced expression of GFP in *Arabidopsis* (Haseloff and Amos, 1995). A modified *gfp* with mutated cryptic splice sequences has been synthesized to improve expression (Haseloff and Amos, 1995). The cryptic splice site is probably recognized not only in *Arabidopsis*, but also in barley and tobacco, since GFP fluorescence could not be detected after transfection of barley or tobacco protoplasts with wild-type *gfp*, but was observed after transfection with a *gfp* from which the cryptic splice site had been removed (Reichel et al., 1996). The readily detectable GFP fluorescence observed in maize and *Citrus* protoplasts transfected with wild-type *gfp* (Hu and Cheng, 1995; Galbraith et al., 1995; Sheen et al., 1995; Niedz et al., 1995) indicates that splicing at the cryptic site probably does not occur in maize and *Citrus*. These results provide evidence that differences are likely to exist in the RNA splicing machinery of different plants, necessitating the testing and careful selection of GFP variants for use in expression studies.

I.A.4 Vectors

Any of the vectors designed for protein expression can be used to make constructs to express GFP in different cells or organisms, either alone or as a fusion protein. The vectors that have been used include the standard plasmid vectors for tranformation or transfection of given cells or organims. Viral and retroviral vectors have also been used to deliver *gfp* constructs into mammalian cells or tissues, and viral vectors have been used to infect plant protoplasts, leaves, and entire plants with *gfp* or *gfp*-gene fusions. Many vectors contain promoters and other specific regulatory elements to optimize expression in given cells. Different promoters can also be incorporated into constructs to confer cell-type specific or inducible expression.

The vectors that have been used to express GFP in bacteria, yeast, *Dictyostelium* and other slime molds, and *Drosophila* have been the standard plasmid vectors for transformation of these organisms. *Caenorhabditis elegans* researchers have used the TU#60-65 plasmids as reporters for specific promoters in work published to date, but see ftp://cinl.ciwemb.edu for the latest information on *C. elegans* GFP vectors.

Mammalian expression vectors include plasmids such as pcDNA I (Invitrogen), pCR3 (Invitrogen), and pcDL-SRα (Takebe et al., 1988). Many of the plasmids used to express genes in mammalian cells contain the human cytomegalovirus (CMV) immediate early promoter to drive gene expression, and a Kozak translation initiation consensus sequence (Kozak, 1987) and SV40 polyadenylation sequence to increase translation efficiency. These plasmids have been used successfully for expression of GFP in various mammalian cell lines.

Viral vectors that have been used to deliver *gfp* into mammalian cells include adenovirus (Moriyoshi et al., 1996) and adeno-associated virus (Barlett et al., 1995). Vaccinia virus and adeno-associated virus carrying *gfp* have also been used to infect neurons of live *Xenopus laevis* tadpoles (Wu et al., 1995) or rats (Peel et al., 1997). Retroviruses that have been used to transfer *gfp* into mammalian cells include murine stem cell virus (MSCV) (Cheng et al., 1996) and a murine replication-defective retrovirus (Levy et al., 1996). The viral and retroviral vectors may prove to be important in gene therapy for targeting gene delivery to specific cells or tissues, using GFP as a marker for expression.

Vectors for expression of GFP in plants include plasmid vectors for transient expression (Galbraith et al., 1995; Hu and Cheng, 1995; Niedz et al., 1995; Sheen et al., 1995) and binary *Agrobacterium* vectors (Haseloff and Amos, 1995; Chiu et al., 1996) for generation of transgenic plants. Many of these vectors contain the califlower mosaic virus (CaMV) 35S promoter to drive gene expression, together with the *nopaline synthase* (*nos*) transcriptional terminator. Plant viral vectors include tobacco mosiac virus (TMV) (Casper and Holt, 1996; Heinlein et al., 1995) and the related tobacco mosaic tobamovirus (Epel et al., 1996), and potato virus X (PVX) (Baulcombe et al., 1995; Oparka et al., 1995).

Representative constructs that have been used to express GFP in various organisms are compiled in Table A.I.1, together with promoter and other regulatory sequence elements of interest. Some constructs were designed for specific purposes, for example, high-level expression of GFP, while others contain specific sequences that target an expressed GFP fusion protein to a particular cellular compartment or organelle. Vectors for both transient and stable transformation are shown in the table. All of the vectors except the TU#60-64 vectors (see entry for *C. elegans*) and the adeno-associated virus (AAV) vector (see Rats) lack *gfp*. The *gfp* coding region in the other constructs was cloned into the vector as a PCR product or as a restriction enzyme fragment from one of the original *gfp* plasmids, together with the target promoter and/or gene of interest.

A number of *gfp* vectors are now available from commercial manufacturers for expression of GFP in prokaryotic or eukaryotic cells (CLONTECH, Life Technologies, Maxygen, Quantum Biotechnologies, Pharmingen). These plasmids contain *gfp*, either wild-type or mutant, multiple cloning sites for fusing proteins of interest to the N- or C-terminus of GFP and a promoter for driving gene expression. In several of these plasmids the sequence upstream of the GFP initiation codon has been modified to conform to the Kozak consensus sequence (Kozak, 1987) to maximize the efficiency of translation in eukaryotic cells. Sequences flanking the GFP coding sequences in many of the original cDNA clones, which were thought to inhibit GFP expression in some cell types, have been removed during construction of many commerical GFP vectors.

Prior to the availability of GFP antibodies, constructs incorporating the 9 amino acid epitope of hemagglutinin (HA) or the 13 amino acid MYC-derived epitope were made to visualize expression of the chimera using antibodies specific to these epitopes. Experiments comparing the distribution of antibody staining of these epitopes to that of GFP fluorescence can be useful in studying the time course of GFP expression versus fluorescence, as well as intracellular conditions that alter GFP fluorescence. GFP antibodies are now commercially available from CLONTECH and Molecular Probes, Inc., and purified GFP for microinjection and other experiments is available from CLONTECH.

TABLE A.I.1 Vectors and constructs for expression of GFP

Cells/Organism	Vector	Promoter	Other Regulatory/ Sequence Elements	Purpose	References
Bacteria					
E. coli	pET3a	T7 $\Phi 10$ promoter		Demonstrate GFP expression (TU#58), mutagenesis of *gfp*, a high-expression derivative of this plasmid has been recovered (see GFP purification protocol in Protocol I.E.)	Chalfie et al., 1994; Delagrave et al., 1995
	pKEN2	*tac* promoter		High-level GFP expression, selection of mutants by FACS	Cormack et al., 1996
	pTrcHis-C	*trc* promoter	*lac* operator, 6 histidines at N-terminus	Purification of GFP	Inouye and Tsuji, 1994
Bacillus subtilis	pER82, pOR100, pBluescript SKII(–)	*sspE-2G* (high efficiency), *csfB* (low efficiency), *cotE* (controlled by RNA polymerase containing s^E), and *gerE* (controlled by RNA polymerase containing s^K) promoters		Cell-specific expression, protein localization	Webb et al., 1995
Mycobacterium species	pMV206, pMV261	*ahpC (alkyl hydroperoxide reductase)*, *mtrA* (*M. tuberculosis* response regulator), *hsp60* (*heat shock* 60), and *tbprc3* (a newly isolated M. *tuberculosis*) promoters		Demonstrate GFP expression, analyze promoter expression in individual cells, FACS analysis	Dhandayuthapani et al., 1995
Pseudomonas putida	pUTKm	T7 $\Phi 10$ promoter	Tn5 IRs (inverted repeats), Km^R	Monitor plasmid transfer during conjugation	Christensen et al., 1996

Salmonella typhimurium	pBR322	*lac* promoter		Demonstrate GFP expression, FACS analysis	Kain et al., 1995; Valdivia et al., 1986)
Yersinia pseudotuberculosis	pBR322	*lac* promoter		Demonstrate GFP expression, FACS analysis	Valdivia et al., 1996
Yeast					
Saccharomyces cerevisiase	pCLUC	*CUP1* (copper inducible) promoter		Demonstrate protein aggregation after induction	Patino et al., 1996
	pTS210 derivatives	*ACT1* (*actin1*) promoter	*ACT1* transcriptional terminator	Monitor movement of cortical actin patches	Doyle and Botstein, 1996
	pYES2	*GAL1/GAL10* promoter	*Swi6* NLS (nuclear localization signal)	Monitor protein localization through the cell cycle	Sidorova et al., 1995
Schizosaccharo-myces pombe	pDB248, pSK248	*dis1*$^{+}$ [spindle pole body (SPB) protein] promoter	*S. cerevisiae leu2*, *S. pombe ARS1* (*ori*)	Follow protein localization between SPB and spindle microtubules	Nabeshima et al., 1995
Slime molds					
Dictyostelium discoideum	pDdGal-15	*actin-15* promoter		Follow protein relocalization dynamics in phagocytosis	Maniak et al., 1995
Polysphondylium pallidum (cellular slime mold)	pDdGal56	*actin-6*, *actin-15* and *ecmB* promoters	Kozak translation start sequence	Demonstrate GFP expression, FACS analysis	(Fey et al., 1995)
C. elegans	TU#60-64 (vectors contain *gfp*)	Various *C. elegans* promoters		Demonstrate GFP expression, reporter for specific promoters	Chalfie et al., 1994; Sengupta et al., 1994; Treinin and Chalfie, 1995; Troemel et al., 1995
Drosophila melanogaster	pCaSpeR vectors	Various *D. melanogaster* promoters		Demonstrate GFP expression as a fusion protein, localize proteins, monitor development, follow dynamics of WT and mutant proteins in mitosis, reporter for specific promoter	Wang and Hazelrigg, 1994; Barthmaier and Fyrberg, 1995; Kerrebrock et al., 1995; Endow and Komma, 1996; Plautz et al., 1996

(*continued*)

TABLE A.I.1 (continued)

Cells/Organism	Vector	Promoter	Other Regulatory/ Sequence Elements	Purpose	References
	pUAST	5 *GAL4* UASsJ (upstream activating sequences)		Reporter for *GAL4* enhancer trap lines	Yeh et al., 1995
	pCaSpeR4	*polyubiquitin* (*PUb*) promoter	5′ *PUb* intron, 9 aa NLS, 3′ *hsp27* UTR (untranslated region) + poly-A signal	Nuclear localization, reporter for gene expression	Davis et al., 1995
Vertebrates					
Mice	pCAGGS	Chicken β actin promoter	CMV-IE (cytomegalovirus immediate early) enhancer, rabbit β globin poly-A signal, 5′ intron, Kozak translation start sequence	Demonstrate GFP expression in transgenic mice	Ikawa et al., 1995
Rats	pTR_{BS}-UF2 [adeno-associated virus (AAV) shuttle vector carrying *gfp*]	NSE (neuron-specific enolase) or PDGF (platelet-derived growth factor) β-chain promoter	AAV 145 bp IR, neo^R controlled by HSV (herpes simplex virus) TK (thymidine kinase) promoter with polyoma enhancer, bovine growth hormone poly-A signal, 'humanized' *gfp*	Gene targeting to CNS in transiently transgenic rats	Peel et al., 1997
X. laevis	pSC65 (vaccinia virus shuttle vector)	Strong synthetic early/late vaccinia virus promoter		Gene targeting to amphibian CNS	Wu et al., 1995

Zebrafish	pXex	*Xenopus ef1α* promoter	*Xenopus ef1α* enhancer + 5′ UTR, SV40 poly-A signal, rabbit β-globin IVS2 (intervening sequence 2)	Demonstrate GFP expression in stably transgenic zebrafish	Amsterdam et al., 1995
	pMLC	MLC1 (myosin light chain 1) promoter	840 bp of small intron + poly-A signal of SV40 T antigen, 920 bp of MLC enhancer	Gene targeting to muscle cells in transiently transgenic zebrafish	Moss et al., 1996
Mammalian Cells					
HeLa cells	pcDNA I	Human CMV IE promoter	SV40 *ori*, polyoma *ori*, SV40 poly-A signal, N-terminal 31 aa's of cytochrome c oxidase subunit VIII precursor (mitochondria targeting sequence)	Targeting of GFP fusion protein to mitochondria	Rizzuto et al., 1995
	pcDNA I	Human CMV IE promoter	SV40 *ori*, polyoma *ori*, SV40 poly-A signal, TK leader + aa's 407–794 of rat GR (glucocorticoid receptor), HA1 (hemagglutinin 1) tag	Targeting of GFP fusion protein to nucleus	Rizzuto et al., 1996
	pcDL-SRα	SRα [SV40 promoter + R segment and part of U5 sequence of HTLV1 (human T-cell leukaemia virus 1) LTR] promoter	SV40 late gene splice junction, SV40 poly-A signal, protein-specific signal sequences (murine Golgi α-mannosidase II, human galactosyl transferase), 13 aa MYC-derived epitope	Targeting of GFP fusion proteins to Golgi	Cole et al., 1996

(*continued*)

TABLE A.I.1 (continued)

Cells/Organism	Vector	Promoter	Other Regulatory/ Sequence Elements	Purpose	References
	pcDNA I	Human CMV IE promoter	SV40 *ori*, polyoma *ori*, SV40 poly-A signal, protein-specific signal sequences (murine Golgi α-mannosidase II, human galactosyl transferase), 13 aa MYC-derived epitope		
COS-1	pCMX	human CMV IE promoter	Kozak translation start sequence, 3′ SV40 small t intron + poly-A signal	Monitor temperature dependence of promoter activation	Ogawa et al., 1995
HLtat or human 293 cells	pBsrev, pBsrevM10 BL	HIV-1 (human immunodeficiency virus–1) 5′ LTR (U3 sequence + R segment up to nt +80)	Rev response element (RRE) binding domain and nuclear/nucleolar localization signal, SV40 poly-A signal	Analyze trafficking and interactions of WT and mutant proteins	Stauber et al., 1995
BOSC23 fibroblast cells (derived from SV40 T antigen-transformed 293 HEK cells), NIH3T3 cells, and PA317 murine fibroblast cells	MSCVneoEB (MSCV retroviral vector)	MSCV LTR	600 bp IRES (internal ribosome entry sequence) or neo^R controlled by PGK-1 promoter	Gene transfer using a retroviral vector, FACS analysis	Cheng et al., 1996

Plants					
Arabidopsis thaliana	pBI121	CaMV (cauliflower mosaic virus) 35S promoter	Mutated cryptic splice junctions, altered codon usage to decrease AU content of mRNA	Demonstrate GFP expression in transgenic plants	Haseloff and Amos, 1995
Citrus sinensis (sweet orange) protoplasts	pBI221	CaMV 35S promoter	*nos (nopaline synthase)* transcriptional terminator	Demonstrate GFP expression	Niedz et al., 1995
Glycine max (soybean) cultured cells		CaMV 35S promoter	*gfp* without cryptic splice site	Demonstrate GFP expression	Plautz et al., 1996
Hordeum vulgare (barley) protoplasts	pRTL2 GUS/NIa[Δ] Bam	CaMV 35S promoter	Duplicated CaMV transcriptional enhancer, *Shrunken-1* exon1/intron1 sequences, CaMV poly-A signal, *gfp* with mutated cryptic splice site	Demonstrate expression of GFP variants	Reichel et al., 1996
Nicotiana benthamiana (tobacco)	pOb (tobacco mosaic tobamovirus vector)	Ob *mp (movement protein)* and *cp (coat protein)* gene promoters		Protein localization	Epel et al., 1996
	pU3/12-RO-PL, pU3/12 (TMV vectors)	TMV *mp* and *cp* promoters		Protein localization	Heinlein et al., 1995
Nicotiana benthamiana, Nicotiana clevelandii	pPC2S (PVX vector)	PVX *cp* promoter		Reporter for viral gene expression	Baulcombe et al., 1995

(continued)

TABLE A.I.1 (continued)

Cells/Organism	Vector	Promoter	Other Regulatory/ Sequence Elements	Purpose	References
Zea mays (maize) protoplasts	BlueCATKS	CaMV 35S promoter, *Arabidopsis* HSP81-1 promoter	*nos* 3′ flanking region	Demonstrate transient and heat-inducible GFP expression in maize cells	Hu and Cheng, 1995
	35SC4ppp dkl-CAT	35SC4ppdk promoter (hybrid promoter consisting of the 35S CaMV enhancer fused to maize C4PPDK gene promoter + 5′ UTR)	*nos* 3′ terminator	FACS analysis	Galbraith et al., 1995

I.B GFP Mutants

Wild-type GFP of *Aequorea victoria* absorbs ultraviolet (UV) and blue light with a major peak of absorbance at 395 nm and a minor peak at 475 nm, and emits green light at 508 nm with a shoulder at 540 nm. There are many mutants of *A. victoria* GFP that are now available with altered fluorescence properties relative to wild type. Mutants of interest, together with their excitation and emission characteristics, are listed in Table A.I.2.

1.B.1 Red-shifted and Blue-Shifted GFP Mutants

GFP has been mutagenized and screened extensively for mutants with altered spectra or enhanced fluorescence (Heim et al., 1994; Delagrave et al., 1995; Ehrig et al., 1995; Heim et al., 1995; Cormack et al., 1996; Crameri et al., 1996; see Chapter 5). Many of the mutants that have been identified alter the *A. victoria* GFP chromophore at residues 64–69 and many mutant proteins have excitation spectra shifted towards red or blue of the visible spectrum. The mutant that has proven most useful so far has been the red-shifted S65T mutant (Heim et al., 1995).

The S65T GFP is about six-fold brighter than wild type, and has been reported to undergo post-translational oxidation four-fold faster than wild type. The S65T mutant GFP has also been reported to be more resistant to photobleaching than wild type when excited at 280 nm (Cubitt et al., 1996). However, in cultured mammalian cells the S65T mutant has been found to photobleach slightly faster than wild-type GFP when excited at 488 nm, even after accounting for increases in wild-type GFP fluorescence due to photoisomerization (Patterson and Piston, unpublished results).

The excitation peak of the S65T red-shifted mutant corresponds more closely than wild type to the excitation wavelength of available FITC filter sets, resulting in a signal that is much brighter than that of wild-type GFP. In addition, the argon ion laser used for FACS and laser scanning confocal microscopy emits at 488 nm, so excitation of the S65T mutant GFP is much more efficient using these detection systems than excitation of wild-type GFP.

Mutants of GFP that excite or fluoresce at wavelengths longer or shorter than that of wild type are being sought for their potential use in double-labeling and fluorescence resonance energy transfer (FRET) experiments. Some mutants with clearly distinguishable excitation and emission peaks relative to wild type have been reported, but their usefulness is limited due to weak fluorescence or rapid rates of photobleaching. The blue mutants reported to date, for example, fluoresce dimly and photobleach within seconds. These mutants can be of value when used for some types of analysis, such as FACS with its millisecond detection times, but are not generally useful for microscopy. Dual color FACS analysis has recently been carried out with mammalian cells expressing both wild-type and Y66H GFP (Ropp et al., 1996). A detailed discussion of available mutants of GFP and their characteristics can be found in Chapter 5.

TABLE A.I.2 Green fluorescent protein mutants

GFP Type	Absorbance peak (nm)	Emission peak (nm)	Color	Special Characteristics	References
Wild type					
A. victoria WT GFP	395 (475)	508 (540)	Green		
Renilla reniformis GFP	498	508	Green		
Brighter green fluorescent mutants					
cycle 3 mut: F99S, M153T, V163A[a]	395 (475)	508 (540)	Green	(i) 45-fold brighter whole cell fluorescence relative to wild-type GFP when excited by UV light (ii) 4-fold less bright whole *E. coli* cell fluorescence relative to *EGFPmut1* when excited at 488 nm	Crameri et al., 1996
EGFPmut1: F64L, S65T	488	507	Green	35-fold fluorescence intensity increase relative to WT for equal amounts of soluble protein from *E. coli*	Cormack et al., 1996
EGFPmut2: S65A, V68L, S72A	481	507	Green	19-fold fluorescence intensity increase relative to WT for equal amounts of soluble protein from *E. coli*	Cormack et al., 1996
EGFPmut3: S65G, S72A	501	511	Green	21-fold fluorescence intensity increase relative to WT for equal amounts of soluble protein from *E. coli*	Cormack et al., 1996
RSGFP4: F64M, S65G, Q69L	490	505	Green	24-fold brighter than WT by FACS	Delagrave et al., 1995; Yang et al., 1996; Cheng et al., 1996
S65T	488	511	Green	(i) 6-fold greater peak amplitude of *E. coli* expressed protein relative to WT (ii) 18-fold brighter than WT by FACS (ii) 4-fold faster oxidation (iv) No photoisomerization (v) Very slow photobleaching when excited at 280 nm	Heim et al.,1995; Cheng et al., 1996
smRS-GFP; S65T, F99S, M153T, V163A	489	511	Green	Solubility changes, S65T mutant background	Davis and Vierstra, 1996

Green fluorescent mutants					
clone 5B,9B: S65T, T203H	512	524	Green		Ormö et al., 1966
clone 6C: S65T, T203Y	513	525	Green		Ormö et al., 1966
clone 10B: F64L, S65G, S72A, T203Y	513	525	Green		Ormö et al., 1966
clone 10C: S65G V68L, S72A, T203Y	513	527	Yellow/ green	Longest wavelength emission peak	Ormö et al., 1966
clone 11: S65G, S72A, T203W	502	512	Green		Ormö et al., 1966
E222G	481	506	Green		Ehrig et al., 1995
H9: S202F, T203I	398	511	Green		Heim et al., 1994
P4-1: S65T, K238E, M153A	504 (396)	514	Green		Heim and Tsien, 1996
P9: I167V	471 (396)	502 (507)	Green		Heim et al., 1994
P11: I167T	471 (396)	502 (507)	Green		Heim et al., 1994
S65A	471	504	Green		Heim et al., 1994
S65C	479	507	Green		Heim et al., 1995
S65L	484	510	Green		Cubitt et al., 1995
T203I	400	512	Green		Ehrig et al., 1995

(continued)

TABLE A.I.2 (continued)

GFP Type	Absorbance peak (nm)	Emission peak (nm)	Color	Special Characteristics	References
Blue fluorescent mutants					
BFP5: F64M, Y66H, V68I	385	450	Blue	Used in FRET assay with *RSGFP4*	Mitra et al., 1996
P4: Y66H	382	448	Blue		Heim et al., 1994
P4-3: Y66H, Y145F	381	445	Blue		Heim and Tsien, 1996
smBFP: Y66H, F99S, M153T, V163A	382	448	Blue	Solubility changes, Y66H mutant background	Davis and Vierstra, 1996
W: Y66W	458	480	Blue		Heim et al., 1994
W2: Y66W, I123V, Y145H, H148R, M153T, V163A, N212K	432 (453)	480	Blue		Heim and Tsien, 1996
W7: Y66W, N146I, M153T, V163A, N212K	433 (453)	475 (501)	Blue		Heim and Tsien, 1996
Y66F	360	442	Blue	Shortest wavelength excitation and emission peaks, less bright than WT	Cubitt et al., 1995
EBFP:F64L, S65T, Y66H, Y145F	380	440	Blue	Brighter signal	Yang et al., 1998

[a] Numbering with reference to wild-type GFP. The mutant GFP was induced in a *gfp* gene that encoded an Ala insertion following the initiator Met, increasing the residue number by +1 relative to wild-type GFP (Crameri et al., 1996).

I.B.2 Enhanced Green Fluorescent Proteins

Additional mutants of GFP that fluoresce more brightly than wild type have been recovered by screening bacteria using FACS (Cormack et al., 1996). These mutants were selected for their enhanced fluorescence, red-shifted absorption spectra, and rapid kinetics of chromophore formation in bacteria. The mutants are described in greater detail in Chapter 5. These enhanced GFP (EGFP) mutants contain double or triple missense mutations in the coding region.

Three enhanced GFP mutants have been described, *EGFPmut1-3*. *EGFPmut1* (F64L, S65T) is commercially available from CLONTECH as EGFP and is the brightest of the three as determined by comparison of total units of soluble protein. EGFP is the most sensitive of the enhanced GFP mutants for protein localization studies. Both *EGFPmut2* (S65A, V68L, S72A) and *EGFPmut3* (S65G, S72A) are not as fluorescent per unit of soluble protein as *EGFPmut1* but they appear brighter when expressed in bacteria, most likely due to higher solubility (Cormack et al., 1996). Flow cytometry analysis of *E. coli* expressing *EGFPmut1-3* from P*tac* and T7 expression systems indicates that *EGFPmut3* achieves the highest fluorescence per bacterium (Cormack et al., 1996). The *EGFPmut3* is therefore the mutant of choice for gene expression studies in bacteria, although *EGFPmut1* and *EGFPmut2* are also suitable.

Unlike conventional bacterial gene reporters such as *lacZ* or *cat*, GFP is not an enzyme. There is therefore no signal amplification derived from multiple substrate cleavage by one molecule of the reporter protein. In general, the amount of fluorescence is directly proportional to the number of soluble GFP molecules present in the bacterial cell. Therefore, "high fluorescence" is obtained either when *gfp* transcription occurs from a single-copy strong promoter or when a weaker promoter is present on a multicopy plasmid. Most reports of GFP expression in bacteria have used medium-copy plasmids (Chalfie et al., 1994; Dhandayuthapani et al., 1995; Kremer et al., 1995; Valdivia and Falkow, 1996; Valdivia et al., 1996). In contrast, other workers have reported successful imaging of GFP–protein fusions driven from single-copy gene fusions in *B. subtilus* (Webb et al., 1995). Enhanced GFP mutants are sensitive enough to allow the imaging of fusions expressed from single-copy plasmids, especially with sensitive imaging detectors such as cooled CCD cameras or laser scanning confocal systems. For flow cytometry analysis, where sample detection time is on the order of milliseconds, promoter strength is best detected and quantitated by driving enhanced *gfp* expression from multicopy plasmids.

I.B.3 Codon Usage Variants

A *gfp* gene with a coding region altered to conform to the codon usage of *Arabidopsis* has been synthesized to improve GFP expression (Haseloff and Amos, 1995). This variant also contains mutated cryptic splice sequences to prevent aberrant splicing of transcripts in *Arabidopsis*. Altered *gfp* genes with codons optimized for *E. coli* (Crameri et al., 1996) or human cells (Zolotukhin et al., 1996; Haas et al., 1996) have also been synthesized. The "humanized" *gfp* variant of Haas et al. (1996) shows brighter fluorescence not only in human cells, but also in plant cells — upon transfection into maize leaf protoplasts, about 20-fold brighter fluorescence is observed compared to wild-type *gfp* (Chiu et al.,

1996). This variant was also successfully expressed in *Arabidopsis* leaves and roots, onion skin epidermal cells, tobacco leaf protoplasts, and transgenic tobacco plants (Chiu et al., 1996). These results can be explained by the almost identical preferred codon usage between humans and maize, and the similarity in codon usage between humans and other higher plants (Wada et al., 1991). The "humanized" backbone described by Haas et al. (1996) is used for all EGFP vectors available from CLONTECH.

I.C Toxicity Due to GFP Expression

There have been published reports that overexpression of GFP may be toxic or interfere with regeneration required to generate transgenic plants (Haseloff and Amos, 1995; Chiu et al., 1996), as well as unpublished reports of toxicity associated with the overexpression of wild-type GFP in bacteria. Greatly overexpressed EGFP in bacteria, for example, from pUC-based vectors, can cause slower growth rates and osmosensitivity (Valdivia, Cormack and Falkow, personal communication). Some bacterial species appear to tolerate high levels of GFP better than others. For example, high levels of GFP are tolerated by *Yersinia* species (Valdivia and Falkow, 1996) but not by *Salmonella* or *Anabaena* species (Valdivia and Falkow, 1996) (Buikema and Haselkorn, unpublished results). Toxicity due to high levels of GFP may also explain the difficulty that some workers have had in obtaining stably transfected mammalian cell lines. The problems that have been encountered with toxic effects associated with GFP overexpression may be a general problem associated with protein overexpression, rather than specifically due to GFP.

I.D Folding and Temperature Sensitivity

The ability of GFP to absorb blue light and emit green light is believed to depend on the formation of a chromophore by cyclization and oxidation of S-Y-G (residues 65–67) (Heim et al., 1994; Cody et al., 1993). The time constant of chromophore formation has been reported to be about 2–4 h for wild type and 0.45 h for S65T mutant GFP (Heim et al., 1994; Heim et al., 1995). The rate-limiting formation of the chromophore may limit the ability to visualize GFP fluorescence in transfected cells or transgenic organisms until a defined time after its expression. For *Drosophila*, the time required for the appearance of GFP fluorescence has been reported to be 3–5 h following expression (Davis et al., 1995). Evidence bearing on this is controversial, however, and is discussed in greater detail in Chapter 9.

Wild-type GFP has been reported to be sensitive to temperature when expressed in mammalian cells, producing a brighter fluorescent signal at 33°C compared to 37°C (Pines, 1995). Temperature sensitivity of the fluorescence of a wild-type GFP fusion protein has also been observed in studies utilizing a GFP–human glucocorticooid receptor (GFP–hGR) fusion construct. These studies showed efficient transactivation of the mouse mammary tumor virus promoter in the presence of dexamethasone at 30°C but not at 37°C (Ogawa et al., 1995), a result demonstrating that the activity of the GFP–fusion protein, as well as its fluorescence, is greatly reduced at higher temperature. These effects may be due to the folding or redox state of GFP in the cell. The studies also showed that cells fluorescing at 30°C continued to fluoresce for at least 48 h upon shifting to 37°C. The time course of GFP–hGR movement from the cytoplasm into the nucleus after induction could be determined by addition of hormone to cells grown at 30°C, followed by incubation for various time periods at 37°C.

Studies in yeast have shown that wild-type GFP and a wild-type GFP fusion protein expressed in *S. cerevisiae* showed markedly reduced fluorescence when cells were grown at ≥30°C, and that fluorescent cells grown at lower temperature retained their fluorescence after a shift to higher temperature (Lim et al., 1995). These observations allowed the workers to monitor relocalization of a GFP–nucleoplasmin fusion protein in a temperature-sensitive mutant of the nucleoporin gene by first culturing cells at 23°C to allow the fusion protein to accumulate, then shifting to 35°C.

These studies illustrate the usefulness of the temperature sensitivity of wild-type GFP given appropriate experimental design. Such temperature sensitivity has not been reported for mutant forms of GFP expressed in mammalian cells. It should be noted, however, that the autocatalytic folding process may be less efficient under certain conditions, and that optimized expression protocols are likely to continue to be developed over the next few years.

I.E Purification of GFP

The original purification of *Aequorea* GFP from photogenic organs of the jellyfish (Morise et al., 1974) is described in Chapter 1. The cloning of the *gfp* gene (Prasher et al., 1992) has permitted expression of GFP in bacteria for biochemical and biophysical studies. Bacterially-expressed GFP has been characterized in clarified induced cell lysates without further purification (Heim et al., 1994) or following purification on a Ni-affinity column of a His_6-tagged GFP containing a 34-residue peptide with six contiguous His residues fused to the N-terminus of GFP (Inouye and Tsuji, 1994). A fusion of GFP to GST (glutathione *S*-transferase) has also been made and purified by glutathione affinity chromatography (Niswender et al., 1995). Purification of bacterially-expressed GFP without an affinity tag has also been achieved and is detailed in Protocol 1.

PROTOCOL 1 Purification of recombinant GFP from bacteria

The following protocol gives high yields of purified GFP from a high-expression strain derived from TU#58 (Chalfie et al., 1994). The yield of purified protein has been as much as 150 mg/L.

This protocol has also been used to purify native GFP directly from the jellyfish, *A. victoria*, and several recombinant wild-type or mutant GFPs. One example is the mutant GFP expressed from the pBAD/*gfp* construct (Crameri et al., 1996). The mutant *gfp* in this vector is under the tight control of the *arabinose* promoter/repressor, *araBAD*, and can be induced continuously with a final concentration of 0.2% L(+) arabinose (w/v) in LB. We have transformed the *E. coli* strain DH5α with pBAD/*gfp* and have grown 10 L of LB under continuous induction at 28°C for 24 h before harvesting. Ultimately, we were able to purify 50 mg of mutant GFP from the 10 L.

The most readily available high GFP expressing construct is TU#60 (Chalfie et al., 1994) sold by CLONTECH as pGFP. The *gfp* gene in pGFP is fused in-frame to the *lacZ* initiation codon from pUC19 (which adds an additional 24 amino acids to the N-terminus of GFP) that allows for high expression from the *lac* promoter. The pGFP vector also contains the *bla* gene for ampicillin selection and is a high copy number plasmid. Suitable *E. coli* strains which can be used to produce the protein include DH5α, JM109, and TB1 (New England BioLabs, Inc.).

The GFP purified according to this protocol is suitable for biochemical and biophysical experiments, including crystallization trials.

Materials

Bacterial strain derived from TU#58 [pET3a/*gfp*, BL21(DE3)]

This strain expresses GFP as a nonfusion protein under the control of the T7 Φ10-s10 promoter fragment. The transcription of the gfp gene is directly controlled by the T7 RNA polymerase. For transcription of gfp, TU#58 must be maintained in E. coli cells lysogenic for the λ phage derivative, DE3 (Studier et al., 1990). The λDE3 lysogen carries the T7 RNA polymerase gene driven by the IPTG-inducible lacUV5 promoter. The pET3a plasmid also contains an amp^R *gene for selection. Cells are maintained continuously on ampicillin selection plates and selected on the basis of fluorescence prior to large-scale culture.*

LB (Miller, 1972)
10 g NaCl
10 g Bacto tryptone (Difco)
5 g Bacto yeast extract (Difco)
Add DW to 1 L, autoclave.
Add ampicillin to 37 μg/mL for cell culture.

LB + ampicillin plates
1 L LB
14 g Bacto agar (Difco)
10 g Lactose
Add ampicillin to 37 μg/mL prior to pouring plates.
This concentration of ampicillin is suitable for the copy number of the plasmid. For pGFP, the optimal ampicillin concentration is 60 μg/mL.

0.5 M IPTG (isopropyl-β-D-thiogalactopyranoside) in DW

Extraction buffer
25 mM Tris-HCl, pH 8.0
1 mM β-mercaptoethanol (Eastman Kodak Co.)

0.1 M Phenylmethylsulfonylfluoride (PMSF) (Sigma) in 2-Propanol

Low ionic strength buffer
10 mM Tris-HCl, pH 8.0
10 mM EDTA

50 mg/mL protamine sulfate (Sigma) in DW
The protamine sulfate will not be in solution at room temperature. Warm the bottle under hot tap water immediately before dispensing.

Ammonium sulfate (solid)

Tris base (solid)

Octyl agarose column buffer
10 mM Tris-HCl, pH 8.0
10 mM EDTA
1.0 M ammonium sulfate

Sepharose column buffer
5 mM Tris-HCl, pH 8.0
0.02% NaN_3

DEAE column buffer
5 mM Tris-HCl, pH 8.0
0.02% NaN_3

Special Equipment

Chromatography columns
Octyl agarose HIC (Hydrophobic Interaction Chromatography) column (Pharmacia Biotech), 2.5 × 12 cm
Pre-equilibrate the column with octyl agarose column buffer.

Sepharose CL-6B (Pharmacia Biotech), 3 × 95 cm
Pre-equilibrate the column with sepharose column buffer.

DEAE Sepharose Fast Flow (Pharmacia Biotech), 2.5 × 17 cm
Pre-equilibrate the column with DEAE column buffer.

1. Select a brightly fluorescent colony from an LB + ampicillin plate and use it to inoculate 50 mL of LB + ampicillin. Grow overnight at 37°C.
   Visually select the most intensely green fluorescent colony from an ampicillin plate by placing the plate on a hand-held long-wave UV lamp (λ_{max}=365 nm).
2. Inoculate a flask containing 1 L of LB + ampicillin with the 50 mL overnight culture. Grow at 37°C to OD_{660} = 0.8, then add IPTG to a final concentration of 0.5 mM. Induce cells at 37°C for 12 h.

No significantly higher production of GFP is obtained by longer incubation. The ideal temperature of expression is 37°C, but reasonable expression (up to 100 mg/L of LB) is achieved at 28°C. In our initial work with the TU#58-derived pET3a construct we obtained lower yields of 1–3 mg/L in LB. We increased the yields to 10 mg/L of LB broth by growing the E. coli cells at 28°C. With improvements in our colony selection technique, we now have cells that are considerably more productive and we are now able to grow these cells at 37°C with excellent GFP yields.

3. Collect the GFP-producing *E. coli* cells by gentle centrifugation at 1000x g for 15 minutes at 4°C.

   Pellets can be stored at this step by freezing after harvesting.

4. Extract the GFP from induced cells by repeated cycles of freezing and thawing.

   Slowly freeze (60 min at –20°C) and slowly thaw (60 min at room temperature) the packed pellets through three cycles (Johnson and Hecht, 1994).

5. Follow this by two to four cold buffer washes (20X pellet volume) with the extraction buffer. Collect the wash supernatant by centrifugation (10,000x g, 15 min) and add PMSF to a final concentration of 1 mM.

   The PMSF will help prevent proteolytic cleavage of the protease-susceptible C-terminal "tail" of GFP. Usually it is necessary to freeze pellets between buffer washes to release all the GFP. While the freeze–thaw process is slower than other methods such as sonication or lysozyme treatment, the freeze–thaw extracts are remarkably clean (low viscosity, low DNA content, and high GFP content – up to 10% of total soluble protein). More than 90% of the GFP can be released from the cells by this method.

6. Treat clarified extracts in low ionic strength buffer at 0–4°C with protamine sulfate to remove residual nucleic acids. Generally 1 mg protamine sulfate per 100 OD_{260} units is sufficient to precipitate most of the DNA, but not the GFP. Remove the precipitate by centrifugation (5000x g, 5 min).

   Add the protamine sulfate dropwise while stirring rapidly, so as not to precipitate GFP in localized regions of the extract. It is advisable to test each batch of GFP by small-scale titration in microfuge tubes to avoid "overshooting" the titration. It is not easy to recover GFP that is inadvertently precipitated by protamine sulfate.

7. Precipitate the protamine-treated and clarified extract with ammonium sulfate (100% of saturation, 697 g/L extract) at 0°C. Add approximately 10 g of solid Tris base per liter of extract during the precipitation step to maintain a pH near 7.0.

   The pH of unbuffered saturated ammonium sulfate is close to 5.5, dangerously close to the low end of the GFP pH stability range. Generally, the precipitation of GFP is rapid. Collect the precipitate by centrifugation (10,000 × g, 30 min) within an hour of precipitation. Expect near quantitative recovery if the GFP concentration in the crude extract is $\geqslant$ 0.2 mg/mL.

8. Dissolve the GFP-containing pellet in a minimal volume (just sufficient to dissolve the GFP) of octyl agarose column buffer containing 1 mM

PMSF. Clarify the dissolved pellet by centrifugation (15,000x g, 20 min).

Pelleted GFP, which appears yellow, will take on the familiar bright green color as it goes into solution and the entire suspension becomes clear.

9. Load the clarified solution, at room temperature, onto an octyl agarose HIC column pre-equilibrated with column buffer. Elute the column stepwise, first with 250 mL of low ionic strength buffer + 0.5 M ammonium sulfate, and then with 250 mL of low ionic strength buffer without ammonium sulfate.

   Complete elution of GFP requires 100–200 mL of the second buffer solution. The column is capable of binding more than 1 g of total protein and can be eluted free of GFP in less than 2 h. Hydrophobicity of octyl agarose columns varies greatly with the length and chemical nature of the spacer arm. We prefer a three-carbon spacer with ether linkage to the agarose beads.

10. Concentrate the GFP sample to a volume of 2–5 mL by ultrafiltration or ammonium sulfate precipitation.
11. Chromatograph at room temperature on a column of Sepharose CL-6B, preequilibrated in column buffer, at a flow rate of 1.4 mL/min.

    GFP elutes from this column at an apparent molecular weight of about 40 kDa, indicating significant dimerization.

    a. Alternate two-step purification
    Reversible dimerization can be used in an alternative purification scheme following gel filtration on Sepharose to achieve ~95% purity.

Additional Materials

Bio-Gel column buffer
10 mM Tris-HCl, pH 8.0
10 mM EDTA
1 M ammonium sulfate
0.02% NaN_3

Additional Special Equipment

Two columns of Bio-Gel P-100 medium resin (Bio-Rad), 10 × 120 cm (~8 L) and 3 × 120 cm (~0.75 L)

1. Run the larger Bio-Gel P-100 column at room temperature with a dilute GFP sample (0.2 mg/mL).

   Partial hydrophobic interaction in high salt causes GFP to elute at an apparent molecular weight of 21 kDa.

2. Then run the second smaller column at room temperature with a very concentrated GFP sample (20–100 mg/mL) in the Bio-Gel column buffer, with or without ammonium sulfate.

   GFP dimerizes at high-protein concentrations and elutes at an apparent molecular weight of 44 kDa. Nearly all contaminants that co-elute with GFP on the first column are removed on the second.

TABLE A.I.3 GFP absorption ratios useful in establishing purity

Protein	Chromophore λ_{max}	ε	λ_1/λ_2	Numerical Ratio
Native *Aequorea* GFP	395 470	27,600 12,000	395/280	1.25
Recombinant wild-type GFP	397 475	27,000 12,000	397/280	1.25
S65T	489	56,000	489/280	2.25
Stemmer mutant *cycle 3 mut*: F99S, M153T, V163A	397 475	27,000 12,000	397/280	1.25
P4: Y66H	382	25,000	382/280	1.23

12. Final polishing to achieve >95% purity is on a DEAE Sepharose Fast Flow column at room temperature. Load the sample in the column buffer by gravity at a flow rate of 2-4 mL/min and elute with a 2.0 L gradient of salt (0 to 0.5 M NaCl) in the same buffer.
Gravity-driven flow rates of 5 mL/min can be achieved with excellent resolution, equaling that achieved with a 4 hour long shallow salt gradient on Pharmacia's Mono Q FPLC column. In fact, minor isoforms of GFP that differ by one charged amino acid are quantitatively removed on DEAE Fast Flow. This column is capable of purifying up to 1 g of GFP.
13. Judge the purity of GFP by the ratio of the absorbance of the chromophore at its λ_{max} to that of the aromatic region of the protein at 280 nm.
Note that, in the chromophore absorption band, the wild-type recombinant GFP, the so-called "Stemmer" mutant [cycle 3 mut (F99S, M153T, V163A)] (Crameri et al., 1996), and native Aequorea GFP all fail to follow Beer's law. For these three types of GFP, as protein concentration increases, the absorbance at 395 nm increases disproportionately while the absorbance at 475 nm decreases disproportionately. The suppression of the 475 nm shoulder upon dimerization (as occurs in cells overexpressing GFP) can be as great as 5-fold, making these forms of GFP exceedingly poor absorbers of blue light (molar extinction coefficient $\leqslant$3,000 for the dimer). Thus, for accurate quantitation, it is necessary to measure absorbance at low protein concentration (0.05-0.20 mg/mL) in a concentration range that obeys Beer's law. Table A.I.3 shows a partial list of GFP forms and spectral characteristics that may be used to judge purity.

CONTRIBUTED by DANIEL G. GONZÁLEZ and WILLIAM W. WARD

REFERENCES

Amsterdam, A., Lin, S., and Hopkins, N. (1995). The *Aequorea victoria* green fluorescent protein can be used as a reporter in live zebrafish embryos. *Dev. Biol.* 171:123–129.

Ausubel, F. M., Brent, R., Kingston, R. E., Moore, D. D., Seidman, J. G., Smith, J. A., and Struhl, K. (1996). Current Protocols in Molecular Biology. Wiley, New York.

Barlett, R. J., Singer, J. T., Shean, M. K., Weindler, F. W., and Secore, S. L. (1995). Expression of a fluorescent reporter gene in skeletal muscle cells from a virus-based vector: implications for diabetes treatment. *Transplantation Proc.* 27:3344.

Barthmaier, P. and Fyrberg, E. (1995). Monitoring development and pathology of *Drosophila* indirect flight muscles using green fluorescent protein. *Dev. Biol.* 169:770–774.

Baulcombe, D. C., Chapman, S., and Santa Cruz, S. (1995). Jellyfish green fluorescent protein as a reporter for virus infections. *Plant J.* 7:1045–1053.

Casper, S. J. and Holt, C. A. (1996). Expression of the green fluorescent protein-encoding gene from a tobacco mosaic virus-based vector. *Gene* 173:69–73.

Chalfie, M., Tu, Y., Euskirchen, G., Ward, W. W., and Prasher, D. C. (1994). Green fluorescent protein as a marker for gene expression. *Science* 263:802–805.

Cheng, L., Fu, J., Tsukamoto, A., and Hawley, R. G. (1996). Use of green fluorescent protein variants to monitor gene transfer and expression in mammalian cells. *Nature Biotech.* 14:606–609.

Chiu, W.-L., Niwa, Y., Zeng, W., Hirano, T., Kobayashi, H., and Sheen, J. (1996). Engineered GFP as a vital reporter in plants. *Curr. Biol.* 6:325–330.

Christensen, B. B., Sternberg, C., and Molin, S. (1996). Bacterial plasmid conjugation on semisolid surfaces monitored with the green fluorescent protein (GFP) from *Aequorea victoria* as a marker. *Gene* 173:59–65.

Cody, C. W., Prasher, D. C., Westler, W. M., Prendergast, F. G., and Ward, W. W. (1993). Chemical structure of the hexapeptide chromophore of the *Aequorea* green-fluorescent protein. *Biochemistry* 32:1212–1218.

Cole, N. B., Smith, C. L., Sciaky, N., Terasaki, M., Edidin, M., and Lippincott-Schwartz, J. (1996). Diffusional mobility of Golgi proteins in membranes of living cells. *Science* 273:797–801.

Cormack, B .P., Valdivia, R. H., and Falkow, S. (1996). FACS-optimized mutants of the green fluorescent protein (GFP). *Gene* 173:33–38.

Crameri, A., Whitehorn, E. A., Tate, E., and Stemmer, W. P. C. (1996). Improved green fluorescent protein by molecular evolution using DNA shuffling. *Nature Biotech.* 14: 315–319.

Cubitt, A. B., Heim, R., Adams, S. R., Boyd, A. E., Gross, L. A., and Tsien, R. Y. (1995). Understanding, improving and using green fluorescent proteins. *Trends Biochem. Sci.* 20:448–455.

Davis, I., Girdham, C. H., and O'Farrell, P. H. (1995). A nuclear GFP that marks nuclei in living *Drosophila* embryos; maternal supply overcomes a delay in the appearance of zygotic fluorescence. *Dev. Biol.* 170:726–729.

Davis, S. J. and Vierstra, R. D. (1996). Soluble derivatives of green fluorescent protein (GFP) for use in *Arabidopsis thaliana.* http://genome-www.stanford.edu/Arabidopsis/ww/.

Delagrave, S., Hawtin, R. E., Silva, C. M., Yang, M. M., and Youvan, D. C. (1995). Red-shifted excitation mutants of the green fluorescent protein. *BioTechnology* 13: 151–154.

Dhandayuthapani, S., Via, L. E., Thomas, C. A., Horowitz, P. M., Deretic, D., and Deretic, V. (1995). Green fluorescent protein as a marker for gene expression and cell biology of mycobacterial interactions with macrophages. *Mol. Microbiol.* 17:901–912.

Doyle, T. and Botstein, D. (1996). Movement of yeast cortical actin cytoskeleton visualized in vivo. *Proc. Natl. Acad. Sci. USA* 93:3886–3891.

Ehrig, T., O'Kane, D. J., and Prendergast, F. G. (1995). Green-fluorescent protein mutants with altered fluorescence excitation spectra. *FEBS Lett.* 367:163–166.

Endow, S. A. and Komma, D. J. (1996). Centrosome and spindle function of the *Drosophila* Ncd microtubule motor visualized in live embryos using Ncd-GFP fusion proteins. *J. Cell Sci.* 109:2429–2442.

Epel, B. L., Padgett, H. S., Heinlein, M., and Beachy, R. N. (1996). Plant virus movement protein dynamics probed with a GFP-protein fusion. *Gene* 173:75–79.

Fey, P., Compton, K., and Cox, E. C. (1995). Green fluorescent protein production in the cellular slime molds *Polysphondylium pallidum and Dictyostelium discoideum*. *Gene* 165: 127–130.

Fire, A., Harrison, S. W., and Dixon, D. (1990). A modular set of lacZ fusion vectors for studying gene expression in *Caenorhabditis elegans*. *Gene* 93:189–198.

Galbraith, D. W., Lambert, G. M., Grebenok, R. J., and Sheen, J. (1995). Flow cytometric analysis of transgene expression in higher plants: green-fluorescent protein. *Methods Cell Biol.* 50:3–12.

Haas, J., Park, E.-C., and Seed, B. (1996). Codon usage limitation in the expression of HIV-1 envelope glycoprotein. *Curr. Biol.* 6:315–324.

Haseloff, J. and Amos, B. (1995). GFP in plants. *Trends Genet.* 11:328–329.

Heim, R., Cubitt, A. B., and Tsien, R. Y. (1995). Improved green fluorescence. *Nature (London)* 373:663–664.

Heim, R., Prasher, D. C., and Tsien, R. Y. (1994). Wavelength mutations and posttranslational autoxidation of green fluorescent protein. *Proc. Natl. Acad. Sci. USA* 91: 12501–12504.

Heim, R. and Tsien, R. Y. (1996). Engineering green fluorescent protein for improved brightness, longer wavelengths and fluorescence resonance energy transfer. *Curr. Biol.* 6: 178–182.

Heinlein, M., Epel, B. L., Padgett, H. S., and Beachy, R. N. (1995). Interaction of tobamovirus movement proteins with the plant cytoskeleton. *Science* 270:1983–1985.

Hu, W. and Cheng, C.-L. (1995). Expression of *Aequorea* green fluorescent protein in plant cells. *FEBS Lett.* 369:331–334.

Ikawa, M., Kominami, K., Yoshimura, Y., Tanaka, K., Nishimune, Y., and Okabe, M. (1995). Green fluorescent protein as a marker in transgenic mice. *Dev. Growth Differ.* 37: 455–459.

Inouye, S. and Tsuji, F. I. (1994). *Aequorea* green fluorescent protein Expression of the gene and fluorescence characteristics of the recombinant protein. *FEBS Lett.* 341:277–280.

Johnson, B. H. and Hecht, M. H. (1994). Recombinant proteins can be isolated from *E. coli* cells by repeated cycles of freezing and thawing. *BioTechnol.* 12:1357–1360.

Kain, S. R., Adams, M., Kondepudi, A., Yang, T.-T., Ward, W. W., and Kitts, P. (1995). Green fluorescent protein as a reporter of gene expression and protein localization. *BioTechniques* 19:650–655.

Kerrebrock, A. W., Moore, D. P., Wu, J. S., and Orr-Weaver, T. L. (1995). Mei-S332, a Drosophila protein required for sister-chromatid cohesion, can localize to meiotic centromere regions. *Cell* 83:247–256.

Kozak, M. (1987). An analysis of 5′-noncoding sequences from 699 vertebrate messenger RNAs. *Nucleic Acids Res.* 15:8125–8148.

Kremer, L., Baulard, A., Estaquier, J., Poulain-Godefroy, O., and Locht, C. (1995). Green fluorescent protein as a new expression marker in myobacteria. *Mol. Microbiol.* 17: 913–922.

Levy, J. P., Muldoon, R. R., Zolotukhin, S., and Link, C. J., Jr. (1996). Retroviral transfer and expression of a humanized, red-shifted green fluorescent protein gene into human tumor cells. *Nature Biotech.* 14:610–614.

Lim, C. R., Kimata, Y., Nomaguchi, K., and Kohno, K. (1995). Thermosensitivity of green fluorescent protein fluorescence utilized to reveal novel nuclear-like compartments in a mutant nucleoporin NSP1. *J. Biochem.* 118:13–17.

Maniak, M., Rauchenberger, R., Albrecht, R., Murphy, J., and Gerisch, G. (1995). Coronin involved in phagocytosis: dynamics of particle-induced relocalization visualized by a green fluorescent protein tag. *Cell 83:915–924.*

Miller, J. H. (1972). Experiments in molecular genetics. Cold Spring Harbor Laboratory, Cold Spring Harbor, New York.

Mitra, R. D., Silva, C. M., and Youvan, D. C. (1996). Fluorescence resonance energy transfer between blue-emitting and red-shifted excitation derivatives of the green fluorescent protein. *Gene* 173:13–17.

Morise, H., Shimomura, O., Johnson, F. H., and Winant, J. (1974). Intermolecular energy transfer in the bioluminescent system of *Aequorea. Biochemistry* 13:2656–2662.

Moriyoshi, K., Richards, L. J., Akazawa, C., O'Leary, D. D. M., and Nakanishi, S. (1996). Labeling neural cells using adenoviral gene transfer of membrane-targeted GFP. *Neuron* 16:255–260.

Moss, J. B., Price, A. L., Raz, E., Driever, W., and Rosenthal, N. (1996). Green fluorescent protein marks skeletal muscle in murine cell lines and zebrafish. *Gene* 173:89–98.

Nabeshima, K., Kurooka, H., Takeuchi, M., Kinoshita, K., Nakaseko, Y., and Yanagida, M. (1995). $p93^{dis1}$, which is required for sister chromatid separation, is a novel microtubule and spindle pole body-associating protein phosphorylated at the Cdc2 target sites. *Genes Dev.* 9:1572–1585.

Niedz, R. P., Sussman, M., and Satterlee, J. S. (1995). Green fluorescent protein: an in vivo reporter of plant gene expression. *Plant Cell Rep.* 14:403–406.

Niswender, K. D., Blackman, S. M., Rohde, L., Magnuson, M. A., and Piston, D. W. (1995). Quantitative imaging of green fluorescent protein in cultured cells: comparison of microscopic techniques, use in fusion proteins and detection limits. *J. Microscopy* 180:109–116.

Ogawa, H., Inouye, S., Tsuji, F. I., Yasuda, K., and Umesono, K. (1995). Localization, trafficking, and temperature-dependence of the *Aequorea* green fluorescent protein in cultured vertebrate cells. *Proc. Natl. Acad. Sci. USA* 92:11899–11903.

Oparka, K. J., Roberts, A. G., Prior, D. A. M., Chapman, S., Baulcombe, D., and Santa Cruz, S. (1995). Imaging the green fluorescent protein in plants—viruses carry the torch. *Protoplasma* 189:133–141.

Ormö, M., Cubitt, A. B., Kallio, K., Gross, L. A., Tsien, R. Y., and Remington, S. J. (1966). Crystal structure of the *Aequorea victoria* green fluorescent protein. *Science* 273:1392–1395.

Patino, M. M., Liu, J.-J., Glover, J. R., and Lindquist, S. (1996). Support for the prion hypothesis for inheritance of a phenotypic trait in yeast. *Science* 273:622–626.

Peel, A. L., Zolotukhin, S., Schrimsher, G. W., Muzyczka, N., and Reier, P. J. (1997). Efficient transduction of green fluorescent protein in spinal cord neurons using adeno-associated virus vectors containing cell-type specific promoters. *Gene Therapy* 4:16–24.

Pines, J. (1995). GFP in mammalian cells. *Trends Genet.* 11:326–327.

Plautz, J. D., Day, R. N., Dailey, G. M., Welsh, S. B., Hall, J. C., Halpain, S., and Kay, S. A. (1996). Green fluorescent protein and its derivatives as versatile markers for gene expression in living *Drosophila melanogaster*, plant and mammalian cells. *Gene* 173:83–87.

Prasher, D. C., Eckenrode, V. K., Ward, W. W., Prendergast, F. G., and Cormier, M. J. (1992). Primary structure of the *Aequorea victoria* green-fluorescent protein. *Gene* 111:229–233.

Reichel, C., Matthur, J., Eckes, P., Langenkemper, K., Koncz, C., Schell, J., Reiss, B., and Maas, C. (1996). Enhanced green fluorescence by the expression of an *Aequorea victoria* green fluorescent protein mutant in mono- and dicotyledonous plant cells. *Proc. Natl. Acad. Sci. USA* 93:5888–5893.

Rizzuto, R., Brini, M., De Giorgi, F., Rossi, R., Heim, R., Tsien, R. Y., and Pozzan, T. (1996). Double labelling of subcellular structures with organelle-targeted GFP mutants *in vivo*. *Curr. Biol.* 6:183–188.

Rizzuto, R., Brini, M., P., P., Murgia, M., and Pozzan, T. (1995). Chimeric green fluorescent protein as a tool for visualizing subcellular organelles in living cells. *Curr. Biol.* 5:635–642.

Ropp, J. D., Donahue, C. J., Wolfgang-Kimball, D., Hooley, J. J., Chin, J. Y. W., Hoffman, R. A., Cuthbertson, R. A., and Bauer, K. D. (1995). *Aequorea* green fluorescent protein analysis by flow cytometry. *Cytometry* 21:309–317.

Ropp, J. D., Donahue, C. J., Wolfgang-Kimball, D., Hooley, J. J., Chin, J. Y. W., Hoffman, R. A., Cuthbertson, R. A., and Bauer, K. D. (1996). *Aequorea* green fluorescent protein: simultaneous analysis of wild-type and blue-fluorescing mutant by flow cytometry. *Cytometry* 24:284–288.

Sambrook, J., Fritsch, E. F., and Maniatis, T. (1989). Molecular Cloning—A Laboratory Manual. Cold Spring Harbor Laboratory Press, Cold Spring Harbor, New York.

Sengupta, P., Colbert, H. A., and Bargmann, C. I. (1994). The *C. elegans* gene *odr-7* encodes an olfactory-specific member of the nuclear receptor superfamily. *Cell* 79:971–980.

Sheen, J., Hwang, S., Niwa, Y., Kobayashi, H., and Galbraith, D. W. (1995). Green-fluorescent protein as a new vital marker in plant cells. *Plant J.* 8:777–784.

Sidorova, J. M., Mikesell, G. E., and Breeden, L. L. (1995). Cell cycle-regulated phosphorylation of Swi6 controls its nuclear localization. *Mol. Biol. Cell* 6:1641–1658.

Stauber, R., Gaitanaris, G. A., and Pavlakis, G. N. (1995). Analysis of trafficking of Rev and transdominant Rev proteins in living cells using green fluorescent protein fusions: transdominant Rev blocks the export of Rev from the nucleus to the cytoplasm. *Virology* 213:439–449.

Studier, F. W., Rosenberg, A. H., Dunn, J. J., and Dubendorff, J. W. (1990). Use of T7 RNA polymerase to direct expression of cloned genes. *Methods Enzymol.* 185:60–89.

Takebe, Y., Seiki, M., Fujisawa, J., Hoy, P., Yokota, K., Arai, K., Yoshida, M., and Arai, N. (1988). SRα promoter: an efficient and versatile mammalian cDNA expression system composed of the simian virus 40 early promoter and the R-U5 segment of human T-cell leukemia virus type 1 long terminal repeat. *Mol. Cell Biol.* 8:466–472.

Treinin, M. and Chalfie, M. (1995). A mutated acetylcholine receptor subunit causes neuronal degeneration in *C. elegans*. *Neuron* 14:871–877.

Troemel, E. R., Chou, J. H., Dwyer, N. D., Colbert, H. A., and Bargmann, C. I. (1995). Divergent seven transmembrane receptors are candidate chemosensory receptors in *C. elegans*, *Cell* 83:207–218.

Valdivia, R. H. and Falkow, S. (1996). Bacterial genetics by flow cytometry: rapid isolation of *Salmonella typhimurium* acid-inducible promoters by differential fluorescence induction. *Mol. Microbiol.* 22:367–378.

Valdivia, R. H., Hromockyj, A. E., Monack, D., Ramakrishnan, L., and Falkow, S. (1996). Applications for the green fluorescent protein (GFP) in the study of host-pathogen interactions. *Gene* 173:47–52.

Wada, K., Wada, Y., Doi, H., Ishibashi, F., Gojobori, T., and Ikemura, T. (1991). Codon usage tabulated from the GenBank genetic sequence data. *Nucleic Acids Res.* 19:1981–1986.

Wang, S. and Hazelrigg, T. (1994). Implications for *bcd* mRNA localization from spatial distriution of *exu* protein in *Drosophila* oogenesis. *Nature (London)* 369:400–403.

Webb, C. D., Decatur, A., Teleman, A., and Losick, R. (1995). Use of green fluorescent protein for visualization of cell-specific gene expression and subcellular protein localization during sporulation in *Bacillus subtilis*. *J. Bacteriol.* 177:5906–5911.

Wu, G.-Y., Zou, D.-J., Koothan, T., and Cline, H. T. (1995). Infection of frog neurons with vaccinia virus permits in vivo expression of foreign proteins. *Neuron* 14:681–684.

Yang, T.-T., Kain, S. R., Kitts, P., Kondepudi, A., Yang, M. M., and Youvan, D. C. (1996). Dual color microscopic imagery of cells expressing the green fluorescent protein and a red-shifted variant. *Gene* 173:19–23.

Yeh, E., Gustafson, K., and Boulianne, G.L. (1995). Green fluorescent protein as a vital marker and reporter of gene expression in *Drosophila*. *Proc. Natl. Acad. Sci. USA* 92:7036–7040.

Zolotukhin, S., Potter, M., Hauswirth, W. W., Guy, J., and Muzyczka, N. (1996). A humanized green fluorescent protein cDNA adapted for high-level expression in mammalian cells. *J. Virol.* 70:4646–4654.

PROTOCOL II SPECIMEN PREPARATION

II.A Transfer of *gfp* into Cells

The standard methods for transformation, transfection, and retroviral or viral infection can be used to transfer constructs for expression of GFP into cells or organisms under study. These conventional methods have been used for many cells and organisms, including bacteria, *Dictyostelium*, *C. elegans*, *Drosophila*, mice, mammalian cells, plants, and plant protoplasts.

In some cases, however, a nonstandard method such as microinjection of DNA, RNA, or protein must be used to deliver *gfp* or GFP into live organisms. Microinjection is particularly useful in cases in which specific tissues are being targeted. For example, microinjection has been used to deliver vaccinia virus carrying *gfp* into the brain of *Xenopus laevis* tadpoles to monitor infected neurons (Wu et al., 1995) and to deliver *gfp*-carrying adeno-associated virus into the central nervous system of rats (Peel et al., 1997). Microinjection has also been used for expression of *gfp* in zebrafish, which have proven difficult to transform stably using other methods. Here, microinjection of *gfp* plasmid DNA constructs (Amsterdam et al., 1995; Moss et al., 1996), *gst-gfp* cRNA (Peters et al., 1995), or purified recombinant GFP protein (Amsterdam et al., 1996) into live embryos has been used to induce or evaluate GFP expression. The use of RNA to obtain expression is not widespread, but has also been used for *Xenopus* where *gfp* RNA was co-injected into embryos as a cell lineage marker, together with a plasmid for expression of a *Drosophila Enhancer of split* basic helix–loop–helix protein (Tannahill et al., 1995). There are some instances in which RNA is routinely used for gene transfer; one of these is for plant RNA viruses. Infectious viral

RNA carrying *gfp* can be transferred into plant cells using conventional methods to inoculate leaves or infect protoplasts (Baulcombe et al., 1995; Heinlein et al., 1995; Casper and Holt, 1996; Epel et al., 1996). Infection of plants with RNA viruses can also be achieved using plasmid DNA (Baulcombe et al., 1995; Oparka et al., 1995).

Microprojectile bombardment, another nonstandard method for transfer of GFP constructs into cells, has been used to deliver *gfp* expression plasmids into neurons and glia in intact rodent brain slices (Lo et al., 1994), as well as into cultured soybean cells (Plautz et al., 1996) and *Arabidopsis* leaves and roots (Sheen et al., 1995) for transient expression studies. Transgenic *Arabidopsis* stably expressing GFP have also been produced using standard methods of regeneration of seedlings from root callus infected with *Agrobacterium* carrying *gfp*-binary plasmids (Haseloff and Amos, 1995).

For transient transfection of mammalian cells, conventional methods include calcium phosphate transfection, liposome-mediated transfection, and electroporation. Commercial reagents are available for transfection of mammalian cells and are used by most workers. Special handling of cells can help to ensure GFP expression, as is true for transfection with any vector. Procedures should be optimized for cell density, amount and purity of DNA, transfection time, and post-transfection interval required for GFP expression. Efficiencies of several transfection protocols should be compared for the cell line of interest. In some instances, it is desirable to allow recovery from transfection for 5 h and then treat cells with 2 mM sodium butyrate to enhance transcription levels.

Many laboratories have had great difficulty in selecting stable GFP expressing cell lines by conventional means (Kain, personal communication), although BHK, HIT, and CHO cells that stably express GFP chimeras have been isolated (Olson et al., 1995; Patterson et al., 1996; Cole et al., unpublished results). Murine C2C12 muscle cell lines have also been reported that stably transmitted and expressed GFP through at least three passages in selection media (Moss et al., 1996). Expression levels in cells with stable GFP gene products are sometimes lower than in transfected cells, possibly due to toxic effects of GFP expression, but for many applications it is useful to have stable cell lines that express GFP even at low levels. Cells with stably integrated plasmids are selected by growing transiently transfected cells in the presence of 50–500 μg/mL G418, a neomycin analogue (Life Technologies), and clonal isolates are obtained by plating cells at limiting dilutions. Stable transformants can then be amplified and maintained in medium containing G418.

For transient expression in plant protoplasts, vectors for GFP expression can be transfected using standard electroporation methods. During preparation of maize protoplasts from etiolated plants, special care should be taken with seedling growth. Disease-free, uniformly yellow leaves should be taken from vigorously growing plants. Maize lines $FR9^{cms}$ × FR37 and FR992 × FR637 (Illinois Foundation Seed, Champaign, IL) are particularly recommended for use (Galbraith et al., 1995). For the electroporation procedure itself, protoplast viability and transfection efficiency should be optimized as a function of protoplast concentration, voltage settings, and pulse lengths. Bacterial contamination can be avoided by using sterile methods or by adding antibiotics (e.g., ampicillin, 100 μg/mL) to the growth medium. Following transfection, protoplasts should be cultured under conditions that ensure adequate oxygenation; shallow tissue culture plates that have been rinsed with 5% calf serum to reduce adhesion to the plastic surface are suitable for this purpose.

II.B Fixed Specimens

Visualization of GFP in fixed cells or tissues has been successfully achieved for many types of cells after fixation with formaldehyde. In some cases, fixation of cells in sodium azide, methanol, ethanol or glutaraldehyde has also proven successful for subsequent visualization of GFP (some fixatives with sodium azide have been problematic). Although glutaraldehyde is a superior fixative to formaldehyde, it is not commonly used in fluorescence microscopy because it causes autofluorescence. At high concentrations, glutaraldehyde will destroy GFP fluorescence, but at low concentrations (e.g., 0.025%), this problem can be partially avoided.

Denaturants such as 1% sodium dodecyl sulfate (SDS) or 8 M urea at room temperature can also be used in fixation procedures with the preservation of GFP fluorescence, but if GFP is fully denatured, or treated with 1% hydrogen peroxide or sulfhydryl reagents (Inouye and Tsuji, 1994), fluorescence is irreversibly destroyed. At high protein concentrations (above 5–10 mg/mL) or in high-salt solutions, GFP has been reported to dimerize, resulting in a 4-fold reduction in absorption.

GFP fluorescence has been reported to be sensitive to some nail polishes used to seal coverslips to slides (Chalfie et al., 1994; Wang and Hazelrigg, 1994). Molten agarose, rubber cement, or VALAP [1:1:1 Vaseline (petroleum jelly): lanolin:paraffin chips, heated until clear] is recommended as a substitute for sealing coverslips. Alternatively, specimens can be viewed without applying a sealant by pressing down firmly on the coverslip to remove any excess mounting media between the slide and coverslip that might cause slippage of the coverslip.

The use of slides marked with grids can aid in relocating cells of interest. Gridded slides available for use contain divisions of several millimeters in a printed Teflon coating (Cel-line Associates, Carlson Scientific), numbered divisions of 200–500 μm ("England finders"; Klarfield Rulings, Inc.), or squares of 55 μm (Eppendorf North America).

The following methods have been successfully used to prepare fixed cells for GFP visualization.

PROTOCOL 1 Bacteria (*E. coli, EGFPmut1-3*)

Materials

E. coli expressing *EGFPmut1-3*

5 mM Sodium azide in PBS (phosphate-buffered saline)

Microscope slides

Coverslips

1. Resuspend bacterial samples in 5 mM sodium azide in PBS to fix cells.
2. Mount cells on a glass slide with a glass coverslip.

   Fluorescence due to soluble EGFP is maintained as long as the cell surface integrity is not compromised. Fixing in sodium azide in PBS has no obvious effect on fluorescence. The cells can be fixed with a variety of crosslinking agents such as paraformaldehyde (2% w/v) and formaldehyde (1% v/v), and will retain some of the fluorescence from soluble EGFP. The GFP fluorescence in the fixed cells is also stable to photobleaching. Detergents and permeabilizing fixatives like methanol will destroy fluorescence. The effect of fixation on the ability to visualize GFP is most pronounced when imaging low levels of EGFP expression.

CONTRIBUTED by RAPHAEL H. VALDIVIA, BRENDAN P. CORMACK, and STANLEY FALKOW

PROTOCOL 2 Yeast (*S. cerevisiae*, TUB4-GFP)

Materials

Yeast expressing Tub4-GFP

Formaldehyde (37%)

Methanol

Acetone

DAPI (4',6-diamidino-2-phenylindole)

Microscope slides

Coverslips

1. Fix cells by adding formaldehyde to a final concentration of 3.7%.
2. Incubate at RT or 30°C 1–2 h.
3. Treat with cold methanol for 5 min, then with cold acetone for 30 s.

   This procedure was performed to help flatten the cells, making the spindles easier to visualize in a single focal plane, rather than specifically for the GFP visualization.
4. To visualize DNA, incubate in 1 µg/mL DAPI for 1 min.
5. Mount cells on a glass slide with a glass coverslip.

   The ability to visualize GFP after fixation with formaldehyde may not pertain to all GFP fusion proteins in yeast. Although Tub4-GFP fluorescence could be observed following formaldehyde fixation, α-tubulin-GFP fusion proteins do not fluoresce visibly after fixation.

CONTRIBUTED by TIM STEARNS (MARSHALL et al., 1996)

PROTOCOL 3 Yeast (*S. pombe*, p93^{dis1}-GFP)

Materials

Yeast expressing p93^{dis1}-GFP

Methanol or 2.5% glutaraldehyde

Microscope slides

Coverslips

1. Treat cells with methanol for 8 min at –80°C or with 2.5% glutaraldehyde at 33°C for 1 h.
2. Mount cells on a glass slide with a glass coverslip.

TAKEN from NABESHIMA et al. (1995)

PROTOCOL 4 Yeast (*Schizosaccharomyces pombe*, GFP)

Protocol for preparation of fixed yeast cells for FACS analysis

Materials

S. pombe cells expressing GFP from an episomal expression vector

EMM medium

Sterile DDW

100% Ethanol

50 mM Sodium citrate

1. Grow cells in EMM to a density of 10^7 cells/mL.
2. Harvest 20 mL of cells by pelleting.
3. Suspend cells in sterile DDW to wash. Pellet again.
4. Resuspend cells in 6 mL sterile DDW.
5. Add 14 ml 100% ethanol.
   The fixed cells can be stored indefinitely at 4°C.
6. Prior to analysis, harvest cells by centrifugation.
7. Wash with sterile DDW.
8. Resuspend in 5 volumes of 50 mM sodium citrate.
9. Sonicate briefly.
10. Analyze by FACS.
    Data was acquired for 20,000 cells for each sample and analyzed by plotting fluorescence against forward scatter.

TAKEN from ATKINS and IZANT (1995)

PROTOCOL 5 *Drosophila (D. melanogaster*, GFP-Exu in egg chambers)

Materials

exu^{Sco2}/*exu*^{Sco2}; *P[Cas,NGE]3/+* females

PBS (phosphate buffered saline)

Fixative (Theurkauf and Hawley, 1992)
8% Formaldehyde
100 mM Potassium cacodylate, pH 7.2
100 mM Sucrose
40 mM Potassium acetate
10 mM Sodium acetate
10 mM EGTA

Strips of Whatman filter paper

50% (v/v) Glycerol in PBS

Microscope slides

Coverslips

1. Collect 0–1 day-old females and place in well-yeasted vials with males.
2. Keep 2 days prior to use, so females are 2–3 days old.
3. Anesthetize females with CO_2.
4. Dissect ovaries in PBS.
5. Place in fixative for 10 min.
6. Wash 3 × 10 min in PBS.
7. Tease ovarioles apart in a drop of PBS on a glass slide.
8. Remove excess PBS with strips of Whatman filter paper, then cover with a drop of 50% glycerol in PBS.
9. Cover with a glass coverslip.

CONTRIBUTED by TULLE HAZELRIGG (WANG and HAZELRIGG, 1994)

PROTOCOL 6 Mammalian cells (HeLa cells, GFP chimeras)

Materials

PBS (phosphate buffered saline)

Fixative
2% Formaldehyde or 0.025% glutaraldehyde in PBS

Fluoromount (Southern Biotechnology)

Microscope slides

Coverslips

1. Place cells in fixative for 10 min at room temperature.
2. Rinse twice in PBS.
3. Mount cells on a glass slide in Fluoromount with a glass coverslip.
   a. Staining of fixed cells with antibodies

Additional Materials

Primary antibody solution
Primary antibody in PBS
10% Bovine serum
0.5% Saponin

10% Bovine serum in PBS

Secondary antibody solution
Rhodamine-labeled secondary antibody in PBS
10% Bovine serum
0.5% Saponin

1. Carry out fixation and PBS wash steps as described above.
2. Incubate cells for 1 h at room temperature in primary antibody solution.
3. Wash cells 3 times over 30 min in 10% bovine serum in PBS to remove unbound antibody.
4. Incubate cells for 1 h at room temperature in secondary antibody solution.
5. Wash cells 3 times over 30 min in 10% bovine serum in PBS to remove unbound secondary antibody.
6. Rinse cells quickly in PBS without serum.
7. Mount cells on a glass slide in Fluoromount with a glass coverslip. Observe using rhodamine and fluorescein filters to determine the distribution of antibody and GFP.

CONTRIBUTED by JENNIFER LIPPINCOTT-SCHWARTZ

II.C Live Specimens

While it is often convenient to fix samples for future analysis, the best imaging of GFP is obtained with live samples. For many cells and organisms, no special preparation of live specimens is needed for visualization of GFP other than mounting in a non-fluorescent medium and prevention of anoxia during observation.

Phosphate-buffered saline (PBS) or other simple salt solutions can be used as a non-fluorescent mounting medium for most cells. Fluorescent compounds that should be omitted from mounting media include flavins [e.g., riboflavin (Ludin et al., 1996)], tryptophan and other aromatic amino acids, and serum. A low-fluorescence defined medium has been developed for yeast cells (Waddle et al., 1996). The GFP-expressing cells used with this medium should be TRP^+ as the medium lacks tryptophan.

For time-lapse imaging of cells such as yeast, cells can be immobilized in a thin layer of agarose on a microscope slide to prevent drift during observation. Motile organisms such as *C. elegans* require anesthesia to prevent movement. Standard anesthetics used for *C. elegans* have been found to lead to a decrease in GFP fluorescence, apparently due to faster photobleaching. For example, phenoxy propanol and sodium azide, which have been used as reversible anesthetics for *C. elegans*, are not suitable for studies with GFP. Levamisole, an acetylcholine agonist, can be used for GFP visualization as a substitute for standard *C. elegans* anesthetics (Chalfie and Fire, personal communication). Levamisole causes hypercontraction of muscle, then eventually leads to relaxation with markedly reduced movement. The effect is reversible, and allows observation and photography of live animals.

Drosophila egg chambers, oocytes, and embryos can be mounted under halocarbon oil to prevent dehydration during visualization of GFP. Dechorionating oocytes and embryos prior to mounting and subsequent visualization results in higher image quality. Oxygenating the halocarbon oil by bubbling with O_2 prior to use can help to prevent anoxia during imaging. Specimens intended for observation on an inverted microscope can be mounted under oil on a large coverslip without a slide to ensure exchange of oxygen. Special microscope chambers have also been designed and used that allow the specimen to be mounted on an O_2 permeable membrane instead of a slide and covered with a coverslip (Davis et al., 1995).

Mammalian and other cultured cells expressing GFP can be observed after washing and mounting in a non-fluorescent buffer. Cells that are normally grown at 37°C will survive at room temperature for several hours, or can be maintained using a temperature-controlled microscope stage or by placing the microscope in a warm room. Cells can be grown on coverslips and maintained in chambers filled with medium for extended periods of visualization. Growth of cells in chambers with coverglass bottoms can help to ensure minimal perturbation when visualization is with an inverted microscope.

Standard methods can be used to prepare live mammalian cells and plant protoplasts for analysis by flow cytometry. Analysis of bacteria by flow cytometry presents special problems because of the small size of bacteria—bacterial cells are close to the limits of detection of available instruments. Some bacteria tend to aggregate, leading to artifacts. For live *Mycobacteria* species, aggregation can be reduced by treatment with detergents and sonication prior to analysis (Dhandayuthapani et al., 1995; Kremer et al., 1995; Ramakrishnan, Valdivia, and Falkow, unpublished results). Appropriate precautions should be taken when handling pathogenic bacteria, especially if there is any danger from aerosol formation (Russell, 1994). Treatment of the flow cytometer with a 0.5% sodium dodecylsulfate (SDS) solution followed by sterilization with 95% ethanol is usually sufficient to remove any adherent organisms (Ramakrishnan, Valdivia, and Falkow, unpublished results).

Gridded slide sources
Carlson Scientific
514 S. Third St.
Peotone, IL. 60468

Cel-Line Associates
PO Box 35
Newfield , NJ 08344
T 609 697-4590

Klarmann Rulings Inc.
PO Box 4795
Manchester, NH 03108
T 800 252-2401

Eppendorf North America
545 Science Drive
Madison, WI 53711
T 800 421-9988

The following section presents protocols for preparation of live specimens for spectrofluorometry, flow cytometry, and time-lapse microscopy using a cooled CCD camera or confocal microscope. Still images of live specimens can also be acquired after preparation of specimens using these methods.

PROTOCOL 1 Preparation of *Mycobacteria* for fluorescence assays using a spectrofluorometer

Untransformed and recombinant *Mycobacterium smegmatis* BCG were grown in microtiter plates to test GFP expression in three different growth media by assaying fluorescence following growth using a spectrofluorometer.

Materials

Untransformed BCG

Recombinant *Mycobacterium smegmatis* BCG expressing GFP

PBS (phosphate buffered saline)

Growth media [Sauton, Dubos (Difco), Middlebrook 7H9 (Difco)]

Sterile 96-well microtiter plates (Nunc)

1. Harvest recombinant *Mycobacterium smegmatis* BCG expressing GFP or untransformed BCG from exponentially growing cultures.
2. Wash cells twice with PBS.
3. Resuspend cells in PBS to a final concentration of 4×10^8 bacilli/mL.
4. Mix 25 μL of bacterial suspensions with 175 μL of each medium in 96-well microtiter plates.
5. Incubate at 37°C.
6. Measure intensity of fluorescence after 24 and 48 h incubation.

TAKEN from KREMER et al. (1995)

PROTOCOL 2 Preparation of *Mycobacteria* for flow cytometry or FACS

Materials

Untransformed BCG

Recombinant *Mycobacterium smegmatis* BCG expressing GFP

PBS (phosphate buffered saline)

Special Equipment

Branson Sonicator Model 450 or equivalent

1. Grow recombinant BCG expressing GFP or untransformed BCG to mid-log phase.
2. Harvest by centrifugation.
3. Wash with PBS.
4. Resuspend in PBS.
5. Sonicate with 3×10 s bursts at minimal constant output.
6. Analyze by flow cytometry.

TAKEN from KREMER et al. (1995)

PROTOCOL 3 Yeast (*S. cerevisiae*, NUF2-GFP)

Using a GFP-tagged centrosomal protein (Nuf2-GFP), the dynamics of mitotic spindle elongation can be observed by time-lapse fluorescence microscopy. This protocol, adapted from Kahana et al. (1995), describes specimen preparation for observing mitotic movements in live yeast cells using a cooled CCD camera.

Materials

Yeast strain JKY85 (Mat α; *ura3-52; his3Δ200; leu2Δ1; nuf2Δ::NUF2-sGFP::URA3*)

This strain constitutively expresses the centrosomal protein, Nuf2p, fused to a mutant (S65T, V163A) of GFP (Kahana and Silver, 1996). *Expression of Nuf2-sGFP is driven by the NUF2 5′ promoter and 3′ UTR. The gene fusion is integrated into the yeast genome at the NUF2 locus. Thus, the only copy of NUF2 is the GFP-fusion. This strain grows at the same rate as its isogenic parent which expresses wild-type Nuf2p.*

Synthetic complete (SC) medium lacking tryptophan (Rose et al., 1990)

Low-fluorescence media (Waddle et al., 1996)
0.9 g/L KH_2PO_4
0.23 g/L K_2HPO_4
0.5 g/L $MgSO_4$
3.5 g/L $(NH_4)_2SO_4$
20 g/L Dextrose
2 g/L Nutrient mix lacking tryptophan (Rose et al., 1990)
10 g/L Agarose
For minimal autofluorescence, this medium lacks the amino acid tryptophan. Therefore, a TRP^+ yeast strain must be used.

VALAP (VAseline, LAnolin, Paraffin)
1:1:1 Vaseline (petroleum jelly):lanolin:paraffin chips
Mix together and heat until clear. Can be cooled and reheated.

Microscope slides

Coverslips (22 mm^2, No. 1 $\frac{1}{2}$)

1. Grow strain JKY85 in SC-Trp medium at 30°C with constant aeration until the cells are in mid-log phase ($\sim 1 \times 10^7$ cells/mL).
2. Prepare the low-fluorescence medium containing 1% agarose. The solution should be heated in a microwave oven until the agarose is fully dissolved.
   Use of solid medium prevents the cells from moving during the experiment.
3. Pipette 1 mL of the medium on top of a microscope slide.
   Surface tension will prevent the medium from running off the sides of the slide.
5. Place a second microscope slide on top of the molten medium and allow to cool for 10 min.
6. Carefully lift the second slide off.
   This will leave an extremely thin (<0.5 mm) layer of solid medium upon the bottom slide.
7. Pipette 3 μL of the yeast culture onto the middle of the slide and cover with a 22 mm^2 coverslip.
8. Cut away the residual solid media with a clean razor blade.

9. Using a cotton-tipped applicator, seal the edges of the coverslip with VALAP.

   This will prevent the medium from drying out during the experiment.

10. Place the slide on the microscope stage for observation using a cooled CCD camera.

    See Protocol IIIB.4.d for imaging protocol.

CONTRIBUTED by PAMELA A. SILVER

PROTOCOL 4 Yeast (*S. pombe*, p93^{dis1}-GFP)

Live yeast cells can simply be mounted on a microscope slide for still image acquisition or recording with a videocamera (Nabeshima et al., 1995).

Materials

Yeast cells expressing p93^{dis1}-GFP

EMM2 minimal media

Microscope slides

Coverslips

1. Grow cells at 33°C in minimal media to log phase.
2. Mount cells on a microscope slide.
3. Cover with a coverslip.

TAKEN from NABESHIMA et al. (1995)

PROTOCOL 5 *Dictyostelium* (*D. discoideum*, pBigGFPmyo)

Dictyosteium cells expressing GFP fused to the N-terminus of cytoplasmic myosin heavy chain are prepared for time-lapse imaging using a cooled CCD camera as described below.

Materials

Dictyostelium cells expressing pBigGFPmyo (Moores et al., 1996)

The pBigGFPmyo plasmid encodes a fusion protein between wild-type GFP *and the N-terminus of the Dictyostelium discoideum myosin II heavy chain. There are no intervening residues between the two peptides in the fusion protein. Expression of the gene fusion is regulated by the Dictyostelium actin-15 promoter. The plasmid encoding the fusion protein was electroporated into a myosin II heavy chain gene null cell line, and transformants were selected based on antibiotic resistance. The only myosin II heavy chain in transformant lines is from the gfp-myo fusion gene.*

Nutrient medium

Lab-Tek chambered coverglasses (Nunc, Inc., Naperville, IL)

Imaging buffer
20 mM MES, pH 6.8
0.2 mM $CaCl_2$
2 mM $MgSO_4$

1. Plate cells expressing the GFP-myosin heavy chain onto Lab-Tek chambered coverglasses and replace the nutrient media with the imaging buffer.

   The use of an imaging buffer is required as the nutrient medium is fluorescent in the visible range used to excite GFP.

2. Place chambers on an inverted microsope stage for imaging.

CONTRIBUTED by JAMES H. SABRY

PROTOCOL 6 *Drosophila (D. melanogaster,* GFP-Exu in egg chambers)

Fusion of GFP to the maternally-expressed Exuperantia (Exu) protein was used to demonstrate the subcellular localization of Exu to particles (Wang and Hazelrigg, 1994). The GFP-Exu fusion protein is expressed from the native *exu* promoter, and the *gfp-exu* transgene rescues the maternal-effect lethal phenotype of exu^{Sco2}, an *exu* null mutant.

Materials

exu^{Sco2}/exu^{Sco2}; *P[Cas,NGE]3/+* females
These females carry 1 copy of the gfp-exu transgene in an exu null mutant background so the only source of Exu is the GFP-Exu fusion protein.

Modified Robb's medium (Theurkauf and Hawley, 1992)
55 mM Potassium acetate
40 mM Sodium acetate
100 mM Sucrose
10 mM Glucose
1.2 mM $MgCl_2$
1.0 mM $CaCl_2$
100 mM HEPES, pH 7.4

Microscope slides

Coverslips

1. Collect 0–1 day old females and place in well-yeasted vials with males.
2. Keep 2 days prior to use, so females are 2–3 days old.
3. Anesthetize females with CO_2.
4. Dissect ovaries in modified Robb's medium and tease apart ovarioles.

5. Mount egg chambers in modified Robb's medium under a coverslip for imaging on an upright microscope.

   For short imaging sessions, the presence of mature oocytes in these preparations provides sufficient support for the coverslip so that younger egg chambers remain intact for several minutes.

CONTRIBUTED by TULLE HAZELRIGG (WANG and HAZELRIGG, 1994)

PROTOCOL 7 *Drosophila (D. melanogaster,* GFP-Exu in egg chambers, time-lapse analysis)

Time-lapse imaging of GFP-Exu in live egg chambers demonstrates that the Exu-associated particles move through the ring canals into the oocyte (Hazelrigg and Theurkauf, unpublished results). The transport of the particles is affected by drugs such as colcemid that affect microtubule stability.

Materials

exu^{Sco2}/*exu*^{Sco2}; *P[Cas,NGE]3/+* females

These females carry 1 copy of the gfp-exu transgene in an exu null mutant background so the only source of Exu is the GFP-Exu fusion protein.

Halocarbon oil

Coverslips (24 × 40 mm)

1. Culture and anesthetize females as described in steps 1–3 of protocol above.
2. Dissect ovaries in a drop of halocarbon oil on a coverslip and tease ovarioles apart.
3. Transfer coverslip to a support on an inverted microscope.
4. Image by laser scanning confocal microscopy. Collect a single slow scan on the Bio-Rad MRC 600 every 10 s.

CONTRIBUTED by TULLE HAZELRIGG and WILLIAM E. THEURKAUF

PROTOCOL 8 Drosophila (*D. melanogaster,* UAS-*gfp* in GAL4-enhancer trap lines)

A *P[Cas,UAS-gfp]* transgene that can be activated to express GFP in the presence of GAL4 was crossed to *gal4*-enhancer trap lines (Yeh et al., 1995). The enhancer trap lines contain insertions of *gal4* that are transcriptionally activated by enhancer sequences at the sites of insertion. Tissues were examined without fixation to determine the patterns of GAL4 expression using GFP as a reporter.

Materials

GAL4-expressing enhancer trap lines (Brand and Perrimon, 1993)

Each of these lines contains an insertion of gal4 that is activated transcriptionally by enhancer sequences at the site of insertion. Previous analysis of these lines using an UAS-lacZ reporter showed that each line produces an interesting GAL4 pattern by β-galactosidase staining.

P[Cas,UAS-gfp] line

This transgene contains five optimal GAL4 binding sequences 5′ to gfp. In the presence of GAL4, the transgene is activated to express GFP.

70% Glycerol (v/v) + 30% 0.1 M Tris, pH 9

3% sodium hypochlorite

DW

Schneider's medium + 10% FCS (fetal calf serum)

Microscope slides

Paper slips (5 × 22 mm)

Vacuum grease

Coverslips

1. Cross *P[Cas,UAS-gfp]* flies to GAL4-expressing enhancer trap lines to obtain F1 embryos, larvae or adults.
   a. Dissect imaginal discs in DW. Mount in 70% glycerol (v/v) + 30% 0.1 M Tris, pH 9.
   b. Dechorionate embryos in 3% Na hypochlorite. Rinse in DW. Mount in 70% (v/v) glycerol + 30% 0.1 M Tris, pH 9.
   c. Dissect ovaries in Schneider's medium + 10% FCS. Dissect stage 8 egg chambers out of the epithelial sheath overlaying ovarioles.
2. Transfer tissue or embryos to a microscope slide with medium.
3. Make coverslip supports of stacked paper slips covered with vacuum grease.
4. Add sufficient medium to immerse the sample. Cover tissue with coverslip.
5. Collect Z series of confocal images every hour for 4 h.

TAKEN FROM YEH et al. (1995)

PROTOCOL 9 ***Drosophila (D. melanogaster,*** **PUbGFP and PUbnlsGFP in early embryos)**

Materials

P[Cas,PUbgfp] *and P[Cas,PUbnlsgfp]* flies

Grape juice plates

Chlorox (dilute 1:1 with DDW)

Permeable membrane (YSI Inc., Yellow Springs, OH 45387)

Custom slide for mounting membrane

Halocarbon oil

Coverslips

1. Collect embryos on grape juice plates.
2. Wash embryos.
3. Dechorionate in 50% bleach.
4. Wash embryos. Blot dry briefly.
5. Mount embryos on a permeable membrane in halocarbon oil.
6. Cover with a glass coverslip.

TAKEN FROM DAVIS et al. (1995)

PROTOCOL 10 ***Drosophila (D. melanogaster,*** **Ncd-GFP in early embryos, time-lapse analysis)**

The Ncd microtubule motor protein is associated with spindle fibers throughout the early cleavage divisions of the embryo. Fusion of Ncd to wild-type or S65T mutant GFP permits the analysis of mitotic spindle dynamics in early embryos (Endow and Komma, 1996). The fusion protein is expressed from the *ncd* promoter, and the *ncd-gfp* transgene rescues the ca^{nd} null mutant of *ncd* for chromosome segregation and embryo viability. Live embryos are prepared for time-lapse imaging using laser scanning confocal microscopy, as described below.

Materials

ca^{nd}/ca^{nd} flies homozygous for 1 or 2 independent *ncd-gfp* insertions
These flies carry two or four copies of ncd-gfp in an ncd null mutant background so the only source of Ncd is the Ncd-GFP fusion protein.

Grape juice agar plates

Yeast paste
Mix dried baker's yeast with DDW to form a paste.

Light halocarbon oil (Halocarbon Oil 27, Sigma Chemical Co.)

Double-stick tape (Scotch, 3MM)

Microscope slides (Clay Adams, Gold Seal)

Coverslips (Corning, No. 1, 22 mm sq)

1. Transfer ~200–500 females and ~50 males to a pint *Drosophila* bottle containing standard cornmeal/yeast/agar food. Place in the dark at room temperature for 3–4 days prior to collecting embryos.

   Light and loud noise inhibit the flies from laying eggs.
2. Allow grape juice agar plates to warm to room temperature. Place a dab of yeast paste in the center of each plate.
3. Collect embryos in the dark at 30 min intervals by inverting bottles of *Drosophila* onto yeasted grape juice agar plates.

   The plates can be taped to the bottle for added stability. Transfer plates in a darkened room to ensure that the flies continue laying eggs.
3. Age embryos on the collection plates for 1.5 h at room temperature to allow development to syncytial blastoderm stage.
4. Transfer embryos to a piece of double-stick tape on a microscope slide with a brush or fine-tipped forceps. Remove chorions manually by gently rolling the embryos on the tape.

   Discard cellularized embryos, embryos damaged during dechorionation, and abnormally developing embryos. Cellularized embryos can be recognized by a refractile layer just below the vitelline membrane, damaged embryos exude drops of cytoplasm at the micropyle, and the cytoplasm of abnormally developing embryos contains dark or light areas and does not appear homogeneous.
5. Transfer dechorionated embryos to a drop of light halocarbon oil on the slide.

   To help prevent anoxia during observation, the halocarbon oil can be bubbled briefly with O_2 prior to use. Set up a size E oxygen tank fitted with a 1/4–7 L min^{-1} regulator and attach 2 mm i.d. tygon tubing to the outlet. Firmly seat a yellow pipetteman tip onto the end of the tubing and place the open end of the tip at the bottom of a 1.5 mL microfuge tube containing ~1 mL of halocarbon oil. Open the regulator valve to the lowest setting. Allow O_2 to bubble into the oil for 20–30 s and then cap the tube tightly until used.
6. Mount a coverslip fragment onto two layers of double-stick tape placed on either side of the embryos. Stage the embryos under visible light on a microscope using a 63X objective. Precellularized embryos with pole cells or pole cell buds, indicating cycle 8–13, can be examined immediately by conventional fluorescence or laser scanning confocal microscopy.

   Cycle 7 and 8 divisions can also be observed, but these occur further into the interior of the embryo than cycles 9–13 and the spindles are usually not completely in focus.
7. For confocal imaging, collect 2 or 3 Kalman-averaged slow scans every 15 s.

CONTRIBUTED by SHARYN A. ENDOW (ENDOW and KOMMA, 1996)

PROTOCOL 11 Mammalian Cells (HeLa cells, GFP chimeras)

Live cells expressing GFP or GFP chimeras can be grown on glass coverslips and the coverslips mounted on depression slides for visualization using a laser scanning confocal microscope.

Materials

Cells grown on No. 1 glass coverslips

Depression slides

Medium

VALAP
1:1:1 Vaseline (petroleum jelly):lanolin:paraffin chips
Mix together and heat until clear

1. Grow cells expressing GFP or GFP chimeras on sterile No. 1 glass coverslips placed in a 100 mm tissue culture dish.
2. Mount coverslips on depression slides in 200 μL of buffered growth medium.
3. Seal with VALAP.
4. Place slide on a microscope stage for viewing.
   a. Mounting cells for extended observation
   For extended periods of observations, cells are mounted in a special chamber and maintained at 37°C for experimental manipulation and visualization.

Special Equipment

Silicon rubber chamber made by cutting out a small square or circular frame from a silicon rubber sheet (Reiss Corp.) so that the coverslip will fit over it. Press the silicon rubber chamber down on a microscope slide.

1. Grow cells expressing GFP chimeras on No. 1 glass coverslips as above.
2. Mount coverslips on top of a rubber chamber containing 100 μl medium.
3. Seal by pressing gently but firmly on the coverslip to remove all liquid on the edges where the coverslip sits on top of the silicon rubber chamber.
4. Wipe excess fluid with a Kimwipe so the outside of the coverslip and chamber is dry.
5. Place slide on a microscope stage for viewing.

   The microscope stage can easily be heated to 37°C by covering the entire microscope with a plastic sheet and gently blowing in warmed air. For perfusion experiments, solutions containing drugs can be applied to one side of a coverslip mounted on two strips of paraffin or double-stick tape on a standard microscope slide. The solution being applied can then be drawn under and across the coverslip by capillary action using a filter paper wick.

Silicon rubber sheets can be ordered from Reiss Corp., 1 Polymer Place, Blackstone, VA 23824.

CONTRIBUTED by JENNIFER LIPPINCOTT-SCHWARTZ

PROTOCOL 12 Mammalian Cells (HeLa cells, GFP or GFP fusion proteins)

Materials

Medium

PBS

Culture dishes with cover glass bottoms (Mat-Tek P35G-1-14, MatTek Corp., Ashland, MA)

These are normal 35 mm culture dishes with a 14 mm hole in the bottom and a No. 1 coverslip glued onto the bottom of the dish to cover the hole. Since we use an inverted microscope, the coverslip bottom dish is a convenient and efficient way to grow and image cells with minimal perturbation of the cells. The dishes come in sterile (gamma irradiated) packages and we have not observed any problems of toxicity in using them. They are easy to adapt to most perfusion systems, and allow access to localized perfusion and electrophysiology pipettes. For observations with an upright microscope, cells can be grown on a coverslip placed in the bottom of a 35 mm culture dish and mounted over a chamber as described in the previous protocol, or imaged with a water immersion objective without a coverslip.

1. Grow cells on Mat-Tek dishes according to normal protocols.

   For some cells that do not grow well on glass, we find that a coating of polylysine or collagen is required.

2. Replace the media in which the cells have been growing with buffer. Phosphate buffered saline (PBS) can be used with most tissue culture cells.

   It is usually necessary to wash the cells once or twice with buffer to remove any remaining media that may cause fluorescence. If used with a perfusion system, only 1 mL of buffer is needed at the onset of the experiments. If no perfusion will be used, then at least 2 mL of buffer is required to maintain the cells for up to 1 h on the microscope.

3. Place the dish of cells on an inverted microscope stage for viewing.

We use the Adams and List TLC-MI temperature-controlled stage, which heats both the sample dish and the perfusate to maintain the sample at 37°C. See Protocol III for imaging procedures.

CONTRIBUTED by DAVID W. PISTON

REFERENCES

Amsterdam, A., Lin, S., and Hopkins, N. (1995). The *Aequorea victoria* green fluorescent protein can be used as a reporter in live zebrafish embryos. *Dev. Biol.* 171:123–129.

Amsterdam, A., Lin, S., Moss, L. G., and Hopkins, N. (1996). Requirements for green fluorescent protein detection in transgenic zebrafish embryos. *Gene* 173:99–103.

Atkins, D. and Izant, J. G. (1995). Expression and analysis of the green fluorescent protein gene in the fission yeast *Schizosaccharomyces pombe. Curr. Genet.* 28:585–588.

Baulcombe, D. C., Chapman, S., and Santa Cruz, S. (1995). Jellyfish green fluorescent protein as a reporter for virus infections. *Plant J.* 7:1045–1053.

Brand, A. H. and Perrimon, N. (1993). Targeted gene expression as a means of altering cell fates and generating dominant phenotypes. *Development* 118:401–415.

Casper, S. J. and Holt, C. A. (1996). Expression of the green fluorescent protein-encoding gene from a tobacco mosaic virus-based vector. *Gene* 173:69–73.

Chalfie, M., Tu, Y., Euskirchen, G., Ward, W. W., and Prasher, D. C. (1994). Green fluorescent protein as a marker for gene expression. *Science* 263:802–805.

Davis, I., Girdham, C. H., and O'Farrell, P. H. (1995). A nuclear GFP that marks nuclei in living *Drosophila* embryos; maternal supply overcomes a delay in the appearance of zygotic fluorescence. *Dev. Biol.* 170:726–729.

Dhandayuthapani, S., Via, L. E., Thomas, C. A., Horowitz, P. M., Deretic, D., and Deretic, V. (1995). Green fluorescent protein as a marker for gene expression and cell biology of mycobacterial interactions with macrophages. *Mol. Microbiol.* 17:901–912.

Endow, S. A. and Komma, D. J. (1996). Centrosome and spindle function of the *Drosophila* Ncd microtubule motor visualized in live embryos using Ncd-GFP fusion proteins. *J. Cell Sci.* 109:2429–2442.

Epel, B. L., Padgett, H. S., Heinlein, M., and Beachy, R. N. (1996). Plant virus movement protein dynamics probed with a GFP-protein fusion. *Gene* 173:75–79.

Galbraith, D. W., Lambert, G. M., Grebenok, R. J., and Sheen, J. (1995). Flow cytometric analysis of transgene expression in higher plants: green-fluorescent protein. *Methods Cell Biol.* 50:3–12.

Haseloff, J. and Amos, B. (1995). GFP in plants. *Trends Genet.* 11:328–329.

Heinlein, M., Epel, B. L., Padgett, H. S., and Beachy, R. N. (1995). Interaction of tobamovirus movement proteins with the plant cytoskeleton. *Science* 270:1983–1985.

Inouye, S. and Tsuji, F. I. (1994). Evidence for redox forms of the *Aequorea* green fluorescent protein. *FEBS Lett.* 351:211–214.

Kahana, J. A. and Silver, P. A. (1996). Use of the A. *victoria* green fluorescent protein to study protein dynamics *in vivo. In* Current Protocols in Molecular Biology, Ausubel, F. M., Brent, R, Kingston, R. E., Moore, D. E., Seidman, J. G., Smith, J. A. and Struhl, K., Eds., Wiley, New York, 9.6.13–9.6.19.

Kremer, L., Baulard, A., Estaquier, J., Poulain-Godefroy, O., and Locht, C. (1995). Green fluorescent protein as a new expression marker in myobacteria. *Mol. Microbiol.* 17:913–922.

Lo, D. C., McAllister, A. K., and Katz, L. C. (1994). Neuronal transfection in brain slices using particle-mediated gene transfer. *Neuron* 13:1263–1268.

Ludin, B., Doll, T., Meili, R., Kaech, S., and Matus, A. (1996). Application of novel vectors for GFP-tagging of proteins to study microtubule-associated proteins. *Gene* 173:107–111.

Marshall, L. G., Jeng, R. L., Mulholland, J., and Stearns, T. (1996). Analysis of Tub4p, a yeast γ-tubulin-like protein: implications for microtubule-organizing center function. *J. Cell Biol.* 134:443–454.

Moores, S. L., Sabry, J. H., and Spudich, J. A. (1996). Myosin dynamics in live *Dictyostelium cells. Proc. Natl. Acad. Sci. USA* 93:443–446.

Moss, J. B., Price, A. L., Raz, E., Driever, W., and Rosenthal, N. (1996). Green fluorescent protein marks skeletal muscle in murine cell lines and zebrafish. *Gene* 173:89–98.

Nabeshima, K., Kurooka, H., Takeuchi, M., Kinoshita, K., Nakaseko, Y., and Yanagida, M. (1995). $p93^{dis1}$, which is required for sister chromatid separation, is a novel microtubule and spindle pole body-associating protein phosphorylated at the Cdc2 target sites. *Genes Dev.* 9:1572–1585.

Olson, K. R., McIntosh, J. R., and Olmstead, J. B. (1995). Analysis of MAP 4 function in living cells using green fluorescent protein (GFP) chimeras. *J. Cell Biol.* 130:639–650.

Oparka, K. J., Roberts, A. G., Prior, D.A.M., Chapman, S., Baulcombe, D., and Santa Cruz, S. (1995). Imaging the green fluorescent protein in plants—viruses carry the torch. *Protoplasma* 189:133–141.

Patterson, G. H., Magnuson, M. A., and Piston, D. W. (1996). Confocal microscopy studies of secretion using insulin-green fluorescent protein. *Biophys. J.* 70:A426.

Peel, A. L., Zolotukhin, S., Schrimsher, G.W., Muzyczka, N., and Reier, P. J. (1997). Efficient transduction of green fluorescent protein in spinal cord neurons using adeno-associated virus vectors containing cell-type specific promoters. *Gene Therapy* 4:16–24.

Peters, K. G., Rao, P. S., Bell, B. S., and Kindman, L. A. (1995). Green fluorescent fusion proteins: powerful tools for monitoring protein expression in live zebrafish embryos. *Dev. Biol.* 171:252–257.

Plautz, J. D., Day, R. N., Dailey, G. M., Welsh, S. B., Hall, J. C., Halpain, S., and Kay, S.A. (1996). Green fluorescent protein and its derivatives as versatile markers for gene expression in living *Drosophila melanogaster*, plant and mammalian cells. *Gene* 173:83–87.

Rose, M. D., Winston, F., and Hieter, P. (1990). Methods in Yeast Genetics: A Laboratory Manual. Cold Spring Harbor Laboratory Press, Cold Spring Harbor, NY.

Russell, D. G. (1994). Obtaining and maintaining microbial pathogens. *In* Microbes as Tools for Cell Biology, (Russell, D. G., Ed.,) Academic, San Diego, CA, pp. 1–4.

Sheen, J., Hwang, S., Niwa, Y., Kobayashi, H., and Galbraith, D. W. (1995). Green-fluorescent protein as a new vital marker in plant cells. *Plant J.* 8:777–784.

Tannahill, D., Bray, S., and Harris, W.A. (1995). A *Drosophila E(spl)* gene is "neurogenic" in *Xenopus*: a green fluorescent protein study. *Dev. Biol.* 168:694–697.

Theurkauf, W. E. and Hawley, R. S. (1992). Meiotic spindle assembly in *Drosophila* females: behavior of nonexchange chromosomes and the effects of mutations in the nod kinesin-like protein. *J. Cell Biol.* 116:1167–1180.

Waddle, J. A., Karpova, T. S., Waterston, R. H., and Cooper, J. A. (1996). Movement of cortical actin patches in yeast. *J. Cell Biol.* 132:861–870.

Wang, S. and Hazelrigg, T. (1994). Implications for *bcd* mRNA localization from spatial distribution of *exu* protein in *Drosophila* oogenesis. *Nature (London)* 369:400–403.

Wu, G.-Y., Zou, D.-J., Koothan, T., and Cline, H. T. (1995). Infection of frog neursons with vaccinia virus permits in vivo expression of foreign proteins. *Neuron* 14:681–684.

Yeh, E., Gustafson, K., and Boulianne, G. L. (1995). Green fluorescent protein as a vital marker and reporter of gene expression in *Drosophila. Proc. Natl. Acad. Sci. USA* 92:7036–7040.

PROTOCOL III VISUALIZATION OF GREEN FLUORESCENT PROTEIN

III.A Fluorescence

Since the original report of cloning of the GFP gene (Prasher et al., 1992) and the demonstration that GFP fluoresces when expressed in foreign cells (Chalfie et al., 1994), fluorescence of GFP has been measured and visualized using many different techniques. Fluorescence is the re-emission of an absorbed photon at a wavelength different from the incident, or excitation, light. Ideally, all of the fluorescent signal would come from the molecule of interest (in this case, from GFP) but some excitation light is always scattered, causing background. In biological samples, there are also naturally occurring fluorescent molecules which fluoresce with characteristics similar to those of GFP. These substances contribute to the background fluorescence, or autofluorescence, of the specimen.

III.B Detection of Fluorescence

Each technique that is used to detect or visualize fluorescence has advantages and limitations that require special consideration. In addition, many GFP mutants with altered spectral characteristics are now available. The combination of these variables has resulted in an explosion of confusing information. For any detection method that is used, however, one basic principle governs the visualization of GFP, and that is to maximize the GFP signal over existing background. Since the background varies from specimen to specimen, optimization of the GFP signal over background is accomplished by different means depending on the specimen. These may include selecting an excitation wavelength that minimizes autofluorescence of the specimen or an emission barrier filter that maximizes the collection of GFP signal, or using a GFP variant that is brighter or optimized for the specific organism under consideration. The basis for making these choices is described both in the individual chapters and in these protocols. In particular, considerations related to instrument selection and set-up are described in this protocol.

The light microscope is the most common instrument used for visualizing biological specimens. For fluorescence microscopy, consideration should be given to selection of the filter cube, excitation light source, detector, and objective. Since fluorescence is a different color, or wavelength, than the excitation light, the two can be separated by the combination of an exciter filter, dichroic mirror, and emitter filter. These three elements are generally combined in a "filter cube." The choice of an exciter filter depends on the light source to be used for the fluorescence excitation. For example, a mercury lamp has a series of bright blue lines between 450 and 490 nm, so a fairly broad excitation filter is usually used (e.g., 470/40 with transmission of 450–490 nm), while an argon ion laser has a single blue line at 488 nm, permitting the use of a very narrow exciter filter (e.g., 490/10).

The mercury lamp and argon ion laser are the two most commonly used light sources for GFP excitation, but two other light sources, the halogen lamp and the xenon arc lamp, are also used. Both of these sources offer the advantage of a broad spectrum thoughout the visible and near-UV range, rather than series of specific lines. Halogen lamps are not very bright and are generally useful for fluorescence imaging only with the brightest samples. The xenon lamp, however, is quite bright and often requires attenuation by neutral density filters to avoid problems with photobleaching. For certain GFPs (e.g., wild-type GFP excited at 395 nm or the Y66H blue-shifted mutant), a xenon arc lamp will give a much better signal than a mercury lamp, which is not very bright around 400 nm.

The choice of emitter filter depends on the fluorescence of the particular GFP being used, and this is described in detail in this protocol. Finally, the choice of dichroic mirror is defined by the exciter and emitter filters chosen.

The process of choosing filters is illustrated in the visualization of GFP-Exu in vitellogenic egg chambers of *Drosophila*, where the yolk autofluorescence can interfere with the GFP signal. This problem can be overcome with a long pass FITC filter set (Zeiss 09, 450–490 exciter filter, LP520 barrier filter), which causes the yolk to appear more yellow than GFP, which appears sharply green. The GFP signal and autofluorescence can be further resolved using a filter cube with a UV excitation filter (405 ± 6 nm), and blue and green emission filters (460 ± 11 and 525 ± 15 nm), in which case the GFP appears green and yolk, blue. However, the GFP signal is not as bright and bleaches more rapidly with this filter set (Hazelrigg, personal communication).

Detection of fluorescence can be accomplished either by eye, film, or an electronic detector [such as a videocamera, a charge-coupled device (CCD), or a photomultiplier tube (PMT)]. Film recording and CCD detectors (Protocols 3 and 4 in this protocol) are the two most common methods used in conventional fluorescence microscopy for recording fluorescence, while PMT detectors are used in laser scanning confocal microscopy (Protocols 5–8 in this protocol), as well as flow cytometry (Protocols 1 and 2 in this protocol).

Thermoelectric cooling of a CCD detector greatly improves sensitivity by reducing electronic noise, thus yielding essentially zero background, high dynamic range (16-bit = 2^{16} shades of gray) and high detection efficiency (40–80%). The absence of real color detection (as with the eye or color film) in CCD detectors, though, means that filters must be carefully chosen to separate the GFP signal from autofluorescence. To take advantage of the dynamic range of a CCD detector, long integration times may be required, especially in dim samples. In living samples, usable integration times are limited by both the dynamics under observation and sample photodamage. The use of a cooled CCD does permit quantitation of GFP signals, especially in thin samples such as tissue culture cells and monolayers. In order to obtain quantitation by CCD imaging in thicker or more heavily expressing samples, where out-of-focus background may be a problem, it is often necessary to use image deconvolution as described at the end of Protocol III.B.4. PMT detectors also lack real color detection, so in confocal microscopy or flow cytometry, filters must be used to separate the GFP signal from autofluorescence. When used in confocal microscopy, PMTs usually give lower dynamic range (8-bit = 2^8 shades of gray) and detection efficiency (15–30%) than a cooled CCD. However, the excellent linearity of the PMT response coupled with the background rejection properties of confocal microscopy make this combination near ideal for quantitative imaging of GFP fluorescence, as described in Protocol III.B.5.

The final and perhaps the most important consideration is the objective lens. Objective lenses are defined by three parameters: magnification, numerical aperture (N.A.), and lens design. For fluorescence microscopy, magnification is the *least important* of these parameters. It determines how large an object will appear in the eyepiece or detector, but not the resolution (i.e., the smallest detail that can be clearly observed) or brightness (the amount of fluorescence signal that will be collected). Both of these parameters are a function of the N.A., which is the *most important* criterion for selection of an objective. The N.A. defines the amount of light that will enter the objective lens. The larger the N.A., the higher the resolution and the brighter the GFP signal will be. In general, the highest N.A. lens available should be used. Increasing the objective lens magnification reduces the field of view, proportionately reducing the amount of fluorescence collected. A larger specimen will therefore appear less bright than when viewed with a lower magnification objective of the same N.A.

For the lens type, a Fluor or NeoFluar design (consisting of three or four elements) will give a brighter fluorescence image than a more complicated design such as a Plan-Apochromat that may contain as many as eight elements. Lenses that have been used successfully for GFP visualization include the Zeiss Plan-Neofluar oil immersion lenses (40X/1.3 N.A., 63X/0.7-1.3 N.A. iris, and 100X/1.3 N.A.) and Nikon Plan-Fluor oil immersion lenses (40X/1.3 N.A., and 100X/1.3 N.A.). Further details about objective lenses (Keller, 1995) and fluorescence microscopy in general (Rost, 1992) can be found elsewhere.

In this protocol, several aspects of GFP visualization are described together with techniques that have been used to visualize, record and measure GFP fluorescence. The utility of each method and the issues that should be considered to optimize GFP visualization are discussed. Two recently developed methods of microscopy that promise to prove useful for studies with GFP, two-photon microscopy and image deconvolution, are also described.

III.B.1 Fluorescence Properties of GFP

Wild-type GFP exhibits two absorption maxima, one in the near-UV (395 nm) and one in the blue (475 nm). The second absorption peak, although smaller than the 395 nm peak, is more generally useful for fluorescence studies using GFP. This usefulness arises from the similarity between GFP excited with blue light and fluorescein (commonly used in the form of FITC) in fluorescence emission spectra. This similarity allows GFP to be readily detected using normal fluorescein filter combinations on a flow cytometer or fluorescence microscope. Many of the mutants of GFP, notably the S65T mutant (Heim et al., 1995) and the F64L, S65T (*EGFPmut1*) mutant (Cormack et al., 1996) show an enhanced blue absorption peak, increasing the use of standard fluorescein filter sets for detecting GFP. Other mutants, such as the Y66H, Y145F (*P4-3*) mutant with a blue-shifted emission spectrum (Heim and Tsien, 1996), permit double labeling with the S65T mutant (Rizzuto et al., 1996) but require special filter sets for visualization of GFP fluorescence. For visualization, the Y66H, Y145F mutant requires a UV excitation filter. As noted in Protocol I, the *P4-3* and other blue mutants of GFP fluoresce dimly and photobleach within seconds, so they may be of limited usefulness for imaging applications.

The general rule of thumb for GFP visualization is that if the level of GFP expression is high enough, GFP can be easily visualized by eye or conventional photography using a standard fluorescein filter set. Localized subcellular distribution can compensate for low levels of expression. As the level of GFP expression decreases due to a weak promoter, or if GFP is dispersed throughout the cell, it becomes more important to optimize detection by using higher efficiency detectors such as cooled CCDs or devices with photomultiplier tubes (confocal scanning detectors, flow cytometers) and filters specifically designed for GFP. For low levels of GFP, autofluorescence may seriously contaminate images recorded by digital cameras or detectors with photomultiplier tubes. For these detectors, special filters that preferentially select the GFP signal should be used to minimize this contamination. Still, the most common limitation in the use of GFP is insufficient expression of the protein. Methods to optimize GFP expression in various organisms and cells are discussed in Chapters 6–12 and Protocol I.

III.B.2 Photobleaching, Photoactivation, Photodamage, and pH Dependence of GFP

Photobleaching of GFP has been reported to be slow (Chalfie et al., 1994; Niswender et al., 1995), certainly much slower than fluorescein under similar conditions. For instance, continuous observation in a confocal microscope for 20 min only reduced GFP intensity to one-half its original value (Niswender et al., 1995). Although GFP is resistant to photobleaching, it has proven very useful in fluorescence photobleaching recovery experiments (described below and in Chapter 12) because of its ability as a fusion protein to be targeted to specific organelles.

Some problems with photobleaching of GFP have been reported. For instance, photobleaching with 395–440 nm light is accelerated by some agents used to anesthestize *C. elegans*, such as 10 mM NaN_3, and another anesthetic agent, phenoxypropanol, has been reported to quench GFP fluorescence (Chalfie et al., 1994). An alternative anesthetic agent for use in visualizing GFP in live *C. elegans* is described in Protocol II. Some mutant forms of GFP may also be more susceptible to photobleaching. For imaging in cultured mammalian cells, 10 μM Trolox has been added to live cells when visualizing Y66H, Y145F (*P4-3*) GFP to reduce the rapid photobleaching of this mutant GFP (Rizzuto et al., 1996).

Contrary to early reports, the photobleaching rate of wild-type GFP is close to the same whether it is excited in the UV or the blue, but the decrease in UV-excited fluorescence appears more rapid because of photoisomerization (Cubitt et al., 1995). Wild-type GFP has two absorption peaks, 395 and 475 nm, which are thought to be related via rotation about a bond (isomerization) within the chromophore structure. The isomerization can be induced by irradiation of wild-type GFP with either 395 or 490 nm light and the kinetics of this photo-induced reaction have recently been measured (Chattoraj et al., 1996). The photoisomerization affects the brightness of wild-type GFP during visualization, but does not occur in either the S65T or F64L, S65T (*EGFPmut1*) mutants of GFP. The recent solving of the crystal structures of wild-type GFP and the S65T mutant (described in Chapter 4) shows that the chromophore is tilted within the protein structure differently in the two proteins, This difference in tilting angle is likely the origin of the photoisomerization observed for wild-type GFP.

The photoisomerization can also be used to photoactivate wild-type GFP. That is, the fluorescence properties of GFP change after irradiation. When wild-type GFP is irradiated with either UV (~395 nm) or blue light (488 nm), photoisomerization occurs and causes an increase in the 475 nm peak and a decrease in the 395 nm peak (Cubitt et al., 1995). In this manner, UV pre-exposure can be used to increase the blue excitation brightness of wild-type GFP (Chalfie et al., 1994). A detailed discussion of the chromophore structure and its photo-induced reactions is presented in Chapters 4 and 5.

Since the photobleaching of GFP is low, cellular photodamage arising from GFP can also be expected to be low. Only a few extended studies have been reported to date, but photodamage associated with using GFP has not been identified as a problem. Anecdotal evidence suggests that photodamage associated with GFP fusion proteins is less than that found with labeling with other fluorophores. For example, fluorescence photobleaching recovery has been used to examine the diffusional mobilities of Golgi-targeted GFP chimeras (Cole et al., 1996), and in neither approach used was there evidence of cellular photodamage or photo-induced cross-linking of the GFP chimeras. Multiple bleaches of the same spot did not affect the measured diffusion coefficient, and staining with antibodies after photobleaching revealed intact membrane structures. Moreover, the mobile fraction did not decrease with successive bleaches, which would be expected if incomplete recovery was due to photo-induced immobilization of a fraction of the labeled molecules. Evidence such as this suggests that GFP will be excellent for time-lapse and four-dimensional imaging. The resistance of GFP to photobleaching may be due to the protection of the chromophore by a tightly packed barrel of β sheets [referred to as a β-can structure by Yang, et al. (1996)], as revealed by the crystal structures (Ormö et al., 1966; Yang et al., 1996).

Still, the intense light typically used for excitation of fluorescence can generate free radicals that in turn can damage cellular proteins. Excitation with UV light can also cause cross linking and breakage of DNA, producing further detrimental effects. As with any fluorescence microscopy study of living samples, it is important to monitor sample viability during and after the experiment. One measure of viability is the ability of irradiated cells to divide further with the same doubling or mitotic cycle time as unirradiated cells. During time-lapse confocal imaging of mitotic spindles in live *Drosophila* embryos visualized using the Ncd motor protein fused to wild-type or S65T mutant GFP, photodamage caused the spindles in the irradiated region of the embryo to become delayed relative to the unirradiated region, resulting in asynchronous divisions (Endow, unpublished results). The cause of this damage was not determined (it could be caused by the GFP or by interactions between the incident light and endogenous absorbers), but lowering the excitation intensity to a level that reduced or eliminated the photodamage still gave a strong GFP signal. Delayed mitotic divisions or cell arrest has also been observed in *Dictyostelium* (Maniak et al., 1995) and yeast (Kahana and Silver, unpublished results) upon over-irradiation of cells.

In general, addition of free radical scavengers (Mikhailov and Gundersen, 1995) or antioxidants to the medium may aid in the imaging of GFP in live specimens with bright light for long periods of time. Oxyrase (Oxyrase Inc. P.O. Box 1345, Mansfield, OH 44901) (0.3 U/mL) and ascorbic acid (0.1–1.0 mg/mL) are two antioxidants that have been shown to reduce photodamage when added to the medium of living cells. Where possible, increasing *gfp* gene dose, or the use of brighter GFP variants is advantageous in imaging live specimens, permitting greatly reduced levels of exposure to fluorescent or laser light.

Finally, the brightness of some GFP mutants appears to be sensitive to pH, although detailed studies have not yet appeared in the literature. As an example, wild-type GFP shows relatively even brightness from pH 5 to 10 (Ward, 1981), while the S65T mutant is 2-fold brighter at pH 7 than at pH 6 (Patterson and Piston, unpublished results). A similar fall-off in brightness at lower pH is exhibited by the F64L, S65T (*EGFPmut1*) double mutant.

III.B.3 Filter Sets for GFP

Standard filter sets can be used to visualize GFP with available equipment, and at least one of the following sets should be available on most fluorescence microscopes (see Tables A.III.1 and A.III.2 for specific filters and filter sets). For wild-type GFP, either a UV filter cube set with longpass emission filter or FITC filter cube set with either bandpass or longpass emission filter can be used. For the blue excitation enhanced S65T and F64L, S65T (*EGFPmut1*) mutants, only the FITC filter sets should be used, while with the UV enhanced Y66H, Y145F (*P4-3*) mutant, a UV filter cube is needed.

If GFP is being used regularly, or if it is in a low expression system, filters designed specifically for GFP (Chroma Technology Corp.) (Niswender et al., 1995; Endow and Komma, 1996) will enhance GFP signal collection. Although some of the custom GFP sets were originally designed for a specific purpose, for example, epifluorescence or confocal microscopy, these filter sets can be used for visualization of the indicated GFP mutants using many different types of microscopy. The choice of dedicated filter sets for use with GFP is governed by two basic principles. First, for qualitative visualization, usually performed using detection by eye or film, long pass emission filters can be used to visualize the sharp green of GFP above a greenish-yellow or greenish-orange background of autofluorescence (autofluorescence generally exhibits a very broad spectrum, while the GFP emission is relatively narrow). Second, for methods in which color information is not collected with the GFP fluorescence, performed using a digital detector such as in flow cytometry or confocal microscopy, or for visualization by digital cameras such as a cooled CCD camera, narrow bandpass filters are needed to increase the GFP signal versus the collected autofluorescence. The latter considerations also apply to methods used for quantitative measurement of GFP such as spectrofluorometry.

Tables A.III.1 and A.III.2 show filter sets that are useful for visualizing wild-type GFP and several of the more useful GFP mutants using detection methods that retain color information (Table A.III.1) or digital and other methods in which color information is not collected with the GFP signal (Table A.III.2). The entries in the table are general suggestions and should be useful for almost every imaging situation. However, visualization of GFP for a particular application in a particular organism may still be enhanced by use of other filters and should be evaluated on a case by case basis.

	Exciter	**Dichroic**	**Emitter**
Chroma No. 11004	D365x	380 DCLP	GG420LP
Chroma No. 31013	D365x	400DCLP	D460/50
Chroma No. 31019	D425/40	460DCLP	D500/40
Chroma No. 32001	D425/60	470DCXR	E480LP
Chroma No. 41001	HQ480/40	Q505LP	HQ535/50
Chroma No. 41012	HQ480/40	Q505LP	HQ510LP
Chroma No. 41014	HQ450/50	Q480LP	HQ510/50
Chroma No. 41015	HQ450/50	Q480LP	HQ485LP
Chroma No. 41017 (Endow set)	HQ470/40	Q495LP	HQ525/50
Chroma No. 41018 (Endow set)	HQ470/40	Q495LP	HQ500LP
Chroma Bright Blue BP No. 31021	D390/22	420DCLP	D460/50
Chroma Bright Blue LP No. 31022	D390/22	420DCLP	E430LP
DWPiston Custom Blue BP	D370/40	400DCLP	D445/60
Leica A cube	340-380 BP	RKP 400	LP 430
Leica D cube	BP355-425	RKP455	LP 470
Leica K3 cube	BP470-490	RKP510	LP 520
Leica L4 cube	BP450-490	RKP510	BP 515–560
Nikon UV-2A	BP330-380	400	LP 420
Nikon V-2B	BP380-425	430	LP 460
Nikon B-2A	BP450-490	510	LP 520
Nikon B-2E	BP450-490	510	520–560
Olympus U-MU	BP330-385	DM400	BA 420(LP)
Olympus U-MNV	BP400-410	DM455	A 455(LP)
Olympus U-MWIB	BP460-490	DM505	BA 515(LP)
Olympus U-MWIBA	BP460-490	DM505	BA 515–550(BP)
Zeiss Set 02	G365	FT395	LP 420
Zeiss Set 05	BP395-440	FT460	LP 470
Zeiss Set 09	BP450-490	FT510	LP 520
Zeiss Set 10	BP450-490	FT510	BP 515–565
Bio-Rad BHS	488/10	510LP	OG 515LP
Bio-Rad GR2/T3	488/10	T3 (trichroic)	522/32BP
Typical Confocal FITC LP	488/10	FT488	LP515
Endow Confocal GFP	488/10x	Q498LP	HQ518/40BP
Niswender Confocal GFP	488/10[a]	Q498LP	HQ512/27BP

[a]Standard excitation filter in the Zeiss LSM410 confocal microscope filter wheel.

III.B.4 Methods for Visualization of GFP

Methods of fluorescence detection that have been used with GFP include use of a hand-held UV source, spectrofluorometry, flow cytometry, or fluorescence activated cell sorting (FACS), cooled CCD cameras, confocal microscopy, and two-photon excitation microscopy. Protocols for each of these methods are described below, together with a description of deconvolution of images for 4-dimensional imaging.

III.B.4.a Detection Using a Hand-Held UV Illuminator GFP expressed in *E. coli* was detected in the initial report of expression in foreign host cells by irradiation with a hand-held long-wave UV illuminator (Chalfie et al., 1994). The standard UV illuminators available from laboratory suppliers are suitable for use. GFP fluorescence has subsequently been detected in other bacteria, such as *Mycobacteria*, by irradiation of colonies on Petri dishes using either a hand-held long-wave (366 nm) or short-wave (312 nm) UV source (Kremer et al., 1995). This method has also been used to detect GFP fluorescence in transformed plants such as *Arabidopsis* (Haseloff and Amos, 1995) and *Nicotiana* (Baulcombe et al., 1995; Casper and Holt, 1996).

The new Leica fluorescence illuminator, designed for use with the Wild MZ12 and other Wild stereomicroscopes, should prove useful for visualization of GFP in bacterial plates or larger specimens that have until now required the use of a hand-held UV illuminator. The fluorescence illuminator is also suitable for visualizing GFP in specimens the size of *Drosophila* and *C. elegans*. The light source is either a 100 W mercury or 75 W xenon lamp, and the illuminator will soon be supplied with filter cubes that can be fitted with specified filters available from Chroma Technology Corp.

III.B.4.b Spectrofluorometry

Measurement of GFP Fluorescence in Cells by Spectrofluorometry

Special Equipment

Labsystems spectrofluorometer or equivalent

One of the simplest ways to measure total fluorescence from a cell population is with a spectrofluorometer. Fluorescence is measured directly from a cell suspension placed in a cuvette and thus very little special preparation is needed. For some mammalian cells that are particularly adherent, a stir bar must be used in the cuvette. Most commercial spectrofluorometers have accessories for stirring the sample. Many samples can be processed in parallel in readers that can measure 96-well plates. For example, this system has been used for the rapid testing of drug susceptibility of *Mycobacterium bovis* BCG (Kremer et al., 1995), and to assay GFP transfections in mammalian cells (Ropp et al., 1995; Niswender et al., 1995). An added advantage of this system is that the optical density of the cultures in the plates can be measured and fluorescence can be normalized to cell number by using a standard cell counting technique (such as a Coulter counter). For quantitative measurements, samples should not be observed at optical densities greater than 1.0.

CONTRIBUTED by RAPHAEL H. VALDIVIA, BRENDAN P. CORMACK, STANLEY FALKOW and DAVID W. PISTON

TABLE A.III.1 Filter sets for detection of GFP by color

GFP Type	Excitation Wavelength: UV	Excitation Wavelength: Blue
Wild-type GFP	*Sets designed for DAPI (or other UV dyes) with longpass emission filter* Zeiss Set 02, Set 05 Leica A cube, D cube Nikon V-2B Olympus U-MNV *Sets designed for GFP with longpass emission filter* Chroma No. 32001	*Sets designed for FITC with longpass emission filter* Zeiss Set 09 Leica K3 cube Nikon B-2A Olympus U-MWIB Chroma No. 41012 *Sets designed for GFP with longpass emission filter* Chroma No. 41015 Chroma No. 41018
S65T	NA	Same as wild type
F64L, S65T (*EGFPmut*1)	NA	Same as wild type
Y66H, Y145F (P4-3)	*Sets designed for DAPI (or other UV dyes) with longpass emission filter*[a,b] Zeiss Set 02 Leica A cube Nikon UV-2A Olympus U-MU Chroma No. 11004 *Filter sets for P4-3 with longpass or bandpass emission filter*[a] Chroma Bright Blue LP Set Chroma Bright Blue BP Set	NA

[a] For use in double-labeling with any of the GFP variants listed in Table A.III.1, a bandpass emission filter should be used instead of a longpass filter to isolate the Y66H, Y145F (P4-3) signal from that of the other GFP.

[b] Excitation with wavelengths less than or equal to 380 nm using a mercury light source may cause more photobleaching and more background autofluorescence than excitation with 390 nm light.

TABLE A.III.2 Filter sets for GFP detection without color (digital methods)

	Excitation		
GFP Type	UV	Blue (450–490 nm)	Laser (488 nm)
Wild-type GFP	*UV sets* Chroma No. 31019	*FITC sets* Chroma No. 41001 Leica L4 Nikon B-2E Olympus U-MWIBA Zeiss Set 10 *GFP sets* Chroma No. 41014 Chroma No. 41017	*FITC sets* Bio-Rad BHS Bio-Rad GR2/T3 Leica, Nikon, Zeiss[a] *GFP sets* Niswender confocal GFP set Endow confocal GFP set
S65T	NA	Use FITC and GFP sets above	Use FITC and GFP sets above
F64L, S65T (*EGFPmut1*)	NA	Use FITC and GFP sets above	*FITC sets* Use FITC sets above *GFP sets*[b] Same as WT GFP, except HQ523/50BP emission filter can be used
Y66H, Y145F (P4-3)	*UV sets* Chroma No. 31013 *Filter sets for P4-3 with bandpass emission filter*[c,d] DWPiston Custom Blue BP Set[e]	NA	NA

[a] Bio-Rad MRC 1024, Leica, Nikon, and Zeiss confocal microscopes allow exciter, dichroic, and emitter filters to be chosen separately. A typical FITC set is shown in the list below.
[b] The filters used for WT and S65T also work well for this EGFP but because this GFP is brighter than other GFPs, a broader emission filter can be used and the collected EGFP signal will still be increased over background autofluorescence.
[c] Excitation with wavelengths less than or equal to 380 nm using a mercury light source may cause more photobleaching and more background autofluorescence than excitation with 390 nm light.
[d] The 390/20 excitation filter (Chroma Bright Blue sets) is preferable to avoid NADH absorption (a major source of cellular autofluorescence), which is maximal at 355 nm. The 460/50 emission filter (Chroma Bright Blue BP set) is best for visualization by eye since the eye is not very sensitive below 450 nm, while the 445/60 (DWPiston Custom Blue BP set) should be better for film and digital detection.
[e] Use with a xenon light source for maximum signal.

Detection of *gfp* Mutants Using Digital Imaging Spectroscopy

Special Equipment

CyberDIS digital imaging spectrofluorometer (KAIROS Scientific, Inc., Santa Clara, CA 95054) or equivalent

Digital imaging spectroscopy is a newly developed technique that can be used for screening many mutants in parallel. Colonies on Petri dishes are illuminated by light of different wavelengths and images of fluorescence are captured using a cooled CCD detector and processed using specially designed software. This method of spectroscopy permits rapid screening of colonies using positional and spectral information. Because this technique can use fluorescence detection, it is readily adaptable to the study of *gfp* mutants (Delagrave et al., 1995). The instrumentation and software for this method are highly specialized and have been detailed recently (Youvan et al., 1995). The CyberDIS digital imaging spectrofluorometer is a commercial, turn-key instrument that is now available.

III.B.4.c. Flow Cytometry Flow cytometry or FACS is a laser-based detection system that allows large populations of cells to be assayed and sorted (Shapiro, 1995). By measuring the way a cell scatters light and emits fluorescence, the flow cytometer records and stores a description of each particle that passes through the sample stream. It can determine the size, density, and, depending on the type of fluorescence stain used, DNA content and protein levels for each individual cell. While this technique is not exactly "visualizing" GFP, it is particularly useful for separation of transiently transfected and nontransfected populations of cells. Other applications, such as recording the dynamics of *gfp* gene induction and screening for mutants with enhanced GFP fluorescence, are emerging.

A remarkable feature of the electronic capabilities of a FACS system is that the fluorescence information is stored for each individual cell in the sample analyzed. This information can be displayed as histograms, dot plots, contour plots, and three-dimensional plots [see Parks et al. (1989) for a review]. This results in a multiparameter snapshot of the physical characteristics and genetic behavior of every individual cell within the population. The statistical programs in most FACS software analysis packages permit quantitation of fluorescence intensities and percentages of cells that exhibit specified combinations of characteristics. Filters for flow cytometry or FACS should be selected using the criteria in the table above for detection of GFP by digital devices, since a photomultiplier tube is used for detection of the fluorescence.

Although GFP has some clear advantages over more conventional reporters, it does require the use of expensive, specialized equipment. For many *in vitro* applications, *lacZ, cat,* and *lux* are far more convenient reporters of gene expression. The advantage of GFP lies in the study of development, cell biology, and pathogenesis, where more conventional reporter genes are limited by their inability to measure single-cell gene expression in live cells.

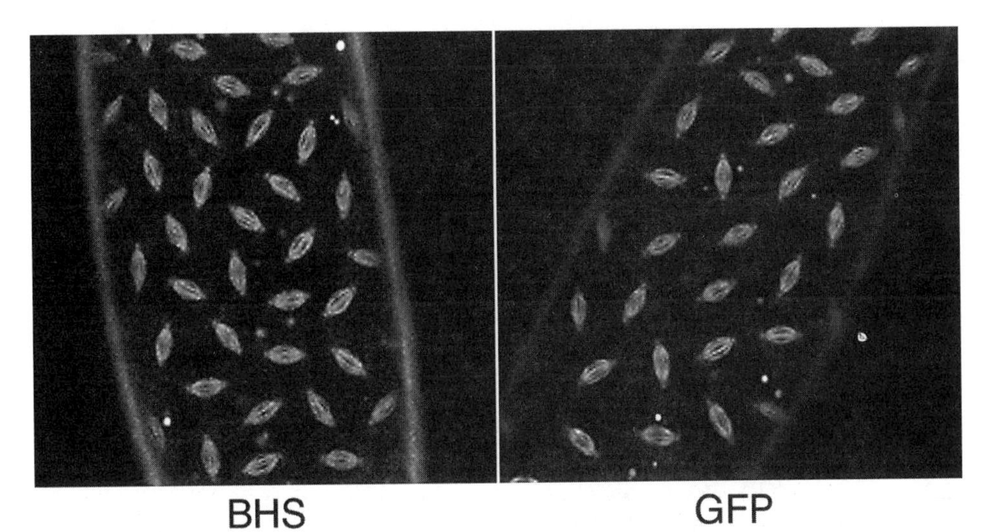

FIG. A.III.1 Suppression of autofluorescence by use of a bandpass emission filter for confocal microscopy. Embryos of *Drosophila* expressing four copies of *ncd-gfp*, encoding the Ncd microtubule motor protein fused to wild-type GFP, were imaged using a Bio-Rad MRC 600 laser scanning confocal detector, a BHS or custom confocal GFP filter set (Endow and Komma, 1996), and a 10% transmission neutral density filter. The pinhole settings were $\frac{2}{3}$ and $\frac{1}{2}$ open for the BHS and custom GFP sets, respectively. A linear contrast adjustment was applied to the images and brightness was adjusted identically for both images. The image acquired using the BHS filter set shows pronounced autofluorescence of the vitelline membrane, which is suppressed by the custom GFP filter set. The BHS filter set contains a long-pass emission filter, while the custom GFP set contains a bandpass emission filter (HQ518/40BP; Chroma Technology Corp.). A similar bandpass emission filter (HQ512/27; Chroma Technology Corp.) has been reported to reduce autofluorescence for confocal and two-photon excitation imaging of mammalian cells (Niswender et al., 1995) (Fig. A.III.4). The figure was provided by S. A. Endow (unpublished).

PROTOCOL 1 Detecting and measuring GFP fluorescence in bacteria in a flow cytometer

Using a flow cytometer, a bacterial culture bearing a test gene fusion to GFP or EGFP can be scanned and the level of fluorescence measured. Bacterial cells, unlike yeast and mammalian cells, are at the lower end of the detection capabilities of most commercially available flow cytometers. Several modifications to a standard FACS protocol for mammalian cells must be made in order to measure and sort them. It is important to determine that bacterial-sized particles can actually be detected by the flow cytometer in use.

Special Equipment

Becton Dickinson FACScan or FACStar^{Plus} equipped with an argon ion laser tuned to 488 nm, a bandpass filter centered around 510–515 (e.g., 515/40 or 530/30), and a flow cell with an 80 µm bore.

There are a variety of commercially available flow cytometers, including the newer FACSCalibur™ and FACSVantage (Becton Dickinson), that can perform at least equally well if not better for bacterial analysis (Shapiro, 1995). The laser power need not exceed 100 mW (higher laser power contributes to higher noise levels in the forward scatter detectors).

1. To begin to define your sample population, increase the voltage of both the forward (FSC) and side scatter (SSC) photomultiplier (PMT) detectors and collect the information with logarithmic amplifiers.

   This increases the dynamic range of detection such that most small particles can be detected. Unfortunately, doing this also increases the amount of noise present. For eukaryotic cells, linear amplifiers would be used.

2. Define the sample population in terms of measurable and constant parameters such as their FSC and SSC characteristics by looking at a third parameter such as fluorescence.

   This is especially important in a low signal population such as small bacteria in order to distinguish them from any extraneous noise.

   a. Define an upper and lower limit of FSC (size) versus fluorescence.

   For example, if a fluorescence-positive (fluorescein-labeled or EGFP-positive) and -negative bacterial culture are analyzed as a function of FSC (size) versus fluorescence, a distinct population among all particles present in the dot plot will shift in fluorescence intensity (Fig. A.III.2(a and A.III.2b. By using the FSC upper- and lower limit of that particular fluorescence-shifted population, a window (gate) can be set that defines the sample population by size.

   b. Repeat the limit definition as a function of SSC (density and light polarization) versus fluorescence.

   The similar window can be used to define bacteria by virtue of their density (Fig. A.III.2c and A.III.2d). Combined with the size-correlated signal, a particular population can be focused on.

3. Adjust the PMT detectors to exclude particles not in the desired population definition.

   The electronics on most flow cytometers will allow you to draw tight boundaries (Fig. A.III.2e) such that only events within those gates are recorded.

4. After defining the sample population, adjust the voltage to the fluorescence PMT detectors such that the desired fluorescence spectrum (i.e., lowest versus highest fluorescence expected) is within range.

5. Scan a culture bearing a test gene fusion to GFP and measure the level of fluorescence. The size of the sample analyzed will vary depending on the objective of the experiment.

Typically when looking at a homogeneous bacterial culture, measuring the fluorescence from 5×10^3 individual bacteria gives a reliable quantitation of the levels of gene expression in a population. However, if the goal of the experiment is to find rare bacterial subpopulations that exhibit altered levels of fluorescence, samples as large as 5×10^5 might be required. This type of analysis is particularly useful when screening for mutants with altered gene expression profiles (see Chapter 6 for further details).

6. Once the parameters are set for a particular organism, analysis is simple, rapid and very reproducible.

   Depending on the sample density, collection number, and flow rate, the analysis can take as little time as 2–3 s.

There are problems using flow cytometric analysis that are specific to some bacterial species. Particularly frustrating are micro-organisms that tend to aggregate. Since aggregates are present even after treating samples with detergents and sonication prior to analysis, it is crucial to have a positive and a negative control in order to define the smallest particles that will shift in fluorescence intensity. By gating on this population, one can bias measurements away from clumps and focus on monodispersed bacteria.

The example shown in Fig. A.III.3 illustrates the use of flow cytometry to measure gene induction from an *Egfp* gene fusion in response to an environmental stimulus. In this case, a gene construct is shown in which a *Salmonella typhimurium* acid-inducible promoter drives the synthesis of EGFP. Gene induction (as measured by green fluorescence) is shown at 0 and 60 min after exposure to low pH (L broth equilibrated to pH 4.5 with HCl). GFP is particularly well suited for this type of analysis because the chromophore fluorescence, unlike other reporter proteins such as β-galactosidase, is highly resistance to pH extremes (Ward, 1981).

CONTRIBUTED by RAPHAEL H. VALDIVIA, BRENDAN P. CORMACK and STANLEY FALKOW

PROTOCOL 2 Separation of bacteria expressing GFP by FACS (fluorescence activated cell sorting)

Modern flow cytometers are equipped with fluorescence detection capabilities and have the ability to separate and collect cells, including bacteria. The decision as to whether a cell is collected or not can be set to be dependent on any of the different parameters described above. Most FACS instruments (and their operators) have been calibrated for the analysis of mammalian cells. To use FACS for bacterial sorting, one must be sure of the sorting efficiency of the machine for the particular micro-organism of interest. A significant amount of time may be needed to set up the sorting conditions for each particular bacterial system.

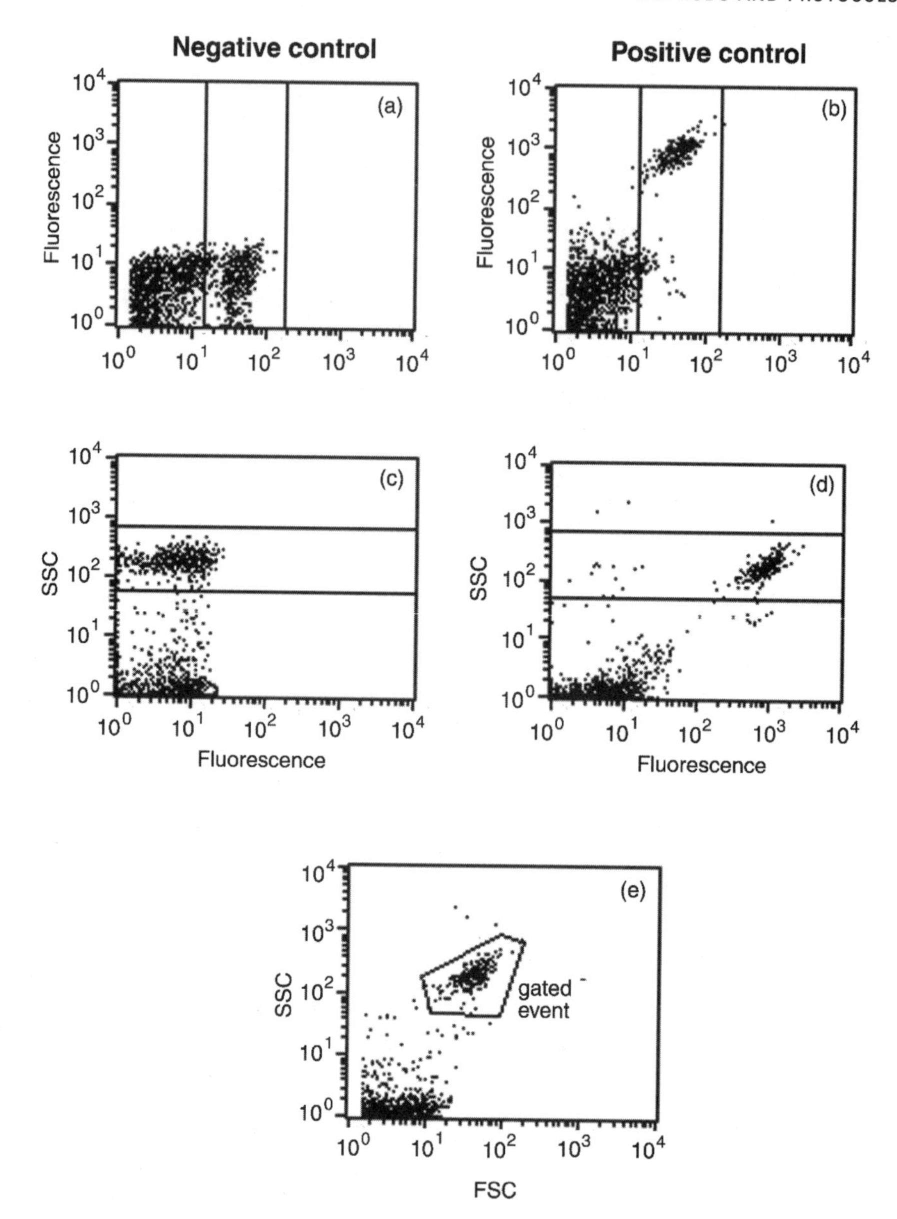

FIG. A.III.2 Bacterial detection by flow cytometry. Non-fluorescent (negative control) and EGFP-labeled bacteria (positive control) were analyzed in a FACScan (Beckton Dickinson). The voltage to the forward scatter (FSC) and side scatter (SSC) photomultiplier (PMT) detectors was raised to 80% of the maximum voltage. The information collected from a scan of 10^3 particles is shown as a dot plot. Both a and b show fluorescence intensity of all particles detected by the flow cytometer with FSC detectors and c and d show fluorescence detected by the SSC detectors. The population within the boundaries defines particles whose fluorescence shift is linked to the expression of *Egfp*. A composite of both boundaries (e) provides a framework (gated events) through which only particles that have light scattering characteristics consistent with bacterial cells are measured.

Special Equipment

Becton Dickinson FACStar^{Plus} or equivalent, as described in previous protocol.

1. Set the FACS machine to trigger (i.e., tell cell sorter electronics what is a cell and what is not) on the SSC signal.

   For eukaryotic cells one normally triggers on FSC. Unfortunately, bacteria give poor forward scatter signals, especially during sorting, and thus it is hard to trigger based on FSC. This difficulty is because extraneous signals introduced by the vibrations from the drop-forming oscillator and stray light from the exposed optics raise the background levels of noise. The SSC signal is affected less by these exogenous factors and thus we routinely trigger on SSC. Alternatively, a stronger third signal can be also be used to trigger for bacterial cells (e.g., DNA stains or bacteria-specific phycoerythrin-labeled antibodies). While the triggering on a third signal is more accurate, it also involves more manipulations and defeats the purpose of looking at dynamic genetic processes. The small FSC and SSC signals produced by bacteria can be partially compensated for by removing the neutral density filters that are usually placed in front of the scatter photodiode detectors.

2. To determine the accuracy of the sorting procedure, mix overnight cultures of GFP labeled and unlabeled bacteria at a ratio of 1:1. Use the same gating process described in the preceding protocol to separate these two known standard preparations.

   For example, in this case, we would use the 1:1 mix to set the fluorescence boundaries that will separate our fluorescent organisms from our negative control. The cell sorter measures the fluorescence of every individual bacterium that passes through the sample stream. If the fluorescence of that particular bacterium is within the gates that we have established, it is separated and collected. The sorting mechanism uses an oscillator that generates fixed-size droplets from the sample stream after it has crossed the laser sensing area. If a droplet contains a bacterium with the desired fluorescence characteristics, it is charged and deflected into a collection tube by charged deflection plates. All other droplets are aspirated into a waste collector. The purity of sorts is dependent on the accuracy of the detection sensors and on "coincidence" logic. Coincidence logic prevents any droplet from being collected if there is the possibility that an unwanted cell is also present in the same droplet or in adjacent droplets. For this reason, it is particularly useful to use dilute bacterial cultures (~10^7 organisms/mL) and to slow down the flow rate of the sample stream. While this compromises the sorting speed, it does increase the purity of the collected fractions.

3. Determine the sorting efficiency. Collect approximately 2×10^6 fluorescent events for the 1:1 mix into a collection tube half-filled with PBS, and then reanalyze the sorted population by FACS.

   If the sorting conditions are working correctly, 90–95% of the events collected should lie within the fluorescence gates used for sorting. For some applications, it is also helpful to determine the sorting efficiency of non-fluorescent bacteria

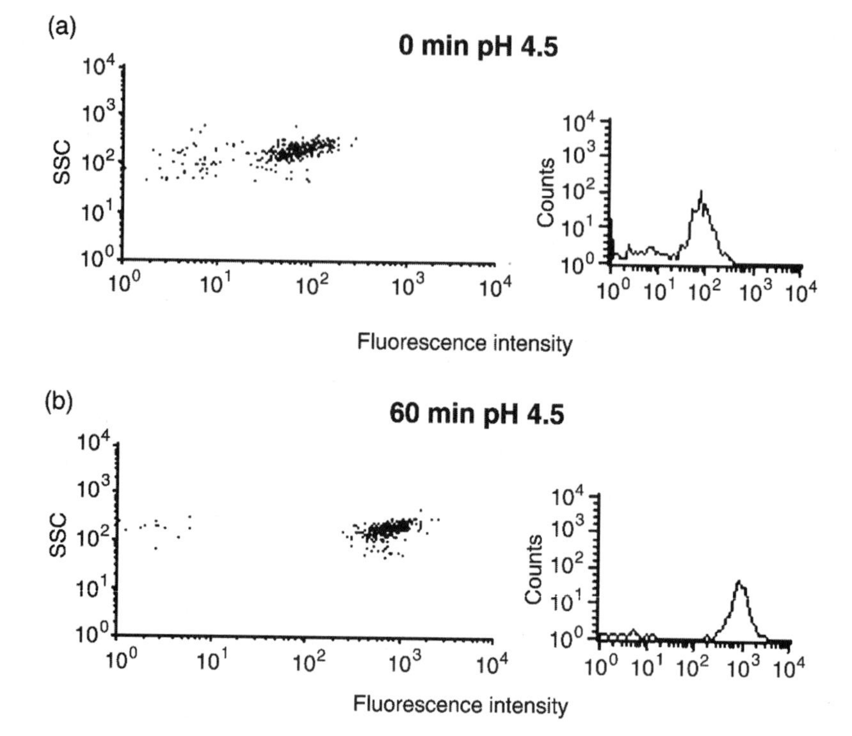

FIG. A.III.3 Quantitation of fluorescence induction. *Salmonella typhimurium* expressing *Egfp* from an acid-inducible promoter P_{aas} was used to demonstrate the application of flow cytometry to the measurement of bacterial gene expression (Valdivia and Falkow, 1996). In this case, bacteria bearing *Egfp* were exposed to L broth equilibrated to pH 4.5 for 0 min and for 60 min. Only events within the gates defined in Fig. A.III.2*e* were measured. The information collected is displayed both in a dot plot format and as a histogram.

(collect 2×10^6 non-fluorescent events from the 1:1 mix and reanalyze as described above).

4. Separate bacteria on the basis of their fluorescence.

   After all parameters are optimized, a FACS machine will separate populations at a rate of 10^3–10^4 per second. While this is technically a genetic screen, the high processivity of the sorter makes this separation essentially a selection. This selection process has two major advantages over traditional prokaryotic gene selection schemes: (1) The selection is independent of the absolute levels of fluorescence intensity. This independence means that one can separate bacteria bearing independent gfp gene fusions with as a little as a two- to threefold difference in expression levels. (2) Selections can be done in response to transient or poorly defined stimuli (e.g., acid shock, host–cell environment).

CONTRIBUTED by RAPHAEL H. VALDIVIA, BRENDAN P. CORMACK, and STANLEY FALKOW

Flow Cytometry and Sorting of Mammalian Cells Commercial flow cytometers are designed for use with large cells, such as yeast and cultured mammalian cells, and the standard fluorescence detection capabilities of these instruments make them especially suitable for sorting transiently transfected cells (Ropp et al., 1995). FACS has been used to sort transiently transfected 293 cells using UV (351 nm), deep blue (407 nm), and blue (488 nm) laser line excitation. Emission was collected from 500 to 550 nm. The autofluorescence in the 500–550 nm range that could contaminate any GFP signal was found to be lowest with 407 nm excitation. The 407 nm excitation was found to be very useful for sorting cells transfected with wild-type GFP, which can be excited with 407 nm light, but it was difficult to detect cells transfected with S65T GFP at this wavelength. However, the increased brightness of the S65T GFP made it readily detectable over autofluorescence with 488 nm excitation (Ropp et al., 1995). FACS has also been used to separate the green fluorescence of wild-type GFP from the blue fluorescence of Y66H GFP. Again, it was found that 407 nm laser line excitation minimized autofluorescence in both cases, and provided the best means for sorting cells transfected with Y66H GFP from those transfected with wild-type GFP (Ropp et al., 1996). Since mammalian cells expressing GFP can be sorted without modification of usual protocols for these cells, we do not present detailed methods here.

III.B.4.d Imaging with a Cooled CCD Camera Cooled CCD cameras offer advantages over other methods of fluorescence detection, for example, the eye, film, or confocal microscopy, in sensitivity (high quantum efficiency, high dynamic range) and ease of use. The quality of images acquired with a cooled CCD detector can be very high, and for some applications a cooled CCD camera is superior to confocal microscopy. High quality CCD cameras are available from manufacturers that include Dage, Hamamatsu, Photometrics, and Princeton Instruments. In purchasing a CCD camera, one of the most important considerations is the CCD chip, which may be the same in cameras made by different companies. A currently popular choice that is available in cameras from several manufacturers is the Kodak KAF-1400 chip, whose 1317 × 1035 array of small pixels (6.8 × 6.8 μm) make it suitable for high resolution images of subcellular structures.

Cooled CCD cameras are usually capable of acquiring 12- or 16-bit (or sometimes higher bit) images. A 12-bit image specifies shades of gray using a 12-digit binary number, with 0 = black and 4095 = white. To ensure that the full dynamic range of the camera is being used, the integration time (i.e., camera exposure time) should be adjusted to obtain pixel values of greater than 0 and less than 4095, that cover as much as possible of the full 0–4095 range. Most software imaging packages include a pre-exposure mode that will set the integration time automatically for each sample. Most image acquisition software packages also contain a histogram function that can be used to display the pixel values in the image, and a min/max display that gives the minimum and maximum pixel reading. These functions can be used to avoid exposures that saturate the pixels (i.e., give readings = 4095 for a 12-bit chip). The gain setting for some cameras can also adjusted to extend the sensitivity of the camera.

As noted previously, it is often not possible to optimize exposure times with respect to the camera's dynamic range for imaging live specimens because of time constraints due to the dynamics of the process under observation or photobleaching/photodamage to the specimen. In these cases, the optimal exposure times for acquiring images of good to acceptable quality over the observation time should be determined empirically. The following two protocols are for time-lapse imaging of GFP in live specimens.

PROTOCOL 3 Time-lapse imaging of live yeast cells using a cooled CCD camera

Time-lapse microscopy of live *nuf2-gfp* cells, expressing a centrosomal protein fused to GFP, was used to follow mitotic spindle elongation using a cooled CCD camera. The use of a sensitive detector permits very short exposure times to fluorescent light, resulting in minimal photobleaching and photodamage to the cells.

Special Equipment

Microscope equipped for epifluorescence with a 100 W mercury or 75 W xenon lamp (Nikon TMD300 or equivalent)

100X/1.4 N.A. PlanApo oil immersion objective (Nikon No. 85025 or equivalent)

Use of a high N.A. objective is required for maximal fluorescence brightness.

Photometrics CH250/KAF1400 cooled CCD camera or equivalent

This camera features pixels that are physically 6.8 × 6.8 μm². Coupled with 100X total magnification, the images are acquired at ~14.7 pixels/μm of the sample. By using an "iso C" mount to connect the camera to the microscope, the use of an intermediate eyepiece lens is avoided for maximal optical efficiency.

Electronically controlled shutter for the epifluorescence source (Uniblitz No. D122 or equivalent)

The shutter is synchronized with the camera so the samples are irradiated with excitation illumination only when the camera is taking an exposure. By doing so, photodamage and photobleaching are minimized.

FITC filter set (Excitation 460–500 nm, Dichroic 505 nm; Barrier 510–560 nm; Chroma Technology No. 41001 or equivalent).

This standard FITC filter set is particularly appropriate for "shifted" GFP isolates with the S65T mutation.

Electronic motorized focus control (Nikon No. 79588 or equivalent)

In order to make the extremely fine focus adjustments required for observing dynamic organelles, a focusing system capable of 0.1 μm movements is required.

Computerized image acquisition system (Universal Imaging Inc. MetaMorph 2.5 or equivalent)

The software should be capable of controlling the camera, shutter, and focus motor. Furthermore, it should perform real-time contrast enhancement and automatic time-lapse acquisition features. While this protocol includes specific references to the MetaMorph software, these instructions should be easily extended to most image acquisition software packages.

1. Place the slide containing live yeast cells on the microscope stage and focus on the cells using transmitted light. In order to observe mitosis, focus upon a mother-daughter pair in which the bud is about 50% the size of the mother.

   Do not use fluorescence to initially focus the microscope because it will damage the cells.
2. Set the software to perform "autoscaling" contrast adjustment.

   Using MetaMorph 2.5, the autoscaling setting is made in the "Acquire from Digital Camera: Define Acquisition Settings" menu. Autoscaling is used for highlighting dim signals when there is little or no background fluorescence. The use of autoscaling enables the use of extremely low exposure times which is critical to prevent photodamage. Most image acquisition software packages are capable of performing autoscaling.
3. Set the software to take 150 ms exposures at 10 s intervals.

   Using MetaMorph 2.5, the exposure setting is made in the "Devices: Acquire from Digital Camera: Define Acquisition Settings" menu, and the interval is set in the "Stack: Acquire Timelapse" menu. Using this scheme, the cells should only be irradiated with excitation illumination for 150 ms every 10 s. Under these conditions, the amount of photobleaching and photodamage should be negligible.
4. Start the timelapse.

   Using MetaMorph 2.5, select "OK" in the "Stack: Acquire Timelapse" menu. The image should appear in a window labeled "Timelapse Temporary Image." The initial image of the mitotic cells should feature two dots roughly 1–2 μm apart located at, and pointing into, the bud neck.
5. Adjust the focus of the microscope inbetween 10 s intervals using the electronic focus motor.

   This adjustment will take a bit of practice. Generally, make 0.2-0.3 μm movements at a time. One or two occasional additional manual exposures may be taken during the 10 s pauses without causing significant damage to the sample. We generally take 50–100 total exposures during an experiment. The 10–20 min encompass the total time for anaphase spindle elongation.
6. When anaphase is completed, stop the time lapse and save the images to disk.

   Metamorph 2.5 can save the images in a number of file formats either as they are being collected or after the experiment is concluded.
7. After completing the time-lapse imaging, continue observing the cells by eye using transmitted light microscopy to confirm that the cells continue through additional rounds of cell division.

This is to show that the cells were not severely damaged by the conditions of the experiment. Generally, the cells will begin to bud within 100 min of the last fluorescence observation.

CONTRIBUTED by PAMELA A. SILVER

PROTOCOL 4 Time-lapse microscopy of *Dictyostelium* expressing GFP-myosin

A GFP-myosin fusion protein consisting of GFP fused to conventional myosin heavy chain was expressed in *Dictyostelim dicoideum* and live cells were assayed for myosin activity (Moores et al., 1996).

Special Equipment

Zeiss Axiovert inverted microscope with a standard 100 W Hg lamp

Planapochromat 63X/1.4 N.A. phase objective

Cells were imaged with this objective combined with a 1.6X optovar lens.

HiQ FITC/GFP filter set (Chroma Technology Corp., Brattleboro, VT)

Princeton Instruments TE/CCD-512 cooled CCD camera

This is a thermoelectrically cooled CCD camera containing a thinned, back-illuminated SITe (formerly Tektronix) 512D (now S1502BA) chip, consisting of a 512 × 512 array of 24 μm pixels. The sensitivity of the CCD extends across the visible range into the UV. The CCD has the highest quantum efficiency (~75–80%) of any CCD in the visible wavelength range. This, combined with a low readout noise (10 e^- at 100 kHz), results in an extremely sensitive camera. This has allowed us to reduce our exposure times considerably and hence minimize photodamage. The camera was mounted on a basement port of the microscope via a 1X adaptor (Diagnostic Instruments, Inc., Sterling Hts, MI). The use of the basement port allows the fluorescent signal to travel to the adaptor-camera unit with no intervening mirrors or glass elements, increasing throughput by approxminately 30%.

Image1/Metamorph 2.0 (Universal Imaging Corp, West Chester, PA)

1. Place the chambered coverglass containing cells on the microscope stage.

   Cells are prepared as described in Protocol 5 (Protocol II.C) and imaged at room temperature.

2. Identify cells using phase illumination to avoid fluorescence-induced photodamage.

   Cells about to undergo cytokinesis can be identified by their lack of coarse granular material under phase illumination.

3. Take images of live cells every 5 s using a 30 ms exposure and illumination from a 100 W Hg lamp attenuated by a 50–70% neutral density filter.

   The use of a neutral density filter was necessary to prevent light-induced damage to the cells. A 50-70% neutral density filter was applied using a Zeiss attoarc controllable Hg lamp. We found that more frequent, more intense, or longer exposures resulted in cell death as evidenced by cessation of pseudopod movement and cell rounding. Using this protocol, cells remained healthy while being imaged for at least 1 h.

4. Image acquisition, storage and analysis was carried out using Image1/Metamorph 2.0.

CONTRIBUTED by JAMES H. SABRY

III.B.4.e Confocal Microscopy Confocal microscopy offers significant advantages for viewing fluorescence in tissues and optically thick samples. Due to the excitation wavelengths of available lasers, confocal imaging of GFP is best used with the red-shifted (fluorescein-like) mutants (see Chapter 5. and Protocol I.B). Confocal microscopy is suited for quantitation because of the excellent linearity of PMT detectors, and can be extended to other types of experiments, such as fluorescence photobleaching and photo-inactivation applications.

PROTOCOL 5 Time-lapse confocal imaging of early *Drosophila* embryos

The Ncd microtubule motor protein localizes to mitotic spindles in early embryos of *Drosophila*. A fusion of Ncd to wild-type or S65T mutant GFP permits spindle dynamics to be monitored during the early mitotic divisions.

Special Equipment

Zeiss Axioskop microscope with a 63X/1.4 N.A. PlanApochromat objective
This is an oil immersion objective. Water immersion objectives (Zeiss 40X/0.9 N.A. PlanNeofluar with adjustable iris) have also been used to image Drosophila embryos, as well as other types of thick specimens in which the GFP is located internally, such as Arabidopsis seedling root tips (Nikon PlanApochromat 60X/1.2 N.A.) (Haseloff and Amos, 1995). The use of a water immersion objective can result in sharper images when the object of focus is relatively deep within the cytoplasm.

Bio-Rad MRC 600 laser scanning confocal detector equipped with a krypton/argon laser

Custom GFP filter block (488/10X excitation, Q498LP dichroic, HQ518/40BP emission filters) (Chroma Technology Corp.)

This filter set was custom designed for imaging wild-type and S65T mutant GFP. For S65T GFP, the GFP filter set transmits ~1.3X as much fluorescence as the Bio-Rad BHS and ~1.8X as much as the Bio-Rad GR2/T3 filter sets (Endow and Komma, 1997).

1. Place the filter block into the detector. Set laser filter wheel to position 0. Set neutral density filter to position 1 (10% transmission) or 2 (3% transmission). Turn on laser and scanning unit.

   The custom GFP filter set contains a sharp cut-on excitation filter that is used instead of an excitation filter in the filter wheel. The laser wheel should therefore be set to position 0 (no excitation filter). A 3% transmission neutral density filter is typically used to attenuate the laser for scanning embryos with 2 copies of S65T GFP and a 10% neutral density filter is used for embryos with 2 or 4 copies of wild-type GFP.

2. Turn on the computer and open COMOS. Select "63X" microscope objective setting and PMT1. Under Collection Filter in the Image Collection window, select "Kalman" and enter 2 or 3 for the number of scans. Select "Z-Series\Time Lapse" in the Collect Menu and set the number of images to 60. Select "Save to File" and "Contrast Stretch", then click on "OK". Type in a file name to divert collected images to the appropriate directory, click on "OK" to start the scan and save the settings, collect an image, then click on "Cancel" to stop the scan.

   COMOS is the image collection and enhancement software provided by Bio-Rad. The collection settings should be set prior to observing the embryos. The early cleavage divisions occur every 10–12 min so there is usually little time after finding an embryo at the appropriate stage and beginning image collection. Kalman averaging reduces background noise in the image. Each Slow Kalman-averaged scan requires 5 s to complete. The "Contrast Stretch" function converts the minimum pixel value in the collected image to 0 and the maximum value to 255 and applies a linear conversion to pixel values inbetween. This has the effect of applying a linear contrast enhancement to each image as it is collected. For quantitation of images, this function should be deselected. The images are saved to an IBM Pentium PC as stack files of 60 images.

3. Place the slide for aligning the laser on the microscope stage. Focus on fluorescent drops on the slide using white light. Divert the laser to the slide by pulling out the slider on the microscope and removing the fluorescence filters from the light path. Deselect "Auto" controls on detector for PMT1 (Channel 1) gain and black level. Turn gain up all the way and open PMT1 pinhole fully.

4. Start image collection using Normal scan speed. Adjust the focus on a fluorescent drop. Center the laser beam by replacing the 63X objective with the "bull's eye" device and using the Allen alignment tools to adjust the M1 mirror while the laser is scanning continuously. Center the beam in the "bull's eye."

Centering of the laser is necessary whenever the filter block is changed.

5. Replace the "bull's eye" device with the 63X objective. Allow the laser to scan continuously. Adjust the gain so the fluorescent drop appears soft gray at the brightest part. Using the Allen tools, adjust the M4 mirror to obtain the brightest image of the drop. Reduce the pinhole opening to $\sim\frac{1}{3}$, adjust the gain so the fluorescent drop appears soft gray, then repeat the adjustment of the M4 mirror using the Allen tools.

   Careful alignment of the laser will result in better images. The laser scan speed can be set to "Slow" for the final adjustments to the M4 mirror.

6. Open the pinhole to $\frac{2}{3}$, turn the gain all the way up, select "Kalman" and collect a sample scan. The "Min" pixel value displayed on the scan should be above 0, in the range of 5–10. If necessary, adjust the black level and repeat test scan to bring the reading into the correct range.
7. Replace the focusing slide with the specimen slide.

   The embryos are prepared just prior to observation using methods described in Protocol II.C.

8. Using white light, focus on the posterior end of an embryo to determine the approximate cleavage division by the developmental state of the pole cells. Select an embryo at the correct stage and focus on the vitelline membrane. Divert the laser to the specimen and quickly scan on 1X zoom.
9. If the embryo is the appropriate stage, quickly change to the desired zoom, adjust focus, gain, and black level if necessary, and set image collection options to "Slow" and "Kalman". Select "Z-Series\Time Lapse" in the Collect Menu, type in the time between images (15 or 20 s) and file name, and begin collection.

   Carefully monitor the first few images for correct black level ("Min">0) and white level ("Max"<255). If the image is saturated ("Max"=255), try turning down the gain or using a higher neutral density filter, and restart the time-lapse sequence.

CONTRIBUTED by SHARYN A. ENDOW

PROTOCOL 6 Observation of GFP in living mammalian cells using confocal microscopy

The protocol described here is adapted from Niswender et al. (1995), which describes confocal microscopic observations of GFP in living mammalian cells.

Special Equipment

- Zeiss LSM410 confocal microscope or equivalent

  To take advantage of the confocal microscope's capability of three-dimensional imaging, a motorized focus control is essential. The software packages included with all confocal microscopes are capable of time-lapse and three-dimensional image acquisition, and most of the image processing required. While this pro-

tocol includes specific references to the Zeiss LSM software, similar software is available from all manufacturers.

Adams and List TLC-MI temperature-controlled stage or equivalent

This stage heats both the sample dish and the perfusate (if used) to maintain the sample at 37°C.

40X/1.3 N.A. Plan-Neofluar oil immersion objective (Zeiss 44-04-50)

Use of a high NA objective is required for maximal signal collection. The Plan-Neofluar design offers considerably better throughput than the PlanApo design. For confocal microscopy, lateral resolution is given by the N.A. of the objective lens (not the magnification), so a 40X lens, which gives the widest possible field of view, is superior to higher magnification objectives.

GFP filter set (Dichroic Q498LP, Barrier HQ512/27BP filter; Chroma Technology)

This GFP filter set is appropriate for both wild-type GFP and the S65T mutant. For the F64L, S65T (EGFPmut1) double mutant, a wider bandpass filter (e.g., 500–550 nm) may be preferable (see Protocol 5 above).

1. Place the dish of cultured cells on the microscope stage and focus on the cells using transmitted light. The use of phase contrast or Nomarski DIC is often helpful in locating the cells.

   Do not use fluorescence to initially focus the microscope because it will damage the cells.
2. After focusing on the cells of interest, quickly observe the cells with epifluorescence.

   This step is required for transient transfections where only a small percentage of cells contain GFP. It is imperative to perform this step quickly to minimize photobleaching and photodamage.
3. Before starting the laser scanning, set the laser power down, the contrast (or PMT gain) up, choose a fast scan time (typically 1 s per scan), and adjust the pinhole setting to its optimum value.

   For the Zeiss LSM410, start with laser power 20 (out of 100), contrast 325, 1 s per scan, and a pinhole of 60 for the 40X objective. This pinhole value maximizes the signal-to-noise in the confocal (Sandison et al., 1995).
4. Start laser scanning and increase laser power until an appropriate signal is collected.

   To obtain the best images of bright samples it may be necessary to lower the contrast setting, and for dimmer samples it may be necessary to use a longer scan time or average several successive images. On the Zeiss system, the F9 key will automatically adjust the contrast and brightness to give a low noise image, but it does not give appropriate values for quantitative measurements.

The image at this point will be of a single 1 μm thick optical section of the specimen.

5. Once the appropriate instrument parameters have been determined, time-lapse images can easily be acquired (select "Time Series" from the "Time" menu) or images of successive optical slices through the sample (select "Z Sectioning" from the "Z" menu).

   Using the Zeiss confocal software, images can be stored directly into memory. This feature is convenient when no time lapses between successive images are desired.

6. After a series of images has been acquired, save them to disk immediately. It is also possible to save images as they are acquired.

   We store images in the TIFF image format for maximum portability between Macintosh and IBM PCs and programs such as NIH Image and Adobe Photoshop.

CONTRIBUTED by DAVID W. PISTON

III.B.4.f Fluorescence Photobleaching Techniques Fluorescence recovery after photobleaching (FRAP; also known as FPR, fluorescence photobleaching recovery), is a long-established method [see Edidin (1992) for review] for quantitating lateral diffusion in cell membranes, and a powerful tool for investigating the environment of membrane proteins. This method is based on the property that fluorophores can be irreversibly inactivated, or "photobleached." The behavior of populations of hundreds of thousands of labeled molecules, on a scale of micrometers can be analyzed by this method. Since recovery of fluorescence into the area being photobleached occurs only as a result of diffusional exchange between bleached and unbleached fluorophores, the diffusional mobility of the fluorescent molecules can be determined.

In practice, a high intensity laser beam photobleaches a predefined region in a fluorescently labeled membrane. An attenuated beam (usually $<$1000-fold less intense than the bleaching beam) is used to monitor fluorescence in the region before and after a proportion of the fluorophores is bleached by a short pulse of laser light at full intensity (Edidin, 1994). The half-time to reach maximum recovery is proportional to the area bleached and the diffusion coefficient, D, of the labeled species. The FRAP measurements also yield a second parameter of the fluorescent proteins, the fraction of fluorescent proteins that is not free to diffuse in the time examined, referred to as the immobile fraction. Several types of interactions that constrain the lateral diffusion of proteins in the plasma membrane have been studied by this technique (Edidin, 1992).

New applications of FRAP have been found using GFP. The advantage of using GFP in FRAP studies is that labeling with extrinsic fluorophores has previously limited researchers to studying the plasma membrane. Even dyes that "target" organelles still go to the wrong places, or are internalized or otherwise mislocalized over time, thus complicating the interpretation of FRAP results. GFP fusion proteins are excellent probes for FRAP studies (Storrie and Kreis, 1996) because they can be targeted to specific organelles, and have been used to measure the diffusional mobility of several membrane proteins that reside within the Golgi complex (Cole et al., 1996). The two protocols presented here have been adapted from the work of Cole et al. (1996).

PROTOCOL 7 Fluorescence recovery after photobleaching (FRAP) using a confocal microscope

Photobleaching recovery experiments are easy to perform on a confocal microscope. Sample preparation is described in Protocol II.C and microscope set-up follows steps 1–5 of the Niswender et al. (1995) confocal microscopy protocol (Protocol 6 above). After the cells of interest are identified in the confocal microscope, the following protocol can be used. In order to change settings quickly, a command macro (available on request) must be used to run the following procedure.

1. Define a small box within the specimen by increasing the zoom.

   The ratio of bleaching intensity to monitoring intensity necessary for these experiments can be obtained using zoom 8 at this step.

2. Set the laser intensity to maximum, i.e., 100% laser power and 100% transmission (no neutral density filters).

3. Photobleach the region within the box by scanning it 3–5 times with the high laser energy.

   The time required to bleach a given sample must be determined empirically by iterations of steps 2 and 3. Once this time has been determined for a given preparation, it is usually consistent between cells.

4. Decrease the zoom to 1 and the laser intensity to 10% power and 3% transmission (1.5 OD neutral density filter).

   The combination of reduced zoom, lower laser power, and decreased transmission gives a ratio of bleaching to imaging radiation of more than 1000.

5. Monitor the recovery of fluorescence into the photobleached region by imaging the photobleached box and surrounding area at a low energy for different times following the bleach.

   Image collection should be timed so that approximately 10 images are acquired before the recovery is completed. Because recovery is a diffusional process, early time points are often more important than later ones, so a logarithmic time spacing of images may be preferable,

6. The diffusion constant, D, can be estimated from the half time of the recovery, $t_{1/2}$, by $D = \beta A/4t_{1/2}$ where β is related to the percent photobleached, and A is the bleached area (Yguerabide et al., 1982). This estimate assumes that recovery can occur along the entire border of the bleached area. To calculate mobile fraction of fluorophore, the ratio of fluorescent intensities for two regions of interest, one within the bleaching zone and the other outside of it, are compared before photobleaching and after recovery.

   For intracellular structures like the Golgi complex, where a photobleached box is frequently a significant portion of the entire organelle, the absolute fluores-

cence associated with the organelle is frequently lower after photobleaching than before photobleaching because photobleaching removes a significant proportion of total fluorescence. This method for calculating mobile fraction is insensitive to this loss.

CONTRIBUTED by JENNIFER LIPPINCOTT-SCHWARTZ

Fluorescence Recovery after Photobleaching Using a Dedicated FRAP Instrument Dedicated FRAP instruments usually couple a laser light source and a photomultiplier detector to a standard microscope. One advantage of the high-energy dedicated laser microscope system for FRAP is that a beam-splitting device is used to produce measuring and bleaching beams, so recovery measurements can be made in a fraction of a second after photobleaching (Edidin, 1992). Time resolution is a problem for the confocal microscope, which can take up to 1–3 s before the first postbleach image is acquired. Since most FRAP instruments have been constructed in individual laboratories, a detailed protocol would not be widely applicable. In general, a laser beam is focused on a 1–2 μm spot or a 1–2 μm stripe, which defines a region in a fluorescently labeled membrane. An attenuated beam is used to monitor fluorescence in the region before and after a proportion of the fluorophores is bleached by a short pulse (~20 ms) of laser light at full intensity (the bleaching intensity should be >1000-fold more intense than the attenuated monitor beam) (Edidin, 1994).

Quantitative FRAP experiments of GFP chimeras have been performed on such a dedicated FRAP system using a stripe photobleaching pattern (Cole et al., 1996). Since the stripe extended across the entire depth of sample, diffusion was into and out of a line bounded on its side. Recovery, therefore, was due to one-dimensional diffusion. One problem with this method of approximating the diffusion constant, D, is that some intracellular organelles, such as the Golgi complex and ER (endoplasmic reticulum), consist of extensively interconnected tubules and cisternae which have a meshlike character. This geometry can introduce errors in the calculation of D. However, a tortuous diffusion path has been shown to reduce the apparent D (Wey et al., 1981), so conventional approaches to calculating D can be assumed to be an underestimation of the true D.

PROTOCOL 8 Fluorescence loss in photobleaching (FLIP)

A powerful photobleaching technique for examining the extent of continuity of a compartment (whether it be ER, Golgi, or cytoplasm) is fluorescence loss in photobleaching, or FLIP (Cole et al., 1996). This technique examines the extent to which regions outside a photobleached area can contribute to recovery of fluorescence at a bleached site. Fluorescence loss in photobleaching can also be applied to many different systems for examining the extent of their continuity and for defining compartmental boundaries.

Using this method, a narrow box across a structure is bleached repeatedly and the fluorescence loss in photobleaching outside the box is followed over time. The extent to which areas outside the box lose fluorescence describes the boundaries over which the fluorophore is capable of diffusing. Much of the following procedure is similar to the FRAP experiments performed using a confocal microscope, described above.

1. Define a small box within the specimen by increasing the zoom, and set the laser intensity to maximum, that is, 100% laser power and 100% transmission (no neutral density filters).

   The ratio of bleaching intensity to monitoring intensity necessary for these experiments can be obtained using zoom 8 at this step.

2. Scan the region within the box for 30 s with the high laser energy.

   Since this method requires depletion of the fluorophore throughout any connected membranes, the bleaching times are considerably longer than in Protocol 7 above for FRAP.

3. Decrease the zoom to 1 and the laser intensity to 10% power and 3% transmission (1.5 OD neutral density filter).

   As with the FRAP protocol, the FLIP protocol uses a ratio of bleaching to imaging radiation of more than 1000 to ensure that bleaching does not occur during the monitoring step.

4. Image the entire field of view to assess the extent fluorescence outside the box is lost as a consequence of bleaching within the box. Monitor the recovery of fluorescence into the photobleached region by imaging the photobleached box and surrounding area at a low energy for different times after the bleach.

5. Repeat steps 1–4 until photobleaching can no longer be detected.

   At this point, the specimen fluorescence will be the same as background.

6. Fix the cells.

   Two controls can be performed on the fixed cells after performing a FLIP experiment.

   (a) Stain cells with antibodies to the structure being bleached and observe the morphology of the structure.

   This staining confirms that structures inside the bleached zone are not damaged during exposure to the intense light. See Protocol II.B for fixation and antibody staining methods.

   (b) Repeat FLIP on the fixed cells.

   This is to rule out the possibility that regions on the edge of the illuminated zone are not progressively bleached by light leakage during FLIP.

CONTRIBUTED by JENNIFER LIPPINCOTT-SCHWARTZ

III.B.4.g Two-Photon Excitation Microscopy Two-photon excitation microscopy (TPEM) is a new method that is a superior alternative to confocal microscopy. Use of this technique eliminates out-of-focus photodamage, avoids chromatic aberrations, and provides high resolution three-dimensional measurements. Two-photon excitation arises from the simultaneous absorption of two photons, each of which is one-half of the energy required for the transition to the excited electronic state of the fluorophore. In practice, two-photon excitation is made possible by the very high local instantaneous intensity that is provided by a combination of diffraction-limited focusing of a single laser beam in the microscope and the temporal concentration of a subpicosecond mode-locked laser. Resultant instantaneous peak excitation intensities are 10^6 times greater than typical in confocal microscopy, but the pulse duty cycle of 10^{-5} maintains the average input power at less than 10 mW, which is only slightly greater than typically used in conventional confocal microscopy. Currently, the cost of state-of-the-art lasers which are required for TPEM has limited this technique to only a few laboratories. However, the field of ultrafast lasers is undergoing rapid development and as the price of lasers decreases over the next 10 years, TPEM may become a widely used method of optical sectioning microscopy.

Three properties of two-photon excitation provide the significant advantages over conventional optical sectioning microscopies for the study of UV excitable fluorophores in thick samples:

1. The excitation is limited to the focal volume due to the second-order dependence of the two-photon excitation on intensity and the decrease in intensity with the square of the distance from the focal plane. The excitation localization yields three-dimensional discrimination equivalent to an ideal confocal microscope without requiring a confocal spatial filter. Absence of the need to descan the fluorescence to pass a confocal aperture enhances fluorescence collection efficiency. Confinement of fluorophore excitation to the focal volume minimizes photobleaching and photodamage associated with UV illumination—the ultimate limiting factors in fluorescence microscopy of living cells and tissues.
2. Two-photon excitation allows imaging of UV fluorophores with conventional visible light optics in the scanning and detection systems because both the red excitation light (~700 nm) and the blue fluorescence (400 nm) are within the visible spectrum.
3. Red light interacts much less strongly than UV light with most living cells and tissues (aside from plant cells) since fewer biological molecules absorb at longer wavelengths and red light is scattered less than shorter wavelengths. This nearly eliminates out-of-focus photodamage and background, and allows most of the input power to reach the focal plane. The relative transparency of biological specimens at 700 nm permits deeper sectioning than would be possible with UV excitation.

A limitation of TPEM is that the two-photon excitation spectrum may bear little resemblance to the one-photon absorption profile. GFP is easy to use in TPEM, however, since the two-photon excitation spectra of both wild-type and the S65T mutant have been shown to overlap twice the one-photon absorption spectra, with the same relative absorption cross sections (Xu et al., 1996). This overlap means that the major absorption peak of 395 nm for wild-type GFP is found at 2×395 nm (790 nm) with two-photon excitation. Preliminary results show a similar behavior for two-photon excitation of the F64L, S65T (*EGFPmut1*) double mutant, and also that it is one of the strongest two-photon absorbers ever measured (Xu, Albolta, and Webb, personal communication). These findings are consistent with the extremely large one-photon absorption of this double mutant. Preliminary results indicate that the F64L, S65T (*EGFPmut1*) mutant is actually brighter than wild type even at 790 nm (2×395 nm) excitation (Patterson and Piston, unpublished observation). Using wild-type GFP, useful two-photon excited images have been obtained using ~790 nm (Niswender et al., 1995) and ~900 nm (Potter et al., 1996) excitation from a Ti:Sapphire laser.

The protocol for two-photon excitation microscopy is identical to that of confocal microscopy (Protocol 6 above) except for the use of a different laser and elimination of the confocal pinhole. Currently, the laser of choice for TPEM is an ultrafast mode-locked Ti:Sapphire (e.g., Coherent Mira or Spectra-Physics Tsunami) pumped by either an argon ion laser or an all solid-state, frequency-doubled Nd:YAG laser that has recently achieved the power levels necessary for this application. The mode-locked Ti:Sapphire laser produces ~100 fs pulses at a repetition rate of ~100 Mhz, and can be tuned over a broad range of output wavelengths from 700 to 1100 nm, allowing two-photon excitation of most UV and visibly excited fluorophores. The ability to adjust laser wavelength, though, makes this laser considerably more complicated than the "turn key" lasers used in confocal microscopy.

In the future, ultrafast lasers yielding a single wavelength, or a small group of selected wavelengths, may allow the development of easy-to-use and relatively inexpensive two-photon excitation microscopes. Since the pinhole is not needed to obtain optical sectioning with TPEM, various detection schemes are possible. Most two-photon excitation systems have been built around a confocal laser scanning microscope. The pinhole on these instruments should be opened fully or removed when using two-photon excitation. However, one can use a CCD camera or full-field photomultiplier detector to increase the fluorescence collection efficiency and still acquire optical sections from thick samples. At this point, use of alternate detection schemes requires extensive modification of commercial laser-scanning systems or construction of a new dedicated two-photon excitation microscope. The choice of lasers and detection schemes have been described in detail elsewhere (Denk et al., 1995). Further information about TPEM can be found at http://www.mc.vanderbilt.edu/vumcdept/mpb/piston/files/tpfe.html.

III.B.4h Image Deconvolution Image deconvolution, where a mathematical algorithm is used to deblur a three-dimensional data set collected with a CCD camera (Agard et al., 1989; Carrington et al., 1990; Holmes et al., 1995), is not a new technique, but the advent of fast desktop computers has allowed it to be useful to a wide variety of laboratories. Deconvolution techniques can be used with some biological systems to obtain results similar, or sometimes superior, to those from confocal imaging. These methods allow better signal collection than confocal microscopy, since a high-quality cooled CCD camera has a higher quantum efficiency (i.e., it collects more light) than a photomultiplier tube, and all the fluorescence is collected (there is no pinhole to exclude any fluorescence).

However, deconvolution is less useful on densely fluorescent samples. A "densely" fluorescent sample is one where fluorescence arises from many places

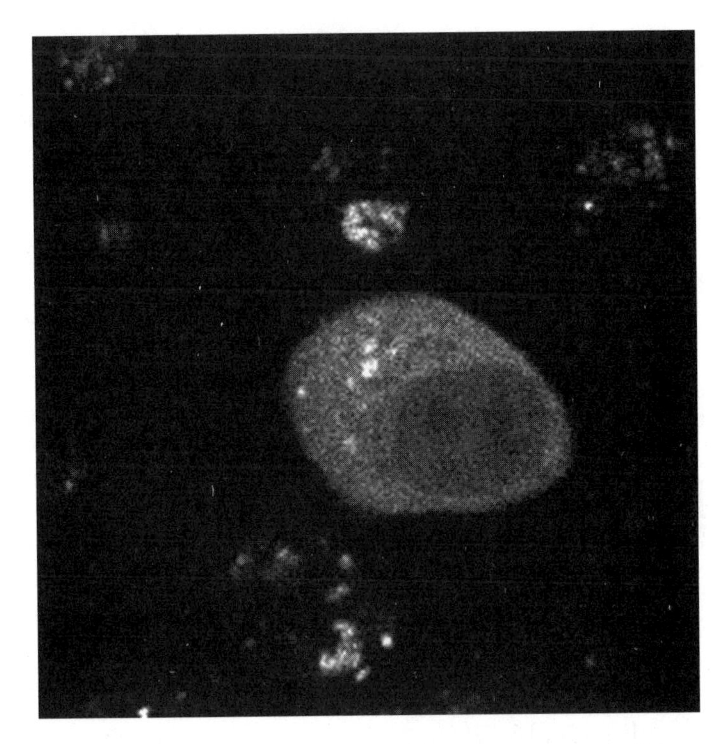

FIG A.III.4 Visualization of GFP in mammalian cells using two-photon excitation microscopy. Two-photon excitation of a HeLa cell transfected with a construct encoding a fusion protein between pancreatic glucokinase (GK) and wild-type GFP. Excitation was at 780 nm and fluorescence was collected through a HQ512/27 bandpass emission filter to minimize background cellular autofluorescence. The GK-GFP fusion protein targets primarily to the cytoplasm, while the GFP alone (not shown) partitions equally throughout the cytoplasm and nucleus. The GFP fluorescence is significantly increased above background autofluorescence. The narrower bandpass emission filter (compared to the HQ518/40BP used for *Drosophila* embryos in Fig. A.III.1) is preferable for use in imaging mammalian cells. For further details, see Niswender et al. (1995).

in the cell or sample (i.e., not just the DNA or a single organelle). For instance, cytoplasmic staining, or microtubule staining of an embryo may look like a blur in a widefield microscope, whereas a confocal image may show sharp detail. There is a point, for any given number of collected photons, where deconvolution techniques can no longer find the details in the blurry image and deconvolve them. This point, however, has never been strictly determined and each research team using deconvolution usually establishes its own "rule of thumb" about which samples are appropriate.

In order to use image deconvolution, a computer-controlled fluorescence microscope is equipped with a high-quality cooled CCD camera. An accurate and reproducible motorized focus control is required. The microscope set-up is similar to those used in the cooled CCD protocols described above. The computer requirements for deconvolution depend mainly on the number of sections in the data set, but Power Macintosh and Pentium computers with sufficient memory can be used to deconvolve most data sets in under an hour. Several commercial software packages (listed below) can be used to process the data sets once they are acquired. The PSF (point spread function) of the microscope—how the microscope images a single point of fluorescence—must generally be measured using small (<0.2 μm) fluorescent latex beads. Some new algorithms, however, do not require a priori knowledge of the PSF (Holmes et al., 1995).

Optical sections are taken throughout the sample at 0.1–0.5 μm intervals, and between 4 and 128 sections are used to perform the deconvolution. To achieve the best results, the acquired data set must be optimized for every sample in terms of spacing and number of sections. It is also possible to use image deconvolution to improve the resolution of data sets acquired with a confocal microscope, but reports of this application to biological samples have yet to appear.

Software packages for three-dimensional image deconvolution, and further information about the image set requirements for each package are available from the vendors listed below.

VayTek, Inc.
P.O. Box 732
305 W. Lowe Avenue
Fairfield, IA 52556
T (515) 472-2227
F (515) 472-8131

Scanalytics
A Division of CSPI
40 Linnell Circle
Billerica, MA 01821
T (800) 882-6247 x1300
F (508) 663-0150

Applied Precision
8505 SE 68th St
Mercer Island, WA 98040
T (206) 236-0704
F (206) 232-4184

AutoQuant Imaging Inc.
1223 Peoples Avenue
Troy, NY 12180-3590
T (518) 276-2138
F (518) 276-3069

III.B.5 Quantitation

A powerful aspect of using GFP in fluorescence microscopy is the capability to perform quantitative measurements of expression in cells without many of the problems that plague immunochemical methods. Quantitative imaging can be easily performed on a confocal microscope, as well as with a cooled CCD camera, but using a CCD can be complicated due to nonuniformities in cell thickness. In some cases, these complications can be eliminated through image deconvolution.

The basic strategy used to quantitate GFP images is to perform parallel spectroscopy and microscopy experiments on a standard solution of fluorescein (Molecular Probes No. F-1300, extinction coefficient = 90,000 $cm^{-1}M^{-1}$ at 488 nm excitation, quantum yield = 85%) and a purified GFP solution, and compare these results to images of GFP-labeled cells (Niswender et al., 1995). Methods for GFP purification are described in Chapter 3 and Protocol I. Once a purified GFP sample is obtained, its concentration can be determined by absorption and fluorescence spectroscopy using the known extinction coefficient and quantum yield of the particular form of GFP being used (Heim and Tsien, 1996). For example, the S65T mutant has an extinction coefficient of ~40,000 $cm^{-1}M^{-1}$ at 488 nm, with a quantum yield of ~66% (Heim and Tsien, 1996). The concentration of the GFP sample is then calculated from the absorption relative to fluorescein, e.g., if the absorption at 488 nm of the S65T sample is the same as a 1 μM fluorescein sample, the S65T concentration is [S65T] = [fluorescein] (extinction coefficient of fluorescein/extinction coefficient of S65T) = 1 μM × (90/40) = 2.25 μM.

The known GFP sample can now be compared with GFP-labeled cells imaged by the confocal microscope. Here we refer to gain and black level on the Bio-Rad MRC 600, which correspond to contrast and brightness, respectively, on the Zeiss LSM410. The standard samples of fluorescein or GFP can be measured on the confocal microscope using a deep-well slide or a spacer between the slide and coverslip. To begin, set the gain so that the image of the desired cell is not saturating (i.e., no pixel values = 255). Once the correct gain is set, it should not be changed.

It is now important to set the microscope offset so that zero pixel value corresponds to zero fluorescence signal. This can be done using the fluorescein sample as follows. Image four concentrations (in a ratio of 1:2:3:4) of fluorescein that each give a good signal but for which the maximum pixel value is <255 at the gain used for the GFP imaging. Determine the mean pixel value of each image using the histogram command, and plot the mean pixel value versus the concentration. The resulting plot should be a straight line, and the Y intercept will go through zero when the black level is correctly set. If the Y intercept is positive, the offset should be reduced, and if the Y intercept is negative the offset should be increased. Image the GFP-labeled cell again. It may require several trials to achieve the optimum gain/offset combination for your sample. Dilute or concentrate the GFP sample to match the signal obtained in the cell GFP image, and determine the concentrations of GFP in the cell over the linear range of the image data (pixel values between 0 and 255). Note that with thick samples such as tissue slices the fluorescence signal may decrease as the focal plane moves deeper into the sample. In this case, it is more difficult to interpret the quantitative amount of GFP in the cells.

CONTRIBUTED by DAVID W. PISTON

REFERENCES

Agard, D. A., Hiraoka, Y., Shaw, P., and Sedat, J. W. (1989). Fluorescence microscopy in three dimensions. *Methods Cell Biol.* 30:353–377.

Baulcombe, D. C., Chapman, S., and Santa Cruz, S. (1995). Jellyfish green fluorescent protein as a reporter for virus infections. *Plant Jl.* 7:1045–1053.

Carrington, W. A., Fogarty, K. E., and Fay, F. S. (1990). 3D fluorescence imaging of single cells using image restoration. *In* Non-invasive Techniques in Cell Biology Foskett, J. K., and Grinstein, S., Eds., Wiley-Liss, New York, pp. 53–72.

Casper, S. J. and Holt, C. A. (1996). Expression of the green fluorescent protein-encoding gene from a tobacco mosaic virus-based vector. *Gene* 173:69–73.

Chalfie, M., Tu, Y., Euskirchen, G., Ward, W. W., and Prasher, D. C. (1994). Green fluorescent protein as a marker for gene expression. *Science* 263:802–805.

Chattoraj, M., King, B. A., Bublitz, G.U., and Boxer, S. G. (1996). Ultra-fast excited state dynamics in green fluorescent protein: multiple states and proton transfer. *Proc. Natl. Acad. Sci. USA* 93:8362–8367.

Cole, N. B., Smith, C. L., Sciaky, N., Terasaki, M., Edidin, M., and Lippincott-Schwartz, J. (1996). Diffusional mobility of Golgi proteins in membranes of living cells. *Science* 273:797–801.

Cormack, B. P., Valdivia, R. H., and Falkow, S. (1996). FACS-optimized mutants of the green fluorescent protein (GFP). *Gene* 173:33–38.

Cubitt, A. B., Heim, R., Adams, S. R., Boyd, A. E., Gross, L. A., and Tsien, R. Y. (1995). Understanding, improving and using green fluorescent proteins. *Trends Biochem. Sci.* 20:448–455.

Delagrave, S., Hawtin, R. E., Silva, C.M., Yang, M. M., and Youvan, D. C. (1995). Red-shifted excitation mutants of the green fluorescent protein. *Bio/Technology* 13:151–154.

Denk, W., Piston, D. W., and Webb, W. W. (1995). Two-photon excitation in laser scanning microscopy. *In* The Handbook of Biological Confocal Microscopy, Pawley, J., Ed., Plenum, New York, pp. 445–458.

Edidin, M. (1992). Patches, post and fences: proteins and plasma membrane domains. *Trends Cell Biol.* 2:376–380.

Edidin, M. (1994). Fluorescence photobleaching and recovery, FPR, in the analysis of membrane structure and dynamics. *In* Mobility and Proximity in Biological Membranes, Damjanovich, S., Edidin, N., Szollosi, J., and Tron, L., Eds., CRC, Boca Raton, FL, pp. 109–135.

Endow, S. A. and Komma, D. J. (1996). Centrosome and spindle function of the *Drosophila* Ncd microtubule motor visualized in live embryos using Ncd-GFP fusion proteins. *J. Cell Sci.* 109:2429–2442.

Endow, S. A. and Komma, D. J. (1997). Spindle dynamics during meiosis in *Drosophila* oocytes. *J. Cell Biol.* 137:1321–1336.

Haseloff, J. and Amos, B. (1995). GFP in plants. *Trends Genet.* 11:328–329.

Heim, R., Cubitt, A. B., and Tsien, R. Y. (1995). Improved green fluorescence. *Nature (London)* 373:663–664.

Heim, R. and Tsien, R. Y. (1996). Engineering green fluorescent protein for improved brightness, longer wavelengths and fluorescence resonance energy transfer. *Curr. Biol.* 6:178–182.

Holmes, T. J., Bhattacharyya, S., Cooper, J. A., Hanzel, D., Krishnamurthi, V., Lin, W., Roysam, B., Szarowski, D. H., and Turner, J. N. (1995). Light microscopic images reconstructed by maximum likelihood deconvolution. *In* The Handbook of Biological Confocal Microscopy, Pawley, J., Ed., Plenum, New York, pp. 389–402.

Keller, H. E. (1995). Objective lenses for confocal microscopy. In The handbook of biological confocal microscopy, Pawley, J., Ed., Plenum, New York, pp. 111–126.

Kremer, L., Baulard, A., Estaquier, J., Poulain-Godefroy, O., and Locht, C. (1995). Green fluorescent protein as a new expression marker in myobacteria. *Mol. Microbiol.* 17:913–922.

Maniak, M., Rauchenberger, R., Albrecht, R., Murphy, J., and Gerisch, G. (1995). Coronin involved in phagocytosis: dynamics of particle-induced relocalization visualized by a green fluorescent protein tag. *Cell* 83:915–924.

Mikhailov, A. V. and Gundersen, G. G. (1995). Centripetal transport of microtubules in motile cells. *Cell Motil. Cytoskel.* 32:173–186.

Moores, S. L., Sabry, J. H., and Spudich, J. A. (1996). Myosin dynamics in live *Dictyostelium* cells. *Proc. Natl. Acad. Sci. USA* 93:443–446.

Niswender, K. D., Blackman, S. M., Rohde, L., Magnuson, M. A., and Piston, D. W. (1995). Quantitative imaging of green fluorescent protein in cultured cells: comparison of microscopic techniques, use in fusion proteins and detection limits. *J. Micros.* 180:109–116.

Ormö, M., Cubitt, A. B., Kallio, K., Gross, L. A., Tsien, R. Y., and Remington, S. J. (1966). Crystal structure of the *Aequorea victoria* green fluorescent protein. *Science* 273:1392–1395.

Parks, D. R., Herzenberg, L. A., and Herzenberg, L. A. (1989). Flow Cytometry and Fluorescence-Activated Cell Sorting. *In* Fundamental Immunology, Paul, W. E., Ed., Raven, New York, pp. 781–802.

Potter, S. M., Wang, C.-M., Garrity, P. A., and Fraser, S. E. (1996). Intravital imaging of green fluorescent protein using two-photon laser-scanning microscopy. *Gene* 173:25–31.

Rizzuto, R., Brini, M., De Giorgi, F., Rossi, R., Heim, R., Tsien, R. Y., and Pozzan, T. (1996). Double labelling of subcellular structures with organelle-targeted GFP mutants *in vivo*. *Curr. Biol.* 6:183–188.

Ropp, J. D., Donahue, C. J., Wolfgang-Kimball, D., Hooley, J. J., Chin, J. Y. W., Hoffman, R. A., Cuthbertson, R. A., and Bauer, K. D. (1995). *Aequorea* green fluorescent protein analysis by flow cytometry. *Cytometry* 21:309–317.

Ropp, J. D., Donahue, C. J., Wolfgang-Kimball, D., Hooley, J. J., Chin, J. Y. W., Hoffman, R. A., Cuthbertson, R. A., and Bauer, K. D. (1996). *Aequorea* green fluorescent protein: simultaneous analysis of wild-type and blue-fluorescing mutant by flow cytometry. *Cytometry* 24:284–288.

Rost, F. W. D. (1992). Fluorescence microscopy. Cambridge University Press, Cambridge, UK.

Sandison, D. R., Piston, D. W., Williams, R. M., and Webb, W. W. (1995). Resolution, background rejection, and signal-to-noise in widefield and confocal microscopy. *Appl. Opt.* 34:3576–3588.

Shapiro, H. M. (1995). Practical Flow Cytometry. Wiley-Liss, New York.

Storrie, B. and Kreis, T. E. (1996). Probing the mobility of membrane proteins inside the cell. *Trends Cell Biol.* 6:321–324.

Valdivia, R. H. and Falkow, S. (1996). Bacterial genetics by flow cytometry: rapid isolation of *Salmonella typhimurium* acid-inducible promoters by differential fluorescence induction. *Mol. Microbiol.* 22:367–378.

Ward, W. W. (1981). Properties of coelenterate green fluorescent proteins. *In* Bioluminescence and Chemiluminescence, DeLuca, M. A. and McElroy, W. D., Eds., Academic, San Diego, pp. 225–234.

Wey, C. L., Cone, R. A., and Edidin, M. A. (1981). Lateral diffusion of rhodopsin in photoreceptor cells measured by fluorescence photobleaching and recovery. *Biophys. J.* 33:225–232.

Xu, C., Zipfel, W., Shear, J. B., Williams, R. M., and Webb, W. W. (1996). Multiphoton fluorescence excitation: new spectral windows for biological nonlinear microscopy. *Proc. Natl. Acad. Sci. USA* 93:10763–10768.

Yang, F., Moss, L. G., and Phillips, G. N., Jr. (1996). The molecular structure of green fluorescent protein. *Nature BioTech.* 14:1246–1251.

Yguerabide, J., Schmidt, J. A., and Yguerabide, E .E. (1982). Lateral mobility in membranes as detected by fluorescence recovery after photobleaching. *Biophys. J.* 40:69–75.

Youvan, D. C., Goldman, E., Delagrave, S., and Yang, M. M. (1995). Digital imaging spectroscopy for massively parallel screening of mutants. *Methods Enzymol.* 246:732–748.

PROTOCOL IV DOCUMENTING GREEN FLUORESCENT PROTEIN

IV.A Documentation Methods

GFP can be documented by photography, digital imaging methods, videotape recordings, and QuickTime movies. Among these methods, photography is still the most common method for documentation but is rapidly becoming replaced by digital methods, which offer greater ease of image manipulation. Photography does offer advantages over most commonly used digital methods in the ability to resolve and record colors, and to record large specimens. These advantages can be combined with those of digital imaging by using negative scanners to convert negatives to digital images or by using commercial film processing services to transfer images to CD-ROM. For organisms in which autofluorescence can be distinguished from the GFP signal by color, such as C. *elegans* (Chalfie et al., 1994), the conversion of color negatives to digital images could prove to be a method of choice.

One of the most valuable features of GFP use is the ability to follow dynamic processes in living cells over time. Until recently, the difficulty of labeling specific structures in living cells and the dependence on relatively sophisticated and expensive equipment for recording and displaying time-lapse images have limited real-time analysis to only a few laboratories. The methods available to most biologists for documenting cellular fluorescence have relied primarily upon still images of fixed cells acquired by photography or confocal microscopy. Now, due to the ability to visualize dynamic processes in living cells and the advent of relatively inexpensive fast desktop computers, real-time imaging and video technology are rapidly becoming part of the standard repertoire of all biologists.

High-quality videotape recordings can be made from time-lapse images collected with a cooled CCD camera or confocal microscope using currently available high-speed personal computers. Time-lapse images can also be made into QuickTime movies and then recorded to videotape, although these recordings are generally of lower resolution and lower quality than videotapes made directly from full-size files. The small QuickTime movies are suitable for posting on Web pages for downloading and viewing, however. Images can also be animated and played continuously on Web pages as GIF animations. The methods for creating GIF animations are not covered here, but those interested in making one should visit the GifBuilder Web site at http://iawww.epfl.ch/Staff/Yves.Piguet/clip2gif-home/GifBuilder.html.

The new video technology provides a new dimension to previous methods of data presentation, but also presents a challenge to journals using conventional methods of publication which cannot display videotapes or movies in published papers. Some journals are beginning to meet this challenge by issuing CD-ROM collections of journal volumes that include video documentation that was not published with the original article. Other journals are now providing internet sites that allow the display of supporting video data or links to sites with GFP movies from detain papers published in the journal. Finally, an increasing number of scientists are taking the initiative of establishing their own homepages with a movie page that allows visitors to download and play their GFP movies.

IV.A.1 Photography

Photography can provide a simple and effective method for recording GFP fluorescence in fixed or living cells. GFP fluorescence in bacteria was documented in the initial report of expression in foreign host cells by photography onto commercial film (Chalfie et al., 1994). Black and white film (Kodak T-MAX 400 ASA, 8–15 s exposure) has been used to record GFP fluorescence from live yeast cells expressing GFP fused to cytoskeletal proteins (Doyle and Botstein, 1996). Kodak Ektachrome 400 ASA color film (Niswender et al., 1995; Kain et al., 1995; Ogawa et al., 1995) has been used to record GFP fluorescence in mammalian cells; in at least one case, the film was processed to enhance sensitivity (Ogawa et al., 1995). Kodak Ektachrome P1600 color reversal film was used to record expression of a yeast-GFP fusion protein in live cells (Hampton et al., 1996); images were transferred to CD-ROM and printed as digital images.

Photography is also an important method for documenting transformation of plants, which are often too large for visualization by microscopy. Expression of GFP can be easily visualized using a hand-held UV light source and the plant can be photographed using commercial film. Leaves of *Nicotiana* infected with tobacco mosaic virus (TMV) carrying *gfp* were photographed with Kodak Ektachrome 100 using a yellow-green filter and illumination from a hand-held UV source to document the sites of infection (Casper and Holt, 1996), and *Nicotiana* plants infected with a potato virus vector carrying *gfp* were photographed onto Kodak Ektachrome 400 ASA film through a yellow (Wratten 8) or green (Wratten 58) filter, using exposure times of up to 70 s (Baulcombe et al., 1995).

IV.A.2 Digital Images

Digital images documenting GFP fluorescence can be collected using a cooled CCD camera or laser scanning confocal microscope and manipulated in the same manner as digital images acquired in other ways. Two procedures that are useful for GFP file manipulation are (1) conversion of images obtained using an IBM PC to Macintosh format, and (2) conversion of grayscale to color images.

PROTOCOL 1 Conversion of digital images from IBM to Macintosh format

Images collected and saved as PICT or TIFF files on an IBM PC can be opened on a Macintosh computer using PC Exchange, a control panel that enables the Macintosh to recognize DOS or Windows disks and files, and an image processing program that can open PICT or TIFF files, such as Adobe Photoshop or NIH Image. Files saved in a proprietary format can often be opened in Adobe Photoshop as a "Raw" file by specifying the dimensions of the file. For example, a full-sized Bio-Rad MRC 600 confocal image saved to an IBM PC can be opened on a Macintosh computer using Photoshop by entering dimensions of 768 × 512 pixels for the width and height, and a header size of 76 bytes. The header size must be correct for the image to open correctly. Click "OK" to the program prompt that the "specified image is smaller than the file; open anyway?" to open the file. For Adobe Photoshop v 2.5.1, use the "Open As" command in the File Menu to access the Raw Options dialogue box. For Adobe Photoshop v 3.0.4 or 4.0, use the "Open" command in the File Menu and click on "Show All Files" and "Open" to display the Raw Options dialogue box.

PROTOCOL 2 Conversion of grayscale to color images

Grayscale images, collected without color information other than knowledge of the fluorochrome providing the signal, can be pseudocolored to regain the color information using Adobe Photoshop or NIH Image. The pseudocoloring is most easily accomplished by converting the image to a RGB or indexed color image. The following protocol can be used to convert a grayscale image to a green (GFP) image in RGB format with Adobe Photoshop. Variations on the protocol can be used to merge rhodamine or DAPI dual images with a GFP image.

1. Open the GFP fluorescence image. Select "RGB Color" in the Image Mode Menu.

   This selection will change the grayscale image to a RGB color image and permit the subsequent manipulations to be carried out.
2. Open the Channels box by selecting "Palettes" and/or "Show Channels" in the Window Menu.
3. Click on the Channels tab and select "Split Channels" from the menu.

   The image will split into 3 identical images with Red, Green, or Blue appended to the title of the windows.

4. Click on the Blue window to make it active. Type "command a" (a) to select the entire image in the window. Press the "delete" key to delete the image.

   The background color should be set to K=100%. The window will appear black.

5. Repeat step 4 with the Red window.
6. Select "Merge Channels" *or the RGB image* in the Channels box menu. The merged image will be green.

   To merge a rhodamine or DAPI dual image with a GFP image, copy the images and paste them into the Red or Blue, and Green windows, respectively, of a new file of the appropriate size. Any unused windows should appear black. Then merge the Red, Green and Blue windows to obtain the final image.

CONTRIBUTED by SHARYN A. ENDOW

IV.A.3 Videotape Recordings

Methods for recording real-time sequences to videotape have in the past been limited to relatively few laboratories. The major limitations have been the ability to convert digital computer images to analogue images and output the images at video rates of 30 fps (frames per second) for recording to videotape. This has required second-party digital-to-analogue conversion boards, which are of limited quality and can be expensive ($10,000 or more) to purchase, and the use of an analogue device such as an OMDR (optical memory disk recorder), which is expensive ($12,000–$25,000) but capable of high speed video-rate output. The availability of reasonably priced high speed Macintosh computers with built-in AV (digital-to-analogue) boards capable of video-rate output should make video technology available to a broader range of scientists. This ability is especially advantageous for documenting GFP in living cells or organisms, one of the most promising and exciting applications of the new GFP technology.

PROTOCOL 1 Recording time-lapse laser scanning confocal images to videotape

This protocol is for recording to videotape time-lapse images acquired with the Bio-Rad MRC 600 laser scanning confocal detector, using the capabilities of available Macintosh PCs. The protocol can be modified for time-lapse images acquired using other methods of confocal or fluorescence microscopy, including cooled CCD cameras.

Special Equipment

Hardware

Power Macintosh 8500/120 AV with 150 MB of RAM and 2 GB hard drive, or equivalent

The PowerMac 8500/120 AV is a high-speed personal computer specifically designed for video input and output. The computer has a high-capacity rapid internal hard disk drive available in 1 or 2 GB. The models being shipped at the time of writing have faster processors (150 or 180 Mhz, cf. 120 Mhz), increasing their usefulness for video applications. The computer is presently available with a maximum of 32 MB RAM and 2 MB VRAM. For work with time-lapse images, it is necessary to increase the RAM to 50 MB or more by installing additional 60 ns DIMMS. DIMMS can be purchased from a second party [e.g., MicroTech International, 158 Commerce St, East Haven CT 06512, T (800) 666-9689, F (203) 469-3926]; prices are currently very low. For maximum performance, DIMMS should be installed in the PowerMac 8500 in pairs (2 × 16 MB, 2 × 32 MB, or 2 × 64 MB) to take advantage of the computer's interleaving capability. Increasing VRAM to 4 MB will permit simultaneous display of the computer screen on both an analogue and digital monitor. For those not wishing to purchase a computer, AV capability can be added to other Power Macintosh PCs. One of the computers in use in my laboratory is a PowerMac 8100/100 (80 MB RAM/1 GB hard drive) with a Radius VideoVision Studio II AV board. The Radius board provides good to high quality video output (but medium to low-quality video capture). The video capture/output quality of current AV PowerMacs exceeds that of medium quality second party AV boards; these PCs are recommended over the purchase of presently available second party boards for installation in a nonAV Macintosh computer.

Sony 17SF multiscan monitor or equivalent

This 17" monitor has multiple resolution capability, suitable for viewing a full-size Bio-Rad MRC 600 confocal image. The monitor contains a Trinitron tube for superior color display. The AppleVision 1710 or 1710AV monitor is a comparable monitor with multiscan capability that contains a Sony Trinitron tube and offers brilliant color representation. Macintosh monitors and computers can be purchased direct from Apple Computer [2420 Ridgepoint Drive, Austin TX 78754, T (800) 998-2775] for educational discounts and no-cost shipping within the United States.

Pinnacle Micro Sierra 1.3 GB magneto-optical (MO) drive, or equivalent

A high capacity storage device is necessary for archiving time-lapse images. The high-speed Sierra drive can also be used to drive image output. The drive uses two-sided MO disks with a capacity of 650 MB on each side. Larger capacity MO drives are now available from Pinnacle Micro and other manufacturers of MO drives. These drives should be tested for speed with the computer in use. Video sequences that show "jerkiness" (as in early motion pictures) are caused by slow or uneven rates of video output, which can result from a slow computer hard drive or storage device drive, or the two combined. The PowerMac 8100/100 performs better using video output driven by the Sierra drive, than without it. Drives for other forms of high capacity storage for digital images, such as tape and CD (compact disk), are not recommended for output to videotape but could serve as reasonably priced storage devices.

Panasonic AG-7350 VCR or equivalent

The Panasonic AG-7xx0 series VCRs are high-quality, heavy-duty VCRs used professionally for recording to videotape. The AG-7350 and other models in this series have the capability of S-VHS (Super-VHS) recording, and can play either VHS or S-VHS recorded tapes. S-VHS is a higher quality video signal than VHS that is recognized by some but not all VCRs. Tapes recorded using S-VHS have higher resolution and higher clarity than those recorded using VHS. Some VCRs can play but not record S-VHS. Many VCRs cannot play or record S-VHS. To ensure that your videotape can be played on other VCRs, it is necessary to make both a VHS and a S-VHS tape, or record using VHS.

Sony PVM-135 monitor

This high-resolution 13" black-and-white monitor is used to display videotaped sequences. The Sony PVM 1354Q monitor is a comparable color monitor that contains a Trinitron tube for superior color display.

Software

NIH Image v 1.61/ppc

This public domain program for image analysis was written and is updated periodically by Wayne Rasbaud. The program can be obtained from the NIH Web Site (http://rsb.info.nih.gov/nih-image/) or by anonymous ftp from zippy.nimh.nih.gov/pub/nih-image and on-line help is provided by e-mail to Wayne Rasbaud (wayne@helix.nih.gov). The above version is specially adapted for use with the Power Macintosh.

Adobe Premiere v 4.2

This program can be used to add professional looking titles and legends to videotape sequences, including special effects, such as fade-in or fade-out titles and transitions. Audio (music, voice) can also be added to videotapes using this program. The program has presets and time base for both NTSC and PAL. NTSC (National Television Standards Committee, 29.97 fps) is the US standard video format and PAL (Phase Alternating Lines, 25 fps) is the UK standard. European countries use either PAL or SECAM video format. Videotapes made in NTSC format can be converted to other formats, and vice versa, with some loss of quality. A few VCRs are available that have the capability of recording in one of several formats (NTSC, PAL, SECAM), but none of these has the capability of recording S-VHS.

Materials

Time-lapse files acquired using the Bio-Rad MRC 600 laser scanning confocal detector

Stack files of 60 images, collected using a Bio-Rad MRC 600 laser scanning detector, are stored upon collection in an IBM PC. Each file is 23 MB (384 KB/image × 60 images). The number of images per file can be changed but

should not be increased so large that transfer and manipulation of files becomes cumbersome. Each file of 23 MB requires about 26–28 MB or more of RAM to open. The size of the files is therefore limited by the RAM of available computers.

Maxell XR-S 120 Black Magnetite videotapes

These are high-quality S-VHS videotapes that can record up to 120 min. Maxell ST-126BQ (NTSC) and SE-180BQ (PAL SECAM) Broadcast Quality S-VHS tapes are of comparable or higher quality. Other brands of high quality videotapes are also available. For videotape recording using S-VHS, the videotapes must specify S-VHS.

1. Turn off File Sharing in the Sharing Setup control panel and make Apple Talk inactive in the Chooser. Turn off Virtual Memory in the Memory control panel.

   These settings are necessary to ensure optimal performance with QuickTime applications.
2. Connect the Sony SF17 monitor to the monitor port on the Radius board in the PowerMac 8100 or to the PowerMac 8500 monitor port.

   Turn the computer off prior to making these connections. The computer will recognize the monitor connected to the Radius board or the computer port as the primary monitor and the Apple desktop will be displayed on the monitor, permitting control of the computer.
3. Connect the S-video OUT port on the Radius outlet or PowerMac 8500 computer to the S-video IN port on the VCR using a cable with S-video connectors at each end.

   S-video is a high-quality video signal that can be output by Power Macintosh and input to many (but not all) VCRs. Recording to videotape using S-video will not limit the ability of VCRs to play the tape. Unlike S-VHS, videotapes recorded using S-video can be played on any VCR, even those that cannot input the S-video signal for tape recording.
4. Connect a video OUT port on the VCR to a video IN port on the Sony PVM-135 monitor using a cable with BNC connectors at each end. Turn on the Sony monitor and select the appropriate line (A or B) to display input from the VCR.
5. Set the VCR input to S-video on the front panel by removing the small panel cover to access the "Input" control switch and changing to "S-video".
6. Set the VCR to record S-VHS ("S-VHS ON") for the highest quality tape recording and black-and-white (Video Mode B/W), and turn "Auto Backspace" on.

   This is done on the Panasonic AG-7350 using the internal control panel. Access the internal control panel by pressing the "Screen Display" and "Down" buttons at the same time. The internal control panel will be displayed on the monitor screen. Press the "Data" button on the VCR to select "ON" or "OFF" or another option, the "Down" button to move down to the next item, and the "Shift" button to change to the next screen. Press the "Screen Display" button to close the control panel and return to a normal screen.

Use of the "Auto Backspace" function helps prevent static between sequences when multiple sequences are recorded to the same tape. When this function is on, pressing the "Pause" then "Rec" buttons causes the VCR to pause. Releasing the "Pause" button causes the VCR to rewind 3 s, play back the tape for 1 s, then record. This eliminates static between sequences caused by starting the sequence playback on the computer, or the VCR.

7. Transfer time-lapse files to the Macintosh computer.

   Transfer of files between computers is best accomplished using available networks. The transferred files should be viewed prior to discarding the original files to ensure that the files were not corrupted during transfer.

8. Increase the amount of memory allocated to NIH Image.

   The memory available to NIH Image determines the number of images that can be opened in the program. The amount of memory allocated to the program can be specified by clicking on the program icon to highlight it prior to launching the program, then selecting "Get Info" from the File Menu. The preferred size of the program can then be entered into the "Memory Requirements" box. The preferred size should be at least 3–5 MB larger than the stack files (i.e., 26–28 MB for a stack file of 23 MB) and is limited by the total RAM in the computer. At least 10 MB of RAM should be reserved for running the system software. A computer with 50 MB of RAM should therefore have a maximum of 40 MB allocated to NIH Image and will be capable of opening full-size confocal stack files of up to 90–95 images. After entering the preferred size, close the Info window.

9. Open NIH Image, select "Load Macros" in the Special Menu and load "Input/Output Macros" from the "Macros" folder, then open the stack file using the "Import BioRad MRC 600 Z Series" command in the Special Menu.

   "Import Bio-rad MRC 600 Z Series" is a macro written specifically for opening Bio-Rad MRC 600 Z series or time-lapse stack files saved in IBM PC format. After opening, files can be saved as stacks in TIFF or PICT format. Stacks can also be converted to single files and saved as individual files numbered sequentially from 001 to xxx, where xxx is the number of images in the stack. Files numbered in series and saved in the same folder can be opened sequentially using the "Open" command in the File Menu by selecting one file from the list of files in the folder and "Open All" in the window. Other stack macros provided with NIH Image permit contrast enhancement or cropping of the entire stack of images. Image montages can also be made for display of time-lapse sequences as figures of sequential images by selecting the "Make Montage" command in the Stacks Menu when a stack file is open.

10. Animate the files using the "Animate" command in the Stacks Menu.

    The speed of the animated sequence playback can be changed using the number keys on the Apple keyboard and the number of fps will be displayed in the Info window. Click anywhere in the image to stop the animation.

11. Adjust image contrast, if necessary, by selecting "Enhance Contrast" in the Process Menu.

    The contrast of the entire stack will be enhanced. To avoid saturation of images in the stack, select the "Enhance Contrast" command when the image of brightest GFP fluorescence is present in the active window. Image contrast can also be adjusted by clicking and dragging in the LUT (look-up table) window using the LUT tool (the tool with the double-headed arrow). Click and drag near the top or bottom of the LUT strip to adjust constrast, or near the middle to adjust brightness. The changes will be applied to the entire stack of images and will be saved with the stack.

12. Go to the last image in the sequence.

    Type "." (period) to step forward one frame at a time to the end of the sequence.

13. Select the Eraser tool in the Tools palette and hold down the "option" key on the Apple keyboard while clicking on the image background.

    This changes the Eraser tool to the same shade black as the image background. Clicking on the LUT strip when the Eraser tool is highlighted will change the Eraser to the shade selected on the LUT strip. The Eraser tool setting will be saved with the image stack.

14. Select "Add Slice" in the Stacks Menu. Repeat four times to add five frames of black background to the end of the sequence.

    This defines the end of the sequence for playback and recording.

15. Save the images as a stack file to a Sierra 1.3 GB MO disk.

    The stack files can be archived on the Sierra MO disk. The Sierra drive can also be used to drive playback of a sequence, resulting in a more constant speed than for the computer alone.

16. Divert the computer screen to the analogue (Sony PVM-135) monitor.

    For the PowerMac 8100 with a Radius board, open the Monitors control panel, click on the "Options" button to open the Radius monitors control panel, then move the slider to "Composite". When the control panel windows are closed, the Apple desktop will disappear from the computer monitor screen and appear on the Sony PVM-135 monitor. For the PowerMac 8500 with 4 MB of VRAM, the Apple desktop will appear on both the computer monitor screen and the Sony PVM-135 monitor when the VCR and Sony PVM-135 monitors are turned on. Open the Sounds and Displays control panel, select "Arrange Displays" and drag the monitors on top of one another to make duplicate displays. Close the window and select "Television" in the Displays box and "Open Settings" in the Configure Menu to open the television settings window, then click in the box to turn flicker control on. This helps reduce the flicker on the composite monitor. For the PowerMac 8500 with 2 MB of VRAM, turn off the computer and computer monitor, and disconnect the computer monitor from the computer. Turn on the VCR, Sony PVM-135 monitor, and computer. The Apple desktop will appear on the Sony PVM-135 monitor. Open the Sounds and Displays control panel and turn the flicker control on.

17. Open the stack file and center the stack in the monitor screen using the mouse to drag the window. The Eraser tool should appear the same shade of black as the image background. Animate the stack by holding down the "option" and "command" () keys and typing "=". The stack will appear with a border the same shade black as the image background and the Apple desktop will no longer be visible. Select the speed or speeds for recording to videotape using the number keys on the Apple keyboard. Type "." (period) to stop the sequence at the frame desired for the start of the videotaped sequence.

Type "." (period) to step forward by single frames through the stack, or "," (comma) to step backward one frame at a time. Click with the mouse to make the Apple desktop reappear under the image. A playback speed corresponding to number key 4 (7.5 fps) or 5 (10 fps) is usually a good speed to record the sequence to videotape. The playback speed should not be confused with the optimal 29.97 fps NTSC video output rate needed to produce smooth video motion.

18. Insert a videotape into the VCR. Press "Play", "Pause", then "Rec" (both the "Pause" and "Rec" buttons should be lighted). Press "Pause" again to release the pause button, wait 3 s, then press the keyboard number key corresponding to the desired speed for the sequence. This will animate the sequence at the desired speed and the sequence will be recorded to videotape.

Sequences can be recorded first with a fast playback speed (number key 4 or 5) followed by a slower speed (number key 2, 2 fps). The speed can be changed by pressing the number key corresponding to the slower speed at any time while the sequence is playing. Recorded transitions appear smoothest if the change is made at the end of the sequence while the black frames are playing. If the "," (comma) key has been pressed prior to the number key to start the sequence, the sequence will run backwards. Care should be taken to avoid this, unless this is desired.

19. Titles made in Adobe Premiere can be displayed and recorded to videotape by selecting "Print to Video" from the File Menu when a project window containing the title is open and the title has been dragged into the construction window. Enter 6 s for the time to blank screen prior to displaying the title. Click on "OK", wait 1 s for the screen to darken, press "Play", "Pause" and "Rec" on the VCR, then release the "Pause" button to record the title to videotape.

"Print to Video" in Adobe Premiere adds a sequence of black frames preceding and following the title, which can serve as a transition to the animated sequence. The titles are displayed surrounded by a black border for the time (duration) displayed in the construction window. Eight to ten seconds is sufficient time to display a title in a videotaped sequence. Titles can also be made in graphics or word processing programs, saved as PICT files, and opened in NIH Image. NIH Image has a "PhotoMode" in the Special Menu which displays the image with a border around it, removing the Apple desktop. The color or black level of the border can be changed using the Eraser tool as described above for the animated stack border.

CONTRIBUTED by SHARYN A. ENDOW

PROTOCOL 2 Transfer to an OMDR and output to videotape

Special Equipment

Sony LVR-3000AN CRV, Panasonic TC-3038F B&W or LQ-3031T Video Disk Recorder/Player, or equivalent

These are models intended for professional/industrial use. The features of the above models should be compared prior to purchase. Current prices range from $12,000–15,000 for the Panasonic recorder/players and $18,600 for the Sony. Other models, such as the Panasonic LQ-4000T Re-Recordable Video Disk Recorder/Player ($25,600), are also available.

Digital images can also be converted to analogue images using a D/A (digital-to-analogue) board and transferred to an OMDR (optical memory disk recorder) for recording high quality sequences of time-lapse images to videotape. The OMDR is an analogue device that can be connected directly to a VCR and has the capability of variable (video-rate) image output. MetaMorph (Universal Imaging Corp.), a Windows-compatible digital imaging system for use with the IBM PC, provides conversion of digital images to video through a VGA to S-VHS video converter for transfer to an OMDR.

IV.A.4 QuickTime Movies

Time-lapse sequences of dynamic processes in living cells visualized using GFP can also be made into QuickTime movies. These movies are suitable for recording to videotape, although there is usually a loss of quality and resolution relative to videotapes made by animating sequences using NIH Image. The play back of animated sequences using NIH Image is smoother than for QuickTime movies, and NIH Image can animate stack files at full size (768 × 512 pixels). By comparison, the computer time required for making full-size (640 × 480 pixels) QuickTime movies (several hours) and the large memory requirements of the movies (333 MB for a 66-frame movie) make practical only movies reduced to quarter size or smaller. The small QuickTime movies are convenient for display of data obtained using GFP fluorescence, however, and can be mounted at Web sites (e.g., http://genome-www.stanford.edu/group/botlab/people/doyle.html http://www2.uchc.edu/~terasaki/flip.html http://abacus.mc.duke.edu/moviepage.html) for viewing by those interested.

PROTOCOL 1 Making QuickTime movies from time-lapse laser scanning confocal images

This protocol is for making QuickTime movies from stack files acquired using the Bio-Rad MRC 600 laser scanning confocal detector, but can also be applied to image stacks collected using other types of confocal or fluorescence microscopy, including cooled CCD cameras.

Special Equipment

Power Macintosh PC and computer monitor as described in previous protocol

Adobe Premiere v 4.2

This software program and others can be used to make QuickTime movies.

1. Open the stack file in NIH Image and add two black frames to the end of the sequence.

   The steps for opening Bio-Rad MRC 600 stack files saved to an IBM PC are described in the previous protocol together with the steps for adding black frames to the end of the sequence.

2. Crop the stack of images to make the width:height ratio of the images 4:3 (e.g., 560 × 420 pixels).

   If the images are not in a 4:3 width:height ratio, Adobe Premiere will try to fit them into a 4:3 window. This can result in distortion of the images. Use the NIH Image Rectangular Selection tool in the Tools palette to outline the region of interest in one image of the stack file, then step through the stack by typing "." (period) to make certain that the object of interest is centered in each of the images. The size of the rectangle is displayed in the Info window and the rectangle can be moved within the image using the arrow keys on the Apple keyboard. After positioning the rectangle, select "Crop and Scale-Fast" from the Special Menu. If this command is not listed in the menu, load the "Stacks" macros from the Macros folder in the NIH Image program folder. Save the file after cropping.

3. Adjust image contrast, if necessary, by selecting "Enhance Contrast" in the Process Menu or by using the LUT tool.

   Methods for adjusting image contrast of a stack file are described in the previous protocol.

4. Create a folder for the files, select "Stack to Windows" in the Stacks Menu to convert the stack file to single image files, then select "Save As" from the File Menu and save the files to the folder as PICT files.

   The files will be saved as a series of files numbered sequentially from 001 to xxx.

5. Open Adobe Premiere. Create a title by selecting "New" and "Title" in the File Menu. Type the title in the window and save the file. Import the title into the project window by selecting "Import" and "File" in the File Menu. Click on the title in the project window to highlight it, then select "Duration" from the Clip Menu and enter 0:00:02:00 (2 s).

6. Import the folder of images into the project window by selecting "Import" and "Folder" in the File Menu. Click on the folder in the project window to highlight it, then select "Speed" from the Clip Menu and enter 500%.

The default time for display of each frame by Adobe Premiere is 1 s. Increasing the speed to 500% will reduce the display time for each frame to 0.2 s. The duration of a movie of 60 images or frames will then be 12 s.

7. Drag the title and folder of images into the construction window.

   The construction window will show the title and images in the sequence that they will appear in the movie.

8. To add lettering (titles, labels, arrows) to the images in the movie, select "New" and "Title" in the File Menu, and type the label in the (approximate) position desired in the image. Change the lettering to white on a white background. Save the file, then import the file into the project window. Drag the file into the construction window onto one of the "S" (Superimpose) tracks under the frames in which you wish the label to appear. Set the desired duration by clicking on the file in the project window to highlight it, then selecting "Duration" from the Clip Menu and entering the desired time.

   The position of the label in the image can be previewed and altered if desired, prior to making the movie (see below).

9. Select the lettering clip in the construction window, then select "Transparency" in the Clip Menu. In the Transparency Settings box, select "White Alpha Matte" for the Key Type.

   The label superimposed on the image will appear in the Sample box when the "Page Peel" icon (3rd from the left) is selected. The image can be magnified using the Magnifying Glass icon and returned to original size by pressing the "option" key on the Apple keyboard and clicking on the image in the Sample box while the Magnifying Glass icon is highlighted.

10. Select the settings for the QuickTime movie.

    Select "Presets" in the Make Menu. Select "Time Base" and enter 30 fps. Click "OK" to return to the "Presets" window. Select "Compression", then select "Graphics" and "Grayscale" in the Compressor box, slide the "Quality" slider to 100%, and select 30 fps in the Motion box. Click "OK" to return to the "Presets" window. Select "Output Options", then select "Entire Project" and "QuickTime Movie", and enter 240 × 180 in the Size box. Deselect "Audio" if no sound will be added to the video sequence. Click "OK" and select "Preview Options" in the "Presets" window. Select 30 fps in the Rate box, then click "OK" to return to the "Presets" window. Save the settings by selecting "Save" in the window. Close the window by clicking on "OK."

11. Select "Movie" in the Make Menu to make a QuickTime movie.

    This will require about 1–5 minutes, depending on the number of frames and the size of the movie. Save the movie to disk and play by selecting "Print to Video" in the File Menu. The movie can also be opened and played using Apple MoviePlayer or Sparkle 2.4.5 (available from the Macintosh Shareware Sites at the Apple Web site at http://www.apple.com), and saved using the MPEG compression option of Sparkle to conserve disk space. To mount the movie in a downloadable form on a Web page, replace the ".M1V" suffix added by Sparkle to the MPEG-compressed movie with ".mpeg" to allow the Web browser to recognize the file as an MPEG movie.

Netscape 3.0 also has a QuickTime plug-in that recognizes QuickTime movies mounted as ".mov" files and will play the movie on the page instead of requiring it to be downloaded to be played.

CONTRIBUTED by SHARYN A. ENDOW

ACKNOWLEDGMENTS

Research in our laboratories is supported by grants from the NIH to S.A.E. and D.W.P., and from the Whitaker Foundation to D.W.P. D.W.P. is a Beckman Young Investigator of the Arnold and Mabel Beckman Foundation. We thank the contributors to the methods sections, especially J. Lippincott-Schwartz, for providing detailed methods, protocols, and observations on the use of GFP. Special thanks to P. Millman for discussion of GFP filter design and to Chroma Technology Corp. for manufacture of custom GFP filter sets. S.A.E. also thanks C. Regen of Bio-Rad Lab. for providing a confocal filter block and GFP filter set, and arranging the mounting of the filter set.

REFERENCES

Baulcombe, D. C., Chapman, S., and Santa Cruz, S. (1995). Jellyfish green fluorescent protein as a reporter for virus infections. *Plant Jl.* 7:1045–1053.

Casper, S. J. and Holt, C. A. (1996). Expression of the green fluorescent protein-encoding gene from a tobacco mosaic virus-based vector. *Gene* 173:69-73.

Chalfie, M., Tu, Y., Euskirchen, G., Ward, W. W., and Prasher, D. C. (1994). Green fluorescent protein as a marker for gene expression. *Science* 263:802-805.

Cole, N. B., Smith, C. L., Sciaky, N., Terasaki, M., Edidin, M., and Lippincott-Schwartz, J. (1996). Diffusional mobility of Golgi proteins in membranes of living cells. *Science* 273:797-801.

Doyle, T. and Botstein, D. (1996). Movement of yeast cortical actin cytoskeleton visualized *in vivo*. *Proc. Natl. Acad. Sci. USA* 93:3886–3891.

Endow, S. A. and Komma, D. J. (1996). Centrosome and spindle function of the *Drosophila* Ncd microtubule motor visualized in live embryos using Ncd-GFP fusion proteins. *J. Cell Sci.* 109:2429–2442.

Hampton, R. Y., Koning, A., Wright, R., and Rine, J. (1996). *In vivo* examination of membrane protein localization and degradation with green fluorescent protein. *Proc. Natl. Acad. Sci. USA* 93:828–833.

Kain, S. R., Adams, M., Kondepudi, A., Yang, T.-T., Ward, W. W., and Kitts, P. (1995). Green fluorescent protein as a reporter of gene expression and protein localization. *BioTechniques* 19:650–655.

Niswender, K. D., Blackman, S. M., Rohde, L., Magnuson, M. A., and Piston, D. W. (1995). Quantitative imaging of green fluorescent protein in cultured cells: comparison of microscopic techniques, use in fusion proteins and detection limits. *J. Microsc.* 180:109–116.

Ogawa, H., Inouye, S., Tsuji, F. I., Yasuda, K., and Umesono, K. (1995). Localization, trafficking, and temperature-dependence of the *Aequorea* green fluorescent protein in cultured vertebrate cells. *Proc. Natl. Acad. Sci. USA* 92:11899–11903.

APPENDIX. GREEN FLUORESCENT PROTEIN ON THE WEB

A Academic Web Sites Related to GFP

gfptub—LivingDictyostelium cells expressing GFP-tubulin.
–http://bioc.rice.edu/~hostos/gfptub.html

green fluorescent protein for expression in plants.
–http://brindabella.mrc-lmb.cam.ac.uk/IndexGFP.html

Mark Terasaki Home Page—Department of Physiology, University of Connecticut Health Center Farmington, CT
—http://www2.uchc.edu/~terasaki/

GFP IMAGES AT YANG'S PAGE—The Green Fluorescent Protein Images.
—http://chronic.dartmouth.edu/yang/gfppix/gfp.html

GFP Studies—Using the Green Fluorescent Protein (GFP) in the Labeling of Cellular Compartments.
—http://www.physiol.arizona.edu/CELL/Department/Llab/Projects/GFP html

GFP Transfection—The following neuron has been calcium phosphate transfected with pCA-GAP-GFP (S65A).
—http://www.life.uiuc.edu/craig/gfp.html

GFP Research—Green Fluorescent Protein.
—http://www.mc.vanderbilt.edu/vumcdept/mpb/piston/files/gfp.html

GFP—GFP: A sequence finishing support tool.
—http://stork.cellb.bcm.tmc.edu/gfp/

Soluble derivatives of green fluorescent protein (GFP) for use in Arabidopsis
—http://genome-www.stanford.edu/Arabidopsis/ww/

The construct GFP-KDEL
—http://uchc.edu/htterasaki/gfpkdelsequence.html

C&C Vislab Project Gallery—Green Fluorescent Protein by Mike Moser.
—http://www.hs.washington.edu/locke/vislab/proj/gfp.html

Endow Movie Page—Department of Microbiology, Duke University Medical Center.
—http://abacus.mc.duke.edu/moviepage.html

gfp—GFP Applications page — Wallace Marshall
—http://util.ucsf.edu/sedat/marsh/gfp_gateway2.html

MEI-S332–GFP localization in Drosophila spermatocytes.
—http://www.wi.mit.edu/orr-weaver/fig1.html

We targeted Green Fluorescent Protein to the ER lumen of starfish eggs.
—http://util.ucsf.edu/sedat/marsh/mito.html

Mitochondrial Dynamics and Transmission in Yeast Visualized with Green Fluores.
—http://util.ucsf.edu/sedat/marsh/mito.html

Transgenic frogs—Transgenic Xenopus.
—http://vize222.zo.utexas.edu/Marker_pages/transgenic.html

Bargmann Lab Home Page—Bargmann Lab Home Page.
—http://devbio-mac1-ucsf.edu/

George Stickney—George Stickney. Click Here to Read About My Project.
—http://www.physiol.arizona.edu/CELL/Department/Llab/George.html

Arabidopsis web site MRC-LMB Cambridge UK
—http://brindabella.mrc-lmb.cam.ac.uk/

B Company Web Sites with GFP Products

Aurora Biosciences: http://www.aurorabio.com/

Chroma Technology: http://www.chroma.com/

CLONTECH: http://www.clontech.com

Invitrogen: http://www.invitrogen.com/

KAIROS Scientific: http://www.www.kairos-scientific.com/

Life Technologies Inc. (LTI): http://www.lifetech.com/

Maxygen: http://www.maxygen.com/

Molecular Probes: http://www.probes.com/

Packard Instruments: http://www.packardinst.com/

PharMingen: http://www.pharmingen.com/

Quantum Biotech: http://www.qbi.com/

Turner Designs: http://www.turnerdesigns.com

Contributed by Steven R. Kain

Index